ENGINEERING OF DYNAMIC SYSTEMS

ENGINEERING OF DYNAMIC SYSTEMS

WILLIAM R. PERKINS

JOSÉ B. CRUZ, JR.

Department of Electrical Engineering
and Coordinated Science Laboratory
University of Illinois, Urbana

JOHN WILEY & SONS, INC.

NEW YORK · LONDON · SYDNEY · TORONTO

10 9 8 7 6 5 4 3 2 1

Library of Congress Catalogue Card Number: 72-81336

SBN 471 68035 4

Printed in the United States of America

To Julia and Pat

Preface

Systems engineering is concerned with the complex interconnection of devices or components. Extensive use is made of mathematical models, computer-aided analysis and design, simulation, and optimization with respect to quantified goals. This approach has been successfully applied to the design of many large-scale engineering systems, including industrial process control systems, factory automation, resource allocation systems, and spacecraft guidance and control systems. Applications of the systems approach have also occurred in nonengineering fields, including economics, management, biology, and sociology. Some exciting beginnings have been made in applying the systems approach to very complex interdisciplinary societal problems such as environmental pollution, transportation, and urban development. Although systems engineering obviously is not a panacea, it is clear that it will play an increasingly important role in finding solutions to many significant problems. Moreover, it provides an effective method of communication not only between engineers of different disciplines but also between engineers and sociologists, economists, urban planners, and other nonengineers involved in multidisciplinary projects. Thus it is important for all engineers to have an introduction to systems engineering.

The purpose of this book is to provide such an introduction. Since the subject is a vast one, we have decided to emphasize the basic concepts and techniques useful in the design of dynamic systems. Dynamic systems are of central importance in engineering. Moreover, most engineers have some basic exposure to dynamics early in their course work (in basic physics courses, for example), which provides a convenient and useful foundation on which to build a systems course.

An important pedogogical aspect of our presentation is the use of a *theme example*, the attitude control problem. Various aspects of this problem are used throughout the book to motivate and illustrate the material. The student thus sees new concepts interpreted concretely and new techniques applied to a familiar problem. Other theme examples could have been chosen, of course. Attitude control is a system problem of intermediate difficulty, complex enough to exhibit the strengths of the systems approach without overwhelming the student.

Several courses can be organized around this book, according to individual tastes and local requirements. We have used versions of the book in a three-semester-hour senior course, *Introduction to Systems Engineering.* The students in this course were already somewhat familiar with much of the material of Part I, *Modeling and Simulation*, and with Chapter 6, System Response via Transfer Functions. Consequently, these chapters could be covered rather quickly.

Many people assisted in the writing of this book. Special thanks are due the students in our classes, who provided the incentive for this undertaking and whose interest and criticism provided useful feedback. We acknowledge our debt to the stimulating suggestions and penetrating comments of Professors D. B. DeBra (Stanford), L. G. Shaw (Polytechnic Institute of Brooklyn), G. F. Franklin (Stanford), and E. I. Jury (University of California, Berkeley), who reviewed all or part of the manuscript. We express special appreciation to Professor M. E. Van Valkenburg (Princeton) for his indispensable enthusiasm, guidance, and assistance with all parts of this project. We are grateful to our colleague, Professor P. Kokotović (Illinois), for many illuminating discussions on systems, which are reflected in many ways in this book. Thanks are due Dr. W. L. Everitt, Dean Emeritus of the College of Engineering, University of Illinois, Professor E. C. Jordan, Head of the Department of Electrical Engineering and Professor W. D. Compton, Director of the Coordinated Science Laboratory, for their strong support and encouragement. We are indebted also to Mrs. Divona Keel for her superb typing of several drafts of the manuscript.

Urbana, Illinois

W. R. Perkins
J. B. Cruz, Jr.

Contents

Chapter 1. Systems and Systems Engineering 1
- 1.1 Engineering as Design 1
- 1.2 Devices and Systems 2
- 1.3 Introduction to the Attitude Control Problem 4

PART I. MODELING AND SIMULATION 7

Chapter 2. Some Basic Elements for Network Modeling 11
- 2.1 Linear Time-Invariant Mechanical Elements 13
- 2.2 Single-Axis Gas-Jet Attitude Control System 22
- 2.3 Linear Time-Invariant Electrical Elements 27
- 2.4 Nonlinear and Time-Varying Elements 34

Chapter 3. Network Modeling of Devices and Systems 40
- 3.1 Kirchhoff's Laws 40
- 3.2 D'Alembert's Principle 46
- 3.3 Conservation Laws 57
- 3.4 Network Model of a Single-Axis Reaction-Wheel Attitude Control System 68
- 3.5 Equilibrium Points and Linearization 72
- 3.6 Sensitivity Models 79

Chapter 4. System Simulation 95
- 4.1 Building Blocks for Analog Simulation 96
- 4.2 All-Integrator System Simulation 100
- 4.3 Sensitivity Functions via Simulation Diagrams 110

Chapter 5. Modeling Spacecraft Motion 117
- 5.1 Rigid Body Equations of Motion 117

5.2 Representation of Spacecraft Attitude 127
5.3 The Gyro 131
5.4 Modeling a Spacecraft with Gas-Jet Actuator 138
5.5 Modeling a Spacecraft with Reaction-Wheel Actuators 141

PART II. SYSTEM ANALYSIS 147

Chapter 6. System Response Via Transfer Functions 148
6.1 The Transfer Function Concept 148
6.2 Transfer Function Block Diagrams 152
6.3 Mason's Formula for Transfer Functions 156
6.4 Transfer Function Block Diagrams with Multiple Inputs and Multiple Outputs 164
6.5 Frequency Response via Sinusoidal Steady-State Response 169
6.6 Frequency Response of Rational Transfer Functions 173
6.7 Frequency Response from Pole-Zero Diagrams 185
6.8 Sensitivity Functions from Block Diagrams 188

Chapter 7. System Response Via State Variables 206
7.1 State Variable Response of First-Order Systems 212
7.2 The State Plane 214
7.3 State Transition Matrix for Second-Order Systems 217
7.4 Phase-Plane Trajectories in Terms of Eigenvectors 219
7.5 Properties of the State-Transition Matrix—Second-Order Systems 227
7.6 Complete Solution for Second-Order Systems 232
7.7 The State Transition Matrix—*N*th-Order System 238
7.8 The Complete Solution 246
7.9 Relationship of State Description to Transfer Functions 252
7.10 Time-Varying Linear Systems 268
7.11 Linearity and Time-Invariance 275

Chapter 8. Stability of Linear Time-Invariant Systems 287
8.1 Characteristic Equation 287
8.2 Routh Stability Test 291
8.3 Root Locus Method 298
8.4 Nyquist Stability Criterion 308

8.5 Input-Output Stability 313
8.6 Parameter Variations and Stability 315

Chapter 9. Response of Nonlinear Systems 323

9.1 Piecewise-Linear Systems 323
9.2 The Describing Function 338
9.3 Lyapunov's Direct Method 341
9.4 Stability of Linear Systems via Lyapunov's Direct Method 350
9.5 Sensitivity of Nonlinear Systems 354

Chapter 10. Analysis of Discrete-Time Systems 359

10.1 System Modeling Using Difference Equations 359
10.2 Digital Simulation 362
10.3 Response of Discrete-Time Systems via Transfer Functions 368
10.4 Response of Discrete-Time Systems via State Equations 374
10.5 Stability of Discrete-Time Systems 377
10.6 Sampled-Data Systems 380

PART III. CONTROL AND OPTIMIZATION 393

Chapter 11. Feedback and Control 395

11.1 Nature of Control Problems 396
11.2 Controllability 398
11.3 Observability 409
11.4 Controllability and Observability of Single-Axis Reaction-Wheel Attitude Control Systems 413
11.5 Controllability and Observability of Composite Systems 418
11.6 State Feedback for Controllable Systems 421
11.7 Comparison of Open-Loop and Closed-Loop Structures when Plant Is Perturbed 436
11.8 Comparison of Open-Loop and Closed-Loop Structures when Plant Has Disturbance Inputs 442

Chapter 12. Parameter Optimization I 454

12.1 Quadratic Performance Criteria 455
12.2 Optimization Using Parseval's Theorem 459

12.3 The Optimum State Regulator 462
12.4 A Derivation of the Riccati Equation 466
12.5 Design Considerations Using Parameter Optimization and Quadratic Indices 471
12.6 Parameter Optimization of Digital Systems 482

Chapter 13. Parameter Optimization II 488

13.1 The Newton-Raphson Method 488
13.2 The Gradient Method 501
13.3 Use of Sensitivity Functions in Newton-Raphson and Gradient Methods 509

Appendix A. Basic Matrix Algebra 524

Appendix B. Solution of Linear Time-Invariant Differential Equations by Laplace Transforms 527

B-1 Laplace Transform Definition 527
B-2 Partial Fraction Expansions 533
B-3 The Second Order All-Pole Transfer Function 543

Selected References for Collateral Reading 551

Answers to Selected Problems 554

Index 565

ENGINEERING OF DYNAMIC SYSTEMS

CHAPTER 1

Systems and Systems Engineering

Many engineering advances have resulted from the invention and development of new devices to solve specific problems. Early engineering accomplishments, like the water clock, the vacuum tube, and the gyroscope had this device orientation. Developments such as the transistor, the maser, and the laser indicate the continued vitality of this important and useful aspect of engineering.

A different class of problems has arisen, however. Automation of factories, industrial process control, resource allocation, and spacecraft guidance and control are typical of engineering problems wherein the complex interconnection of devices is of basic interest, rather than the devices themselves. Such problems are systems problems. The purpose of this book is to provide an introduction to the engineering of systems.

1.1 Engineering as Design

To talk about systems engineering, we must first talk about engineering. All engineering problems, including system problems, begin with the recognition of a *need.* This need may be phrased in very broad terms. For example, we need to eliminate air pollution, or we need to send some men to the moon (or get them back!), or we need a less expensive lighter more efficient toothbrush. The common feature of all engineering is *design.* A system or device is designed to meet a need. Because the final design of devices and systems is the solution of engineering problems, it represents the ultimate concern of engineers.

It is helpful in clarifying the idea of engineering as design to contrast engineering with science. Science is concerned primarily with investigating the natural universe around us and with developing explanations for natural phenomena. The scientist thus *analyzes* a *given* structure (the natural universe), whereas the engineer *designs* new (man-made) structures (devices and systems). Of course, the boundaries between science and engineering are frequently rather blurred in practice. An engineer often will conduct investigations of the natural world (i.e., scientific investigations) while solving engineering problems. For example, the engineer interested in solid-state devices must understand the physics of the solid state. Similarly, the physicist—especially the experimental physicist—may do device and even system design. And, of course, both endeavors provide mutual enrichment. Engineering problems often stimulate scientific investigations of important natural phenomena, and expanded knowledge of natural phenomena leads to new technological applications and to more rational design procedures. But the fundamental difference in viewpoint remains: whereas science is concerned basically with analysis of the natural world, engineering is concerned with the design of objects (devices or systems) to perform a desired task.

1.2 Devices and Systems

Having described engineering in terms of design, we now divide engineering interests into two camps: *device engineering* and *systems engineering*. It is difficult to define these terms precisely. Intuitively, systems engineering is concerned with the design of entire assemblages of parts, whereas device engineering is concerned with the individual parts themselves. A transistor can be thought of as a device, but an entire radar installation, involving thousands of transistors as well as other devices, is a system. Clearly, a hierarchy of systems can be established. For example, a radar installation can be considered as a component or subsystem in a larger national defence system of many radar installations.

Since the systems engineer is concerned with the design of interconnected components, his interest centers primarily on structure rather than the physical nature of the components involved. He uses mathematics and computers as basic tools. The device engineer, on the other hand, draws extensively on physics.

The role of experimentation in device and systems engineering provides another contrast. The device engineer can experiment readily with his actual devices, but experimentation with systems is very difficult and costly. Thus the systems engineer must be fairly certain of the overall performance of his system before it is built. Computer studies and computer simulation of proposed systems are his experiments.

An important aspect of systems engineering is its organized approach to design. Because of the complexity of many systems problems, a systematized approach is mandatory. Six aspects of the systems-engineering approach are listed below.

1. Performance specification.
2. Modeling.
3. Simulation.
4. Analysis.
5. Optimization.
6. Physical realization.

We shall discuss each of these aspects in detail.

As noted earlier, systems engineering is basically the design of a system to meet a need or goal. This need is a form of system specification. Also, certain constraints (such as available components and resources) provide additional specifications. An important part of the systems problem is to express these specifications in a form suitable for the following mathematical parts of the problem. Thus the initial step in the systems approach is to identify the real problem, given a usually rather ill-defined need, and to develop performance specifications useful in the evaluation of proposed system designs.

Modeling is the representation of the system and its components in mathematical form. Models can be developed from applications of basic physical principles (for example, Newton's laws, Kirchhoff's laws) or from measurements conducted on components (for instance, frequency responses). The basic skill required in modeling is the ability to represent all the significant behavior of the corresponding physical device without undue complexity. Clearly, experience is invaluable in developing this skill. Certain principles of approach can be learned, however, and these will be illustrated in the text.

Analysis refers to the study of the component and system model characteristics, such as output response to specified inputs. Several analysis methods, including computer-oriented methods suitable for the study of large-scale systems, are discussed in the chapters to follow. Methods applicable to both linear and nonlinear systems are treated.

System synthesis refers to the design of systems using mathematical models and analytical performance measures. Often a performance criterion such as cost or fuel consumption is to be minimized. Designs that extremize a performance measure are said to be *optimum*, that is, they do the best job possible in the sense of the performance index chosen. High-speed large-memory digital computers have made mathematical optimization techniques practical as synthesis methods and large-scale system simulation studies a reality. Hence, systems can be designed using realistic performance measures, and the proposed design can be evaluated by simulation. This combination of optimization and simulation, discussed at length in the following text, is the heart of the systems approach to design.

Finally, of course, the proposed system must be realized, using actual physical hardware. It is thus "tuned up," and its performance is evaluated. The device and systems engineer share in attaining the ultimate goal—a system that meets a specified need.

Clearly, these aspects of system design are interrelated. For example, the type and complexity of the models allowed depend on the analysis tools available. Also, the design procedure is usually iterative. Successively more elaborate versions of the same basic problems are modeled, analyzed, and synthesized before a final design is established.

1.3 Introduction to the Attitude Control Problem

We introduce and illustrate some aspects of systems engineering in the context of a theme example. This approach provides motivation, since each different concept or technique is discussed in terms of a continuing example.

Many space missions require a specified orientation of a satellite or spacecraft, with respect to prescribed references axes, in the presence of disturbances or perturbations. For example, the Orbiting Astronomial Observatory is a 3600-pound satellite that can perform several astronomical observations. Its orientation in a desired direction must be controlled very precisely if accurate observations of the stars are to be made. Another example is spacecraft reentry into the atmosphere. If the entry is too abrupt, excessive heat will be generated, and perhaps the vehicle will be destroyed. Conversely, if the reentry is grazing, the craft may "skip" on the atmosphere like a stone on a pond, rather than reentering as desired. A third example involves the maintenance of a desired orientation of an Earth-Mars vehicle so that (1) the vehicle power source—solar cells—is always illuminated by the sun, and (2) the communication antenna is always pointed toward the earth. These are only a few of the many, many examples that could be mentioned.

The orientation of a vehicle with respect to reference axes is called *attitude*. The function of *attitude control*, therefore, is to obtain a desired vehicle attitude. Desired vehicle attitudes are determined from the nature of the particular space mission being considered.

A spacecraft may be subjected to a variety of disturbances or perturbations that would result in deviations from desired attitudes if control is not used. For example, aerodynamic torques (lift, drag) are very significant during liftoff and reentry. Propellant motion ("sloshing"), as well as motion of astronauts, can exert large torques on the vehicle. Gravity effects, solar radiation pressure, meteoroids, and magnetic torques, although smaller than the torques just mentioned, may have large cumulative effects for long flights. Finally, the geometry of the vehicle itself can produce disturbance torques because of coupling between various motions, that is, alignment of one axis may produce misalignment of other axes. The atitude control system must achieve the desired attitude in the presence of such disturbances.

Clearly, the control system cannot achieve exactly the desired attitude in all situations. Performance criteria must be established that specify not only the desired attitude but also the acceptable accuracy. Transient response criteria are also important. For example, speed of response in carrying out an attitude maneuver may be significant. Another vital consideration is fuel consumption, usually a specification of utmost importance for space missions. In addition, weight, reliability, and state of the art of sensor technology will all have determining roles in the design.

A block diagram of the control configuration is shown in Figure 1.1. The input to the complete attitude control system is the desired attitude; the system output is the vehicle attitude. The vehicle is controlled by applying control torques. The vehicle is also subjected

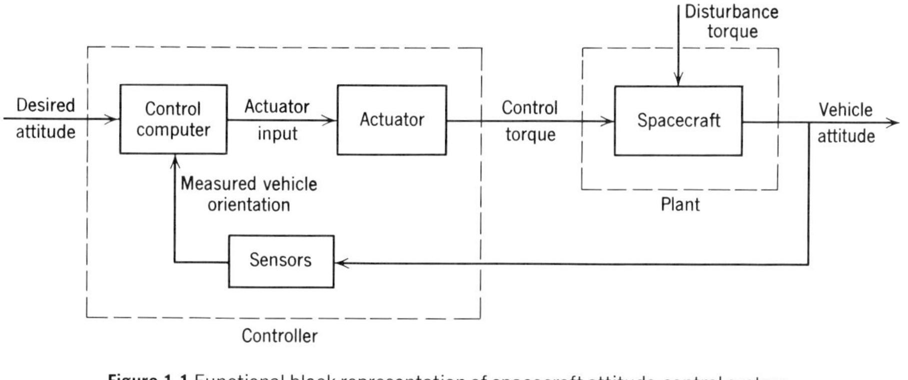

Figure 1.1 Functional block representation of spacecraft attitude-control system.

to disturbance torques. The controller is to be designed to satisfy the performance specifications and includes an "actuator" that applies the control torques to the vehicle. Gas jets and motor-driven flywheels are two commonly used actuators. The controller may be of the "feedback" type, as shown, with the actual attitude sensed and used for generating the control.

There are facets of attitude control not treated in this book. Moreover, some of the illustrations given may not correspond to current practice in the aerospace industry. However, they do provide insight into concepts and methods of systems engineering.

PART I

Modeling and Simulation

Chapter 2. Some Basic Elements for Network Modeling
Chapter 3. Network Modeling of Devices and Systems
Chapter 4. System Simulation
Chapter 5. Modeling Spacecraft Motion

Modeling is one of the important aspects in the engineering of a dynamic system. It involves the mathematical formulation of cause and effect interactions among the significant variables of a system. The mathematical model must enable the designer to predict the performance of the physical system that is being modeled to an acceptable degree of accuracy. The methods of analysis, optimization, and design of engineering systems are based on mathematical models. Thus it is important that the model be accurate but simple enough to be amenable to available mathematical tools of systems engineering.

The first step in the modeling of a physical system is the identification of its functions or operations. Three entities must be identified: *inputs*, *system functional block diagram*, and *outputs*. Inputs are the stimuli that cause the system to produce responses or outputs. The system functional block diagram is then partitioned or decomposed into an interconnection of subsystem functional blocks. Outputs of some subsystems are inputs of other subsystems. The interactions among the subsystems are modeled, using interconnection or topological rules and terminal constraints (for example, the output of one subsystem may be the input of another). Depending on the stage of analysis or design of a specific system, the various subsystem blocks may represent quite elementary devices or quite complex systems by themselves. Although there is great flexibility in the partitioning or decomposition of the whole system, the key idea is that the system model has two levels: (1) a set of models for the various subsystems, and (2) a model for the interconnections of the subsystems. Any subsystem may be decomposed also as an interconnection of simple subsystems.

As a practical matter, the amount of decomposition should be kept small. The smallest level of subsystem beyond which functional block decomposition is not carried out is defined as a *system component*. The determination of the input-output behavior of a system component or a subsystem is facilitated by considering another type of mathematical model called a *network diagram* or simply a *network*. Network analysis is based on a decomposition of the network into an interconnection of basic parts called *network elements*. The interaction among the network elements is governed by topological rules such as Kirchhoff's laws for electrical networks and D'Alembert's principle for mechanical networks. The theory of networks is well developed, and it is very useful in system analysis as well as synthesis.

The mathematical relationships that are obtained by network analysis can be translated into block diagram form for further system analysis or optimization. When the elementary system components correspond to operations in a computer, the resulting block diagram can serve as a guide for computer simulation.

The essential steps in modeling and simulation, as described above, are treated in Chapters 2, 3, 4, and 5. The principles will be studied in the context of modeling and simulation of a specific system. The specific physical dynamic system that will be studied throughout the book is a spacecraft attitude control system.

As far as modeling is concerned, the first step is to identify the inputs and outputs and to form a functional block diagram. For our system, the outputs are attitude angles and angular rates of rotation. Inputs are desired angular positions or rates that may be in the form of radio messages, or signals generated by a computer on board, or signals generated by an astronaut. Figure I.1 shows the system functional block diagram. Note that the attitude control system is not the space vehicle itself but the mechanism by which the desired attitudes are processed to produce the actual attitudes. This functional block diagram may be decomposed into an interconnection of functional subsystems as shown in Figure I.2. The spacecraft equations of motion relate torques to attitude angles. For simple single-axis angular motion, these equations may be obtained from network models described in Chapters 2 and 3. Modeling methods for more complicated three-axis motion is treated in Chapter 5. The sensors are functional subsystems that measure the vehicle attitudes and angular rates of rotation. The outputs are signals

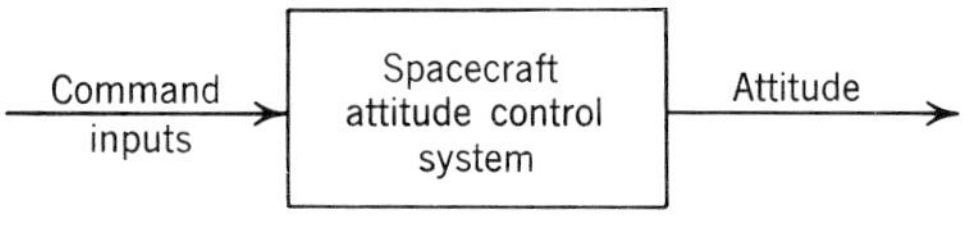

Figure I.1 Functional block representation of spacecraft attitude control system.

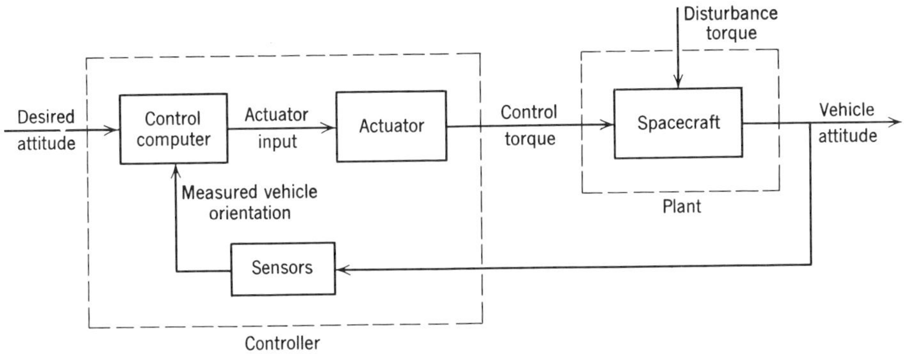

Figure 1.2 Functional block diagram of spacecraft attitude control system with some subsystems identified.

(usually electrical) that represent the measured quantities. Chapter 5 includes the modeling of the gyroscope, a widely used sensor. The subsystem in the dotted lines is called a controller. It processes the input signals, such as desired attitude angles, and the measured vehicle orientation as provided by the sensors to produce the torque that finally controls the vehicle attitudes. The controller is further partitioned into two subsystems, one called the control computer and the other called the actuator. The control computer is discussed in Part III, Control and Optimization. The actuator supplies control torques to the vehicle.

There are two classes of actuators. Control jets and reaction wheels, discussed in Chapters 2 and 3, are examples of *active* torque sources. They rely on on-board energy sources for their operation. *Passive* or *environmental* torque sources, such as gravity gradients, solar radiation, and magnetic torques, make use of energy sources present in the environment. Intensive study is being given currently to a variety of methods of production of control torques. Rather than attempt a survey of all methods, we shall instead consider only reaction wheels and control jets in detail. However, the basic ideas involved in actuation should be well illustrated by those two examples.

Computer simulation of the models is treated in Chapter 4. The computer simulation block diagram is similar to the functional block diagram except that the basic components are elementary mathematical operations such as addition, multiplication by a constant, and integration. Computer simulation of the system is based on the models of the subsystems.

In summary, physical system modeling starts with an identification of input and output quantities and a functional block diagram that symbolizes the processing of input signals to obtain output signals. The block diagram is then decomposed as an interconnection of subsystem functional blocks. The input-output characteristic of each subsystem functional block may be obtained by analyzing the network model for each subsystem. Network models aid in developing block-diagram models that are the basis for computer simulation or further system analysis. The principles of modeling and simulation developed in Chapters 2 to 5 are applied to the attitude control system throughout these chapters.

CHAPTER 2

Some Basic Elements for Network Modeling

The basic elements discussed in this chapter are sometimes adequate models for electrical and mechanical devices. More often, devices can be modeled with sufficient agreement with acutal measurements, using an interconnection of elements. Only the very commonly used elements for electrical and mechanical devices are discussed.

Two types of variables are used in defining network elements: *through variables* and *across variables.* Through variables are associated with flow quantities such as fluid through a pipe, current through a resistor, or force through a mass. Examples of across variables are pressure difference between two points on a pipeline, voltage drop across a resistor, and velocity difference between the ends of a spring. The quantitative use of these variables, including the writing of element equations, requires the adoption of a consistent set of units. Appropriate units for the mechanical and electrical elements discussed in this chapter, together with their dimensions, are given in Tables 2.1 and 2.2.

Table 2.1 Dimension and Units for Mechanical Quantities

Quantity	MKS unit	Dimension in Terms of Force, Length, Time	Dimension in Terms of Mass, Length, Time
Time t	Second (sec)	t	t
Length l	Meter (m)	l	l
Force f	Newton (n)	f	mlt^{-2}
Mass M	Kilogram (kg)	$fl^{-1}t^{2}$	m
Torque τ	nm	fl	$ml^{2}t^{-2}$

Quantity	MKS unit	Dimension in Terms of Force, Length, Time	Dimension in Terms of Mass, Length, Time
Displacement x	m	l	l
Velocity v	m/sec	lt^{-1}	lt^{-1}
Acceleration a	m/sec^2	lt^{-2}	lt^{-2}
Angular displacement θ	Radian (rad)	Dimensionless (arc length/ radius)	Dimensionless (arc length/ radius)
Angular velocity ω	rad/sec	t^{-1}	t^{-1}
Angular acceleration $\dot{\omega}$	rad/sec^2	t^{-2}	t^{-2}
Translational spring stiffness coefficient K	n/m	fl^{-1}	mt^{-2}
Translational damping coefficient D	n/(m/sec)	$fl^{-1}t$	mt^{-1}
Torsional spring stiffness coefficient K	nm/rad	fl	ml^2t^{-2}
Torsional damping coefficient D	nm/(rad/sec)	flt	ml^2t^{-1}
Moment of inertia J	nm/(rad/sec^2)	flt^2	ml^2
Momentum Mv	n sec	ft	mlt^{-1}
Angular momentum $J\omega$	nm sec	flt	ml^2t^{-1}
Work or energy	nm	fl	ml^2t^{-2}
Power	nm/sec	flt^{-1}	ml^2t^{-3}

The element models relate the through and across variables for each element. The equations relating the element variables such as voltage and current associated with a resistor are postulated on the basis of experimental evidence. For example, if the voltage across a

Table 2.2 Dimensions and Units for Electrical Quantities

Quantity	Unit	Equivalent Unit in Newtons, Meter, Second, Ampere	Dimension in Terms of Force, Length, Time, and Current
Time t	Second (sec)	sec	t
Length l	Meter (m)	m	l
Force f	Newton (n)	n	f
Current i	Ampere (a)	a	i
Charge q	Coulomb (cl)	a sec	it
Voltage v	Volt (v)	nm/(a sec)	$fli^{-1}t^{-1}$
Flux linkage λ	v sec	nm/a	fli^{-1}
Resistance R	Ohm, v/a	nm/(a^2 sec)	$flt^{-1}t^{-2}$
Capacitance C	Farad, cl/v	a^2 sec^2/(nm)	$i^2t^2f^{-1}l^{-1}$
Inductance L	Henry, v sec/a	nm/a^2	fli^{-2}
Work or energy	Joule (jl)	nm	fl
Power	Watt, jl/sec	nm/sec	flt^{-1}

resistor is plotted as a function of the current through it, the graph is approximately a straight line over a range of values for current. The resistor model is then defined by a linear equation expressing voltage in terms of a constant times current.

When elements are interconnected, the through and across variables must satisfy equilibrium laws. These laws are described in Chapter 3 where the elements are combined to represent devices and subsystems.

2.1 Linear Time-Invariant Mechanical Elements

Models for characterizing the planar motion of mechanical devices and systems may be constructed from a few types of mechanical elements. Planar motion may be resolved into two categories: rectilinear (or translational) and rotational. That is, planar motion can be represented by a combination of translational motion and rotational motion. Translational elements will be defined first.

Figure 2.1*a* shows a pictorial representation of a mass model for a block of material undergoing translational motion. The block of material is assumed to be sliding on a frictionless horizontal surface. The velocity vector of the mass, relative to the inertial frame of reference, and the force vector acting *on* the mass, are colinear and with positive reference direction indicated by the arrows in Figure 2.1*a*. The component of the velocity vector along the indicated direction is denoted by ν, and the component of the force vector along the same direction is denoted by f. Notice that f and ν are scalar quantities that are the components of the force and velocity vectors. The defining model equation for the mass element based on Newton's law is

$$f = \frac{d}{dt}(M\nu) = M\frac{d\nu}{dt} \tag{2.1}$$

where the constant M is the mass parameter.

The product $M\nu$ is called *momentum*, so that the force acting on the mass is equal to the time rate of change of momentum. Since the mass parameter for the mass model defined by (2.1) is a constant, it factors out of the time differentiation.

The network symbol for mass is shown in Figure 2.1*b*. The network symbol is more useful than the pictorial diagram when studying interconnections of various mechanical component models. First, notice that the force variable f and the velocity variable ν in the network diagram are scalars. Second, observe that M is represented as a two-terminal model. The defining equation for mass, as in (2.1), involves the velocity of the mass with respect to an inertial frame. This reference frame is part of the definition for ν associated with the mass.

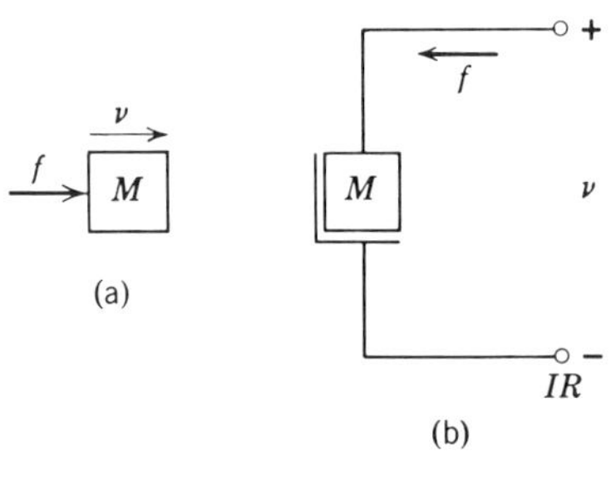

Figure 2.1 (*a*) Pictorial diagram of mass with force and velocity vectors in the same direction. (*b*) Network symbol for mass.

As a reminder that the lower node is the inertial frame of reference, it will be distinguished by the label *IR*. This is especially helpful in remembering that in a mechanical network, *all* mass elements have a terminal in common. Third, notice that for the variable ν in the network diagram, plus and minus reference marks are used and for the variable f, a reference direction is used. By definition, if the velocity and force vectors of the pictorial diagram have the same reference direction, then the plus reference of the ν in the network diagram is placed near the tail of the reference arrow of the f in the network diagram, as in Figure 2.1. The plus and minus marks associated with the across variable ν signify that ν is the velocity of the node with the plus mark with respect to the node with the minus mark. The definition for mass above is a special form of Newton's second law of motion where f is the net force acting on the mass. If the motion is not horizontal, then f includes a component of the force due to gravity.

If f is the input variable and ν is the output variable, then by integrating both sides of (2.1) and solving for $\nu(t)$, we obtain

$$\nu(t) = \nu(t_0) + \frac{1}{M}\int_{t_0}^{t} f(\tau)\, d\tau. \tag{2.2}$$

Furthermore, since ν is the rate of change of position x,

$$\nu = \frac{dx}{dt}, \tag{2.3}$$

then

$$x(t) = x(t_0) + \int_{t_0}^{t} \nu(\xi)\, d\xi. \tag{2.4}$$

Substituting (2.2) in (2.3), position is related to force by

$$x(t) = x(t_0) + (t - t_0)\nu(t_0) + \frac{1}{M}\int_{t_0}^{t}\int_{t_0}^{\xi} f(\tau)\, d\tau\, d\xi. \tag{2.5}$$

In addition to the force function of time from t_0 to t, initial values of position and velocity, $x(t_0)$ and $\nu(t_0)$, are needed in order to determine the position $x(t)$ at time t. The initial quantities $x(t_0)$ and $\nu(t_0)$ are said to specify the *initial state* of the mass. For a general time instant t, $x(t)$ and $\nu(t)$ specify the *state* of the mass.

The next translational element we consider is the *spring*. A pictorial diagram is shown on Figure 2.2*a*. Since any mass of the physical device may be separately modeled by the mass element of Figure 2.1, it is reasonable to exclude the mass of the spring device in order to concentrate on its elasticity, its principal property. Let the force vector acting *on* node a of the spring, the velocity vector of node a, and the velocity vector of node b be colinear and with the same positive direction as indicated in Figure 2.2*a*. Denote the components of these vectors by f, v_1, and v_2, respectively. Note that f, v_1, and v_2 are scalar quantities. Denote the difference between v_1 and v_2 by v

$$v = v_1 - v_2. \tag{2.6}$$

The defining equation for the spring is

$$\frac{df}{dt} = Kv. \tag{2.7}$$

The constant coefficient K is called the *coefficient of stiffness*, and (2.7) is the defining equation for a *linear time-invariant spring*, which relates velocity to time rate of change of force linearly.

Integrating (2.7) from time t_0 to time t,

$$f(t) - f(t_0) = \int_{t_0}^{t} Kv(\tau)\, d\tau. \tag{2.8}$$

Since v is the time rate of expansion of the distance between the terminals of the spring,

$$v = \frac{dx}{dt} \tag{2.9}$$

where x is the distance between the terminals of the spring. Substituting (2.9) in (2.7) and integrating from t_0 to t,

$$f(t) - f(t_0) = K[x(t) - x(t_0)]. \tag{2.10}$$

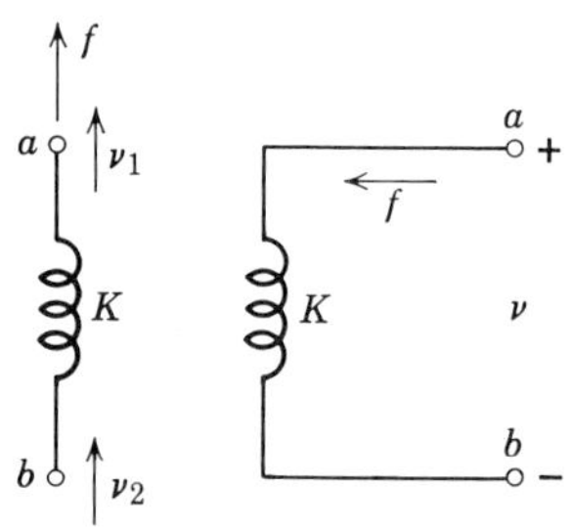

Figure 2.2 (*a*) Pictorial diagram for spring model. (*b*) Network symbol for spring.

Thus the net *change* in force, which tends to elongate the spring, is linearly proportional to the net *change* in the length of the spring. This relationship is known as *Hooke's Law*.

Figure 2.2*b* shows the network symbol for the spring. As in the mass network symbol, f and v are not space vectors, and the actual space orientation must be obtained from the pictorial diagram. The plus and minus marks associated with the across variable v signify that v is the velocity of the node with the plus mark with respect to the node with the minus mark. In Figure 2.2*b*, v is the velocity of node *a* with respect to node *b*. The force f is the force *on* the spring at node *a*.

Our third model is based on viscous frictional forces that depend on velocity. In many instances, devices called dampers or dashpots are built to provide the frictional forces. To depict such frictional phenomena, we introduce a damper model for a dashpot shown pictorially on Figure 2.3*a*. The piston velocity of the dashpot is indicated by the motional variable v_1. Similarly, the motional variable v_2 indicates the velocity of the cylinder. As the piston moves relative to the cylinder, the oil moves from one end of the cylinder to the other through the orifices. The force acting on the piston at *a* is approximately linearly proportional to the relative velocity of the piston with respect to the cylinder and, as a simple model, we define the *linear time-invariant damper* to have the characteristic

$$f = Dv \tag{2.11}$$

where f is the component of the vector force of the piston, v is the component of the relative velocity vector of the piston with respect to the cylinder, which is $v = v_1 - v_2$, and D is a constant parameter called the *damping constant*. The position of the dashpot in Figure 2.3*a* may be reversed so that the force vector is acting on the cylinder, the velocity component of the cylinder is v_1, and the velocity component of the piston is v_2; (2.11) still applies. As in the previous two models, the force f and the velocity v are colinear and with the same positive direction as indicated in Figure 2.3*a*. The force f is acting on the dashpot at node *a*, and v is the velocity of node *a* with respect to node *b*. The network symbol is shown on Figure 2.3*b*. The plus and minus symbols associated with v signify that v is the velocity on the node with the plus mark with respect to the node with the minus mark.

We also introduce a force source model and a velocity source model. The symbols for the f source and v source are shown in Figure 2.4*a* and 2.4*b*, respectively. The force source is defined by the property

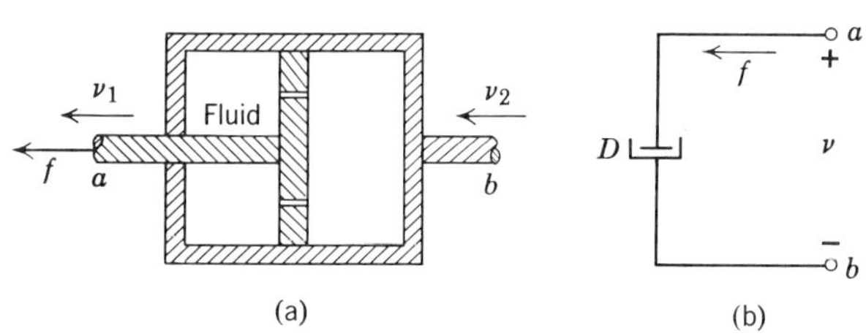

Figure 2.3 (*a*) Pictorial diagram for dashpot model. (*b*) Network symbol for dashpot model.

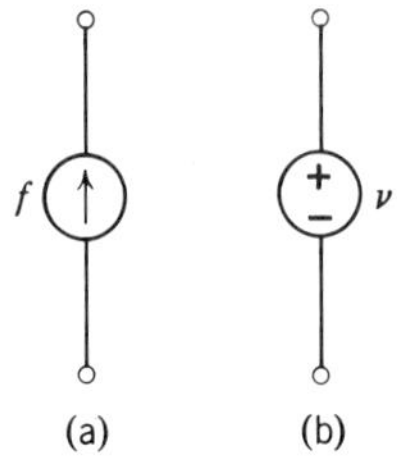

Figure 2.4 (*a*) Symbol for force source. (*b*) Symbol for velocity source.

that the force is independent of the velocity associated with it. It is called an independent force source if it is independent of the nature of the rest of the mechanical models connected ot it. Thus the independent force source is not affected by the velocity of the node resulting from the application of f. Similarly the independent velocity source has the property that the velocity is independent of the forces that develop as a result of the application of ν.

For rotational motion in a plane, we introduce three elements analogous to M, K, and D, and two sources analogous to f and ν. The through and across variables are torque τ and angular velocity ω. The source symbols are the same as those in Figure 2.4 except that instead of f we have τ and instead of ν we have ω.

Instead of mass, we have rotational inertia whose pictorial diagram is shown in Figure 2.5*a*. Note that the torque is acting *on* the rotational inertia. The defining equation for *rotational inertia* is

$$\tau = \frac{d}{dt}(J\omega) = J\frac{d\omega}{dt} \tag{2.12}$$

where J is a constant parameter called the *moment of inertia*.

The product $J\omega$ is defined as the *angular momentum* of the rotational inertia rotating at angular velocity ω with respect to the inertial reference frame. Thus, torque is the time rate of change of angular momentum. Since J is assumed to be a constant, it factors out of the time differentiation. It is assumed that the rotational motion is about the center of mass, and that τ and ω have the same sense for positive direction. Furthermore, ω is the angular velocity with respect to the

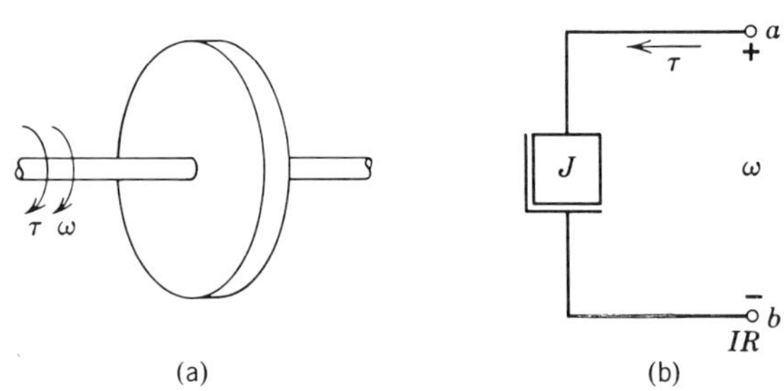

Figure 2.5 (*a*) Pictorial diagram for rotational inertia. (*b*) Network symbol for J is the same as for M except that the network variables are τ and ω instead of f and ν.

inertial reference frame. The network symbol is shown in Figure 2.5*b*. Notice that τ and ω are the components of the torque and angular velocity vectors whose orientations are shown in Figure 2.5*a*. Hence these are scalar quantities. As in the mass element, one terminal of the rotational inertia network element symbol is distinguished by the label *IR* as a reminder that the terminal represents the inertial reference frame. In a mechanical network, *all* rotational inertia elements have one terminal in common. The plus and minus symbols associated with ω on the network diagram have the same significance as before.

Instead of the translational spring, we have torsional or rotational spring as another model for rotational motion in one coordinate. A pictorial diagram is shown in Figure 2.6*a*. A torque on the shaft causes it to twist, so that there may be a difference between ω_1 and ω_2. Denote $\omega_1 - \omega_2$ by ω,

$$\omega = \omega_1 - \omega_2. \tag{2.13}$$

Thus ω is the angular velocity of the shaft at cross-section *a* with respect to the shaft at cross-section *b*. Note that the torque τ and the relative angular velocity ω have the same sense of positive direction, and that τ is the torque acting *on* the torsional spring at cross-section *a*. The defining equation for the linear time-invariant *torsional spring* is

$$\frac{d\tau}{dt} = K\omega \tag{2.14}$$

where K is a constant parameter called the *torsional stiffness coefficient*. The network symbol is shown on Figure 2.6*b*. The plus and minus symbols associated with ω in the network diagram signify that ω is the angular velocity of node *a* with respect to node *b*.

Analogous to translational forces due to viscous friction, we have rotational forces or torques due to viscous friction. To model this type

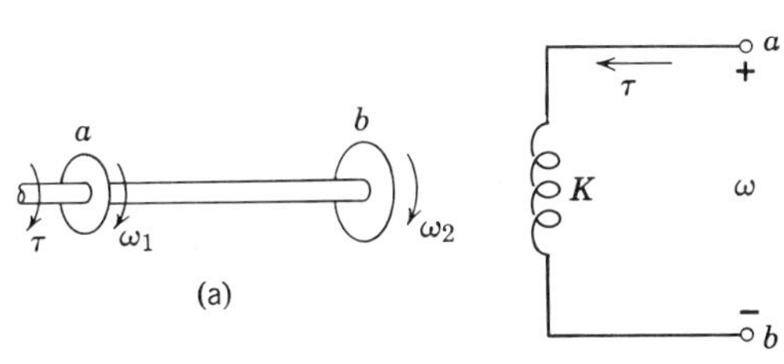

Figure 2.6 (*a*) Pictorial diagram for rotational or torsional spring. (*b*) Network symbol for rotational spring or torsional spring.

of phenomenon, we introduce an element for rotational or torsional damping whose pictorial diagram is shown in Figure 2.7*a*. The larger cylinder is connected to the left-hand side shaft and the smaller cylinder inside the larger cylinder is connected to the right-hand side shaft. Between the two cylinders is some viscous fluid. The torque acting on the larger cylinder is related to the relative velocity of the left-hand side shaft with respect to the right-hand side shaft. As a simple model, we have the defining equation for a *torsional* or *rotational damper*

$$\tau = D\omega \tag{2.15}$$

where $\omega = \omega_1 - \omega_2$ is the component of the relative angular velocity vector of the larger cylinder with respect to the smaller cylinder and torque is applied to the larger cylinder at a. As in the linear or translational damper, the position of the shafts in Figure 2.7*a* may be reversed so that the torque is now acting on the smaller cylinder. Equation (2.15) still applies. The constant parameter D is called the *torsional* or *rotational damping constant*. Figure 2.7*b* shows the network symbol for D. As before, τ, ω_1, and ω_2 are scalar quantities, since they are just the components of the vectors whose orientations are shown in Figure 2.7. The angular velocity $\omega = \omega_1 - \omega_2$ is the angular velocity of shaft a with respect to shaft b. The torque τ and the angular velocity ω have the same sense of positive direction. The torque τ acts on the torsional damper at shaft a. The plus and minus marks associated with ω in the network diagram have the same meaning as before.

The elements mass, spring, damper, rotational inertia, torsional spring, and torsional damper, defined in this section, relate the associated through and across variables (or their integrals) by linear homogeneous algebraic equations. The coefficients of proportionality, which are the element parameters, are also independent of time. For these reasons, the elements are called *linear* and *time-invariant*. Some nonlinear models will be considered in Section 2.3.

Table 2.3 summarizes the terminal relations for the through and

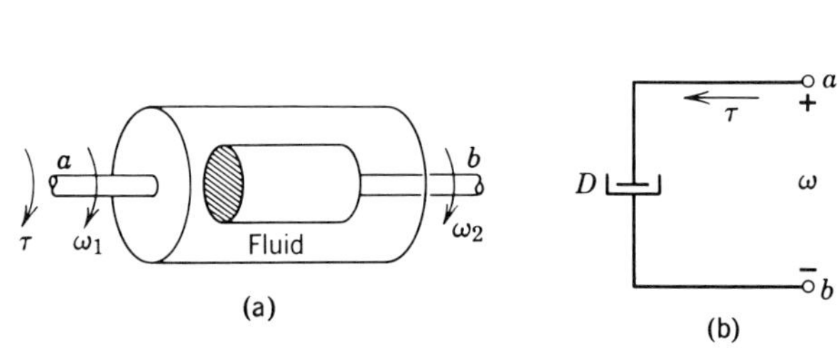

Figure 2.7 (*a*) Pictorial diagram for torsional or rotational damping. (*b*) Network symbol for torsional or rotational damping.

Table 2.3 Linear Time-Invariant Mechanical Elements

Translational			Rotational		
Element	Symbol	Equation	Element	Symbol	Equation
Mass *M*	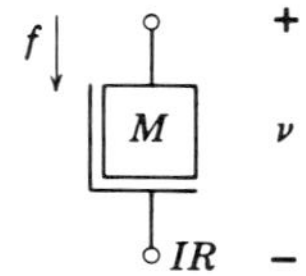	$f = M\frac{d\nu}{dt}$	Rotational inertia *J*	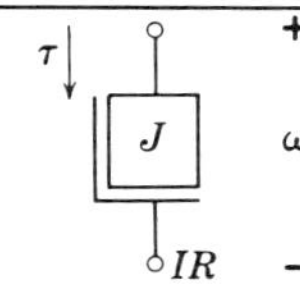	$\tau = J\frac{d\omega}{dt}$
Translational spring *K*	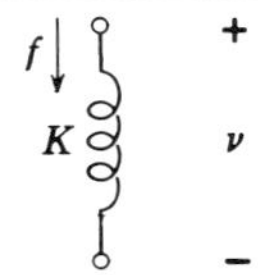	$\frac{df}{dt} = K\nu$	Torsional spring *K*	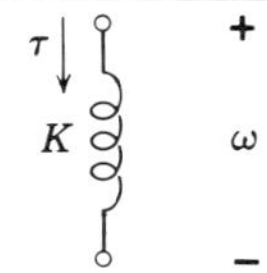	$\frac{d\tau}{dt} = K\omega$
Translational damper *D*	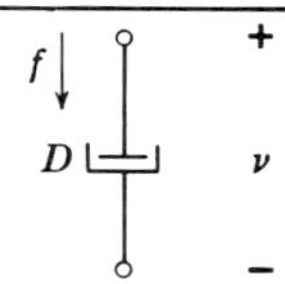	$f = D\nu$	Torsional damper *D*	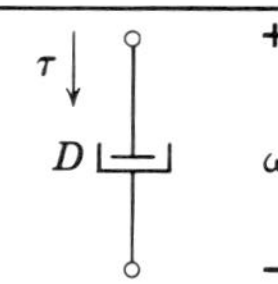	$\tau = D\omega$
Force source *f*	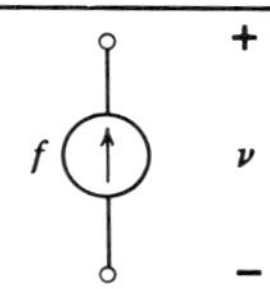	$f = f(t)$ independent of ν	Torque source τ	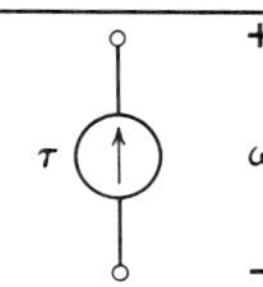	$\tau = \tau(t)$ independent of ω
Velocity source ν	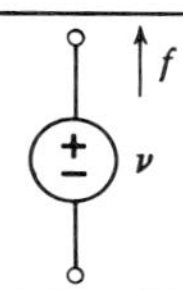	$\nu = \nu(t)$ independent of *f*	Angular velocity source ω	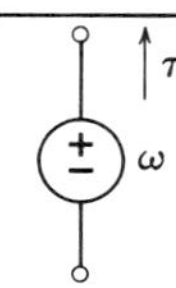	$\omega = \omega(t)$ independent of τ

across variables, associated with linear time-invariant mechanical elements, for translational and rotational motion. The network symbols are also shown.

2.2 Single-Axis Gas-Jet Attitude Control System

Let us model the angular motion about a single axis of a space vehicle controlled with gas jets. Consider a simplified pictorial representation as in Figure 2.8*a* for a space vehicle where the attitude orientation from a reference axis, called the roll angle, is to be controlled by an appropriate firing of gas jets. The system block diagram is shown in Figure 2.8*b*.

Let us model the actuator first. Gas jets produce control action by mass expulsion. Control jets operate in the same manner as propulsion jets, except that the thrust requirements are much smaller. The physical actuator includes a container for the propellant, feed lines to transport the propellant, and valves and thrust chambers to permit propellant ejection.

Three basic types of propellants are in use: cold gases, monopropellants, and hypergolic bipropellants. Cold gas systems depend on the kinetic energy of the stored gas to supply forces. Nitrogen at 298°K is a feasible cold gas propellant. Monopropellants depend on

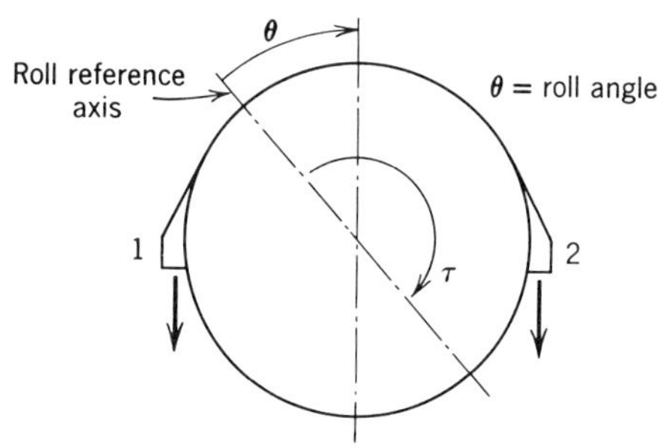

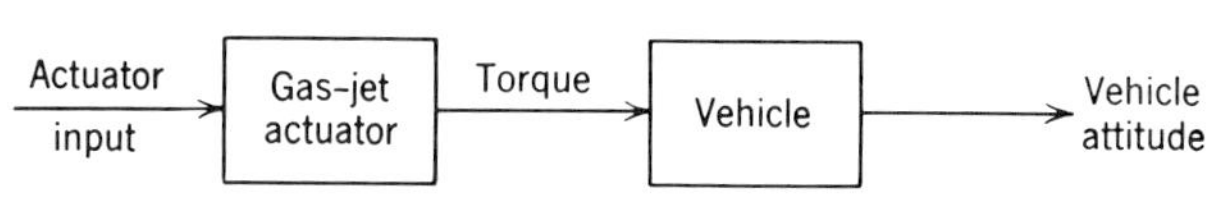

Figure 2.8 (*a*) Simplified pictorial end view of spacecraft single-axis attitude control by gas jets. (*b*) Block diagram of the system.

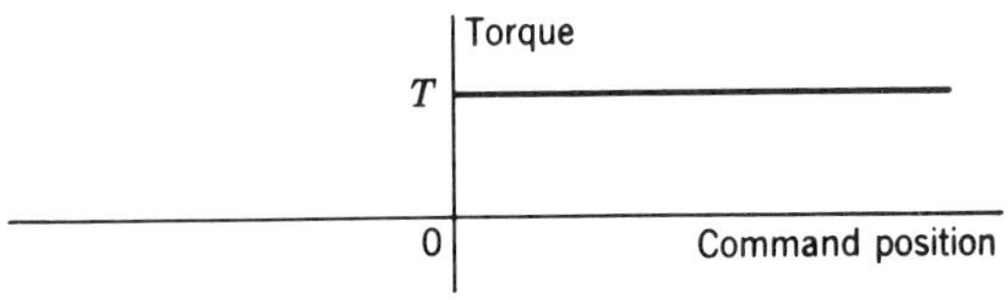

Figure 2.9 Gas-jet input-output characteristic.

chemical reactions for force generation. The monopropellant decomposition must be initiated by glow plug or catalyst bed. Hydrogen peroxide is a typical monopropellant. Hypergolic (self-igniting) bipropellant systems consist of two gases: an oxidizer and a fuel. These are stored separately and come into contact in the thrust chamber where self-ignition takes place.

Gas-jet actuators are basically on-off or "relay" devices, that is, the valves are either closed or open. Because of the small valve openings, it is difficult to obtain "proportional" flow by partially opening a valve. In practice, electronic drivers control the valve position. The device has an input-output characteristic as shown in Figure 2.9, where the command signal is the driver input signal.

The jets typically are mounted to produce rotations about the vehicle body axes in either direction. The spacecraft cross section shown in Figure 2.8 indicates one possible arrangement of gas jets for roll torques in either roll direction, firing one jet for roll in each direction. The torque vs. command signal characteristic for this configuration is shown in Figure 2.10.

In actual operation, the system purposely may be made to depart from the ideal on-off characteristic of Figure 2.10. For example, Figure 2.11 shows a jet actuator response having *deadband*. The deadband is introduced by electronic circuitry.

The characteristic in Figure 2.11 will be taken as the input-output model for the actuator. The output is the actual control torque. Thus the torque can have three values, $-T$, 0, and $+T$ depending on the valve position. The command signal or actuator input may be manipulated as a function of time. If command signal as a function of time

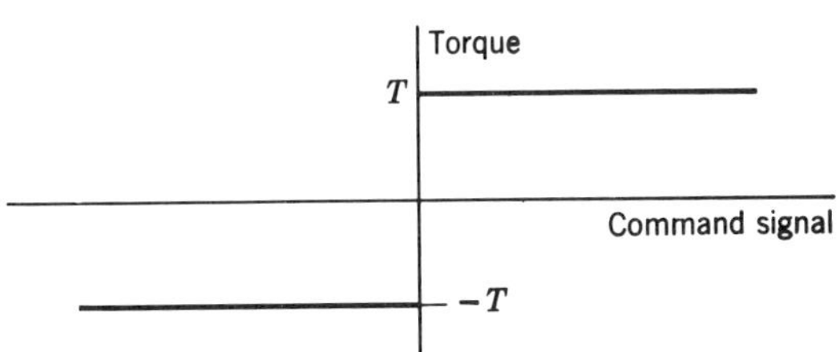

Figure 2.10 Input-output characteristic for two jets.

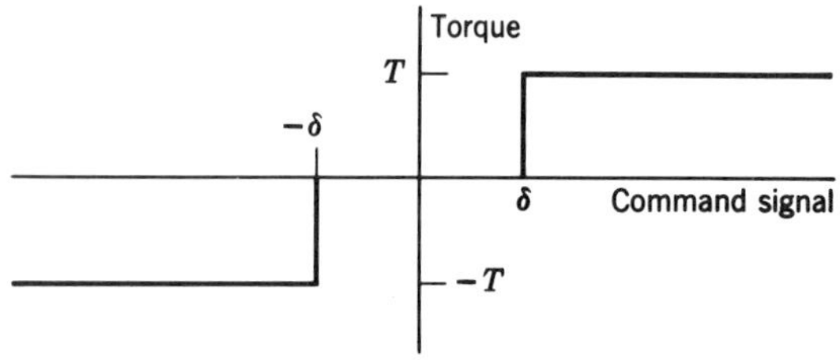

Figure 2.11 Two-jet actuator input-output characteristic with deadband.

is denoted by $p(t)$, then, from Figure 2.11, the input-output characteristic is given by

$$\begin{aligned} \tau(t) &= +T \quad &\text{for} \quad & p(t) \geqslant \delta \\ &= 0 \quad &\text{for} \quad & |p(t)| < \delta \\ &= -T \quad &\text{for} \quad & p(t) \leqslant -\delta. \end{aligned} \tag{2.16}$$

Next we model the subsystem that relates the torque to the attitude angle θ. A network diagram for this simple subsystem is shown in Figure 2.12.

The equation of motion for the space vehicle is given by (2.12), where τ is the torque supplied by the gas jets, J is the moment of inertia of the space vehicle, and ω is the angular rate $d\theta/dt = \theta$. Let us derive an expression for $\theta(t)$ in terms of the torque time function $\tau(t)$. Integrating (2.13) from t_1 to t,

$$\dot{\theta}(t) = \dot{\theta}(t_1) + \frac{1}{J}\int_{t_1}^{t} \tau(\xi)\, d\xi. \tag{2.17}$$

Integrating (2.17) from t_1 to t,

$$\theta(t) = \theta(t_1) + \dot{\theta}(t_1)(t - t_1) + \frac{1}{J}\int_{t_1}^{t}\int_{t_1}^{\eta} \tau(\xi)\, d\xi\, d\eta. \tag{2.18}$$

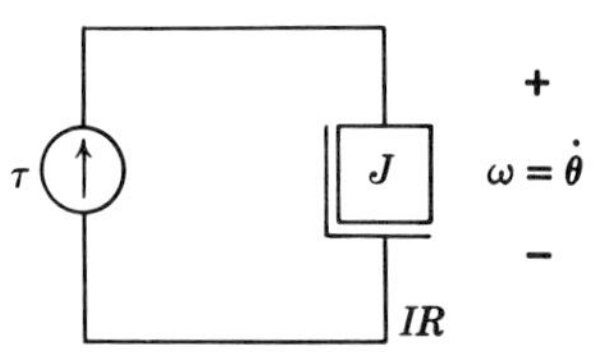

Figure 2.12 Network diagram for subsystem that relates torque τ to angle θ.

From (2.18) it is clear that in order to obtain the attitude angle θ at time t, we need the initial angle at time t_1, the initial angular velocity at time t_1, and the torque function from time t_1 to t.

For any given command signal as a function of time, the torque function of time may be obtained from the actuator characteristic of Figure 2.11. This torque function may be substituted in (2.18) to obtain $\theta(t)$. The desired system model consists of an actuator with the input-output characteristic of Figure 2.11 and a subsystem whose network representation is that shown in Figure 2.12. The system equations are (2.16) and (2.18), where $p(t)$ is the input and $\theta(t)$ is the output.

Example 2-1. Let us compute the angle $\theta(t)$ for a specific simple choice for τ.

Suppose that the initial states $\theta(t_1)$ and $\dot{\theta}(t_1)$ are both zero at the initial time $t_1 = 0$. If a constant torque T (for any $p(t) > \delta$) is then applied at $t = 0$, then from (2.17), the angular rate $\dot{\theta}(t)$ is

$$\dot{\theta}(t) = \frac{T}{J}t \tag{2.19}$$

which is linearly increasing with time, and the angle $\theta(t)$ from (2.18) is

$$\theta(t) = \frac{1}{2}\frac{T}{J}t^2 \tag{2.20}$$

which is parabolically increasing with respect to time. At some time t_2, the angle and angular rate can be evaluated from (2.19) and (2.20) as $\theta(t_2) = \frac{1}{2}(T/J)t_2^2$ and $\dot{\theta}(t_2) = (T/J)t_2$. Suppose that at time t_2, the torque is removed ($\tau = 0$). From (2.17), using t_2 as initial time,

$$\dot{\theta}(t) = \dot{\theta}(t_2) = \frac{T}{J}t_2 \tag{2.21}$$

for all $t \geq t_2$. Thus the spacecraft will continue to rotate at a constant angular velocity. From (2.18), the angle $\theta(t)$ becomes

$$\theta(t) = \frac{1}{2}\frac{T}{J}t_2^2 + (t - t_2)\frac{T}{J}t_2^2 \tag{2.22}$$

for $t \geq t_2$. Thus θ increases linearly with time starting with $t = t_2$. Thus, for torque which is equal to $+T$ from $t = -$ to $t = t_2$ and zero thereafter, the attitude is given by (2.20) from $t = 0$ to $t = t_2$ and given by (2.22) for $t \geq t_2$. This torque can be produced by any command signal $p(t)$ that satisfies

$$p(t) \geq \delta \qquad \text{for} \qquad 0 \leq t \leq t_2$$

and

$$|p(t)| < \delta \qquad \text{for} \qquad t > t_2. \blacktriangle$$

Example 2.2. Suppose that after θ attains the value $\theta(t_2)$ at time t_2 in Example 2-1, it is desired to stop the motion. In order to accomplish this, the angular velocity must be reduced to zero. But at time t_2, the angular velocity $\dot{\theta}(t_2)$ from (2.19) or (2.21) is $\dot{\theta}(t_2) = (T/J)t_2 > 0$. To slow it down we must apply a reverse torque or $\tau = -T$ for $t > t_2$. Using (2.17) with t_1 replaced by t_2,

$$\begin{aligned}\dot{\theta}(t) &= \dot{\theta}(t_2) + \frac{1}{J}\int_{t_2}^{t} \tau(\xi)\, d\xi \\ &= \frac{T}{J}t_2 - \frac{T}{J}(t-t_2), \end{aligned} \tag{2.23}$$

for $t \geqslant t_2$. The angular rate is linearly decreasing. To solve for the instant of time when $\dot{\theta}(t) = 0$, we set (2.23) to zero and obtain

$$\frac{T}{J}t_2 - \frac{T}{J}(t-t_2) = 0$$

or

$$t = 2t_2. \tag{2.24}$$

Denote this instant by t_3.

If the negative torque is not removed after t_3, the angular velocity $\dot{\theta}(t)$ will continue to decrease as given by (2.23). If the torque is removed for $t \geqslant t_3$, then $\dot{\theta}(t)$ will be zero for $t \geqslant t_3$ and the attitude angle will remain fixed at the value attained at $t = t_3$.

The angle for $t \geqslant t_2$ may be obtained by integrating (2.23) from t_2,

$$\theta(t) - \theta(t_2) = \frac{T}{J}t_2(t-t_2) - \frac{T}{2J}(t-t_2)^2. \tag{2.25}$$

But $\theta(t_2)$ from (2.20) which holds for $t_1 \leqslant t \leqslant t_2$ is $\frac{1}{2}(T/J)t_2^2$. Hence (2.25) becomes

$$\theta(t) = \frac{1}{2}\frac{T}{J}t_2^2 + \frac{T}{J}t_2(t-t_2) - \frac{T}{2J}(t-t_2)^2 \tag{2.26}$$

for $t \leqslant t \leqslant t_3$. At $t_3 = 2t_2$, (2.26) yields

$$\theta(t_3) = \frac{T}{J}t_2^2. \tag{2.27}$$

For $t \geqslant t_3$,

$$\theta(t) = \frac{T}{J}t_2^2. \tag{2.28}$$

Figure 2.13 shows the time histories for torque, $\dot{\theta}(t)$, and $\theta(t)$. It is clear that the $\dot{\theta}$ curve is the area under the τ curve divided by J, since

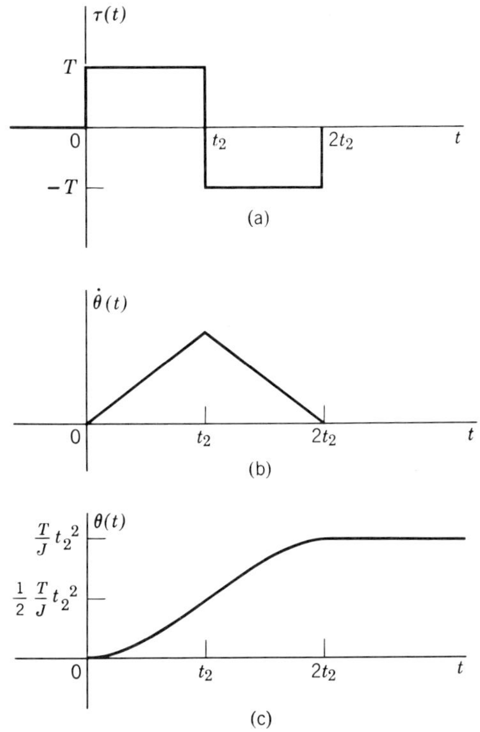

Figure 2.13 Torque, angular velocity, and attitude as functions of time.

the initial value of $\dot{\theta}(t_1)$ is zero. Similarly, it is seen that $\theta(t)$ is the area under the $\dot{\theta}(t)$ curve, since $\theta(t_1)$ is zero. A command signal function $p(t)$ that can generate the torque function in Figure 2.13 is easily obtained. ▲

2.3 Linear Time-Invariant Electrical Elements

Most representations or models for electrical devices involve the elementary models called resistor model, inductor model, and capacitor model. These three elements are useful not only for representing physical electrical components called resistor, inductor, and capacitor, but also for representing more complex devices by appropriate combinations of these three elements.

The *resistor* element is based on the assumption that the voltage and current associated with a device with two terminals may be related by an algebraic equation. The simplest such relationship is given by

$$v = Ri \tag{2.29}$$

where v and i have the reference directions shown on Figure 2.14 and R is a constant parameter called *resistance*. The symbol for the element is shown in Figure 2.15*a*.

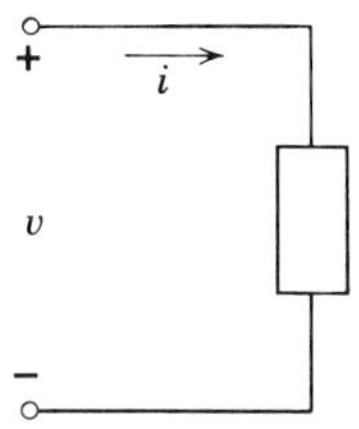

Figure 2.14 Reference directions for network variables used in defining elementary electrical models.

For another element called the *capacitor*, the charge (integral of current) and voltage are related by an algebraic equation. The simplest such representation is

$$q = Cv \tag{2.30}$$

where q is the charge, v is the voltage across the model, and C is a constant parameter called *capacitance*, which is the slope of the straight line $q = Cv$. Differentiating (2.30) with respect to time, we have

$$\frac{dq}{dt} = i = C\frac{dv}{dt}. \tag{2.31}$$

The capacitor element symbol is shown in Figure 2.15*b*.

We may integrate both sides of (2.31) from some time t_0 to t and obtain

$$\int_{t_0}^{t} i(\tau)\, d\tau = C[v(t) - v(t_0)] \tag{2.32}$$

or

$$v(t) = \frac{1}{C}\int_{t_0}^{t} i(\tau)\, d\tau + v(t_0). \tag{2.33}$$

Since the integral of i from t_0 to t is the net charge added to the capacitor from time t_0 to time t, then the integral equation of (2.33) is simply a restatement of (2.30) in the form $v(t) = (1/C)q(t)$ where $q(t)$ is the total charge on C at time t.

The third element we introduce is also a two-terminal model, where the magnetic flux linkage (integral of voltage) and current are related by an algebraic equation. It is called the *inductor*. Again, the linear equation happens to be the simplest one:

$$\lambda = Li \tag{2.34}$$

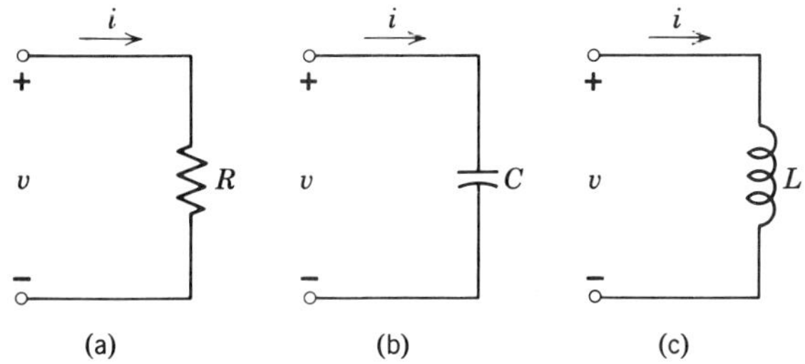

Figure 2.15 Symbols for the resistor, capacitor, and inductor elements are shown in (*a*), (*b*), and (*c*), respectively.

where λ is the magnetic flux linkage, i is the current, and L is a constant parameter called *inductance*, which is the slope of the straight line on the $\lambda - i$ plane. Differentiating both sides of (2.34), we have

$$\frac{d\lambda}{dt} = v = L\frac{di}{dt}. \tag{2.35}$$

The inductor element symbol is shown in Figure 2.15*c*.

We may integrate both sides of (2.35) from t_0 to t to obtain

$$\int_{t_0}^{t} v(\tau)\, d\tau = L[i(t) - i(t_0)] \tag{2.36}$$

or

$$i(t) = \frac{1}{L}\int_{t_0}^{t} v(\tau)\, d\tau + i(t_0). \tag{2.37}$$

The resistor, inductor, and capacitor models described above are *linear* as well as *time-invariant* because the defining equations (2.29), (2.30), and (2.34), which relate the corresponding through and across variables (or their integrals), are linear homogeneous algebraic equations. Furthermore, the parameters R, L, and C are independent of time.

There are two types of models for electrical sources: *voltage source* and *current source*. The symbols for these models are shown in Figure 2.16. The voltage source is defined by the property that the voltage across the terminals of the model is a specified function. It is said to be an independent voltage source if its value is independent of the characteristics of models connected to the voltage source. For example, a

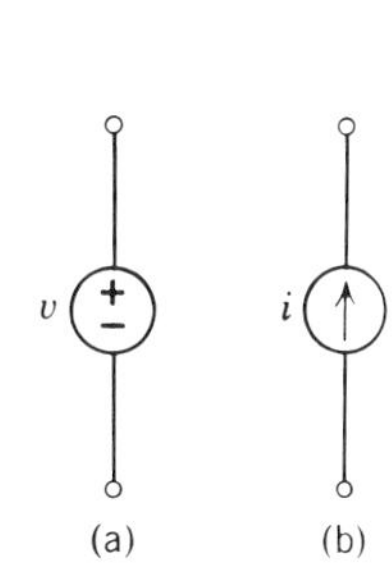

Figure 2.16 (*a*) Symbol for voltage source. (*b*) Symbol for current source.

voltage source v might be a constant, say 10 volts, regardless of what is connected across it (except a short circuit which is not admissible). Similarly, a current source is defined by the property that the current through the terminals of the model is a specified function, and it is called an independent current source if its value is independent of the characteristics of models connected to the current source. As an example, a current source may have a value of 3 amperes. This value is not affected by any network characteristic of other models connected to it (except an open circuit which is not admissible).

If the voltage across a voltage source is specified in terms of other variables in the network, the source is called a dependent or controlled source. For example, a voltage source may have a voltage v_2 which is equal to K times v_1, where v_1 is a voltage somewhere in the network. This is called a voltage-controlled voltage source. Likewise, a current source may be a controlled or dependent current source. Figure 2.17 shows a few examples of controlled sources. Similarly, for Figure 2.17*b*, the voltage-controlled current source $i_2 = Kv_1$ is dependent on the voltage v_1 that appears somewhere in the network. Figure 2.17*c* shows a current-controlled voltage source and Figure 2.17*d* shows a current-controlled current source.

Finally, we mention the magnetically coupled inductor pair as an example of a twoport as shown in Figure 2.18. The defining equations are

$$\lambda_1 = L_1 i_1 + M i_2 \tag{2.38}$$

$$\lambda_2 = L_2 i_2 + M i_1 \tag{2.39}$$

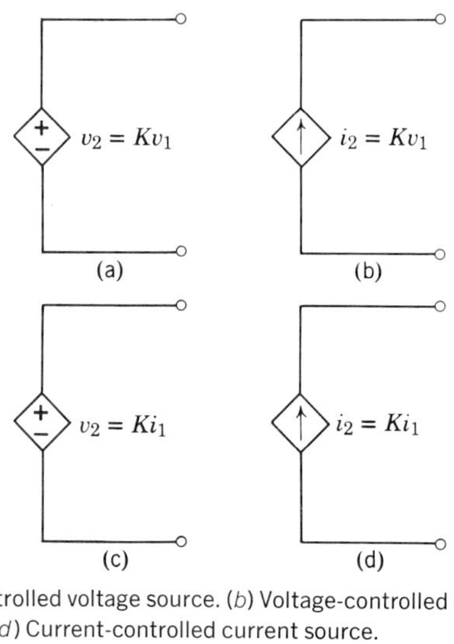

Figure 2.17 (*a*) Voltage-controlled voltage source. (*b*) Voltage-controlled current source. (*c*) Current-controlled voltage source. (*d*) Current-controlled current source.

where λ_1 and λ_2 are magnetic flux linkages which are time integrals of voltage. The constant M is called the *mutual inductance* parameter. The constants L_1 and L_2 are the usual inductance parameters called *self-inductance* to distinguish them from M. Differentiating (2.38) and (2.39),

$$v_1 = L_1 \frac{di_1}{dt} + M \frac{di_2}{dt} \tag{2.40}$$

and

$$v_2 = L_2 \frac{di_2}{dt} + M \frac{di_1}{dt}. \tag{2.41}$$

The sign of the second terms may be negative depending on the relative position of the dot marks and the reference directions of v and i. The rule for determining the signs is as follows. Choose the current reference directions so that the tails of the arrows are at the plus terminals of the voltage references, as in Figure 2.18. If the arrows both leave or both enter the dotted terminals, the signs are both plus. Otherwise, they are both minus (see Figure 2.19).

Table 2.4 summarizes the terminal relationships for the linear time-invariant electrical elements. The network symbols are also shown.

The elements described in this section and those in Section 2.1 for mechanical networks have similar mathematical descriptions. The through and across variables are physically different but if in the terminal equation relating the through and across variables for mechanical elements, the through variable force (f) is replaced by the through variable current (i) and the across variable velocity (ν) is replaced by the across variable voltage (v), then the mass model equation becomes the capacitor model equation if M is replaced by C. With this correspondence of force and current, the capacitor is said to be the electrical *analog* of mass. Similarly, resistor is the analog of

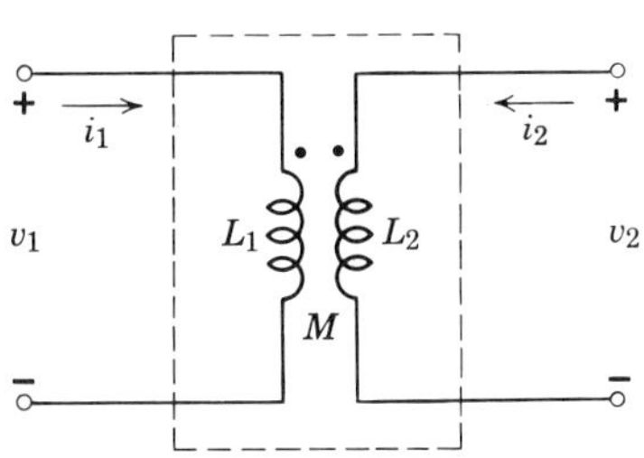

Figure 2.18 Magnetically coupled inductor pair model.
$v_1 = L_1 \frac{di_1}{dt} + M \frac{di_2}{dt}; \quad v_2 = L_2 \frac{di_2}{dt} + M \frac{di_1}{dt}$

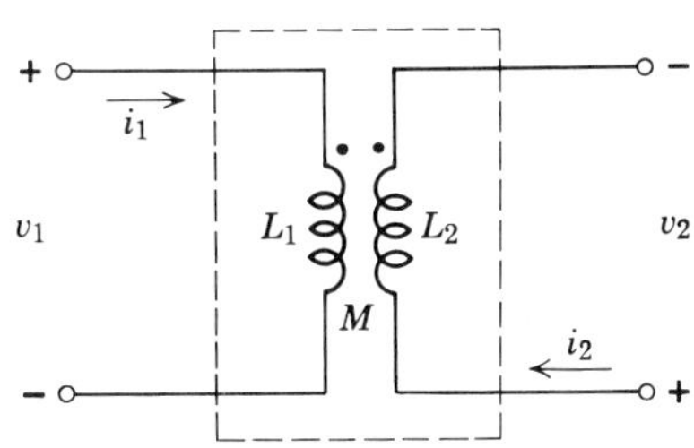

Figure 2.19 For reference and dots shown,
$v_1 = L_1 \frac{di_1}{dt} - M \frac{di_2}{dt}; \quad v_2 = L_2 \frac{di_2}{dt} - M \frac{di_1}{dt}$

Table 2.4 Linear Time-Invariant Electrical Elements

Element	Symbol	Equation
Resistor R	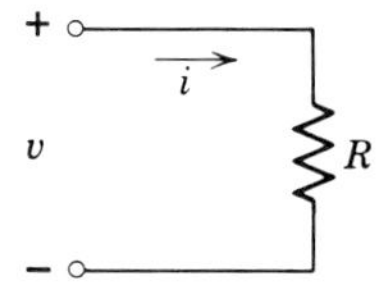	$v = Ri$
Inductor L	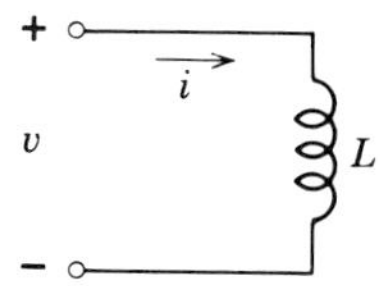	$v = L\frac{di}{dt}$
Capacitor C	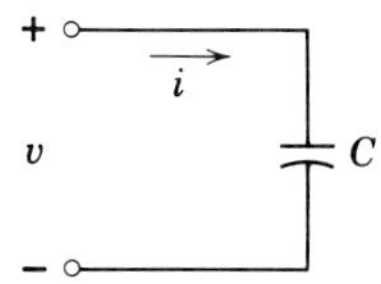	$i = C\frac{dv}{dt}$
Coupled inductors L_1, L_2, M	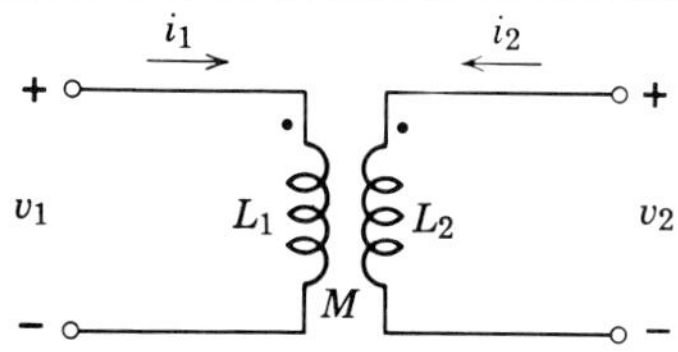	$v_1 = L_1\frac{di_1}{dt} + M\frac{di_2}{dt}$ $v_2 = M\frac{di_1}{dt} + L_2\frac{di_2}{dt}$
Voltage source v	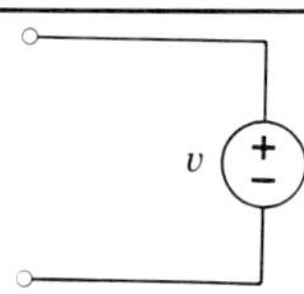	$v = v(t)$, independent of i
Current source i	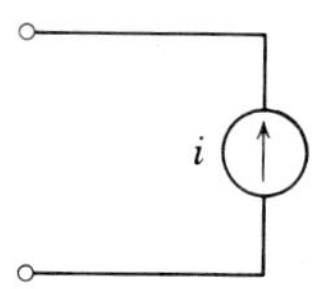	$i = i(t)$, independent of v
Controlled voltage source v	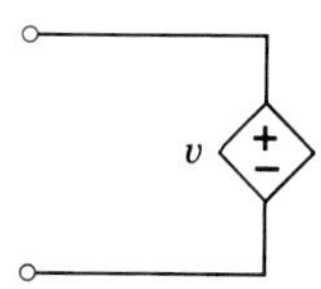	$v =$ controlled by another variable
Controlled current source	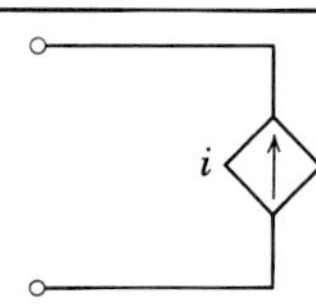	$i =$ controlled by another variable

damper, and inductor is the analog of the spring. The parameter M corresponds to the parameter C, D corresponds to $1/R$, and $1/K$ corresponds to L. The reciprocal of spring stiffness coefficient, $1/K$, is called *compliance*. The reciprocal of resistance, $1/R$, is called *conductance*. For a more symmetrical analogy, compliance may be used instead of stiffness coefficient and conductance may be used instead of resistance so that the parameters will directly correspond. Similarly, if the through variable torque is made to correspond with the through variable current, then angular velocity will correspond to voltage. The corresponding analog elements are rotational inertia and capacitor, torsional spring and inductor, and torsional damper and resistor. The corresponding parameters are J and C, D and $1/R$, and $1/K$ and L. The reciprocal of torsional spring stiffness coefficient, $1/K$, is called *torsional compliance*. Table 2.5 summarizes the analogies for convenient reference. In Chapter 3, we shall see that the equilibrium laws are analogous also. Analogies are useful because knowledge and experience gained in the study of one type of network (say electrical network) can be transferred in connection with the study of another type of network (say mechanical network). However, it should be noted that although an electrical network analog can always be found for any mechanical network, it is not always possible to have a mechanical network analog for an arbitrary electrical network. This is

Table 2.5 Mechanical-Electrical Element Analogies

Electrical		Translational		Rotational
Current	~	Force	~	Torque
Voltage	~	Velocity	~	Angular velocity
Capacitor	~	Mass	~	Rotational inertia
Inductor	~	Spring	~	Torsional spring
Resistor	~	Damper	~	Torsional damper
Conductance	~	Damping	~	Torsional damping
Inductance	~	Compliance	~	Torsional compliance
Capacitance	~	Mass	~	Moment of interia

because all masses and all rotational inertias must have a common terminal, which is the inertial reference node. Thus, mechanical analogs can be found only in electrical networks where all capacitors have a common terminal.

It is also possible to demonstrate an analogy based on a correspondence of velocity to current. However, since such an analogy replaces a through variable by an across variable and an across variable by a through variable, the analog networks will have different topologies or structural arrangement. The force-to-current analogy preserves the network topology, and the analog is simpler to obtain. Topological analogies are discussed in Chapter 3.

2.4 Nonlinear and Time-Varying Elements

Thus far, the elements that we have considered are linear in the sense that the values of the parameters associated with the elements are not dependent on the values of the through or across variables of the element. They are also time-invariant in the sense that the element values do not vary with time. In some applications, these models are not sufficiently accurate. For example, if we have a rheostat (adjustable resistor) where the movable contact is mechanically varied with respect to time and if the motion does not depend on i nor v, a more accurate model is described by

$$v(t) = R(t)i(t) \tag{2.42}$$

where $R(t)$ is a function of time. This represents a *linear time-varying resistor* model. Some devices, such as semiconductor diodes, are represented by nonlinear but time-invariant models. The relationship between voltage and current for a *nonlinear time-invariant* resistor is of the form

$$i = f_R(v) \tag{2.43}$$

or

$$v = g_R(i) \tag{2.44}$$

where f and g are nonlinear functions, as in Figure 2.20. In Figure 2.20*a*, for every v, i is uniquely determined; but, for a given i, v cannot be determined uniquely. Thus, in Figure 2.20*a*, it is not possible to solve for v in terms of i. We say that the inverse of f does not exist. Similarly, in Figure 2.20*b*, it is clear that the inverse of g does not

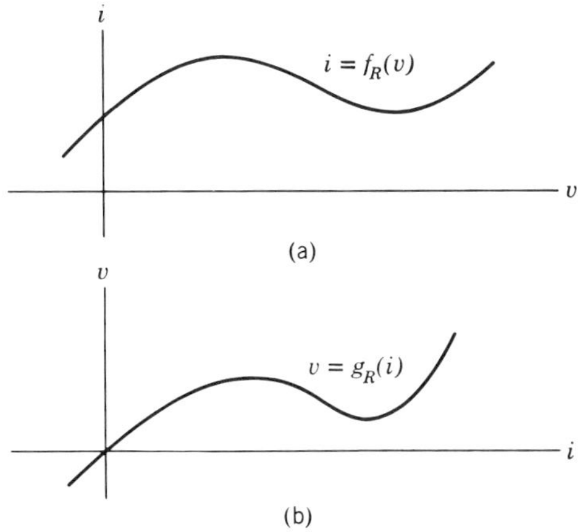

Figure 2.20 (*a*) i as a function of v, (*b*) v as a function of i.

exist. We shall assume that the relationship between v and i can be represented by either (2.43) or (2.44). If (2.43) is the defining equation for the resistor, it is said to be *voltage-controlled.* If (2.44) is the defining equation, it is said to be *current-controlled.* The linear time-invariant resistor defined in the previous section is voltage-controlled and also current-controlled.

A *nonlinear time-varying resistor* is characterized by an equation of the form

$$i = f_R(v, t) \tag{2.45}$$

or

$$v = g_R(i, t) \tag{2.46}$$

depending on whether it is voltage-controlled or current-controlled, respectively.

For a *nonlinear capacitor*, the defining equation relating charge q to voltage v is a nonlinear algebraic equation. If time t is explicitly involved in the equation, the capacitor is also time-varying. Otherwise, it is time-invariant. For a charge-controlled nonlinear time-varying capacitor, the defining equation is of the form

$$v = g_C(q, t), \tag{2.47}$$

and for a voltage controlled nonlinear time-varying capacitor, it is

$$q = f_C(v, t). \tag{2.48}$$

If f and g do not depend on t explicitly, the capacitor is time-invariant. Similarly, for a *nonlinear time-varying inductor*, the defining equation is of the form

$$\lambda = g_L(i, t) \tag{2.49}$$

if it is current-controlled, or

$$i = f_L(\lambda, t) \tag{2.50}$$

if it is flux-controlled. If f and g are not explicit functions of time, the inductor is considered to be time-invariant.

For a nonlinear time-invariant voltage-controlled capacitor,

$$q = f_C(v) \tag{2.51}$$

differentiation with respect to time yields

$$\frac{dq}{dt} = i = \left(\frac{df_C}{dv}\right)\frac{dv}{dt}. \tag{2.52}$$

The coefficient (df_C/dv) in (2.52) is called the *incremental capacitance* of the capacitor. Compare (2.52) with the equation for a linear time-invariant capacitor which is $i = C(dv/dt)$. The incremental capacitance depends on v, and it is the slope of the f_C curve. Similarly, for a current-controlled nonlinear inductor

$$\lambda = g_L(i), \tag{2.53}$$

the time derivative yields

$$v = \frac{d\lambda}{dt} = \left(\frac{dg_L}{di}\right)\frac{di}{dt}. \tag{2.54}$$

The coefficient (dg_L/di) is called *incremental inductance*. It is the slope of the g_L curve, and it varies with i.

The nonlinear mechanical elements may be similarly defined. For example, for a nonlinear spring, the force and spring displacement are related by a nonlinear function

$$f = g_K(x). \tag{2.55}$$

Differentiating with respect to time,

$$\frac{df}{dt} = \left(\frac{dg_K}{dx}\right)\frac{dx}{dt} = \left(\frac{dg_K}{dt}\right)\nu. \tag{2.56}$$

The coefficient (dg_K/dx) is called the *incremental spring stiffness coefficient*. In general, (dg_K/dx) is a function of x, being the slope of the force-displacement curve that is assumed to be nonlinear.

Example 2-3. Suppose that the torque-angular displacement relationship for a torsional spring is

$$\tau = \theta^{2/3}. \tag{2.57}$$

Then

$$\frac{d\tau}{dt} = \left(\tfrac{2}{3}\theta^{-1/3}\right)\frac{d\theta}{dt} = \left(\tfrac{2}{3}\theta^{-1/3}\right)\omega. \tag{2.58}$$

The *incremental torsional stiffness coefficient* is $(\frac{2}{3})\theta^{-1/3}$, which is a function of θ. ▲

Problems

2-1. Draw a functional block diagram, indicating important subsystems, for a forced-air thermostat-controlled home-heating system.

2-2. Repeat Problem 2-1 for a fossil-fuel electric-power generation system.

2-3. Repeat Problem 2-1 for a city sewage-disposal system.

2-4. Repeat Problem 2-1 for a man-machine system composed of a man driving an automobile.

2-5. Repeat Problem 2-1 for some kind of communication system (radio, TV, radar, telemetry, etc.).

2-6. A simplified single-axis model for a spacecraft's angular motion is a linear time-invariant moment of inertia J. A control torque of $+T$ or $-T$ or 0 (where T is a constant) is available for changing the angular position θ and angular velocity $\omega = \dot{\theta}$. Suppose that at $t = 0$, the initial position is θ_0 and the initial velocity is ω_0. Determine the *shortest* time needed to reduce θ to zero, using $+T$, $-T$, and 0 torques.

2-7. Consider the spacecraft model of Problem 2-6. Determine the shortest time needed to stop the spin motion of the vehicle, that is, to make $\omega = 0$. Determine the corresponding θ. Sketch the control torque vs. time.

2-8. A nonlinear inductor is characterized by

$$\lambda = i^3 - 9i^2 + 12i.$$

Determine the range (if any) for $i \geqslant 0$ for which the *incremental inductance* is negative.

2-9. Determine the linear incremental model for the inductor in Problem 2-8 for the operating point $i = 0$.

2-10. Repeat Problem 2-9 for the operating point $i = 3$.

2-11. A voltage pulse

$$v(t) = \begin{cases} V_0, & 0 \leqslant t \leqslant 1 \\ 0, & t > 1 \end{cases}$$

is applied to a linear time-invariant inductor. $i(0) = 0$. Sketch $i(t)$ for $t \geqslant 0$.

2-12. The charge-voltage relationship for a nonlinear capacitor is

$$q = 1 - \epsilon^{-v}.$$

Determine the *incremental capacitance*.

2-13. Suppose $v(t) = Kt$, K a constant, for the capacitor of Problem 2-12. Plot $i(t)$ for $t > 0$.

2-14. Which of the following inductor characteristics are linear? Nonlinear? Time-invariant? Time-varying?

(a) $\lambda = 8i^3$
(b) $\lambda = (8+3t)i$
(c) $\lambda = 6i$
(d) $\lambda = 4ti$
(e) $i = \lambda^{1/3}$
(f) $i = (1-\epsilon^{-t})\lambda$
(g) $i = (1-\epsilon^{-t})^2(\lambda - \lambda^3)$
(h) $i = (1-\epsilon^{-t})^3\lambda$

2-15. Find an equation expressing the output $y(t)$ in terms of the input $x(t)$ for a linear time-invariant moment of inertia J for the following identification of the variables:

(a) $x(t) = \tau$ $\quad y(t) = \theta$
(b) $x(t) = \omega$ $\quad y(t) = \theta$
(c) $x(t) = \alpha$ $\quad y(t) = \theta$
(d) $x(t) = \omega$ $\quad y(t) = \tau$
(e) $x(t) = \theta$ $\quad y(t) = \tau$

where $\tau =$ torque, $\theta =$ angular position, $\omega =$ angular velocity, and $\alpha =$ angular acceleration.

2-16. A linear time-invariant capacitor is connected to a current source as in Figure P2-16*a*. The signal $i(t)$ is sketched in Figure P2-16*b*. $C = 0.1$ microfarad and $v(0) = 0$. Sketch $v(t)$ for $0 \leqslant t \leqslant 4$.

2-17. For the inductor characteristics in Problem 2-14(a), (b), (c), and (d), determine the inductor voltage if the inductor current is $i(t) = \sin t$.

2-18. For the inductor characteristics in Problem 2-14(e), (f), (g), and (h), determine the inductor current $i(t)$ for $t \geqslant 0$ if the inductor voltage is $v(t) = \sin t$ and $\lambda(0) = 0$.

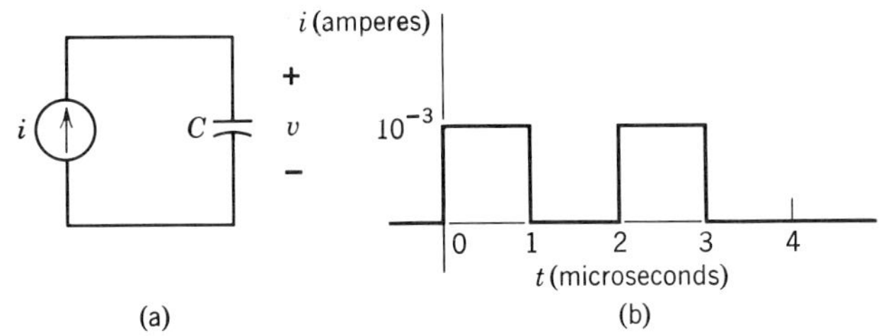

Figure P2-16

2-19. The capacitance of a simple capacitor consisting of two parallel metal sheets separated by dielectric material is given by

$$C = \frac{\epsilon A}{d} \text{ farads}$$

where A is the area of either sheet in meters^2, d is the separation of the sheets in meters, and

$$\epsilon = 8.854 \times 10^{-12} K$$

where K is the relative dielectric constant. For air, $K = 1$. The formula neglects the effects of fringing of the electric field at the edges of the plates. To make the device compact, a second sheet of dielectric is placed over one of the metal sheets, and the combination is wound into a roll such that the metal sheets and dielectric sheets alternate. The formula above is still a useful approximation. Tinfoil is commonly used for the metal sheets. If mica is used for the dielectric ($K = 5$) and the distance between the tinfoil sheets is 10^{-5} meter, determine the area of the sheets for constructing a capacitor of 0.01 microfarad. If the sheets are 0.02 meter wide, determine the total length of one strip.

2-20. For the dashpot model of Figure 2.3a, the velocity of the cylinder ν_2 is shown in Figure P2-20. The damping constant is $D = 1$ newton/(m/sec). If it is desired to hold the piston stationary with respect to inertial reference, what must the force f be?

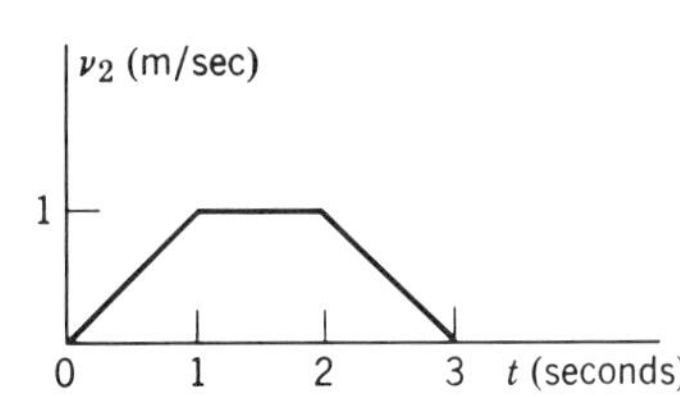

Figure P2-20

CHAPTER 3

Network Modeling of Devices and Systems

In the previous chapter we developed some of the basic network elements used for modeling electrical and mechanical devices and systems. In addition to these elements, we need postulates that model the behavior of interconnections of these elements to form network representations of devices and systems. The postulates will be stated in terms of electrical and mechanical systems, but analogous laws apply to other types of physical systems. Two types of postulates will be discussed in this chapter: the first deals with topological constraints affecting the through and across variables in an interconnected system; the second deals with a set of conservation laws. These postulates will be combined with the element models of the previous chapter to model devices and systems.

3.1 Kirchhoff's Laws

After choosing currents and voltages (or their time integrals) as through and across variables, and element models that relate pairs of currents and voltages, the next step in electric network modeling is to obtain relationships among currents and relationships among voltages in the interconnected network. The equilibrium law for currents is known as Kirchhoff's current law and, for voltages, the equilibrium law is known as Kirchhoff's voltage law.

Kirchhoff's Current Law

Consider a portion of an electrical network as shown in Figure 3.1, where wires DA, BA, and CA join to form a node A. Denote the current from A to B by i_1, the current from A to C by i_3, and the current from D to A by i_2. Notice the arrow directions. In general, the currents are time varying quantities or functions of time. If at a given instant of time i_1 is positive, then the actual positive current in wire AB flows from A to B. If at a given instant of time i_3 is negative, then the actual positive current in wire AC flows from C to A. The arrows indicate the actual direction of current flow when the value of the current is positive.

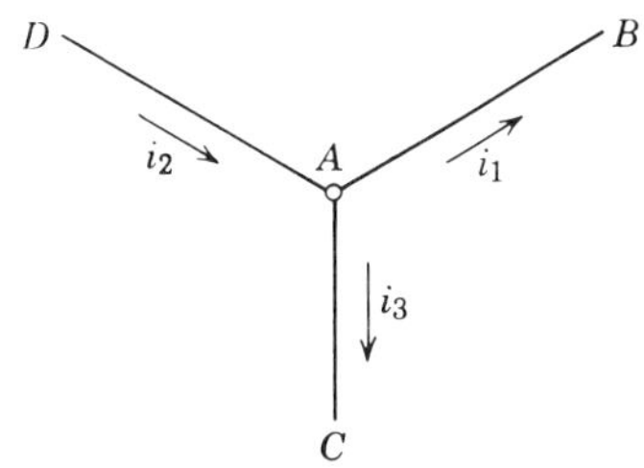

Figure 3.1 Network node for illustrating Kirchhoff's current law. For node A, $i_1+i_3-i_2=0$.

Kirchhoff's current law states that the sum of the currents flowing out of a node minus the sum of the currents flowing into the node is zero. We say that a current is flowing into a node if the reference arrow is toward the node.

For the junction in Figure 3.1 with the indicated current reference directions, Kirchhoff's current law (abbreviated KCL) states that

$$i_1 + i_3 - i_2 = 0. \tag{3.1}$$

Kirchhoff's current law applies to every node of a network at every instant of time. In essence, Kirchhoff's current law is a continuity condition which requires that what flows in must flow out. There can be no accumulation of charge at a node.

Alternatively, Kirchhoff's current law states that

$$\sum_{j=1}^{N_b} a_{kj} i_j(t) = 0, \qquad k = 1, 2, \ldots, N_v \tag{3.2}$$

where there are N_v nodes and N_b branch currents. The coefficient a_{kj} is equal to $+1$ if i_j flows out of node k; it is equal to -1 if i_j flows into node k; and it is equal to zero if i_j does not flow into or out of node k.

Kirchhoff's Voltage Law

Consider a portion of a network as shown in Figure 3.2 where the boxes labelled a, b, and c are electrical devices or subnetworks forming a loop ABC. Let v_1 denote the voltage of A with respect to B, v_2 the voltage of C with respect to B, and v_3 the voltage of A with respect to C. The plus symbol indicates that if the voltage quantity is actually positive, then the voltage of the terminal with the plus symbol with respect to the other terminal is positive. If the voltage quantity is actually negative, then the voltage of the plus terminal with respect to the other terminal is negative. For example, if in Figure 3.2, $v_1 = -10$

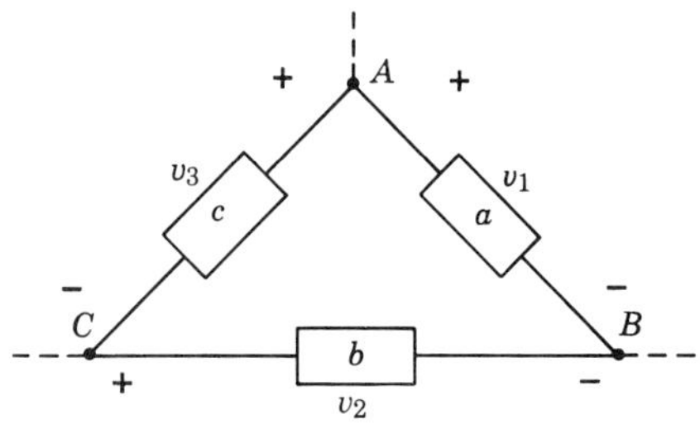

Figure 3.2 Network loop for illustrating Kirchhoff's voltage law. For loop *ABC*, $v_1 - v_2 - v_3 = 0$.

volts at a certain instant of time, then the voltage of *A* with respect to *B* is -10 volts or the voltage of *B* with respect to *A* is $+10$ volts at the same instant of time.

Let us adopt the convention that in tracing through a branch, if we encounter the plus (+) terminal first, there is a drop in voltage. Likewise, if we encounter the minus (−) terminal first before the plus terminal, there is a rise in voltage. Thus, in Figure 3.2, in tracing *AB* from *A* to *B*, we encounter a voltage drop v_1. Similarly, in tracing *AB* from *B* to A, we encounter a voltage rise v_1.

Kirchhoff's voltage law (abbreviated KVL) states that in tracing around a loop, the sum of the voltage drops minus the sum of the voltage rises is equal to zero.

For example, for the loop in Figure 3.2, KVL requires that

$$v_1 - (v_2 + v_3) = 0. \tag{3.3}$$

Alternatively, we state Kirchhoff's voltage law as follows:

$$\sum_{j=1}^{N_b} b_{kj} v_j(t) = 0, \qquad k = 1, 2, \ldots, N_l \tag{3.4}$$

where there are N_l loops and N_b branch voltages. In tracing around loop *k* in a particular direction (the choice is arbitrary) if branch *j* is not in loop *k*, $b_{kj} = 0$; if branch *j* is in loop *k* and if the plus reference of v_j is encountered first, $b_{kj} = +1$; otherwise, $b_{kj} = -1$.

Another convention for reference direction that eliminates the need for the plus symbol is to use double subscript notation. For example v_{AB} means the voltage of node A with respect to node *B*. Similarly, v_{BA} means the voltage of node *B* with respect to node A. Double subscript notation may be used also for current reference direction instead of the arrow. For example, in Figure 3.1, i_{AB} means the current from A to *B*. In the say way, i_{AB} means the current from *B* to A.

Example 3-1. Consider the network in Figure 3.3 which might represent a relay circuit where L is the inductance of the coil for the electromagnet and R is the resistance of the coil. The voltage v is the control variable for producting current. Suppose $v(t)$ is a step function

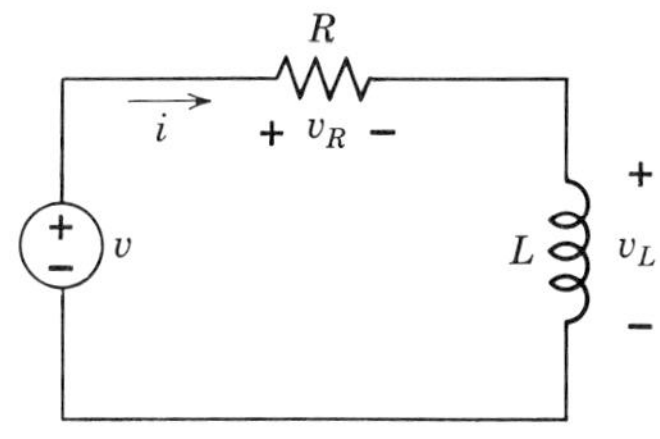

Figure 3.3 Network for Example 3-1.

$$
\begin{aligned}
v(t) &= V_0, \qquad && t \geqslant 0 \\
&= 0, && t < 0.
\end{aligned}
$$

At $t = 0$, the current is zero. Let us determine the resulting current $i(t)$ for $t \geqslant 0$.

Denote the element voltages by v_R, v_L, and v as indicated. From KVL,

$$v_R + v_L - v = 0. \tag{3.5}$$

Assuming linear time-invariant models for the elements, we have

$$v_R = Ri \tag{3.6}$$

and

$$v_L = L\frac{di}{dt}. \tag{3.7}$$

Substituting (3.6) and (3.7) in (3.5), we have

$$L\frac{di}{dt} + Ri = v. \tag{3.8}$$

For $t \geqslant 0$

$$\frac{di}{dt} + \frac{R}{L}i = \frac{1}{L}V_0. \tag{3.9}$$

The simple linear first-order differential equation above may be solved in several ways. The complementary solution is

$$i_c = Ke^{-(R/L)t} \tag{3.10}$$

and the particular solutions is

$$i_p = \frac{V_0}{R}. \tag{3.11}$$

Hence the general solution is

$$i = i_c + i_p = Ke^{-(R/L)t} + \frac{V_0}{R}. \tag{3.12}$$

Since $i(0) = 0$,

$$0 = K + \frac{V_0}{R} \tag{3.13}$$

so that

$$K = -\frac{V_0}{R}. \tag{3.14}$$

Thus the solution for $t \geqslant 0$ that satisfies the zero initial condition is

$$i(t) = \frac{V_0}{R}(1 - e^{-(R/L)t}). \tag{3.15}$$

Figure 3.4 shows the plot of $i(t)$. If L/R is made larger, the response becomes slower, and if L/R is made smaller, the response becomes faster. The ratio L/R is called the time constant of the circuit. Note that the initial slope $di(0)/dt$ equals the time constant. If L is in henrys and R is in ohms, then L/R is in seconds (check the dimension from Table 2.2). ▲

Example 3-2. As another example, let us obtain the network equations for the model shown in Figure 3.5. Let v_1 be the voltage of node 1 with respect to node 3, and let v_2 be the voltage of node 2 with respect to node 3. The initial voltages across C_1 and C_2 and the initial currents for L_1 and L_2 are given at $t = 0$. For the elements connected to node 1, the branch equations are:

$$i_1 = \frac{v_1(t)}{R_1} \tag{3.16}$$

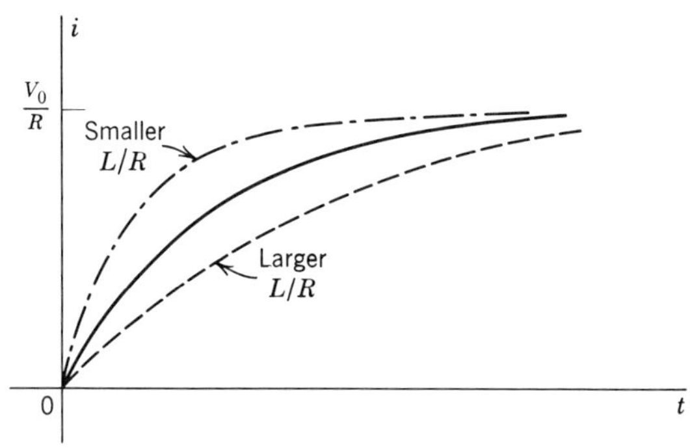

Figure 3.4 Current i for Example 3-1.

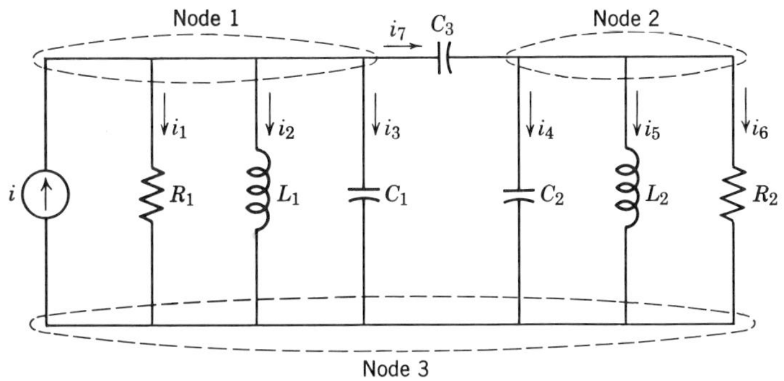

Figure 3.5 A network example with three nodes.

$$i_2(t) = i_2(0) + \frac{1}{L_1}\int_0^t v_1(\tau)\, d\tau \tag{3.17}$$

$$i_3(t) = C_1\frac{dv_1}{dt}(t) \tag{3.18}$$

and

$$i_7(t) = C_3\frac{d(v_1 - v_2)}{dt}. \tag{3.19}$$

But KCL applied to node 1 yields

$$i_1 + i_2 + i_3 + i_7 - i = 0. \tag{3.20}$$

Eliminating i_1, i_2, i_3, and i_7 from (3.16) to (3.20), we have

$$\frac{v_1(t)}{R_1} + C_1\frac{dv_1(t)}{dt} + C_3\left[\frac{dv_1(t)}{dt} - \frac{dv_2(t)}{dt}\right] + i_2(0) + \frac{1}{L_1}\int_0^t v_1(\tau)\, d\tau = i(t). \tag{3.21}$$

Similar application of Kirchhoff's current law to node 2 yields

$$\frac{v_2(t)}{R_2} + C_2\frac{dv_2}{dt} + C_3\left[\frac{dv_2}{dt} - \frac{dv_1}{dt}\right] + i_5(0) + \frac{1}{L_2}\int_0^t v_2(\tau)\, d\tau = 0. \tag{3.22}$$

The voltages $v_1(t)$ and $v_2(t)$ can be obtained by solving (3.21) and (3.22) simultaneously, either analytically or by computer simulation. However, the object of this example is to illustrate how KCL, KVL, and the element model equations may be applied in combination to obtain integrodifferential equations whose solution yields the unknown currents and voltages. Equations (3.21) and (3.22) may be differentiated once to obtain linear ordinary differential equations with constant coefficients, which are valid for $t \geqslant t_0$. ▲

3.2 D'Alembert's Principle

The equilibrium law for through variables in mechanical networks is called D'Alembert's principle. A dual equilibrium law also holds for across variables, but there is no specific name associated with the equilibrium condition. It is sometimes known as the geometrical compatibility condition for across variables.

First consider a mechanical system undergoing translational motion along a given direction, say the x direction. The velocities or motions of bodies may be considered to be those of junctions or nodes connected to the bodies. For example, if a rigid block of material is connected to a spring, then one end of the spring is assumed to have the same motion and velocity as the block which is modeled by mass. D'Alembert's principle states that the sum of the forces in the positive x direction minus the sum of the forces in the negative x direction, for forces acting *on* all bodies connected to a node is zero. The principle applies to every node of the system. In applying D'Alembert's principle, it should be remembered that the force acting *on* a body is the negative of the forces exerted *by* the body. The force reference directions associated with force sources are the directions for forces exerted by the sources. Hence the forces acting on the sources are the negative of those shown with the symbols.

The statement of the above principle is a modification of the original version which, like Newton's second law, is combined with the definition for mass. These two important ideas are separated here. The relationship of the velocity of a mass and the force acting on a mass ($f = M dv/dt$) was discussed in the previous chapter. In the present statement of D'Alembert's principle, we direct our attention to the equilibrium of forces acting on bodies connected to a junction. The equilibrium law is very simple. For one degree-of-freedom motion, the statement is even simpler when cast in the context of a mechanical network. For translational motion, D'Alembert's principle states that in a mechanical network, the sum of the forces flowing out of a node minus the sum of the forces flowing into the node equals zero for every node of the mechanical network. For rotational motion, the statement is the same except that force is replaced by torque. This statement is analogous to Kirchhoff's current law, which is a statement of current equilibrium at a node, without mention of how the currents are related to the branch voltages. The branch relationships are treated separately by means of the defining equations for electrical elements.

A dual equilibrium condition that is concerned only with the topology involves the summation of relative velocities being zero. Pick any sequence of nodes so that the first node is the same as the last node. For example, if we have nodes A, B, and C, pick the sequence $ABCA$. Then the relative velocity of A with respect to B plus the relative velocity of B with respect to C plus the relative velocity of C with respect to A is zero. This condition is analogous to Kirchhoff's voltage law, which also applies to loops of a topology. D'Alembert's principle as stated above, on the other hand, is analogous to Kirchhoff's current law, since it applied to nodes of a topology. The above equilibrium conditions also hold for rotational systems where, instead of force, we have torque and, instead of velocity, we have angular velocity.

In drawing a mechanical network for a translational or a rotational system, the first step is to identify the parts that are rigidly coupled and that move with the same velocity. Each such set of parts is identified as a network node. The masses (and rotational inertias) are the easiest branches to draw, since each mass has a terminal that is connected to the inertial reference frame. The other terminal for each mass is associated with a velocity variable. Dampers and springs can be drawn next, keeping in mind that the velocities associated with each such damper or spring are *relative* velocities and that the appropriate nodes to which they connect can be determined. The sources are then drawn.

Example 3-3. Let us consider a simple example of a mechanical system in linear translational motion along a horizontal line. Figure 3.6 shows a simplified pictorial diagram of two box cars with masses M_1 and M_2 coupled through a spring with spring constant K. The damping constants corresponding to viscous friction are D_1 and D_2 for masses M_1 and M_2 respectively. A force f is applied to M_1 as shown. First we assign reference directions for v_1 and v_2, as shown in Figure 3.6. For

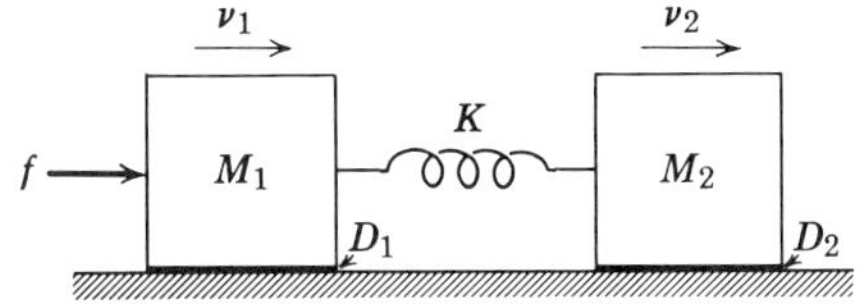

Figure 3.6 Two box cars with masses M_1 and M_2 coupled through a spring with spring constant K, moving on tracks with viscous friction. Damping constants are D_1 and D_2.

all the forces acting on the various bodies, let the reference direction be from left to right. We note that the force, M_1, D_1, and the left end of K can be considered to be connected to a node. M_2, D_2, and the other end of the spring can be considered to be connected to another node. In addition, a node for the inertial frame is needed.

The mass M_1 is connected between the node for v_1 and the reference node. Also connected from the node for v_1 to reference is the damper D_1. This is because there is viscous friction between the first box car and the surface. Since the box car is moving with velocity of v_1 with respect to the inertial reference, the damper model must be connected between the node for v_1 and the reference. Likewise, the left-hand side of the spring is moving with the mass M_1 with velocity v_1. Hence, one end of the spring must be connected to the node for v_1 in the network diagram. The rule here is that terminals corresponding to points that are rigidly connected to each other have the same velocity and, therefore, these terminals must connect to the same node in the network diagram. Similarly, the viscous friction between the second box car and the surface corresponds to a velocity of v_2, so that the damper model must be between the node for v_2 and the reference node. M_2 is also connected between the node for v_2 and the reference node. Finally, the external force f must be connected between the reference node and the node for v_1 because it is counteracted not only by the reaction force of the mass but also by the reaction force of the damper D_1 and the reaction force of K. All these elements have a terminal with velocity v_1. The complete network diagram is shown on Figure 3.7. The positive reference directions for the force components are indicated in Figure 3.7. Notice that f_1, f_2, and f_5 are flows out of the node associated with v_1, but f is a flow into the same node. Hence, D'Alembert's principle for this node yields

$$f_1+f_2+f_5-f=0. \tag{3.23}$$

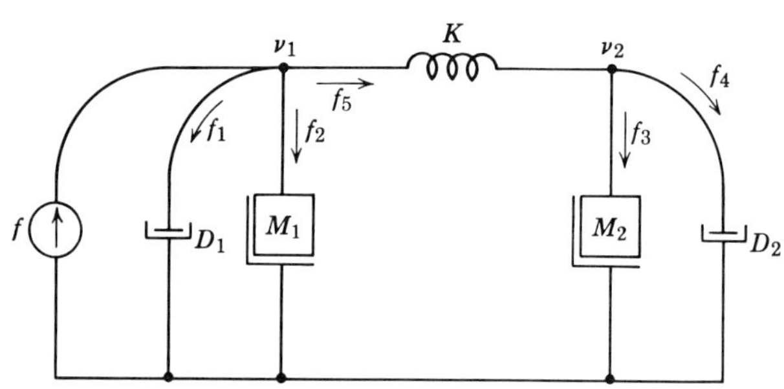

Figure 3.7 Network diagram for mechanical system in Figure 3.6.

Likewise, for the node associated with ν_2, D'Alembert's principle yields

$$f_3 + f_4 - f_5 = 0. \tag{3.24}$$

Assuming linear time-invariant element models the branch equations based on the definitions in Chapter 2 are

$$f_1 = D_1 \nu_1 \tag{3.25}$$

$$f_2 = M_1 \frac{d\nu_1}{dt} \tag{3.26}$$

$$\frac{df_5}{dt} = K(\nu_1 - \nu_2) \tag{3.27}$$

$$f_3 = M_2 \frac{d\nu_2}{dt} \tag{3.28}$$

and

$$f_4 = D_2 \nu_2. \tag{3.29}$$

It is possible to eliminate some of the variables from the seven equations above. For example, if we retain ν_1, ν_2, and f_5 as variables, then substituting (3.25) and (3.26) in (3.23) we have

$$D_1 \nu_1 + M_1 \frac{d\nu_1}{dt} + f_5 - f = 0 \tag{3.30}$$

or

$$\frac{d\nu_1}{dt} = -\frac{D_1}{M_1} \nu_1 - \frac{1}{M_1} f_5 + \frac{1}{M_1} f. \tag{3.31}$$

Substituting (3.28) and (3.29) in (3.24), we obtain

$$M_2 \frac{d\nu_2}{dt} + D_2 \nu_2 - f_5 = 0 \tag{3.32}$$

or

$$\frac{d\nu_2}{dt} = -\frac{D_2}{M_2} \nu_2 + \frac{1}{M_2} f_5. \tag{3.33}$$

Thus (3.31), (3.33), and (3.27) are three first-order differential equations in the variables ν_1, ν_2, and f_5, with f as a known input function.

These equations are readily solved by computer simulation, starting from initial values for $\nu_1(0)$, $\nu_2(0)$, and $f_5(0)$, as shown in the next chapter. ▲

D'Alembert's principle, as noted above, is analogous to Kirchhoff's current law, and the dual equilibrium law for velocities and angular velocities is analogous to Kirchhoff's voltage law. Recall that if the through variables force and current are made analogous, the electrical parameters C, L, $1/R$ are the analogs of the mechanical parameters M, $1/K$, and D, respectively. Furthermore, velocity and voltage are analogous. Similarly, if torque is made analogous to current, then angular velocity and voltage are analogs, and the mechanical parameters J, $1/K$, and D correspond to the electrical parameters C, L, and $1/R$, respectively. Force (or torque) sources are analogous to current sources, and velocity (angular velocity) sources are analogous to voltage sources. For a given mechanical network model for a mechanical system in translational motion, suppose that the mechanical network is replaced with an electrical network such that the topology is the same, and every mass, spring, and damper is replaced by a capacitor, inductor, and resistor, respectively. Let the numerical values of C, L, and $1/R$ be equal to the corresponding M, $1/K$, and D, respectively. Then the equations based on D'Alembert's principle would be the same as those based on KCL except that the through force variables are replaced by the through current variables. Similarly, the topological constraint equations for the velocity variables would be the same as the KVL equations. Furthermore, the element equations would be the same, as discussed in Chapter 2. Hence the solutions for the current and voltage variables of the electrical network would be numerical similar to the solutions for the force and velocity variables of the mechanical network. These networks are said to be analogs of each other.

Example 3-4. Let us construct an electrical network analog for the mechanical network in Figure 3.7, as shown in Figure 3.8. Let the electrical

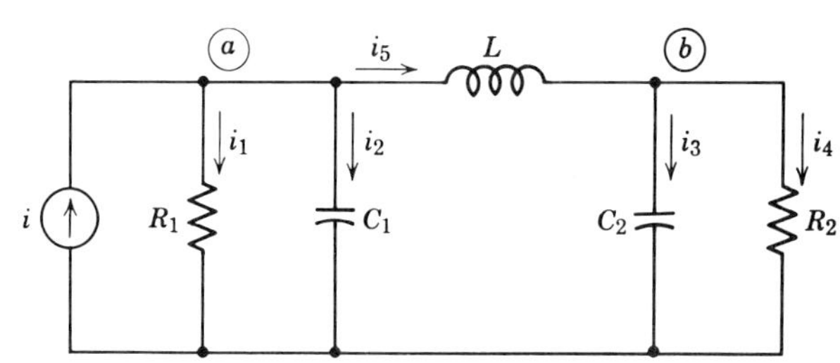

Figure 3.8 Electrical network analog of mechanical network in Figure 3.7 if $R_1 = 1/D_1$, $C_1 = M_1$, $L = 1/K$, $C_2 = M_2$, $R_2 = 1/D_2$, and $i = f$.

element parameters be chosen as follows: $R_1 = 1/D_1$, $C_1 = M_1$, $L = 1/K$, $C_2 = M_2$, $R_2 = 1/D_2$, and $i = f$. Then KCL at node a yields

$$i_1 + i_2 + i_5 - i = 0. \tag{3.34}$$

At node b, KCL yields

$$i_3 + i_4 - i_5 = 0. \tag{3.35}$$

Comparing with (3.23) and (3.24), we see that i_1, i_2, i_3, i_4, i_5, and i are the analogs of f_1, f_2, f_3, f_4, f_5, and f. The element equations are

$$i_1 = \frac{v_1}{R_1}, \tag{3.36}$$

$$i_2 = C_1 \frac{dv_1}{dt}, \tag{3.37}$$

$$\frac{di_5}{dt} = \frac{1}{L}(v_1 - v_2), \tag{3.38}$$

$$i_3 = C_2 \frac{dv_2}{dt}, \tag{3.39}$$

and

$$i_4 = \frac{v_2}{R_2}. \tag{3.40}$$

With the choice of numerical values for the electrical element parameters as indicated above, v_1 and v_2 are the analogs of ν_1 and ν_2, respectively, and the numerical solutions for the analog variables are numerically identical, provided that the corresponding initial conditions are identical. Thus for this example, the differential equations are

$$\frac{di_5}{dt} = \frac{1}{L} v_1 - \frac{1}{L} v_2, \tag{3.41}$$

$$\frac{dv_1}{dt} = -\frac{1}{R_1 C_1} v_1 - \frac{1}{C_1} i_5 + \frac{1}{C_1} i, \tag{3.42}$$

$$\frac{dv_2}{dt} = -\frac{1}{R_2 C_2} v_2 + \frac{1}{C_2} i_5. \; \blacktriangle \tag{3.43}$$

Example 3-5. An *accelerometer* is an instrument for measuring acceleration. It consists of a mass called *test mass*, or *seismic mass*, and a restraining spring mounted on a stable platform, with the mass free to move along one axis. The stable platform is designed to maintain a fixed orientation in inertial space, except for short transients. This is accomplished by fixing the platform rigidly to three mutually orthogonal single-axis gyroscopes (discussed in Chapter 5). For now, let us assume that a platform can be stabilized in a fixed orientation in inertial space. Three such accelerometers are needed to measure the total acceleration. The stretching (or compression) of the spring depends not only on the acceleration of the vehicle but also on the gravitational force along the axis at the point in space where the vehicle is located. For simplicity, let us first consider an accelerometer mounted on a "horizontal" platform, a platform that is perpendicular to the gravity gradient as shown in Figure 3.9.

Since the platform is perpendicular to the gravity gradient, gravitational force does not affect motion in the x direction. It is assumed that the mass M may move only in the $+x$ or $-x$ direction because of mechanical guide constraints. Let x be the position of the platform in the direction shown in Figure 3.9 measured from some fixed point in inertial space. Let the equilibrium point on the mass M, which coincides with the point 0 of the guide when the platform is not accelerated, be denoted by P. Let y be the displacement of P from 0 when the platform is accelerated, where y is measured positively to the left as indicated in Figure 3.9. Figure 3.10 shows the network for the accelerometer. Applying D'Alembert's principle to the left-hand side node,

$$M(\ddot{x}-\ddot{y}) + \frac{D}{2}[(\dot{x}-\dot{y})-\dot{x}] + \frac{D}{2}[(\dot{x}-\dot{y})-\dot{x}] + \frac{K}{2}[(x-y)-x] + \frac{K}{2}[(x-y)-x]=0 \tag{3.44}$$

or

$$\ddot{y}+\frac{D}{M}\dot{y}+\frac{K}{M}y=\ddot{x}. \tag{3.45}$$

The acceleration $\ddot{x}$ is the x component of the acceleration of the space vehicle, which is to be measured. If y, $\dot{y}$, and $\ddot{y}$ are known, $\ddot{x}$ is determined from (3.45). Suppose that the acceleration $\ddot{x}$, which is to be measured, is constant. The solution for y in (3.45) consists of the particular solution and the complementary solution. The particular solution is

$$y_p=\frac{M}{K}\ddot{x}\cdot \tag{3.46}$$

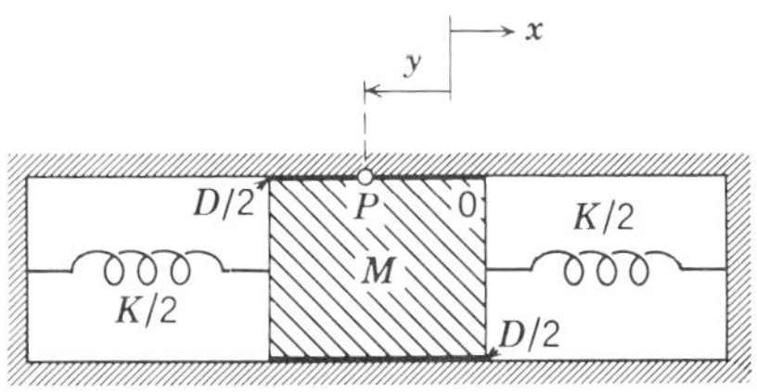

Figure 3.9 Basic accelerometer in "horizontal" direction.

Assuming that

$$\frac{K}{M}-\left(\frac{D}{2M}\right)^2 > 0$$

(see Problems 3-12 and 3-13 for other cases), the complementary solution is

$$y_c = K_1 \exp\left(-\frac{D}{2M}t\right) \cos\sqrt{\frac{K}{M}-\left(\frac{D}{2M}\right)^2}\,t + K_2 \exp\left(-\frac{D}{2M}t\right)\cdot \sin\sqrt{\frac{K}{M}-\left(\frac{D}{2M}\right)^2}\,t. \quad (3.47)$$

The complete solution is

$$y = y_c + y_p \quad (3.48)$$

such that the initial conditions are satisfied. If the initial conditions

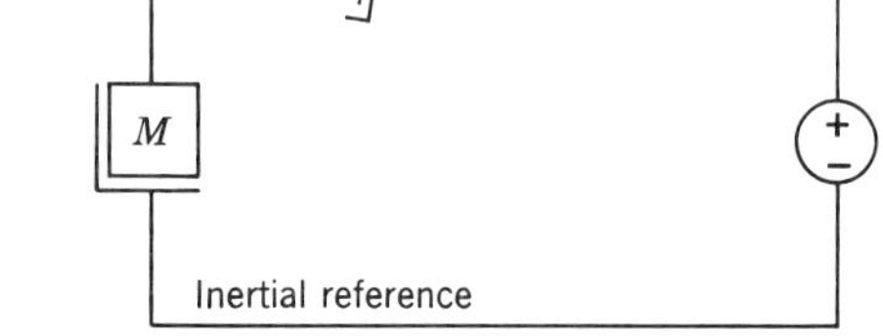

Figure 3.10 Mechanical network for accelerometer.

are known, the constants K_1 and K_2 are completely determined. For $y(0) = 0$ and $y'(0) = 0$, the result for $t \geqslant 0$ is

$$y = \frac{M}{K}\ddot{x}\left[1 - \exp(-\alpha t)\left\{\cos \beta t - \frac{\alpha}{\beta}\sin \beta t\right\}\right] \tag{3.49}$$

where

$$\alpha = \frac{D}{2M} \tag{3.50}$$

and

$$\beta = \sqrt{\frac{K}{M} - \left(\frac{D}{2M}\right)^2} \tag{3.51}$$

Thus for $t > 5/\alpha$, $\exp(-\alpha t) < \epsilon^{-5}$ and, from (3.49), y is approximately equal to $M/K\ddot{x}$. Thus, after waiting for a transient time $t \approx 5/\alpha$,

$$\ddot{x} \approx \frac{K}{M}y. \tag{3.52}$$

The oscillation of y during the transient period has a frequency of $\sqrt{K/M - (D/2M)^2}$ rad/sec and a damping factor of $\exp(-\alpha t)$. Thus by proper design of $D/2M$ and K/M, (3.52) can be used as a basis for measuring $\ddot{x}$ by measuring y. That is, the vehicle acceleration $\ddot{x}$ is approximately proportional to the compression of the left-hand side spring. If $\ddot{x}$ is not constant but varies slowly with time, so that it is approximately constant during a time duration of $5/\alpha$, (3.52) is still a good approximation.

Suppose that the platform is not horizontal (not perpendicular to the gravity gradient), as shown in Figure 3.11. Then there will be a

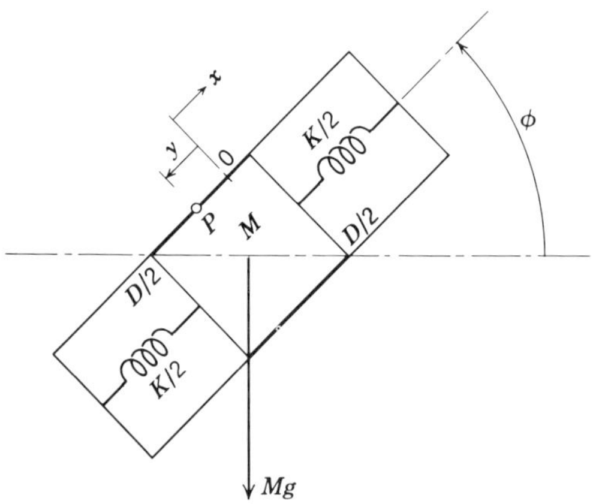

Figure 3.11 Accelerometer tilted by angle ϕ from horizontal.

component of the gravitational force in the y direction. The gravitational force is Mg where g is the acceleration due to gravity corresponding to the point in space where the accelerometer is located. The component of the gravitational force in the y direction is $Mg \sin\phi$ where ϕ is the tilt angle as shown in Figure 3.11. Then Figure 3.10 is modified by adding a force source $Mg \sin\phi$ connected from the reference node to the top left node. Instead of (3.44), we have

$$M(\ddot{x}-\ddot{y})+\frac{D}{2}[(\dot{x}-\dot{y})-\dot{x}]+\frac{D}{2}[(\dot{x}-\dot{y})-\dot{x}]$$
$$+\frac{K}{2}[(x-y)-x]+\frac{K}{2}[(x-y)-x]-Mg\sin\phi=0 \quad (3.53)$$

or

$$\ddot{y}+\frac{D}{M}\dot{y}+\frac{K}{M}y=\ddot{x}-g\sin\phi. \quad (3.54)$$

Since $g \sin\phi$ is a known quantity for any given direction and position in space, the measurement of $\ddot{x}$ is similar to the one in which $g \sin\phi$ is not present.

The actual position signals from the accelerometer may be electrical in nature. Electrostatic, magnetic, or electromechanical pick-offs may be used. We have discussed only the most elementary configuration for accelerometers. High-performance accelerometers are much more elaborate, but the basic principle of operation is the principle discussed above. ▲

Example 3-6. Consider a single-axis reaction wheel for attitude control of a space vehicle. A simplified diagram is shown in Figure 3.12. The axis of revolution is fixed in inertial space. The reaction wheel is a fly-wheel that serves as the rotor of an electric motor. The motor frame is rigidly attached to the vehicle. The motor (to be discussed in the next section) produces torque, driving the wheel. A reaction torque is produced, which acts on the vehicle, and it can be used to reorient the vehicle. Denote the angular velocity of the vehicle by $\dot{\theta}=\omega$ and the speed of the flywheel relative to the vehicle by Ω. For now, we shall assume that the motor is an angular velocity source. In the next

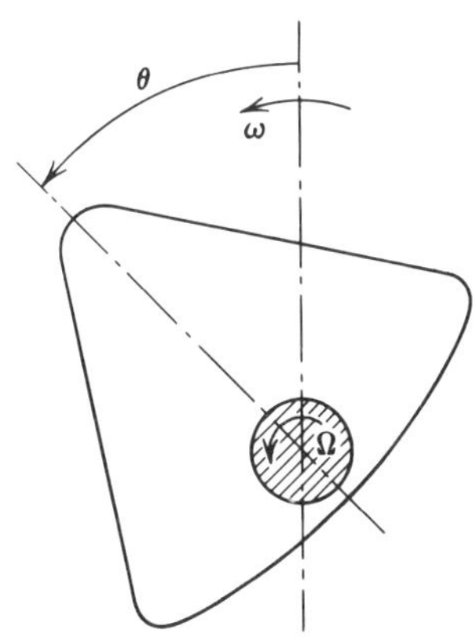

Figure 3.12 Single-axis reaction-wheel system.

section, a more elaborate model for the motor will be developed. The mechanical network for the reaction-wheel attitude system is shown in Figure 3.13, where J_V is the moment of inertia of the vehicle, J_W is the moment of inertia of the wheel, and the angular velocity source represents the motor. Since Ω is the angular velocity of the wheel relative to the vehicle, the angular velocity of the wheel relative to the inertial reference is $\Omega+\omega$. The torque supplied by the motor is denoted by τ_m, and the external disturbance torque is represented by the torque source τ_d. D'Alembert's principle applied to the left node yields

$$J_V\dot{\omega}-\tau_d+\tau_m=0. \tag{3.55}$$

For the right node, D'Alembert's principle yields

$$J_W(\dot{\omega}+\dot{\Omega})-\tau_m=0. \tag{3.56}$$

Adding (3.55) and (3.56), we have

$$(J_V+J_W)\dot{\omega}=\tau_d-J_W\dot{\Omega} \tag{3.57}$$

or

$$\dot{\omega}=\frac{\tau_d-J_W\dot{\Omega}}{J_V+J_W}. \tag{3.58}$$

Thus the motor torque $J_W\dot{\Omega}$ can counteract the disturbance torque τ_d. Furthermore, $\dot{\Omega}$ may be manipulated so as to produce a desired ω. Finally, the rate ω is related to the attitude angle θ by

$$\dot{\theta}=\omega. \tag{3.59}$$

The angular motion of the vehicle is described by (3.58) and (3.59).

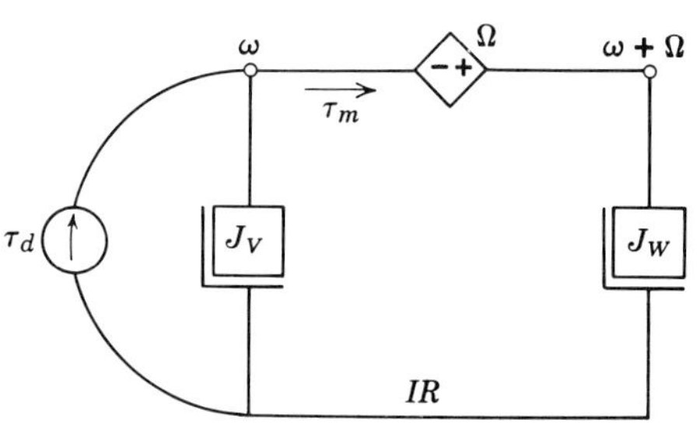

Figure 3.13 Mechanical network of reaction-wheel system.

The angular velocity ω may be obtained by integrating (3.58). Knowing ω, (3.59) may be integrated to obtain θ. ▲

3.3 Conservation Laws

A conservation law for physical devices and systems is a law of invariance of certain physical quantities. For example, the physical quantity charge is conserved in an electrical network or system. The conservation of charge law states that the total charge in a system is a constant. Kirchhoff's current law is a direct consequence of the conservation of charge. Since there is no accumulation of charge at a node, the net current (rate of change of charge) into a node is zero.

A very fundamental conservation law that applies to physical systems is the conservation of energy. (Here energy is broadly interpreted to include the energy equivalent to mass according to Einstein's law if nuclear reactions are involved.) A closely related law is the conservation of mass law. (Here again, the mass is broadly interpreted to include the mass equivalent of energy according to Einstein if nuclear reactions take place.) In this section, we shall illustrate the use of conservation of energy to obtain models for electromechanical devices and systems.

An electromechanical system consists of an electrical portion, a mechanical portion, and a coupling portion. The purely electrical and purely mechanical portions can be modeled using the techniques discussed in Chapter 2 and Sections 3.1 and 3.2. The new feature is the coupling portion. This portion converts electrical energy to mechanical energy or vice versa.

The energy flows in the coupling portion of an electromechanical system are indicated in Figure 3.14. Consideration of these flows will lead to a useful model for the coupling portion. We shall assume that all energy is represented in the diagram: W_e is electrical energy input, W_m is mechanical energy output, W_s is energy stored in the coupling portion (typically as stored energy in electrical or magnetic

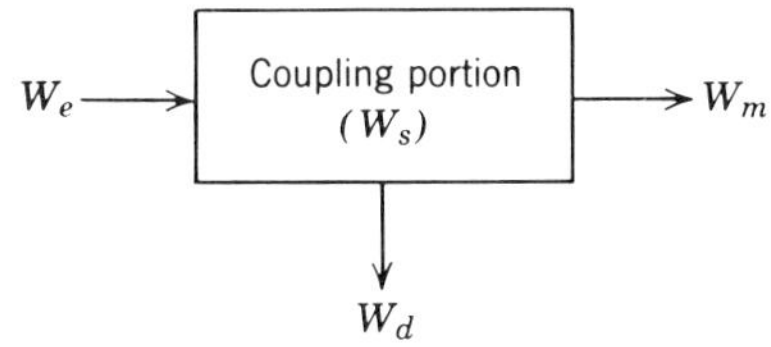

Figure 3.14 Coupling portion of electromechanical system.

fields), and W_d is energy dissipated (loss in resistors, dampers, friction, etc.). We suppose that no mass-to-energy conversions occur, nor is any energy lost from the system by radiation (electromagnetic waves). Notice the reference conventions chosen for energy flow. The quantities shown are positive for "motor action," that is, for a system that actually has electrical inputs and mechanical outputs. For systems having mechanical inputs and electrical outputs, W_e and W_m would be negative. Applying conservation of energy to the system in Figure 3.14, with energy flow references as shown, we obtain

$$W_e = W_m + W_s + W_d. \tag{3.60}$$

In the following, we shall consider only coupling systems for which $W_d = 0$. Such systems are called *conservative* or *lossless*. Notice that we do not assume the entire electromechanical system to be conservative; the electrical and mechanical portions may contain lossy elements. Only the coupling portion must be lossless. (Even some apparently lossy physical coupling systems can be represented by lossless systems, with resistors or dampers externally appended to account for the losses.) For $W_d = 0$, (3.60) becomes

$$W_e = W_m + W_s. \tag{3.61}$$

The energy functions of a conservative system have a particularly useful property as shown in the example below.

Example 3-7. Consider the linear time-invariant capacitor model of Chapter 2,

$$i = C\frac{dv}{dt}, \tag{3.62}$$

where C is a constant. This device stores energy in its electrostatic field. Moreover, $W_m = 0$ (no mechanical motion) and $W_d = 0$ (no dissipation). Electrical energy input is the integral of the instantaneous power input to the capacitor

$$W_e = \int_{t_0}^{t_1} p_e \, dt. \tag{3.63}$$

But electrical power input is

$$p_e = vi,$$

or

$$W_e = \int_{t_0}^{t_1} vi \, dt = \int_{v(t_0)}^{v(t_1)} Cv \, dv. \tag{3.64}$$

Equation 3.64 integrates to

$$W_e = \tfrac{1}{2}Cv^2(t_1) - \tfrac{1}{2}Cv^2(t_0). \tag{3.65}$$

That is, W_e *can be found without knowing the specific time dependence of* $v(t)$ *between* t_0 *and* t_1. W_e depends only on the values of v at the beginning and end of the interval, and not on the nature of the variation of v within the interval. This independence of integration path is a general property of energy functions for conservative systems. ▲

Now consider (3.61). Applying (3.61) at $t = t_0$, we obtain

$$W_{e0} = W_{m0} + W_{s0}, \tag{3.66}$$

where the subscript 0 indicates that the energy function is to be evaluated at $t = t_0$. Similarly, at $t = t_1$,

$$W_{e1} = W_{m1} + W_{s1}. \tag{3.67}$$

Now, because of the conservative property of the system, we can form the difference of (3.67) and (3.66), obtaining correct expressions for the energy *changes* in the interval $\Delta t = t_1 - t_0$:

$$\Delta W_e = \Delta W_m + \Delta W_s, \tag{3.68}$$

where

$$\begin{aligned} \Delta W_e &= W_{e1} - W_{e0} \\ \Delta W_m &= W_{m1} - W_{m0} \\ \Delta W_s &= W_{s1} - W_{s0}. \end{aligned} \tag{3.69}$$

Furthermore, we may consider a differentially small time interval $\Delta t \to dt$, so (3.68) becomes

$$dW_e = dW_m + dW_s, \tag{3.70}$$

which is the differential form of the *energy balance equation* of the conservative coupling system. Equation (3.70) is the basic equation used in modeling the coupling portion of a system, as will be illustrated below.

Example 3-8. As an illustration of the use of (3.70) in modeling electromechanical devices, we shall consider the *solenoid* pictured in Figure

3.15. A solenoid is an inductor with a movable magnetic core. Solenoids are widely used in various devices, including household appliances and many measuring instruments. The opening and closing of attitude control gas jets is done by solenoids. The movable core of mass M is constrained to translation in the x direction only by a sleeve. Friction with the sleeve as well as possible external damping of the motion is represented by the ideal viscous damper D. A spring provides restoring force on the core. We assume a linear spring with spring stiffness coefficient K. The inductor is excited electrically as shown. L represents the self-inductance of the inductor. R represents the resistance of the inductor windings as well as any resistance added externally in the electrical circuit. Motion is measured by the displacement of the center of the movable core from its rest position with respect to the center of the inductor windings. Various mechanical configurations are possible, depending on the use for the solenoid. We shall consider the spring and damper to have one terminal fixed a distance d from the coil center, and the other terminal fastened to the core.

At this point, the electrical portion has been modeled by v, R, and L, and the mechanical portion by K, D, and M. Only the coupling portion remains to be modeled. We would expect, from physical considerations (recalling the familiar action of an electromagnet, for example), that a force of electrical origin should tend to pull the core *into* the coil. Let us investigate this in detail.

The coupling portion of this device consists of the magnetic field produced by the inductor. This coupling system is outlined in Figure 3.15. We shall apply (3.70) to model this portion of the system. For this system,

$$dW_e = v_1 i \, dt. \tag{3.71}$$

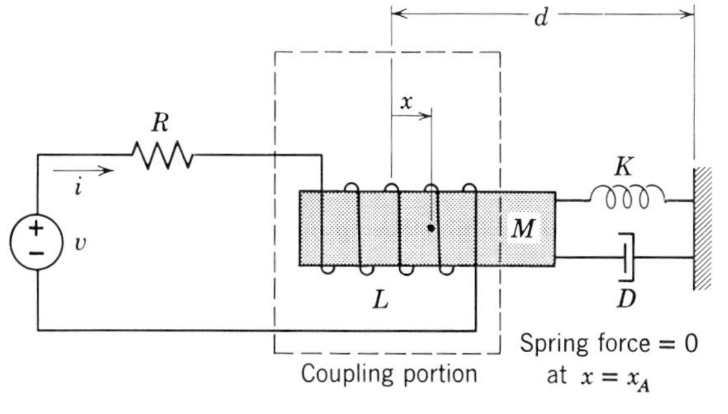

Figure 3.15 A solenoid.

Moreover, work (mechanical energy) is given by force times distance,

$$dW_m = f_e\, dx,$$

where f_e is the mechanical force of electric origin acting on the core. Since, by our reference convention, dW_m is a mechanical *output*, f_e has a *positive reference in the* $+x$ *direction*. Thus, (3.70) becomes

$$vi_1 dt = f_e dx + dW_s. \tag{3.72}$$

Equation 3.72 appears to contain too many unknowns. The force f_e is not known, nor is dW_s. However, we can make use of the conservative nature of the coupling system to determine *both* these quantities from the *single* equation (3.72)! Let us first compute W_s. Recalling that, for a conservative system, W_s is a function of only the initial and final values of the system variables, we may impose any mathematically convenient constraints in calculating the buildup of stored energy, as long as the appropriate initial and final values are achieved. *These constraints need not be the physical constraints imposed on the actual device*! In particular, let us compute W_s for a *fixed value of* x. Because the system is conservative, we shall obtain the correct functional dependence on x even if we hold x fixed for the computation of W_s. With x fixed, $dW_m = 0$, and

$$v_1 i\, dt = dW_s, \tag{3.73}$$

or

$$W_s = \int_{t_0}^{t_1} v_1 i\, dt. \tag{3.74}$$

To proceed, we need more details concerning the inductor. These may be determined experimentally if a physical model of the solenoid is available, or they may be calculated from physical analysis of inductors. Let us suppose that, as a first approximation, it is reasonable to consider the coil to be *electrically linear*, that is,

$$\lambda = L(x)i, \tag{3.75}$$

where the flux linkages are related to the voltage v_1 by

$$v_1 = \frac{d\lambda}{dt}. \tag{3.76}$$

Note that, for fixed x, this is just the usual model for a linear inductor. Thus we are supposing that the motion of the core produces an inductance $L(x)$ that varies with core position. Using (3.75) and (3.76), (3.74) becomes

$$W_s = \int_{\lambda(t_0)}^{\lambda(t_1)} \frac{\lambda}{L(x)}\, d\lambda. \tag{3.77}$$

Since x is fixed,

$$W_s = \frac{1}{2L(x)}[\lambda^2(t_1) - \lambda^2(t_0)]. \tag{3.78}$$

Because only incremental values of W_s appear in (3.70), we may take $\lambda(t_0) = 0$ and $t_1 = t$, the present time, without loss of generality. Thus

$$W_s = \frac{\lambda^2}{2L(x)} = W_s(\lambda, x), \tag{3.79}$$

using λ and x as variables, or

$$W_s = \tfrac{1}{2}L(x)i^2 = W_s(i, x), \tag{3.80}$$

substituting (3.75) and using i and x as variables. This is the correct expression for W_s no matter how λ, i, and x vary (provided, of course, that (3.75) is satisfied), although the derivation was made with x fixed!

We now return to (3.72) to find f_e. Using (3.79) for W_s,

$$i d\lambda = f_e\, dx + d\left[\frac{\lambda^2}{2L(x)}\right]. \tag{3.81}$$

Now we must let x vary, of course. In general, λ may also be allowed to vary, so that we obtain two terms from dW_s:

$$i d\lambda = f_e\, dx + \frac{\lambda\, d\lambda}{L(x)} - \frac{\lambda^2}{2L^2(x)}\, dL. \tag{3.82}$$

Note that, since $Li = \lambda$, the terms involving $d\lambda$ cancel each other no matter what the variation in λ is! Thus we could assume any variation for λ that we find convenient. In particular, λ could be held fixed! In any event, (3.82) yields

$$f_e = \frac{\lambda^2}{2L^2}\frac{dL}{dx} = \tfrac{1}{2}i^2\frac{dL}{dx}. \tag{3.83}$$

The same expression could be obtained using (3.80) for W_s, of course. Let us do this. Equation 3.81 becomes

$$i d\lambda = f_e\, dx + d[\tfrac{1}{2}L(x)i^2]. \tag{3.84}$$

Since x and i both vary,

$$d[\tfrac{1}{2}L(x)i^2] = \tfrac{1}{2}dL\, i^2 + iL\, di. \tag{3.85}$$

Also, using (3.75),

$$d\lambda = L\, di + dL\, i. \tag{3.86}$$

Substituting (3.85) and (3.86) in (3.84), we obtain

$$iL\, di + i^2 dL = f_e\, dx + \tfrac{1}{2}i^2 dL + iL\, di. \tag{3.87}$$

Notice that the terms involving di cancel each other no matter what the variation in i. In particular, we could have held i constant at the outset, shortening the calculation. We obtain

$$f_e = \tfrac{1}{2}\, i^2 \frac{dL}{dx} \tag{3.88}$$

as before.

Thus the force of electrical origin acting on the core is proportional to the square of the inductor current and proportional to the rate of change of the inductance with core position.

To carry the calculation further, we need to know the specific function $L(x)$. This can be determined experimentally or from physical considerations.

The electrical relations of the coupling system are readily determined by noting that

$$v_1 = \frac{d\lambda}{dt} = L(x)\frac{di}{dt} + i\frac{dL}{dx}\nu, \tag{3.89}$$

and $v_1 = v_L + v_m$, where the core velocity $\dot{x}$ is denoted by ν. Thus, there are two contributions to v_1; one due to self-inductance (this is the $L\, di/dt$ term obtained with no core motion), and a second term proportional to the velocity (this is sometimes called a "speed voltage").

The complete electromechanical device can be modeled by the network in Figure 3.16. Controlled sources have been used to show the electrical voltage of mechanical origin and mechanical force of electrical origin of the coupling section.

Applying Kirchhoff's voltage law to the electrical side of the model, we obtain

$$v = iR + L(x)\frac{di}{dt} + i\frac{dL}{dx}\nu. \tag{3.90}$$

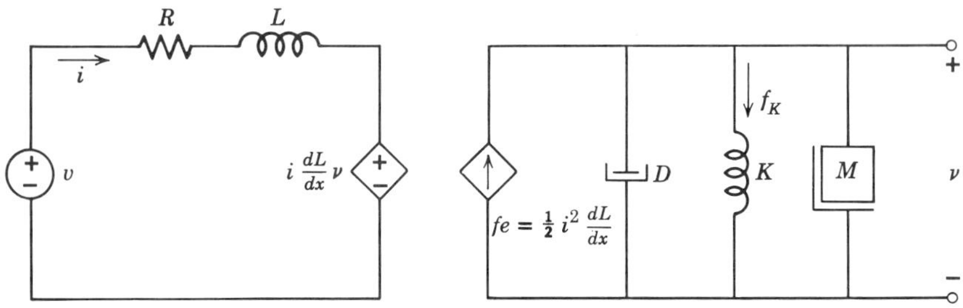

Figure 3.16 Complete network diagram for model of solenoid in Figure 3.15.

Applying D'Alembert's principle to the mechanical side, we obtain

$$\tfrac{1}{2}\,i^2\frac{dL}{dx} = M\frac{d\nu}{dt} + D\nu + K\Big[\int_{t_0}^{t}\nu(\tau)\,d\tau + x(t_0) - x_A\Big]. \qquad (3.91)$$

These two simultaneous equations represent a mathematical model for the electromechanical system. The model is *nonlinear* because of the nonlinear coupling terms. ▲

Example 3-9. The D-C Rotating Machine. Rotating machines (motors and generators) are of tremendous importance in accomplishing the work of the world. They generate most of the electrical and much of the mechanical energy used in our energy-hungry society. They are of special interest in dynamic systems engineering because of their use in instrumentation devices and in automatic control systems. In particular, we shall use rotating machines in the spacecraft attitude-control problem.

The important features of the construction of a typical d-c machine are indicated in Figure 3.17. Basically, the machine consists of two portions, a frame (stator) and a rotating member (rotor). A set of windings on the stator produces one magnetic field when excited. A second set of windings is on the rotor.

An important part of d-c machines is the *commutator*. The commutator is a special switching device that interchanges the voltages applied to the rotor windings so that the angle between the rotor and stator magnetic fields remains fixed at 90° despite the physical rotation of the rotor. The situation is shown pictorially in Figure 3.18. The subscript f on the stator variables denotes "field," the subscript a on the rotor variables denotes "armature." This is standard terminology for d-c machines.

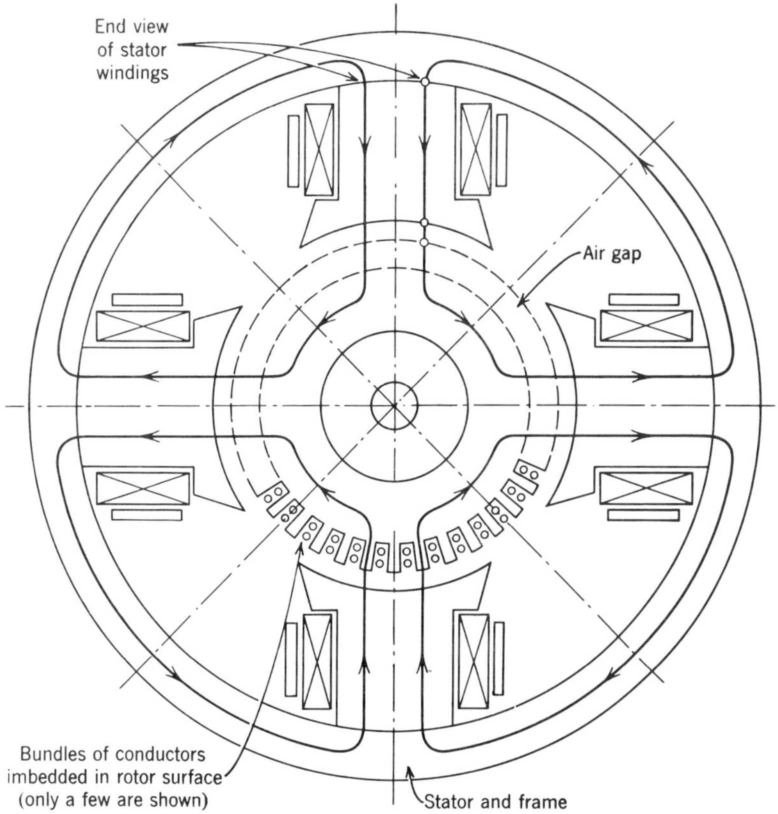

Figure 3.17 End view of a d-c machine.

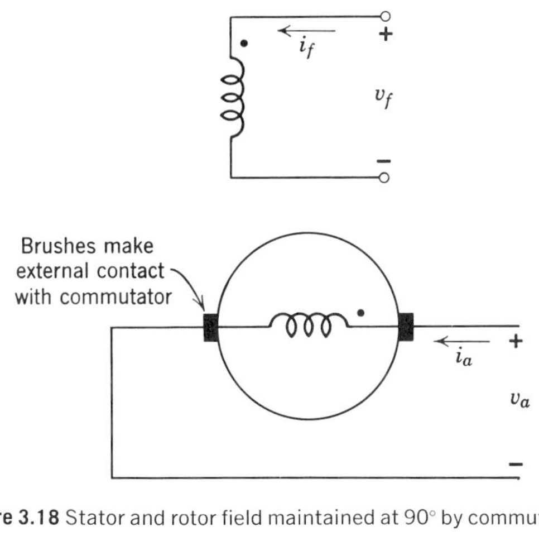

Figure 3.18 Stator and rotor field maintained at 90° by commutator.

We shall develop a simple model for the d-c machine. We shall neglect nonlinear effects (primarily magnetic saturation of the steel) for this development.

The field circuit is modeled easily. The winding possesses a resistance R_f and a self-inductance L_f. Since the field and armature coils are perpendicular, there will be very little mutual coupling. We shall neglect this small amount of coupling. Also, the field structure is stationary with respect to the magnetic field in the machine (remember the commutator maintains a *nonrotating* field structure relative to the stator); hence, there will be no induced voltage in the field windings due to Faraday's law. Thus the field circuit is modeled by

$$v_f = R_f i_f + L_f \frac{di_f}{dt}. \tag{3.92}$$

The armature circuit is slightly more complicated. The winding possesses resistance R_a and self-inductance L_a. Moreover, the mutual coupling with the field winding is negligible, as discussed above. However, there will be an induced voltage term due to Faraday's law, since the rotor windings are rotating at a mechanical speed ω_m with respect to the stator magnetic field. This induced voltage is called a "speed voltage," because it is proportional to ω_m. This voltage is also proportional to the flux, which is proportional to i_f, if saturation is neglected. Thus

$$v_0 = K i_f \omega_m. \tag{3.93}$$

The constant K depends on the geometry of the stator and rotor construction and on the windings. If we were interested in machine design, we would need a more detailed model indicating the effects of construction details. Such modeling would require an investigation of the field and magnetic circuit structure in the machine as a function of geometry. Here, however, we are interested primarily in the terminal characteristics of the machine.

With v_0 given in (3.93), the equation for the armature circuit becomes

$$v_a = R_a i_a + L_a \frac{di_a}{dt} + v_0. \tag{3.94}$$

To model the mechanical portion of the system, we need an expression for the torque of electrical origin. This is readily obtained from energy balance. The *electromechanical coupling portion* of the machine can be characterized by an electrical input of $v_0 i_a\, dt = dW_e$

and a mechanical output of $\tau_e \omega_m \, dt = dW_m$, as indicated in Figure 3.19. The coupling system is conservative ($dW_a = 0$). Moreover, there is no stored energy in the coupling system as we have identified it (recall that the inductances have been included in the external electrical circuits and, thus, should not be included again in the coupling system). Therefore (3.70) becomes

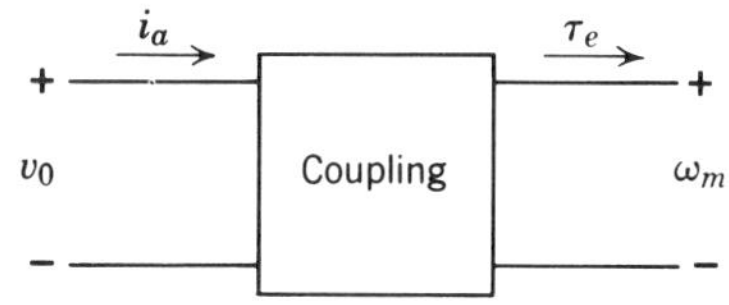

Figure 3.19 Coupling portion of d-c machine.

$$v_0 i_a \, dt = \tau_e \omega_m \, dt. \tag{3.95}$$

Substituting (3.93) into (3.95), we obtain

$$\tau_e = K i_f i_a. \tag{3.96}$$

In addition to this "developed torque," the machine typically has rotor inertia as well as viscous-friction losses ("friction and windage"). These may be modeled by inertia J_m and friction D_m. A *threeport* network model of the d-c machine is given in Figure 3.20. The mechanical portion is not shown because its nature depends on the environmental situation in which the d-c machine is used. For example, in the spacecraft example, if the stator is rotating relative to inertial reference, the moment of inertia of the stator must be modeled

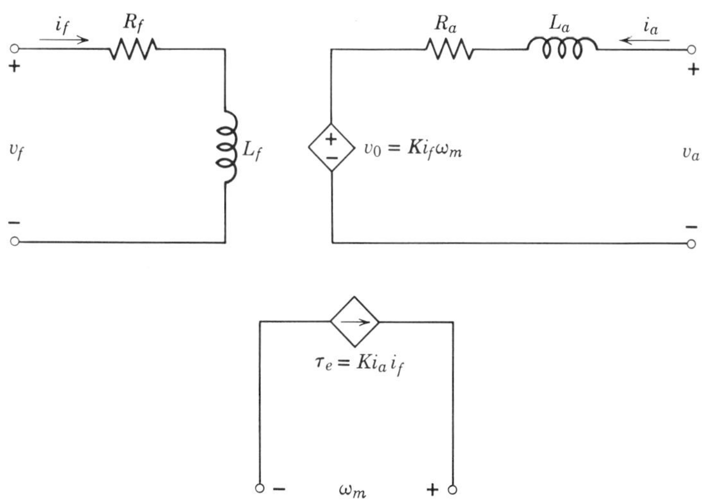

Figure 3.20 Threeport model of d-c- machine.

also. Such an application is discussed in the next section. The d-c machine can be used in a variety of configurations. The model of Figure 3.20 can be used to represent any configuration, provided that consideration of magnetic saturation is not essential. ▲

3.4 Network Model of Single-Axis Reaction-Wheel Attitude Control System

Let us reconsider the single-axis model for a reaction-wheel attitude control system discussed in Example 3-6. For driving the reaction wheel, a variety of motors is available such as a-c synchronous motors, a-c induction motors (including the servomotor), separately excited d-c motors, permanent-magnet d-c motors, and many others. Because the motor must operate in space, the considerations involved in motor selection differ from those governing the usual terrestrial applications. Ease of speed control, together with torque capability as a function of weight and fuel consumption, are involved in the choice of motors. For example, the speed of a-c synchronous motors is difficult to control over a wide range and, thus, such machines are virtually eliminated from consideration. Separately excited d-c machines require excitation for the field even when not operating, an undesirable feature. The permanent magnet d-c motor overcomes this difficulty, but it is used only in small output (millihorsepower) applications. Difficulties with brushes and commutation present the major objections to the use of d-c machines in space. Electronic commutation has been developed to overcome this problem. However, the circuitry contributes additional weight. Single-phase and three-phase a-c induction machines generally are subject to limited speed control. A two-phase induction motor, the servomotor, does have satisfactory speed-control characteristics, however. We shall concentrate on the use of servomotors and armature-controlled d-c motors here.

The d-c machine was discussed and modeled in Example 3-9. The weight of the motor, an important consideration in space applications, depends primarily on its mechanical power-output capabilities. The curve in Figure 3.21 has been obtained by study of typical d-c reaction-wheel motors available. It can be used to estimate the expected weight of a motor, given its output-power requirements. The output power can be determined from a study of the attitude control and maneuvering requirements.

A type of two-phase a-c induction motor, the servomotor, is also widely used. The motor consists of two windings on the stator placed

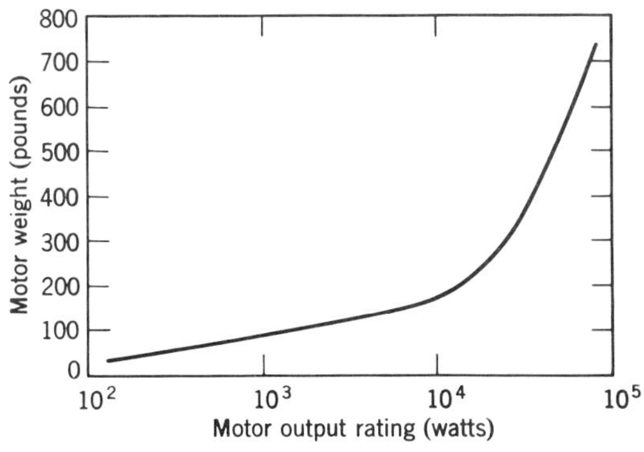

Figure 3.21 Motor weight versus power output for typical d-c machines.

90° apart, as shown schematically in Figure 3.22. An a-c sinusoidal voltage of fixed magnitude and phase is applied to the reference phase. An a-c sinusoidal voltage of variable magnitude and differing 90° in phase from the reference voltage, is applied to the control phase. This combination of excitations produces a field in the air gap that *rotates* at the *synchronous speed*

$$N_s = \frac{120f}{P}\ \text{rpm},$$

where f is the frequency of the a-c excitation and P is the number of poles of the machine. No electrical connections are made to the rotor. Instead, currents are *induced* in the rotor circuit in accord with Faraday's law (hence the name *induction* motor). A torque is developed by interaction of the current-carrying rotor conductors with the rotating stator field.

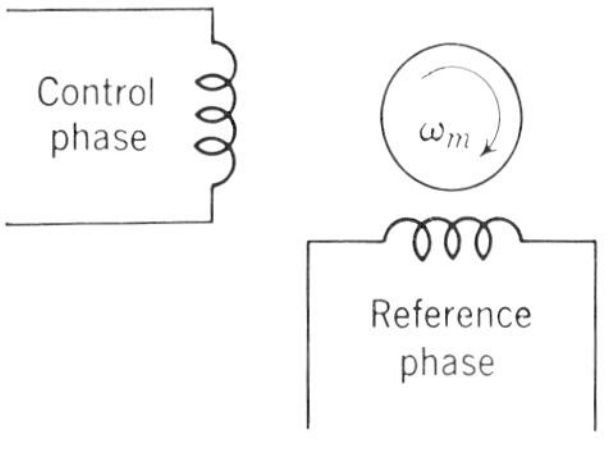

Figure 3.22 Pictorial representation of servomotor.

The weight of the a-c servomotor depends primarily on its output power, as is exhibited in Figure 3.23.

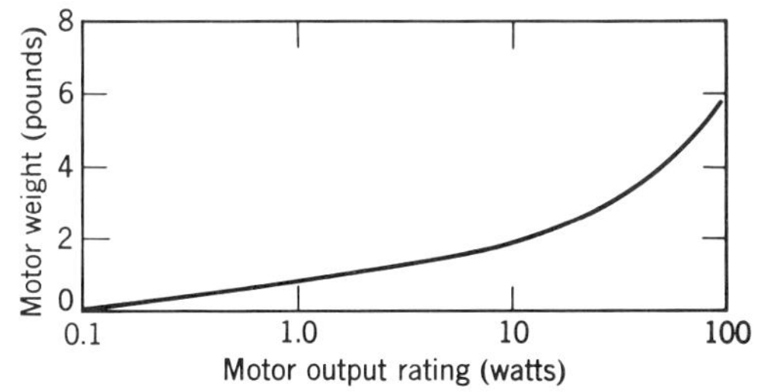

Figure 3.23 Weight versus power output for typical a-c servomotors.

Typical torque-speed characteristics of a servomotor are indicated in Figure 3.24. V_c is the magnitude of the control voltage, ω_m is the speed, and τ is the torque. Although these characteristics are nonlinear, they can typically be linearized meaningfully in the vicinity of an operating point.

Let us now consider the modeling of a single-axis reaction-wheel attitude-control system using a d-c motor for producing the relative speed Ω between the reaction wheel and the vehicle. Instead of a simple velocity source as in Figure 3.13, we shall employ a more detailed model for the motor. The system outputs are the attitude angle θ and the attitude rate $\dot{\theta}$. The system input is the armature voltage on the motor. The system equations will be obtained from a network model.

Figure 3.20 shows the network model for a d-c machine derived in Section 3.3. For our application, we shall employ the machine as a motor so that we have electrical inputs and mechanical outputs. The field is excited by a constant current source, so the field current i_f is a constant. (Often the field is supplied by a permanent magnet, which is equivalent in our model to $i_f =$ constant.) As mentioned above, the armature voltage is the system input. The armature inductance L_a is usually very small, and we shall neglect it in our model here.

Since the motor frame is rigidly connected to the vehicle, the moment of inertia of the motor stator may be included in the moment of inertia of the vehicle. Let this total amount of inertia be denoted by J_V. Similarly, let J_W denote the total moment of inertia of the rotor-flywheel combination. Our network model for the system must include two rotational inertias, J_V and J_W, with one terminal from each connected to the inertial reference frame. The other node of J_V corresponds to the vehicle angular velocity ω with respect to inertial reference. The other node of J_W corresponds to a wheel angular

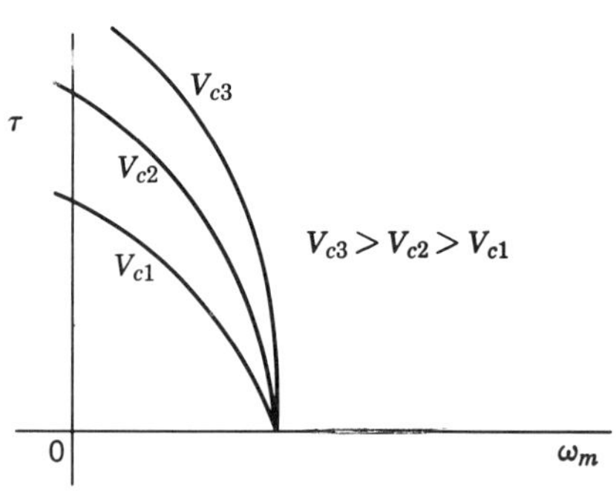

Figure 3.24 Servomotor torque-speed characteristics.

velocity of $\omega + \Omega$ with respect to inertial reference. Between these two nodes, we connect the torque source from the motor.

The complete network diagram is shown in Figure 3.25. Frictional losses of the rotor proportional to the speed Ω are accounted for by means of a damper D as shown in Figure 3.25.

Let us now write the equations of motion for the system from the network of Figure 3.25. From KVL, we may write

$$-V_a + R_a i_a + K i_f \Omega = 0. \tag{3.97}$$

Applying D'Alembert's principle to the top node of J_V we have

$$-\tau_d + J_V(\dot{\omega}) + D[\omega - (\omega + \Omega)] + K i_f i_a = 0. \tag{3.98}$$

For the top node of J_W, D'Alembert's principle yields

$$J_W(\dot{\omega} + \dot{\Omega}) + D[(\omega + \Omega) - \omega] - K i_f i_a = 0. \tag{3.99}$$

We shall assume that i_f is known and constant because V_f is known and constant. Hence the unknowns are Ω, ω, and i_a with V_a as input. We also have

$$\dot{\theta} = \omega \tag{3.100}$$

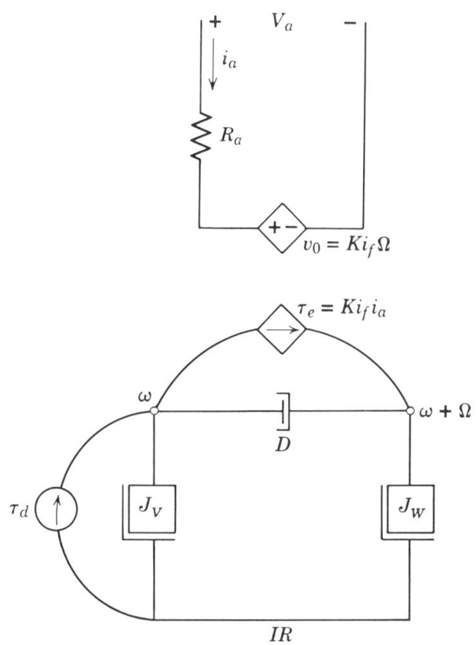

Figure 3.25 Network diagram for single-axis reaction-wheel attitude-control system.

which relates the rate ω to the attitude angle θ. The armature current i_a may be eliminated by solving for i_a in (3.97) and substituting the expression for it in (3.98) and (3.99). Thus (3.98) becomes

$$J_V\dot{\omega} = D\Omega + \tau_d - \frac{Ki_f}{R_a}(V_a - Ki_f\Omega)$$

or

$$\dot{\omega} = \frac{1}{J_V}\left(D + \frac{K^2 i_f^2}{R_a}\right)\Omega + \frac{1}{J_V}\tau_d - \frac{Ki_f}{J_V R_a}V_a. \tag{3.101}$$

Similarly, (3.99) becomes

$$J_W(\dot{\omega} + \dot{\Omega}) + D\Omega - \frac{Ki_f}{R_a}(V_a - Ki_f\Omega) = 0. \tag{3.102}$$

As discussed in Chapter 4, it is convenient for computer simulation to have differential equations written as a set of first-order differential equations where only one derivative appears in each equation. Equations (3.100) and (3.101) are already in such form. However, (3.102) is not. This is easily resolved by substituting $\dot{\omega}$ from (3.101) in (3.102) to obtain

$$\frac{J_W}{J_V}\left(D + \frac{K^2 i_f^2}{R_a}\right)\Omega + \frac{J_W}{J_V}\tau_d - J_W\frac{Ki_f}{R_a}V_a + J_W\dot{\Omega} + D\Omega - \frac{Ki_f}{R_a}V_a + \frac{K^2 i_f^2}{R_a}\Omega = 0$$

or

$$\dot{\Omega} = -\left(\frac{1}{J_V} + \frac{1}{J_W}\right)\left(\frac{K^2 i_f^2}{R_a} + D\right)\Omega + \frac{Ki_f}{R_a}\left(\frac{1}{J_V} + \frac{1}{J_W}\right)V_a - \frac{1}{J_V}\tau_d. \tag{3.103}$$

Thus (3.100), (3.101), and (3.103) are the desired three system equations for our attitude control system. They are first-order differential equations in the variables Ω, ω, and θ. These equations can be simulated on a computer for any desired input function V, initial conditions $\Omega(0)$, $\omega(0)$, and $\theta(0)$, and any disturbance torque τ_d.

Instead of the d-c motor, a two-phase a-c induction motor, the servomotor, is also widely used. Its linearized model is similar to the one considered in this section, and it leads to equations like (3.101), (3.102), and (3.103) except that the constant coefficients have different interpretations.

3.5 Equilibrium Points and Linearization

In the operation of some physical systems, the values of certain signals or variables may be confined to narrow ranges. For example, the attitude angles of a spacecraft during a particular phase of a mission may vary within a fraction of a degree from say 30°. The variation may be caused by small disturbance torques arising from a variety of factors including solar radiation, micrometeorite impact, or minor redistribution of mass within the spacecraft. Although the equations describing the system may be nonlinear, a linear approximation may be adequate for describing the relationship between the *deviation* or change in the input to the *deviation* or change in the output. For example, if the disturbance torque that is added to the control torque in maintaining an attitude of 30° is small, then the deviation of the angle from 30° may be related to the disturbance torque by a linear approximation.

The main advantage in using linear models is that they can be studied analytically. This usually provides greater insight into system behavior and may result in computational advantages as well. The following example illustrates one method of system linearization using Taylor's series.

Example 3-10. Consider the nonlinear series circuit shown in Figure 3.26 where the input is a voltage v. The resistor R is assumed to be constant, but the inductor is nonlinear with λ-i characteristic

$$\lambda = \tanh i. \tag{3.104}$$

This characteristic is sketched in Figure 3.27. From KVL and the network element equations,

$$-v + Ri + \frac{d\lambda}{di}\frac{di}{dt} = 0. \tag{3.105}$$

Since $\lambda = \tanh i$,

$$\frac{d\lambda}{di} = \text{sech}^2\, i. \tag{3.106}$$

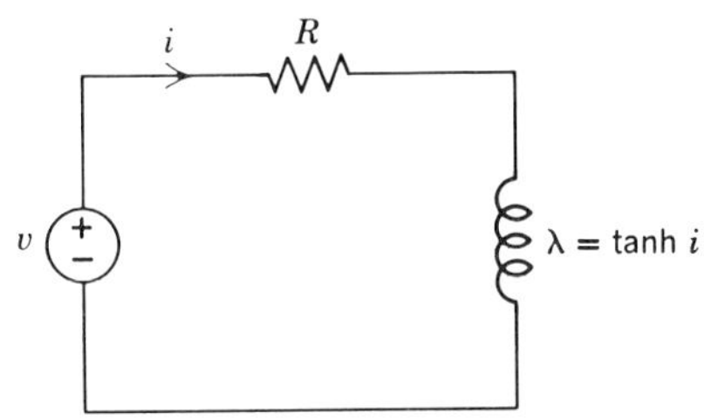

Figure 3.26 Nonlinear series circuit.

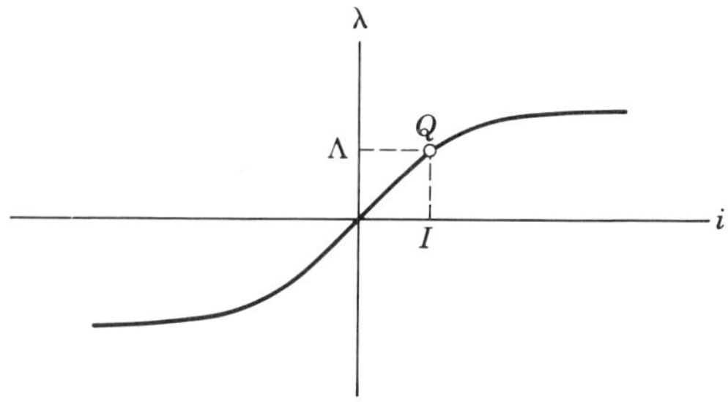

Figure 3.27 Nonlinear inductor linearized at point Q.

Substituting (3.106) in (3.105), we have

$$\frac{di}{dt} = (v - iR) \cosh^2 i. \tag{3.107}$$

We note that, for $v = V =$ constant, a *particular solution* of (3.107) is

$$i = \frac{V}{R} = \text{constant} = I. \tag{3.108}$$

That is, $i = I$ satisfies the differential equation (3.107) when $v = V$, since, of course,

$$\frac{di}{dt} = \frac{dI}{dt} = 0. \tag{3.109}$$

The point $i = I$ is called an *equilibrium point* of the differential equation (3.107). The network response $i(t)$, that is, the *complete solution* of (3.107) with specified initial conditions, is not equal to I in general. However, the *steady-state* value of the network response is equal to I.

We now consider a general input v. Equation 3.107 is

$$\frac{di}{dt} = f(i, v), \tag{3.110}$$

where

$$f(i, v) = (v - iR) \cosh^2 i. \tag{3.111}$$

The right-hand side of (3.110) may be expanded in Taylor's series about the equilibrium point $i = I$ with $v = V$:

$$f(i, v) = f(I, V) + \left.\frac{\partial f}{\partial i}\right|_{\substack{i=I\\v=V}} \Delta i + \left.\frac{\partial f}{\partial v}\right|_{\substack{i=I\\v=V}} \Delta v + \text{higher-order terms}, \tag{3.112}$$

$$\begin{aligned} \Delta i &= i - I \\ \Delta v &= v - V. \end{aligned} \tag{3.113}$$

Since $i = I$ is an equilibrium point for $v = V, f(I, V) = 0$. Furthermore,

$$\frac{di}{dt} = \frac{dI}{dt} + \frac{d}{dt}(\Delta i) = \frac{d}{dt}(\Delta i). \tag{3.114}$$

Thus (3.110) may be written

$$\frac{d}{dt}(\Delta i) = \left.\frac{\partial f}{\partial i}\right|_{\substack{i=I\\ v=V}} \Delta i + \left.\frac{\partial f}{\partial v}\right|_{\substack{i=I\\ v=V}} \Delta v + \text{higher-order terms.} \tag{3.115}$$

If Δi and Δv are small, that is, if we are operating in the vicinity of the equilibrium point, then a good approximation to the actual system behavior may be obtained by solving (3.115) with the higher-order terms omitted:

$$\frac{d}{dt}(\Delta i) \approx \left.\frac{\partial f}{\partial i}\right|_{\substack{i=I\\ v=V}} \Delta i + \left.\frac{\partial f}{\partial v}\right|_{\substack{i=I\\ v=V}} \Delta v. \tag{3.116}$$

Note that (3.116) is a *linear* differential equation with *constant* coefficients. ▲

For a general system with n state variables, instead of the single equation (3.110), we have the system of n equations

$$\begin{aligned} \frac{dx_1}{dt} &= f_1(x_1, x_2, \ldots, x_n, u_1, u_2, \ldots, u_m) \\ \frac{dx_2}{dt} &= f_2(x_1, x_2, \ldots, x_n, u_1, u_2, \ldots, u_m) \\ &\vdots \\ \frac{dx_n}{dt} &= f_n(x_1, x_2, \ldots, x_n, u_1, u_2, \ldots, u_m) \end{aligned} \tag{3.117}$$

where x_i for $i = 1, \ldots, n$ are the state variables of the system and u_1, $u_2, \ldots, u_m$ are the inputs to the system. The first step in the linearization process is to find the equilibrium point (or points) of the system, when the inputs assume their constant or steady values. Let these constant input values be denoted by $U_1, U_2, U_3, \ldots, U_m$. At the equilibrium point, the state variables have constant values so that the time derivatives are all zero. Denoting the equilibrium values for $x_1, x_2, \ldots, x_n$ by $X_1, X_2, \ldots, X_n$, we have

$$\begin{aligned}
0 &= f_1(X_1, X_2, \ldots, X_n, U_1, U_2, \ldots, U_m) \\
0 &= f_2(X_1, X_2, \ldots, X_n, U_1, U_2, \ldots, U_m) \\
&\vdots \\
0 &= f_n(X_1, X_2, \ldots, X_n, U_1, U_2, \ldots, U_m).
\end{aligned} \tag{3.118}$$

The equilibrium values $X_1, X_2, \ldots, X_n$ are obtained by solving the n nonlinear nondynamic equations in (3.118).

The next step is to obtain the linearized equations by a Taylor's series expansion at the equilibrium point, retaining only the linear terms. What is needed is the Taylor's series form for several variables. Denote the input variables by

$$\begin{aligned}
u_1 &= U_1 + \Delta u_1 \\
u_2 &= U_2 + \Delta u_2 \\
&\vdots \\
u_m &= U_m + \Delta u_m
\end{aligned} \tag{3.119}$$

and the state variables by

$$\begin{aligned}
x_1 &= X_1 + \Delta x_1 \\
x_2 &= X_2 + \Delta x_2 \\
&\vdots \\
x_n &= X_n + \Delta x_n.
\end{aligned} \tag{3.120}$$

Now expand the functions in (3.120) in Taylor series at the equilibrium point. For example, for the first equation,

$$\begin{aligned}
\frac{d\Delta x_1}{dt} &= f_1(X_1, X_2, \ldots, X_n, U_1, U_2, \ldots, U_m) \\
&+ \frac{\partial f_1}{\partial x_1}\Delta x_1 + \frac{\partial f}{\partial x_2}\Delta x_1 + \cdots + \frac{\partial f_1}{\partial x_n}\Delta x_n \\
&+ \frac{\partial f_1}{\partial u_1}\Delta u_1 + \cdots + \frac{\partial f_1}{\partial u_m}\Delta u_m + \text{higher-order terms}
\end{aligned} \tag{3.121}$$

where the partial derivatives are evaluated at the equilibrium point. From (3.118), we notice that the first term on the right-hand side is zero, so that if we neglect the higher-order terms we have

$$\frac{d(\Delta x_1)}{dt} = \sum_{i=1}^{n} \frac{\partial f_1}{\partial x_i} \Delta x_i + \sum_{i=1}^{m} \frac{\partial f_1}{\partial u_i} \Delta u_i. \tag{3.122}$$

Similarly for the other equations, we may obtain a linear approximation by a Taylor's series expansion and by neglecting the higher-order terms. Thus

$$\begin{aligned} \frac{d(\Delta x_2)}{dt} &= \sum_{i=1}^{n} \frac{\partial f_2}{\partial x_i} \Delta x_i + \sum_{i=1}^{m} \frac{\partial f_2}{\partial u_i} \Delta u_i \\ &\vdots \\ \frac{d(\Delta x_n)}{dt} &= \sum_{i=1}^{n} \frac{\partial f_n}{\partial x_i} \Delta x_i + \sum_{i=1}^{m} \frac{\partial f_n}{\partial u_i} \Delta u_i. \end{aligned} \tag{3.123}$$

It should be remembered that the partial derivatives are evaluated at the equilibrium point so that these partial derivatives are constants. Hence (3.122) and (3.123) yield a system of ordinary linear differential equations with constant coefficients, relating the deviations of the state variables Δx_i to the deviations of the input variables Δu_i, from the corresponding equilibrium points X_i and U_i.

Example 3-11. Let us obtain a linear model for the electromechanical system of Figure 3.15. The system equations are given by (3.90) and (3.91). Although these equations can be linearized directly, they are not in the state (simultaneous first order) form of (3.117). To apply our linearization technique discussed above, the equations must be converted to first-order form. Dividing by $L(x)$, (3.90) becomes

$$\frac{di}{dt} = -\frac{R}{L(x)} i - \frac{1}{L(x)} \frac{dL}{dx} i\nu + \frac{1}{L(x)} v. \tag{3.124}$$

Conversion of (3.91) is slightly more complicated. We first note that

$$\frac{dx}{dt} = \nu, \tag{3.125}$$

and thus (3.91) can be put into the form

$$\frac{d\nu}{dt} = -\frac{D}{M}\nu - \frac{K}{M}(x - X_A) + \frac{1}{2M} i^2 \frac{dL}{dx}. \tag{3.126}$$

Thus (3.124), (3.125), and (3.126) constitute the desired simultaneous first-order equations. The equilibrium point is obtained by setting the right-hand sides of (3.124), (3.125), and (3.126) to zero with $v = V$, as indicated by (3.118). Denoting the equilibrium point values by uppercase letters, we have

$$\begin{aligned} 0 &= -\frac{R}{L(X)} I - \frac{1}{L(X)} \frac{dL}{dX} I\mathrm{N} + \frac{1}{L(X)} V \\ 0 &= \mathrm{N} \\ 0 &= -\frac{D}{M}\mathrm{N} - \frac{K(X - X_A)}{M} + \frac{1}{2M} I^2 \frac{dL}{dX}. \end{aligned} \tag{3.127}$$

The second equation shows that $\mathrm{N} = 0$, that is, the equilibrium point for velocity is zero. Thus the equilibrium position is a constant. Note that dL/dX is evaluated at this constant equilibrium position. With $\mathrm{N} = 0$, the equilibrium point conditions reduce to

$$\begin{aligned} I &= \frac{V}{R} \\ X &= \frac{1}{2} \frac{dL}{dX} \frac{V^2}{R^2 K} + X_A. \end{aligned} \tag{3.128}$$

Suppose, for example, that

$$L(x) = L_0 + L_1 \,\epsilon^{-K_1 x^2} \tag{3.129}$$

where L_0, L_1 and K_1 are positive constants. Then

$$\left.\frac{dL}{dx}\right|_{x=X} = -2K_1 L_1 X \,\epsilon^{-K_1 X^2}. \tag{3.130}$$

X may be determined by substituting (3.130) in (3.127),

$$X - X_A = \frac{1}{2} \frac{V^2(-2K_1 L_1 X \,\epsilon^{-K_1 X^2})}{KR^2} \tag{3.131}$$

Equation 3.131 is a transcendental equation in X, which may be solved numerically or graphically. The quantities V, K_1, K, L_1, R, and X_A are given.

The small-signal equations are obtained by applying (3.122)–(3.123) to (3.124), (3.125), and (3.126):

$$\frac{d}{dt}(\Delta i) = -\frac{R}{L(X)}\Delta i - \left[\frac{I}{L(X)}\frac{dL}{dX}\right]\Delta \nu + \frac{1}{L(X)}\Delta v$$

$$\frac{d}{dt}(\Delta x) = \Delta \nu$$

$$\frac{d}{dt}(\Delta \nu) = \left[\frac{I}{M}\frac{dL(X)}{dX}\right]\Delta i - \frac{D}{M}\Delta \nu - \frac{K_{eq}}{M}\Delta x, \tag{3.132}$$

where

$$K_{eq} = \left(K - \frac{I^2}{2}\frac{d^2L}{dX^2}\right). \tag{3.133}$$

Thus the linearized system is described by the three simultaneous first-order constant-coefficient linear-differential equations in (3.132). A *linear* network model of the linearized solenoid is given in Figure 3.28.

Note from (3.133) that the equivalent spring stiffness K_{eq} may be negative! This will result in unstable motion of the solenoid. Stability of linear systems is discussed in detail in Chapter 8. ▲

3.6 Sensitivity Models

An important consideration in both analysis and design is the effect of system parameter variations on system performance. There will always be some variation in subsystems that are supposed to be identical. Recognizing that it is impossible to build or measure any system parameter exactly, tolerances often are associated with parameter values, performance measures, and system responses. For example, dimensions on machine parts have associated tolerances. Resistance values are stated with tolerances, such as 1000 ohms ±5%. Desired

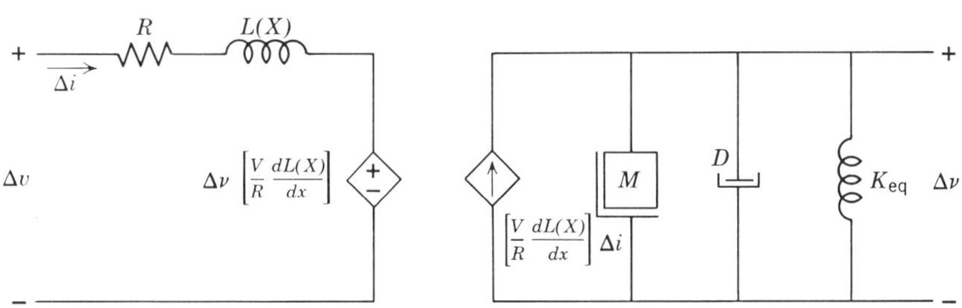

Figure 3.28 Network model of linearized solenoid.

attitude positions are stated in terms of acceptable ranges of values. In design, acceptable tolerances on components must be set as high as possible in the interest of simplicity and economy, but low enough so that the overall system performance is still satisfactory.

In addition, a study of the effects of parameter variation on system responses and performance can be used for the determination of nominal design values for controller parameters. On the basis of a parameter sensitivity analysis, tentative controller parameter values may be adjusted manually or automatically until a satisfactory or even optimum setting of the parameter values is reached. Such adjustments may be performed in the design process before the system is fabricated or "on-line" while the system is operating to adapt to new system goals or environmental conditions.

One straightforward method of studying parameter variation effects is simply to simulate the system and observe responses for various parameter settings. This approach is actually used successfully in some analysis and evaluation situations. However, this approach may fail to provide insight into the roles of critical parameters in system performance. This insight is especially important in design, in contrast with analysis, when parameter adjustments are being made to obtain desired performance.

Several other approaches to the study of parameter effects have been proposed. A common method involves a partial derivative, called a sensitivity function or parameter influence coefficient, to approximate, to first order, effects of small parameter variations. Sensitivity functions have the property of exhibiting clearly, in "magnified" form, the effects of parameter change. Furthermore, the sensitivity functions are useful in several numerical optimization methods, such as the gradient method, as will be discussed in later chapters.

Consider a signal $y(t, q)$ as a function of time t and parameter q. To first-order approximation,

$$y(t, q_0+\Delta q) = y(t, q_0) + \left.\frac{\partial y}{\partial q}\right|_{q=q_0} \Delta q. \tag{3.154}$$

Given a response $y(t, q_0)$ corresponding to a nominal value of parameter q_0 the perturbed response $y(t, q_0+\Delta q)$ may be predicted for any parameter perturbation Δq by using (3.154) if the partial derivative in (3.154) is available. For simplicity in notation, we shall denote the partial derivative evaluated at $q = q_0$ by

$$\frac{\partial y}{\partial q} = \frac{\partial y(t, q_0)}{\partial q} = \left.\frac{\partial y(t, q)}{\partial q}\right|_{q=q_0}.$$

This partial derivative is defined as the *sensitivity function* of y with respect to q.

In this section, we shall demonstrate that if $y(t, q)$ appears as a signal in a network, $\partial y/\partial q$ also appears as a signal in a similar network with the same topology. If the network is linear and time-invariant and if q is one of the element parameters, then the network for determining $\partial y/\partial q$ is the same as the original except that the original independent sources are reduced to zero, and a controlled source is inserted in the branch where q appears. The nature of the insertion is described in detail below. This method of generating sensitivity function is computationally more accurate than forming the difference $y(t, q_0+\Delta q)-y(t, q_0)$ and then scaling by Δq. Furthermore, the same computer network analysis program (using ECAP, for example) can be used both for the original network and for the network generating the sensitivity functions. The method was first enunciated by M. L. Bihovski in "The Accuracy of Electric Network Analyzers," *Izv. AN SSR Otd. Tekhn, Nauk*, No. 8, 1948, in Russian. See also J. V. Leeds, Jr., "Transient and Steady-State Sensitivity Analysis," *IEEE Trans. on Circuit Theory*, Vol. CT-13, No. 3, September 1966, pp. 288–289.

The sensitivity model is easily derived by taking the partial derivative with respect to q of all the equilibrium equations and branch equations describing the original network.

Consider an electrical network whose KCL equations are

$$\sum_{j=1}^{N_b} a_{kj} i_j = 0, \qquad k = 1, 2, \ldots, N_v \tag{3.155}$$

and whose KVL equations are

$$\sum_{j=1}^{N_b} b_{kj} v_j = 0, \qquad k = 1, 2, \ldots, N_l \tag{3.156}$$

where a_{kj} and b_{kj} may be 0, +1, or −1 as discussed in Section 3.1. Taking the partial derivative of (3.155) and (3.156), we have

$$\sum_{j=1}^{N_b} a_{kj} \frac{\partial i_j}{\partial q}, \qquad k = 1, 2, \ldots, N_v \tag{3.157}$$

and

$$\sum_{j=1}^{N_l} b_{kj} \frac{\partial v_j}{\partial q}, \qquad k = 1, 2, \ldots, N_l. \tag{3.158}$$

If a particular current i_j is a source that does not depend on the parameter q, then $\partial i_j/\partial q$ is zero. In general, if i_j is not a source, the partial derivative is not identically zero. Similarly, $\partial v_j/\partial q$ is generally not identically zero unless v_j is a source that does not depend on q. Thus, if $\partial v_j/\partial q$ is regarded as a voltage and $\partial i_j/\partial q$ is regarded as a current, the topology of the associated network must be the same as the original network except that current sources are open-circuited and voltage sources are short-circuited. This is because the KCL and KVL equations are the same, except for the source terms.

Let us consider the branch equations. Suppose the network is linear and time-invariant so that the branch equations are of the form

$$v_j = R_j i_j, \tag{3.159}$$

$$i_k = C_k \frac{dv_k}{dt}, \tag{3.160}$$

$$v_l = L_l \frac{di_l}{dt}. \tag{3.161}$$

Let the parameter q be an element parameter R_1. Then if $j \neq 1$

$$\frac{\partial v_j}{\partial R_1} = R_j \frac{\partial i_j}{\partial R_1} \tag{3.162}$$

$$\frac{\partial i_k}{\partial R_1} = C_k \frac{d}{dt}\left(\frac{\partial v_k}{\partial R_1}\right) \tag{3.163}$$

$$\frac{\partial v_l}{\partial R_1} = L_l \frac{d}{dt}\left(\frac{\partial i_l}{\partial R_1}\right) \tag{3.164}$$

and for $j = 1$

$$\frac{\partial v_1}{\partial R_1} = R_1 \frac{\partial i_1}{\partial R_1} + i_1. \tag{3.165}$$

Thus, with $\partial i_j/\partial R_1$ and $\partial v_j/\partial R_1$ for all j, regarded as current and voltage, respectively, the above equations indicate that the branch characteristics are the same as before except that, instead of the branch R_1, we have a controlled voltage source equal to i_1 in series with R_1. The sensitivity network is the same as the original network except that all of the independent voltage sources are short-circuited ($v = 0$) and all of the independent current sources are open-circuited ($i = 0$), and a controlled voltage source equal to i_1 is inserted in series with R_1.

Example 3-12. Consider the network in Figure 3.29*a* where it is desired to measure the sensitivity of v_2 with respect to R_1. Let the current through R_1 be denoted by i_1 as indicated in Figure 3.29*a*. The sensitivity network is obtained by redrawing the original network, but with v reduced to zero, and inserting a controlled voltage source equal to i, in series with R_1 as shown. The voltage from node 2 to the bottom node of the sensitivity network is equal to $\partial v_2/\partial R_1$. Similarly, $\partial v_3/\partial R_1$ appears as a voltage across nodes 3 and 1 of the sensitivity network. The sensitivity of the current i_2 of Figure 3.29*a* with respect to R_1 is the corresponding current in the sensitivity network. ▲

Sensitivities with respect to other types of elements are determined in a similar way. The KCL and KVL equations are differentiated with respect to the parameter. Then the branch equations are differentiated with respect to the parameter. The resulting equations are the same as the original equations except that, instead of the original currents and voltages, we now have their partial derivatives with respect to the parameter. Furthermore, there is a controlled source term added to the branch equation that contains the parameter explicitly.

Example 3-13. Suppose that for the network of Figure 3.29*a*, it is desired to compute $\partial v_2/\partial L_2$. If we write all the KCL and KVL equations of the network in Figure 3.29*a* and all the branch equations, and then differentiate all the equations with respect to L_2, the result is a set of equations that is like the original set with the following exceptions:

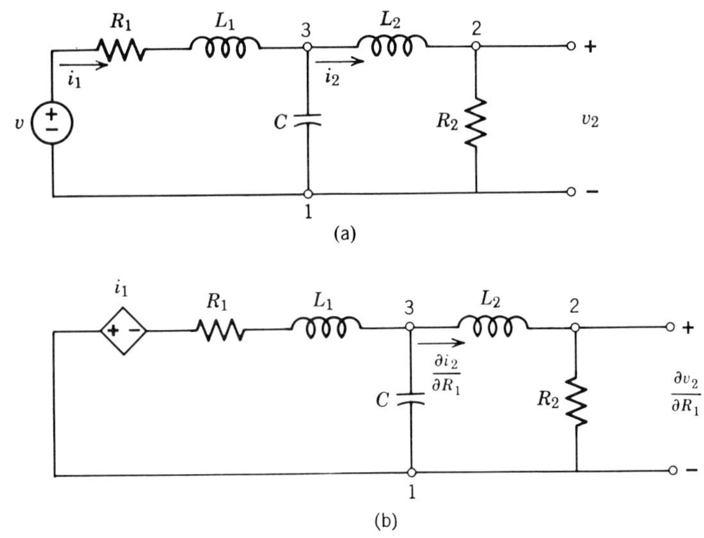

Figure 3.29 (*a*) Network where $\partial v_2/\partial R_1$ is desired. (*b*) Sensitivity network whose output is $\partial v_2/\partial R_1$.

(1) instead of the original set of currents and voltages, we have their partial derivatives; (2) the partial derivative $\partial v/\partial L$ is identically zero; (3) for the branch equation

$$v_{32} = L_2 \frac{di_2}{dt} \tag{3.166}$$

the partial derivative is

$$\frac{\partial v_{32}}{\partial L_2} = \frac{di_2}{dt} + L_2 \frac{d}{dt}\left(\frac{\partial i_2}{\partial L_2}\right). \tag{3.167}$$

Thus the various partial derivatives can be obtained as corresponding currents and voltages of a sensitivity network as shown in Figure 3.30. The voltage v_{21} in Figure 3.30 is equal to $\partial v_{21}/\partial L_2$ for the network of Figure 3.29*a*. The current through R_1 as indicated in Figure 3.30 is equal to $\partial i_1/\partial L_2$ for the network of Figure 3.29*a*. ▲

Example 3-14. For the network of Figure 3.29*a* let us determine the sensitivity network for determining the partial derivatives of currents and voltages with respect to C. There will be a controlled source associated with C in the sensitivity network. The original branch equation for C is

$$i_{31} = C \frac{dv_{31}}{dt} \tag{3.168}$$

and its partial derivative with respect to C is

$$\frac{\partial i_{31}}{\partial C} = \frac{dv_{31}}{dt} + C \frac{d}{dt}\left(\frac{\partial v_{31}}{\partial C}\right). \tag{3.169}$$

Thus we need a controlled current source equal to dv_{31}/dt in parallel with C for the sensitivity network, which is shown in Figure 3.31. The

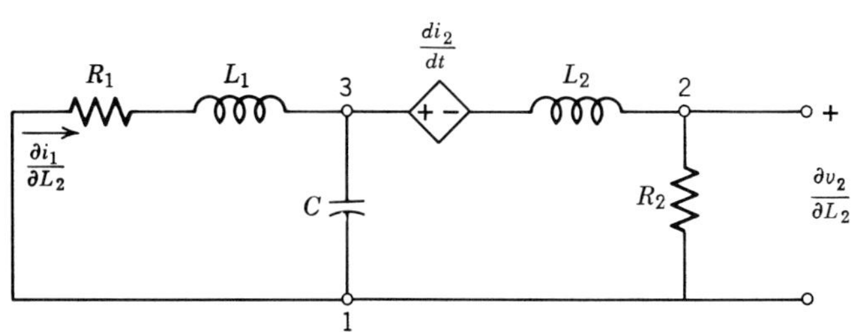

Figure 3.30 Sensitivity network for computing the sensitivity with respect to L_2 of the currents and voltages in Figure 3.29 (*a*).

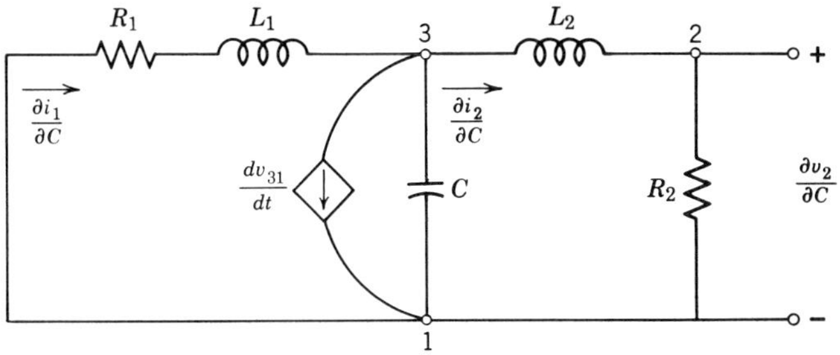

Figure 3.31 Sensitivity network for computing the sensitivity with respect to C of the currents and voltages in Figure 3.29(a).

voltage v_{21} in Figure 3.31 is equal to $\partial v_{21}/\partial C$ for the network of Figure 3.29a. ▲

Notice that in the above examples, the controlled sources depend on currents or voltages in the original network. These currents and voltages correspond to the nominal parameter values and a particular input $v(t)$. The sensitivity functions will be different for different choices of input $v(t)$.

The sensitivity network models may be used as a guide in the experimental determination of the partial derivatives. Furthermore, network analysis method applicable to the original network are also applicable to the sensitivity network. However, the currents and voltages of the sensitivity network are the sensitivity functions of the original network.

The sensitivity network models for mechanical networks are obtained in the same fashion as above. That is, the equilibrium equations based on D'Alembert's principle and the branch equations are differentiated with respect to the parameter. For the linear time-invariant network, the sensitivity network is the same as the original network except for the absence of independent sources and the addition of a controlled source. If desired, the sensitivity network may be the electrical analog of the mechanical network.

Example 3-15. Consider the mechanical network in Figure 3.7. Let us obtain a sensitivity network with respect to the parameter D_1. For the branch equation

$$D_1 \nu_1 = f_1 \tag{3.170}$$

the partial derivative with respect to D_1 is

$$\nu_1 + D_1 \frac{\partial v_1}{\partial D_1} = \frac{\partial f_1}{\partial D_1}\,. \tag{3.171}$$

Hence the sensitivity network will have a controlled force source equal to ν_1, as shown in Figure 3.32. The pictorial diagram for the mechanical sensitivity system is shown in Figure 3.33. Thus, if f is replaced by a force equal to $-\nu_1$, the resulting velocity of M_1 is $\partial\nu_1/\partial D_1$ and the resulting velocity of M_2 is $\partial\nu_2/\partial D_1$. ▲

We conclude this section by forming a sensitivity network for the spacecraft example of Figure 3.25.

Example 3-16. A network diagram for a single-axis reaction wheel attitude control system was given in Figure 3.25, which is reproduced here as Figure 3.34 for convenience, assuming constant field current i_f. Suppose we wish to obtain the sensitivity network with respect to the vehicle moment of inertia J_v. As discussed above, the sensitivity network diagram connections are the same as those in the original network diagram. Since the input V_a does not depend on J_v, $\partial V_a/\partial J_v = 0$, and the input terminals are *shorted* (zero voltage input) in the sensitivity network. For the branch equation

$$\tau' = J_v\omega, \tag{3.172}$$

the partial derivative with respect to J_v is

$$\frac{\partial\tau'}{\partial J_v} = J_v\frac{\partial\omega}{\partial J_v} + \omega. \tag{3.173}$$

Thus the sensitivity network, including the controlled torque source indicated by (3.173), is as shown in Figure 3.35. In Figure 3.35, it is assumed that the external disturbance torque τ_d does not depend on J_v, so $\partial\tau_d/\partial J_v = 0$. ▲

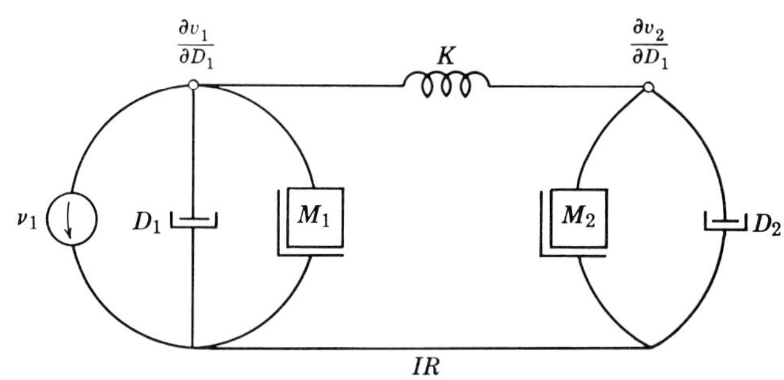

Figure 3.32 Sensitivity network with respect to D_1 for the mechanical network in Figure 3.7.

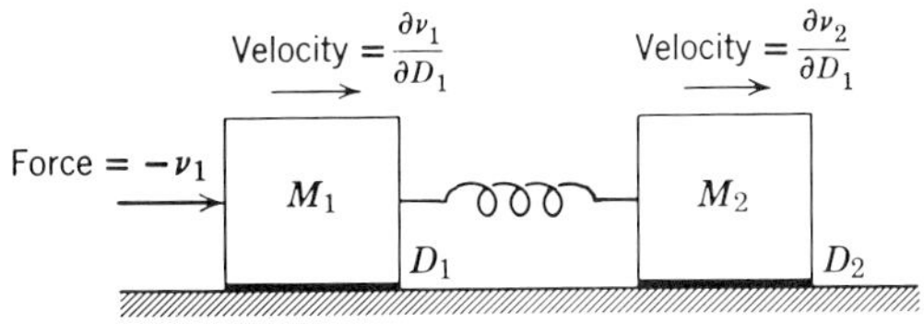

Figure 3.33 Mechanical sensitivity system for computing $\partial \nu_1/\partial D_1$ and $\partial \nu_2/\partial D_1$ of the mechanical system in Figure 3.6.

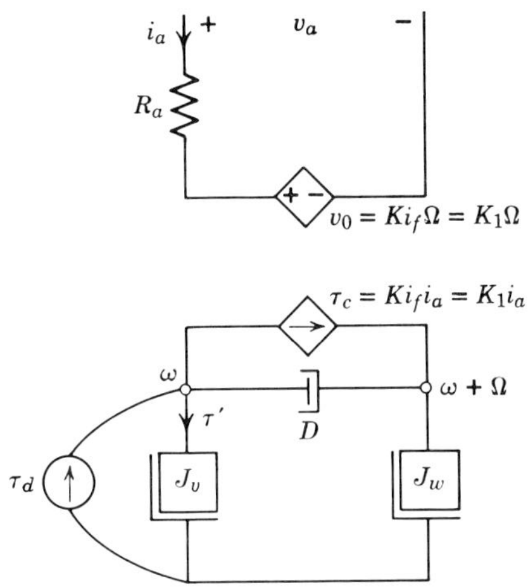

Figure 3.34 Network diagram for single-axis reaction-wheel attitude-control system.

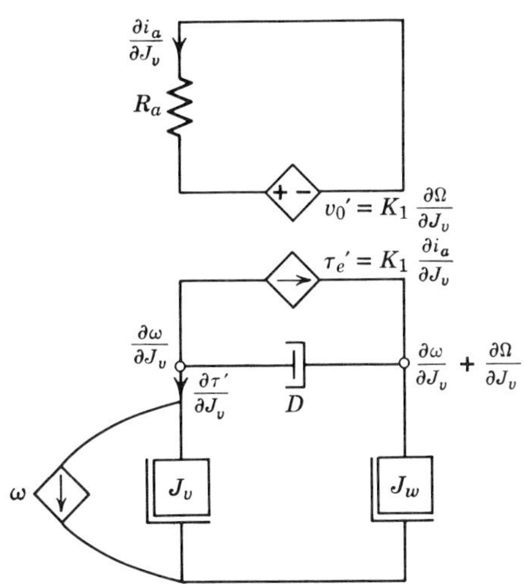

Figure 3.35 Sensitivity network with respect to J_v for single-axis reaction-wheel attitude-control system.

Problems

3-1. Write a set of simultaneous first-order equations for the network shown in Figure P3-1 where v is the input and i is the output.

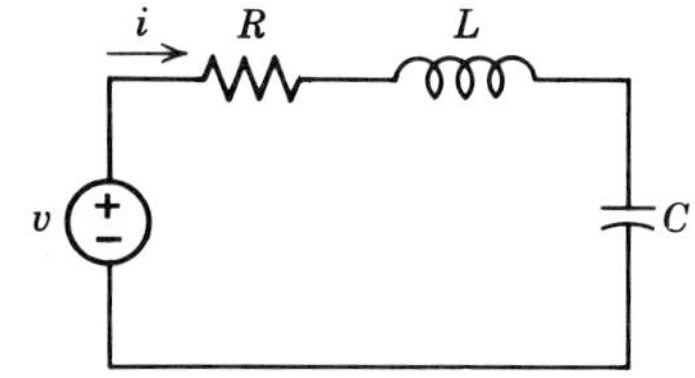

Figure P3-1

3-2. Repeat Problem 3-1 for the network in Figure P3-2. The input is i. The outputs are v_4 and i_6.

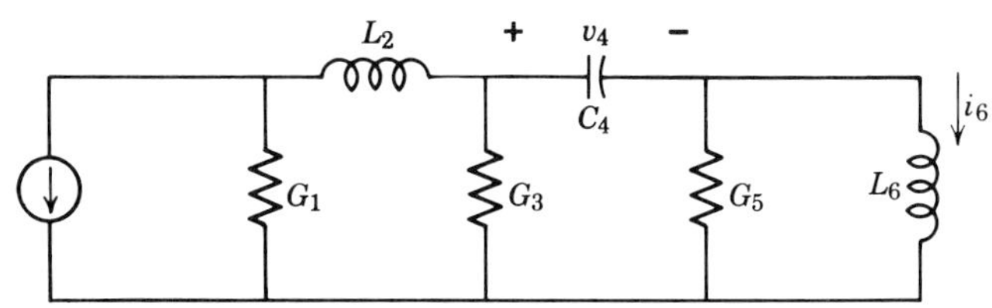

Figure P3-2

3-3. Draw a mechanical network diagram for the mechanical system pictured in Figure P3-3. f_{ext} is an external force source.

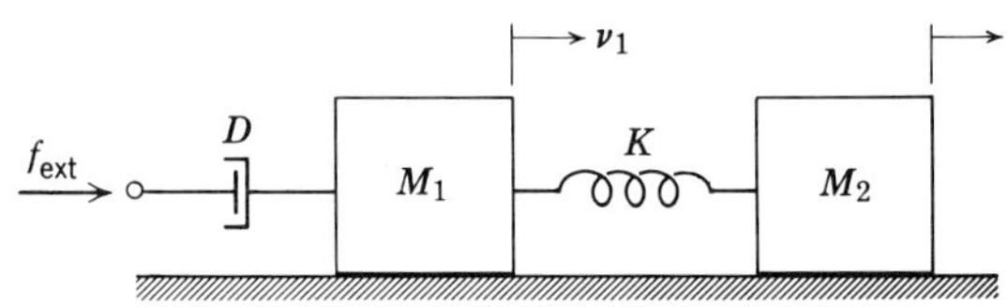

Figure P3-3

3-4. Repeat Problem 3-3 for the system in Figure P3-4.

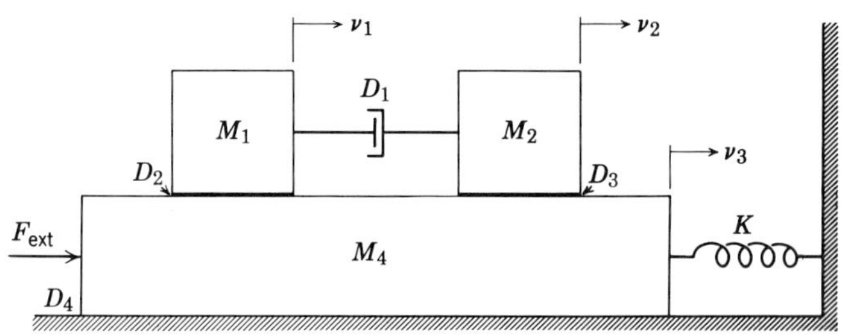

Figure P3-4

3-5. Repeat Problem 3-3 for the system in Figure P3-5. Include effects of gravity. ν_{ext} is an applied external velocity (relative to inertial reference).

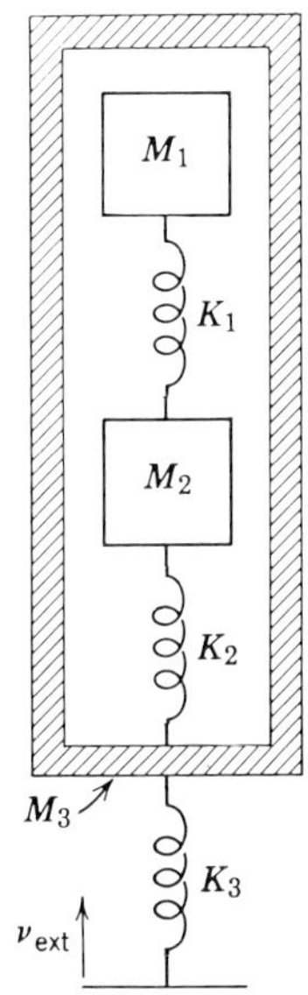

Figure P3-5

3-6. Repeat Problem 3-3 for the system in Figure P3-6. ω_{ext} is the input and Ω is the output.

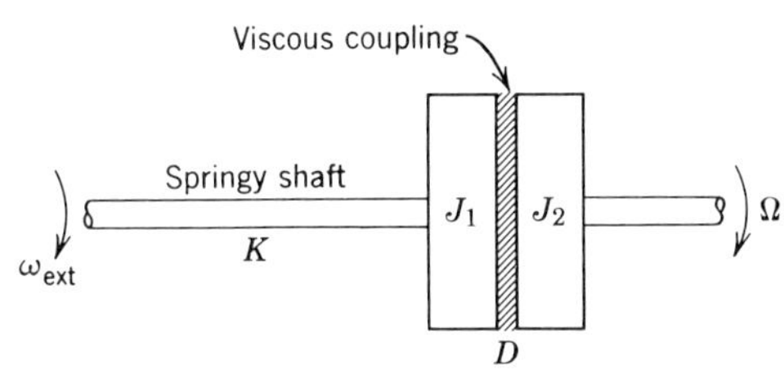

Figure P3-6

3-7. In Chapter 2, the electrical transformer is discussed as an example of a coupling element. Gears are the mechanical rotational counterpart of the transformer. Develop a model for ideal gears analogous to the electrical transformer. What is the mechanical translational analog?

3-8. Write simultaneous equations (not necessarily first-order) from which ν_2 of Problem 3-3 can be determined.

3-9. Repeat Problem 3-8 for the system of Problem 3-4 with f_{ext} as input, and v_1 and v_2 as outputs.

3-10. Repeat Problem 3-8 for the system of Problem 3-5 with v_{ext} as input and the velocity of M_1 *with respect to* M_3 as the output.

3-11. Obtain a single differential equation relating input ω_{ext} to output Ω for the system in Problem 3-6.

3-12. For the accelerometer of Example 3-5, suppose that

$$\frac{K}{M} - \left(\frac{D}{2M}\right)^2 < 0.$$

Find $y(t)$ for $t \geq 0$, for $y(0) = \dot{y}(0) = 0$, and $\ddot{x} =$ constant.

3-13. Repeat Problem 3-12 for

$$\frac{K}{M} - \left(\frac{D}{2M}\right)^2 = 0.$$

3-14. Consider the single-axis reaction-wheel attitude control system of Example 3-6. Suppose the spacecraft is initially motionless $[\dot{\theta}(0-) = 0]$ at $\theta(0-) = \theta_0$. The external torque $\tau_d = 0$. The wheel is rotated with $\Omega(t)$ as indicated in Figure P3-14. Find and sketch $\dot{\theta}(t)$ and $\theta(t)$ vs t.

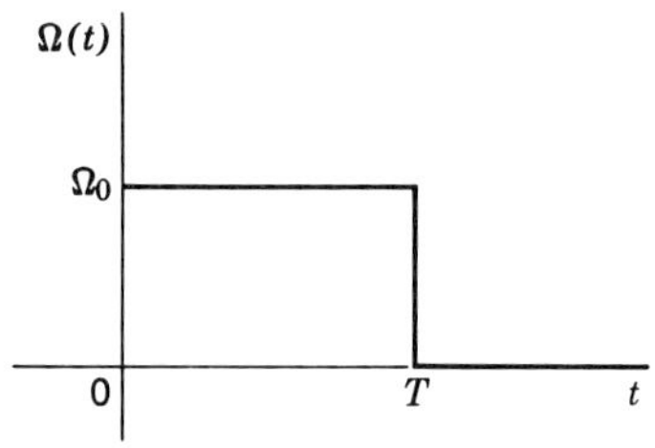

Figure P3-14

3-15. Figure P3-15 shows a movable-plate capacitor. The plate has mass M. Suppose the capacitor is electrically linear, that is,

$$q = C(x)v_1,$$

where q is the charge. Find the mechanical force of electric origin acting on the plate in the $+x$ direction.

3-16. Suppose that

$$C(x) = \frac{1}{x}$$

for Problem 3-15. Draw a network model for the system showing electrical elements, mechanical elements, and controlled source coupling elements.

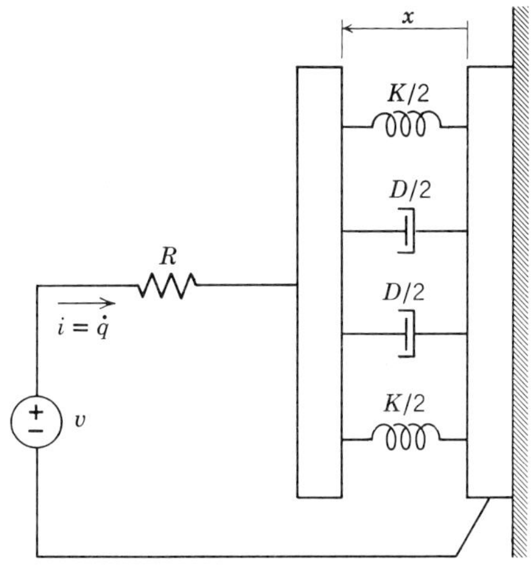

Figure P3-15

3-17. Using the method of Section 3-5, develop linearized equations and a linear network model for the system of Problems 3-15 and 3-16.

3-18. Determine a sensitivity network for the mechanical network of Figure 3.10 from which the sensitivity function of any signal with respect to K can be determined.

3-19. Determine a sensitivity network for the electromechanical network of Figure 3.28 from which the sensitivities with respect to R can be found.

3-20. Repeat Problem 3-18 for determining sensitivities with respect to D instead of K.

3-21. Repeat Problem 3-18 for determining sensitivities with respect to M instead of K.

3-22. Repeat Problem 3-19 for determining sensitivities with respect to M instead of R.

3-23. Repeat Problem 3-19 for determining sensitivities with respect to D instead of R.

3-24. For the tilted accelerometer of Figure 3.11, determine the differential equation for the sensitivity function $\partial y/\partial \phi$ at $\phi = 0$.

3-25. Derive the expression for stored energy in a linear time-invariant inductor in terms of (a) L and i, (b) L and λ.

3-26. Draw an electrical network analog for the mechanical network of Figure 3.10, using a force to current analogy.

3-27. Draw an electrical network analog for the mechanical network of Figure 3.13, using a torque to current analogy.

3-28. Repeat Problem 3-26 for the mechanical system in Problem 3-3.

3-29. Repeat Problem 3-26 for the mechanical system in Problem 3-4.

3-30. Repeat Problem 3-26 for the mechanical system in Problem 3-5.

3-31. Repeat Problem 3-27 for the mechanical system in Problem 3-6.

3-32. Draw a mechanical system whose signals are sensitivity functions for the mechanical system of Figure P3-3, where the parameter subject to perturbation is K.

3-33. Repeat Problem 3-32 for M_1 instead of K as the perturbed parameter.

3-34. Repeat Problem 3-22 for M_2 instead of K as the perturbed parameter.

3-35. Repeat Problem 3-32 for D instead of K as the perturbed parameter.

3-36. Repeat Problem 3-32 for the mechanical system of Figure P3-4, and M_1 as the perturbed parameter.

3-37. Repeat Problem 3-32 for the mechanical system of Figure P3-4 and M_2 as the perturbed parameter.

3-38. Repeat Problem 3-32 for the mechanical system of Figure P3-4 and M_3 as the perturbed parameter.

3-39. Repeat Problem 3-32 for the mechanical system of Figure P3-4 and K as the perturbed parameter.

3-40. Repeat Problem 3-32 for the mechanical system of Figure P3-4 and D_1 as the perturbed parameter.

3-41. Repeat Problem 3-32 for the mechanical system of Figure P3-4 and D_2 as the perturbed parameter.

3-42. Repeat Problem 3-32 for the mechanical system of Figure P3-4 and D_3 as the perturbed parameter.

3-43. Repeat Problem 3-32 for the mechanical system of Figure P3-4 and D_4 as the perturbed parameter.

3-44. Repeat Problem 3-32 for the mechanical system of Figure P3-5 and M_1 as the perturbed parameter.

3-45. Repeat Problem 3-32 for the mechanical system of Figure P3-5 and M_2 as the perturbed parameter.

3-46. Repeat Problem 3-32 for the mechanical system of Figure P3-5 and M_3 as the perturbed parameter.

3-47. Repeat Problem 3-32 for the mechanical system of Figure P3-5 and K_1 as the perturbed parameter.

3-48. Repeat Problem 3-32 for the mechanical system of Figure P3-5 and K_2 as the perturbed parameter.

3-49. Repeat Problem 3-32 for the mechanical system of Figure P3-5 and K_3 as the perturbed parameter.

3-50. Repeat Problem 3-32 for the mechanical system in Figure P3-6 and K as the perturbed parameter.

3-51. Repeat Problem 3-32 for the mechanical system in Figure P3-6 and J_1 as the perturbed parameter.

3-52. Repeat Problem 3-32 for the mechanical system in Figure P3-6 and D as the perturbed parameter.

3-53. Repeat Problem 3-32 for the mechanical system in Figure P3-6 and J_2 as the perturbed parameter.

3-54. Derive an expression for stored potential energy in a linear time-invariant translational spring.

3-55. Repeat Problem 3-54 for a rotational spring.

3-56. Derive an expression for kinetic energy in a linear time-invariant mass.

3-57. Repeat Problem 3-56 for a linear time-invariant rotational inertia.

3-58. A nonlinear control system is described by

$$\dot{y} = \tanh\,[x - y]$$

where y is the output and x is the input.

(a) Determine the equilibrium point (or points) for $x = 1$.
(b) Linearize the above equation about the equilibrium point in (a) and obtain a linear differential equation relating the deviation Δy from the equilibrium point to the deviation Δx from $x = 1$.

3-59. Repeat Problem 3-58 for

$$\dot{y} = 10\tanh\,[x - y].$$

3-60. Instead of obtaining an equilibrium point as described in Section 3.5, it is often simpler to set to zero all the time derivatives of signals that appear in the network element equations and then to solve the resulting set of nondynamic equations. In the case of electrical networks, this is equivalent to replacing all inductors by short circuits and replacing all capacitors by open circuits. The remaining network is resistive, with constant sources. Instead of the linearization in Section 3.5, each network element may be linearized about the equilibrium point. Then the linearized branch equations and linear equilibrium equations may be combined to obtain the same result as in Section 3.5.

Using the above method, develop the linearization about the equilibrium point of a nonlinear series RL circuit similar to that in Figure 3.26 where the nonlinear resistor is given by

$$v_R = f_R(i)$$

and the nonlinear inductor is given by

$$\lambda = f_L(i).$$

Denote the constant input voltage at the equilibrium point by V.

3-61. Write a set of first-order differential equations for the network in Figure P3-61 in the form

$$\frac{di_1}{dt} = a_{11}i_1 + a_{12}i_2 + b_1 v$$

$$\frac{di_2}{dt} = a_{21}i_1 + a_{22}i_2 + b_2 v$$

The network parameters R_1, R_2, L_1, L_2, and M are all constants.

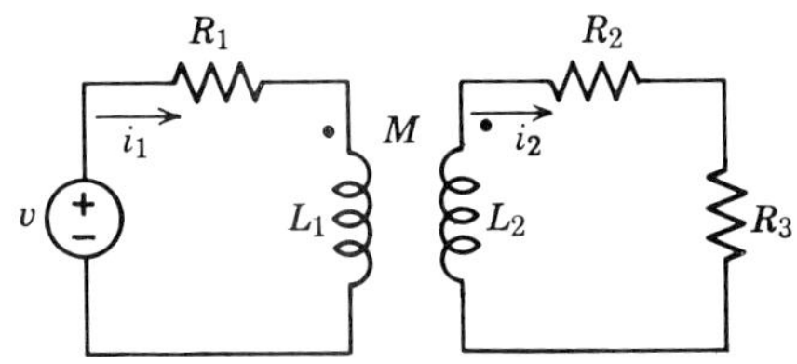

Figure P3-61

3.62. Apply KVL to the two loops of the network in Figure P3-61 and differentiate the equations with respect to the mutual inductance parameter M. Draw a sensitivity network where $\partial i_2/\partial M$ appears as a signal.

3-63. The linearized network model of the solenoid system shown in Figure 3.28 involves parameters that are equilibrium points of the nonlinear model. Determine a network sensitivity model where $\partial(\Delta\nu)/\partial V$ appears as a signal. Note that for the network in Figure 3.28, X and K_{eq} depend on V.

3-64. Draw a sensitivity network diagram for the sensitivity analysis with respect to J_w of the reaction-wheel attitude-control system in Figure 3.34.

3-65. Repeat Problem 3-64 for a sensitivity analysis with respect to D.

3-66. Repeat Problem 3-64 for a sensitivity analysis with respect to R_a.

3-67. Repeat Problem 3-64 for a sensitivity analysis with respect to K_1.

CHAPTER 4

System Simulation

In Chapters 2 and 3, network models were used for representation of systems. In this chapter another modeling technique is discussed: the simulation diagram. This diagram can be considered to be a more detailed development of the functional block diagram discussed previously. It is a representation of a system in terms of interconnected subsystems, as is the functional block diagram. In addition, however, the mathematical operations relating the inputs and outputs of the subsystems are indicated on the simulation diagram.

Considerable insight and "feel" for the system operation can frequently be attained by inspection of simulation diagrams. Moreover, these diagrams lead *directly* to computer simulation of systems if the subsystems are represented by certain elementary operations.

Computers play a crucial role in systems engineering. For example, a mathematical model (equations) representing a system may be programmed on a computer, with many of the variables that appear physically in the actual system also appearing as computer outputs. This programming can be done directly from simulation diagrams. Such a system representation is called a computer simulation. Analog, digital, and hybrid computers may be used for simulation.

The analog computer is an interconnection of electronic circuits that solves differential equations. Since continuous-time dynamic systems are described by differential equations, the analog computer is especially useful in the analysis and design of such systems. The computer is programmed by interconnecting the circuits so that the differential equations describing the electronic circuits are the same as those describing the system to be studied, that is, the electronic circuit and the system are analogs. Inputs and initial conditions are applied, and the response of the circuits is noted, often by pen recording on a strip chart. Parameter values are readily varied by adjusting potentiometers (variable resistors) or other components until the desired response is obtained.

The digital computer is also an interconnection of electronic circuits but, whereas analog computers involve signals that are functions of a continuous time variable, digital computers deal with sequences of numbers. Thus, dynamic systems described by *difference equations* may be simulated directly on the digital computer. Systems described by differential equations cannot be simulated exactly. The differential equations must be approximated by difference equations for digital simulation. A detailed discussion of such approximations is a part of the subject of *numerical analysis*, and will not be undertaken here.

Hybrid computers contain both analog and digital subsections. These computers are useful in simulating hybrid (or "sampled-data") systems. For example, if a continuous-time system, such as a chemical process, is controlled by a special-purpose digital computer, the simulation of the combined system can be done naturally on hybrid computers.

The details of computer programming vary from machine to machine and thus cannot be discussed here. In this chapter, we concentrate on the fundamental ideas of simulation of continuous-time systems. Simulation of other types of systems is considered in later chapters. Basic building blocks for such simulations are discussed, and methods of obtaining simulation diagrams are developed. The actual programming of analog, digital, or hybrid computers can be done from these diagrams.

4.1 Building Blocks for Analog Simulation

To simulate a system, we must be able to represent the basic mathematical operations required by the model of the system. As discussed in detail in Chapters 2 and 3, some dynamic systems are modeled by differential or integrodifferential equations. Thus, our simulation scheme may contain differentiators and integrators as basic building blocks. Moreover, multiplication by a constant, addition, and nonlinear operations such as multiplication and division of functions are often required.

We shall represent these operations pictorially by simulation diagrams. The operation is represented by a notation in the block, and the input and the output signals are indicated by arrows. For example, multiplication of a function by a constant is represented in Figure 4.1.

Multiplication of a function by a time-varying coefficient is represented in Figure 4.2.

Figure 4.1 Multiplication by a constant.

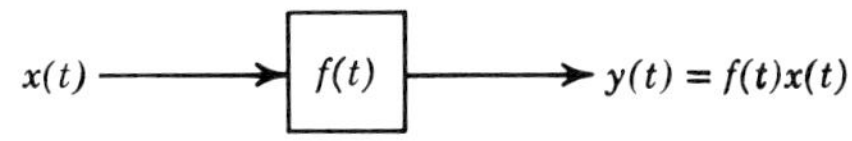

Figure 4.2 Multiplication by a time-varying coefficient.

Addition of signals is represented by a summer, as is shown in Figure 4.3. Any number of inputs may be indicated for a summer.

A block diagram representation of differentiation is shown in Figure 4.4. The input signal w and output signal z are identified, and the block is labeled with the differentiation operator, as indicated.

The same relation between $z(t)$ and $w(t)$ can be expressed in integral form:

$$w(t) = \int_{t_0}^{t} z(\tau)\, d\tau + w(t_0). \tag{4.1}$$

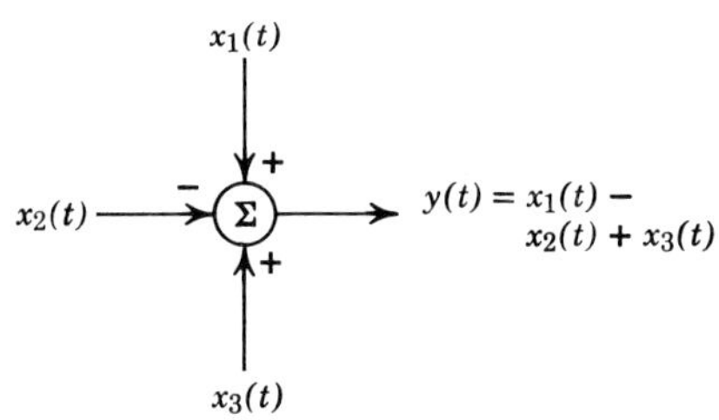

Figure 4.3 Addition of signals by a summer.

Now z is regarded as the input, and w as the output. A corresponding block diagram is Figure 4.5 for the case of zero initial conditions, $z(t_0) = 0$.

The integration with initial condition of (4.1) can be represented as in Figure 4.6.

Operations performed sequentially can be represented by blocks in tandem. Care must be exercised, however, to be certain the blocks appear in the correct order, since the operations represented by the blocks are not commutative, in general. An example is given in Figure 4.7.

Nonlinear operations can be represented easily by block diagrams also. Several examples are given in Figure 4.8.

Any number of inputs to the multiplier are allowed.

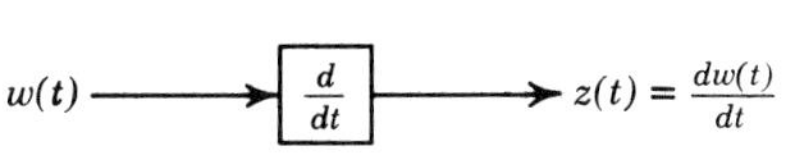

Figure 4.4 Differentiation.

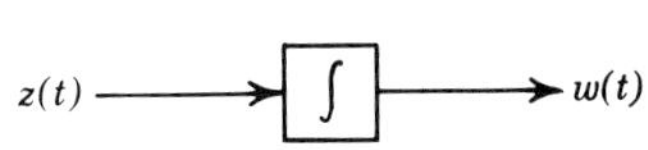

Figure 4.5 Integration with zero initial condition.

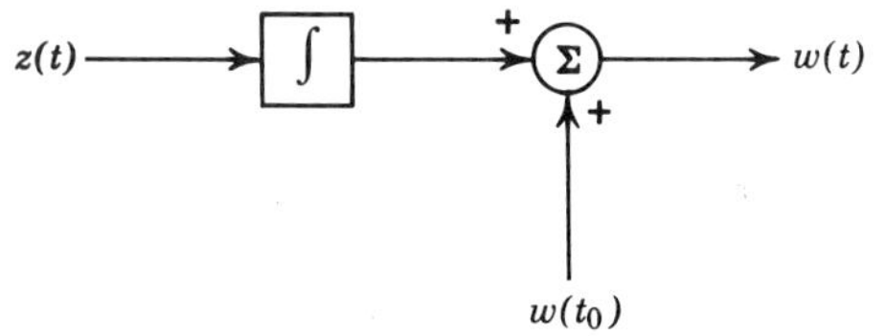

Figure 4.6 Integration with initial condition.

These block diagrams can be interconnected to represent complicated relations among signals, as is illustrated by the following examples.

Example 4-1. This example illustrates a *feedback* relation in which an output $c(t)$ is "fedback" and compared with an input $r(t)$. The difference $e(t)$ is integrated. A block diagram is given in Figure 4.9. We wish to find an equation relating c and r.

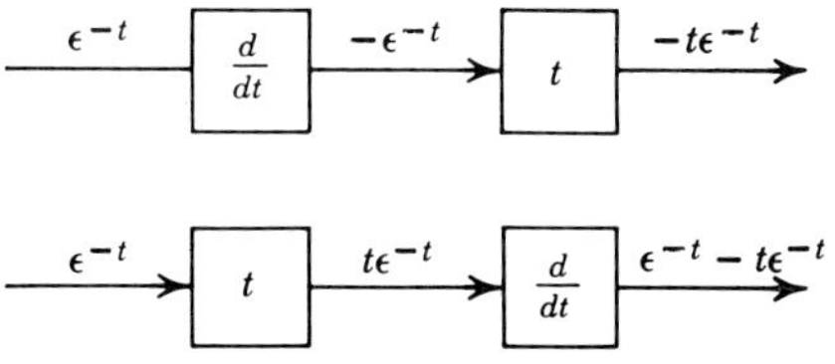

Figure 4.7 Tandem blocks representing noncommutative operations.

We first notice that

$$e(t) = r(t) - Kc(t) \tag{4.2}$$

and

$$c(t) = \int_{t_0}^{t} e(\tau)\, d\tau. \tag{4.3}$$

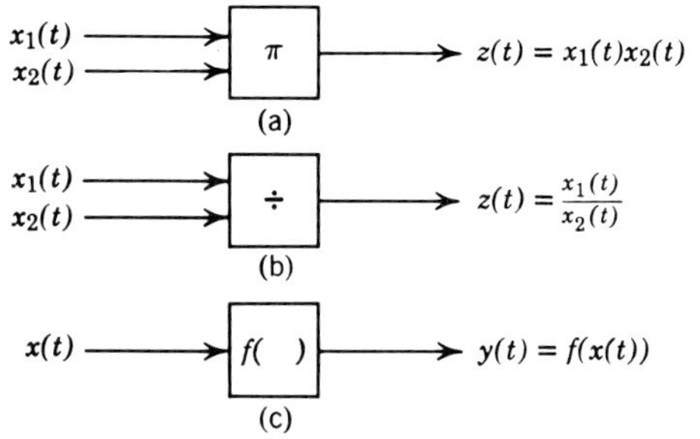

Figure 4.8 Block diagrams for some nonlinear operations: (*a*) Multiplication. (*b*) Division. (*c*) Nonlinear function.

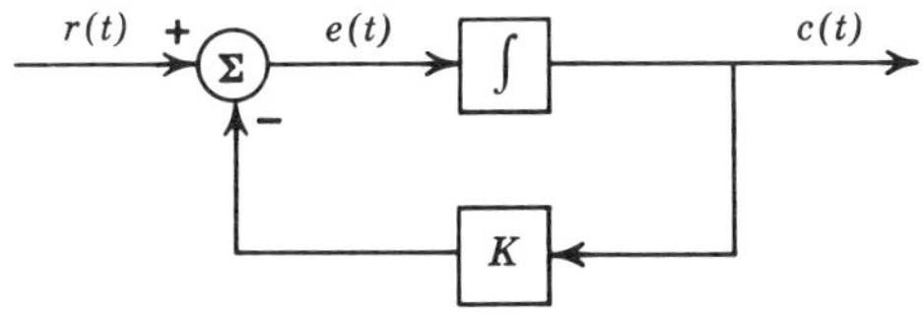

Figure 4.9 A feedback system.

Thus the desired relationship is

$$c(t) = \int_{t_0}^{t} [r(\tau) - Kc(\tau)]\, d\tau \tag{4.4}$$

or in differential form,

$$\frac{dc}{dt} + Kc = r. \tag{4.5}$$

Since no initial conditions are indicated in Figure 4.9, $c(0) = 0$ is the initial condition for (4.5). ▲

Example 4-2. Find an equation relating c and r for the system shown in Figure 4.10.

Writing equations, we have

$$r = e + m_1 c, \tag{4.6}$$

$$e = \frac{dm_1}{dt} = \dot{m}_1, \tag{4.7}$$

$$m_1 = \frac{dc}{dt} = \dot{c}. \tag{4.8}$$

Combining (4.6), (4.7), and (4.8), we obtain, in differential form,

$$\ddot{c} + \dot{c}c = r, \tag{4.9}$$

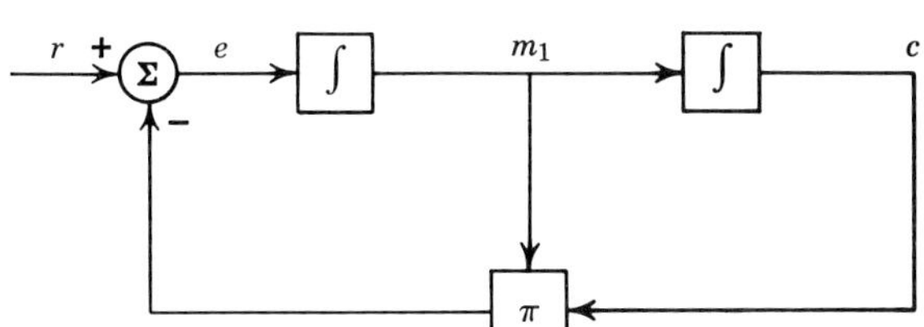

Figure 4.10 Figure for Example 4-2.

where

$$\ddot{c} = \frac{d^2c}{dt^2}. \tag{4.10}$$

Since no initial conditions are shown in Figure 4.10, $c(0) = \dot{c}(0) = 0$. ▲

4.2 All-Integrator System Simulation

The basic building blocks discussed in Section 4.1 can be interconnected to form block-diagram representations for complete systems. The block diagrams often convey a better "feel" for the system behavior than the corresponding set of equations.

For several reasons, it is preferable to use the integrator, rather than the differentiator, as the basic dynamical element. For example, high-frequency noise problems make it very difficult to construct a practical circuit that behaves as a differentiator over a wide range of operating frequencies. Moreover, many of the mathematical results in differential equation theory, useful in systems engineering, are expressed in terms of simultaneous first-order differential equations. Such systems of equations are readily simulated using integrators, as will be seen. To simulate a system, therefore, we wish to use integrators as the basic dynamic building block. There are several ways to do this.

Suppose the system is modeled by a set of simultaneous first-order equations. Then all-integrator simulation can be done by inspection. (This is one reason for the emphasis on this form of modeling in Chapters 2 and 3.) The following example will indicate the ideas involved.

Example 4-3. Suppose a system is modeled by the equations

$$\dot{x}_1 = -6x_1 + x_2 + r \tag{4.11}$$

$$\dot{x}_2 = -8x_1 + x_3 + 4r \tag{4.12}$$

$$\dot{x}_3 = 3r \tag{4.13}$$

$$c = x_1, \tag{4.14}$$

where r is the system input, c the system output, and x_1, x_2, and x_3 are intermediate variables. Such intermediate variables are called *state variables*, and will be quite important in our later work. The equations (4.11), (4.12), and (4.13) are called the *state equations* for the system. Initial conditions will be ignored for the moment.

The basic step in the simulation procedure is to observe that if a state variable, say x_1, is identified as an integrator *output*, then the signal at the integrator *input* must be the time derivative of the state variable, $\dot{x}_1$. Thus the state equations tell us what signals (integrator outputs and system inputs) must be summed to form the integrator input signal. Thus, (4.11) tells us how the input $\dot{x}_1$ to Integrator 1 is made up of a weighted sum of the signals x_1, x_2, and r. The equation is represented in block diagram form in Figure 4.11.

Since x_2 is not a system input, it must be synthesized using (4.12). Figure 4.12 represents (4.12) and (4.11). Finally, x_3 can be synthesized by adding another integrator and using (4.13) to form the integrator input (see Figure 4.13). In Figure 4.13, the *output equation* (4.14) is also represented.

Let us now consider the initial conditions, which have been ignored thus far. The initial conditions must be determined from the physical situation as a part of the same modeling procedure that produced the system equations (4.11) to (4.14). If the variables $x_1(0)$, $x_2(0)$, $x_3(0)$ are known, then it is a simple matter to include these terms at the integrator outputs, as shown in Figure 4.6. The situation is somewhat more complicated if the integrator outputs are not the variables specified at $t_0 = 0$ (suppose $c(0)$, $\dot{c}(0)$, $\ddot{c}(0)$, $r(0)$, $\dot{r}(0)$ are given in this example), but instead are linear combinations of these quantities. The state equations are useful in evaluating the initial integrator outputs $x_1(0)$, $x_2(0)$, and $x_3(0)$ in terms of the known initial conditions. For example, by (4.14),

$$x_1(0) = c(0). \tag{4.15}$$

Moreover, from (4.11) and (4.12),

$$\dot{x}_1(0) = \dot{c}(0) = -6x_1(0) + x_2(0) + r(0) \tag{4.16}$$

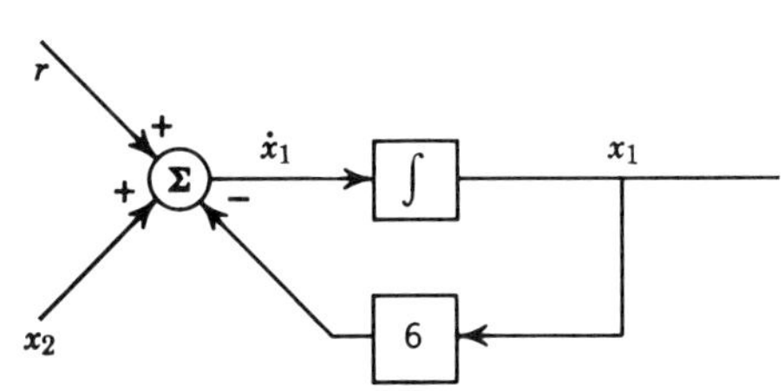

Figure 4.11 Block-diagram representation of (4.11).

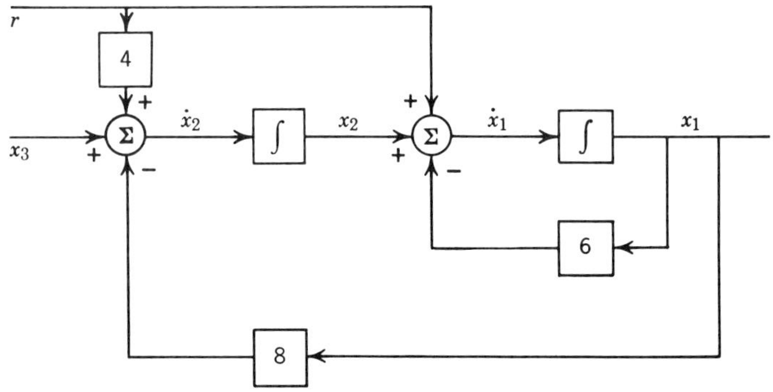

Figure 4.12 Block-diagram representation of (4.11) and (4.12).

or, using (4.15) and (4.16),

$$x_2(0) = \dot{c}(0) + 6c(0) - r(0). \tag{4.17}$$

Finally, from (4.17) and (4.12),

$$\begin{aligned}\dot{x}_2(0) &= \ddot{c}(0) + 6\dot{c}(0) - \dot{r}(0)\\ &= -8c(0) + x_3(0) + 4r(0),\end{aligned}$$

or

$$x_3(0) = \ddot{c}(0) + 6\dot{c}(0) + 8c(0) - 4r(0) - \dot{r}(0). \tag{4.18}$$

Thus it is a straightforward matter to find the initial conditions for the all-integrator flow graph in terms of the given initial conditions. ▲

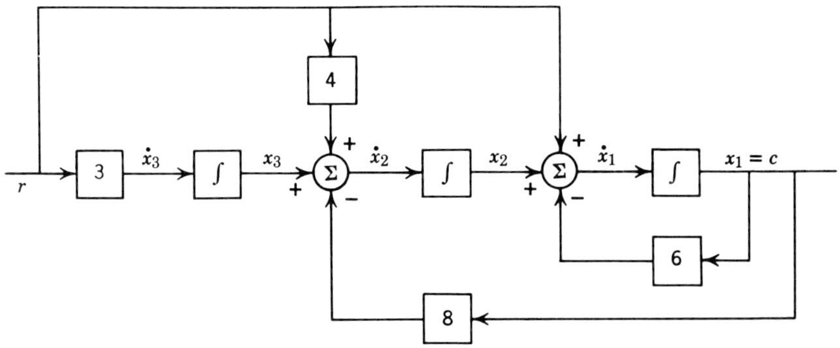

Figure 4.13 All-integrator representation of the system of Example 4-3.

Notice that the initial conditions for the state variables in Example 4-3 involve not only initial conditions for the system output and its derivatives but also the initial values of the system input and its derivative. Thus the initial conditions for the simulation diagram will change depending upon the system input!

Some ambiguity occurs if any signals involved in calculation of initial conditions are discontinuous at $t=0$. In this text, we shall use the signal values at $t=0-$, an instant just before $t=0$, for calculating initial conditions. [This results in some desirable properties of the Laplace Transform, as discussed in Appendix B. Some authors use $0+$ initial conditions. Of course, correct results are obtained by either approach if it is followed consistently. Analog computer simulation requires $0+$ values if we are to avoid "infinite voltages" (impulses) appearing at amplifier inputs.]

Thus we see that modeling a system described by simultaneous first-order equations is quite straightforward. One integrator is needed for each state variable, and the state equations provide the interconnections to the integrator inputs.

The same general procedure can be followed even if the system is modeled by simultaneous time-varying or nonlinear equations, as is indicated by the following example.

Example 4-4. In Chapter 5, we shall show that the rigid body motion of a spacecraft can be modeled by the equations

$$\dot{\omega}_1=\frac{J_2-J_3}{J_1}\omega_2\omega_3+r_1 \tag{4.19}$$

$$\dot{\omega}_2=\frac{J_3-J_1}{J_2}\omega_3\omega_1+r_2 \tag{4.20}$$

$$\dot{\omega}_3=\frac{J_1-J_2}{J_3}\omega_1\omega_2+r_3, \tag{4.21}$$

where J_1, J_2, and J_3 are (constant) moments of inertia, r_1, r_2, and r_3 are input torques, and ω_1, ω_2, and ω_3 are spacecraft rotation rates. The all-integrator simulation can be drawn directly by inspection of the state equations, (4.17) to (4.21). See Figure 4.14. Notice that the equations are nonlinear, because of the product terms, and multipliers are required in the simulation. ▲

If the system is not modeled initially as a set of simultaneous first-order differential equations, the situation is more complicated. Several schemes are available for developing an all-integrator simulation.

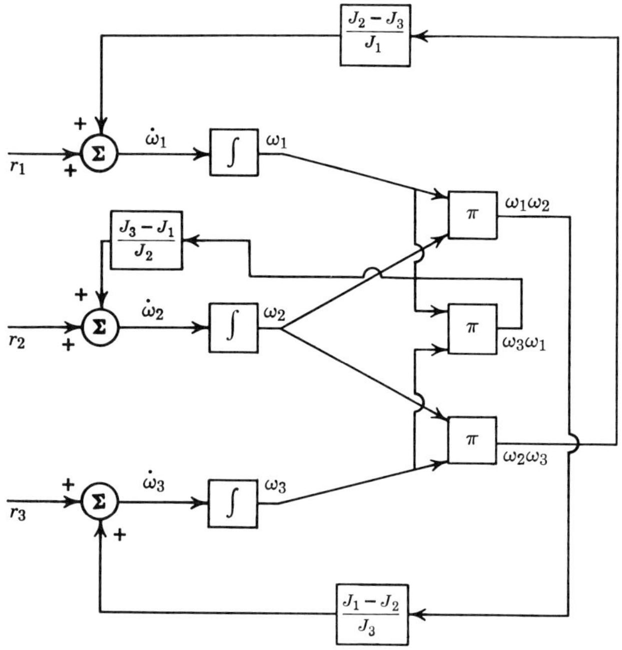

Figure 4.14 Simulation of spacecraft equations of motion.

These are equivalent to techniques for converting general differential equations to state form. Two procedures are indicated by the following two examples.

Example 4-5. Suppose that we wish to simulate a system described by an nth order linear differential equation with constant coefficients. To be specific, suppose that a system is described by

$$\dddot{c} + 6\ddot{c} + 8\dot{c} = \ddot{r} + 4\dot{r} + 3r, \tag{4.22}$$

where $c(t)$ is the output, $r(t)$ is the input, and a dot indicates a differentiation with respect to time. Some initial conditions must also be known. Suppose that we know $c(0)$, $\dot{c}(0)$, and $\ddot{c}(0)$. Also $r(t)$ is given, so $r(0)$ and $\dot{r}(0)$ can be computed.

If the operation d/dt is denoted by the symbol D, d^2/dt^2 by D^2, and d^3/dt^3 by D^3, (4.22) becomes

$$D^3c + 6D^2c + 8Dc = D^2r + 4Dr + 3r. \tag{4.23}$$

We wish to simulate (4.23) using integrators as the dynamic elements.

Consider first the special case of identically zero input:

$$D^3c + 6D^2c + 8Dc = 0. \tag{4.24}$$

If we identify c as an integrator output, then $Dc = \dot{c}$ must be the input to that integrator. But $\dot{c}$ may be identified as the output of a second integrator, having $D^2c = \ddot{c}$ as its input. Thus, if we define

$$\begin{aligned} x_1 &= c \\ x_2 &= \dot{c} \\ x_3 &= \ddot{c}, \end{aligned} \tag{4.25}$$

then we may write

$$\begin{aligned} \dot{x}_1 &= x_2 \\ \dot{x}_2 &= x_3 \\ \dot{x}_3 &= -8x_2 - 6x_3, \end{aligned} \tag{4.26}$$

where the first two equations come from (4.25) and the third equation utilizes (4.24). The system is now in first-order form and may be simulated using the method discussed previously. The simulation is shown in Figure 4.15.

This procedure will not work if for this example r is not zero, since derivatives of r will be required. Thus the simulation is not all-integrator. The procedure can be modified as follows. Group derivatives of the same order in (4.24):

$$D^3c + D^2(6c - r) + D(8c - 4r) = 3r. \tag{4.27}$$

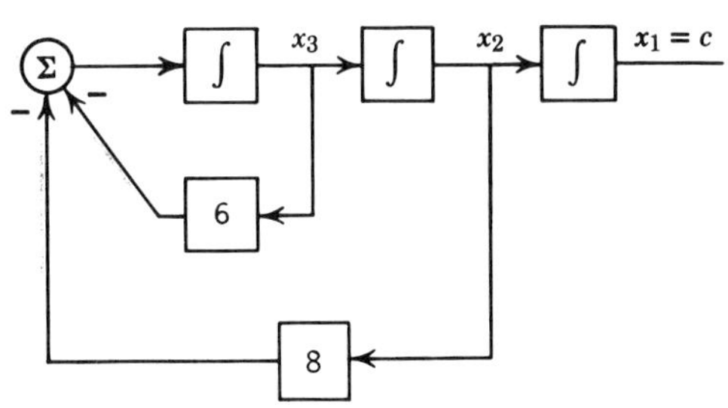

Figure 4.15 Simulation of (4.26).

This may be expressed as

$$D\{D[Dc + (6c - r)] + (8c - 4r)\} = 3r. \tag{4.28}$$

The all-integrator diagram can now be drawn by inspection. We proceed step by step for clarity. We note that the quantity within the braces is the output of an integrator having $3r$ as input. Denoting the brace by x_3,

$$x_3 = \{D[Dc + (6c - r)] + (8c - 4r)\}, \tag{4.29}$$

we have

$$\dot{x}_3 = 3r. \tag{4.30}$$

The corresponding simulation is shown in Figure 4.16*a*.

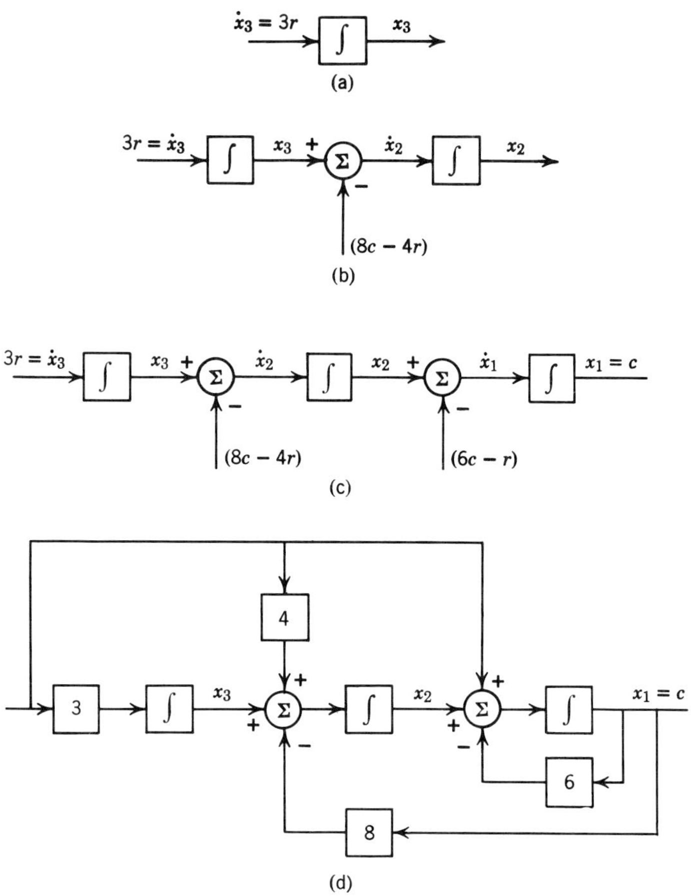

Figure 4.16 (*a*) First step in simulation. (*b*) Second step in simulation. (*c*) Third step in simulation. (*d*) Final step in simulation.

But, identifying the brackets term as x_2,

$$Dc + (6c - r) = x_2, \tag{4.31}$$

(4.29) can be written as

$$\dot{x}_2 = x_3 - (8c - 4r). \tag{4.32}$$

See Figure 4.16*b*.

Finally, identifying

$$c = x_1, \tag{4.33}$$

(4.31) becomes

$$\dot{x}_1 = x_2 - (6c - r). \tag{4.34}$$

See Figure 4.16*c*. To complete the diagram, it is necessary to show only one input, r. Thus $3r$, $(8c-4r)$, and $(6c-r)$, which appear in Figure 4.16*d*, must be synthesized from r and from the output c. This modification is shown in Figure 4.16*d*. ▲

Notice that Figure 4.16*d* is the same as Figure 4.13. Thus, Examples 4-3 and 4-5 deal with the same system.

Example 4-6. Let us illustrate another method using the same differential equation (4.22) as in Example 4-5.

We consider first a similar system having $r(t)$ as the right-hand side; rather than $\ddot{r}+4\dot{r}+3r$,

$$\dddot{y} + 6\ddot{y} + 8\dot{y} = r. \tag{4.35}$$

As we shall see, the output y of this system is closely related to the output c of the original system. The system of (4.35) is readily simulated by noting that

$$\dddot{y} = -6\ddot{y} - 8\dot{y} + r, \tag{4.36}$$

a form readily obtained from a simulation diagram on which y, $\dot{y}$, and $\ddot{y}$ appear as integrator outputs. See Figure 4.17.

To form c, we utilize linearity (superposition). The response of the system (4.22) may be considered to result from the superposition of the responses to three inputs, $\ddot{r}$, $4\dot{r}$, and $3r$. Since y of (4.35) is the response to r, $3y$ is the response to $3r$. Moreover, by differentiating

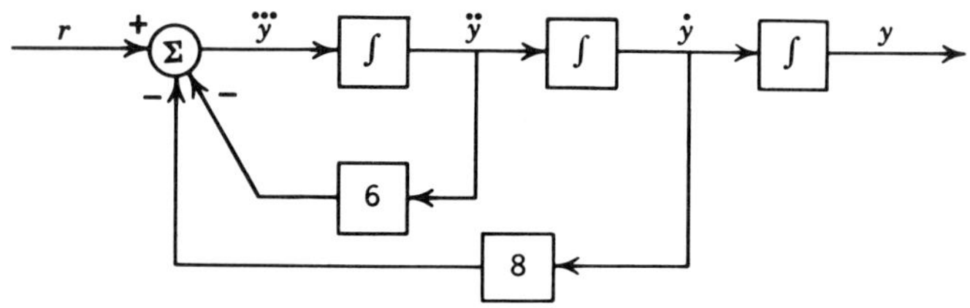

Figure 4.17 Simulation of (4.36).

(4.35), the response to $\dot{r}$ is just $\dot{y}$, and the response to $4\dot{r}$ is $4\dot{y}$. Finally, by differentiating (4.35) twice, the response to $\ddot{r}$ is $\ddot{y}$. Thus the response c due to $\ddot{r}+4\dot{r}+3r$ is by superposition, the sum of the three responses,

$$c = 3y + 4\dot{y} + \ddot{y}. \tag{4.37}$$

This is indicated in Figure 4.18, which is the final simulation diagram for this example. This diagram is different from that in Figure 4.16*d*, although the differential equation relating r to y is the same as in Example 4-5. ▲

Example 4-7. Equations describing a single-axis reaction-wheel attitude-control system were developed from a network model in Chapter 3. These equations, (3.100), (3.101), and (3.103), are repeated here, with Ki_f replaced by K_1 and with $\tau_d \equiv 0$.

$$\dot{\theta} = \omega \tag{4.38}$$

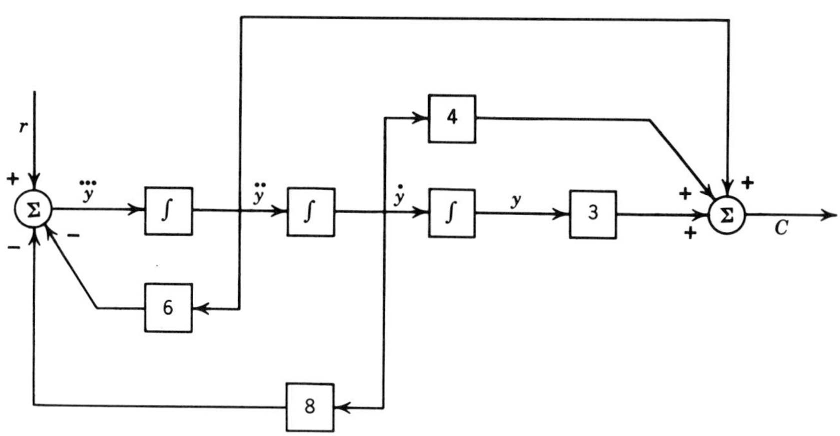

Figure 4.18 Simulation diagram for Example 4-6.

$$\dot{\omega} = \frac{1}{J_v}\left(D + \frac{K_1^2}{R_a}\right)\Omega - \frac{K_1}{J_v R_a} v_a \tag{4.39}$$

$$\dot{\Omega} = -\left(\frac{1}{J_v} + \frac{1}{J_w}\right)\left(\frac{K_1^2}{R_a} + D\right)\Omega + \frac{K_1}{R_a}\left(\frac{1}{J_v} + \frac{1}{J_w}\right) v_a. \tag{4.40}$$

Since these equations are in state form (simultaneous first-order differential equations), a simulation diagram can be drawn readily, using the procedure of Example 4-3. See Figure 4.19. ▲

As noted earlier, simulation diagrams are not unique. In particular, alternative diagrams may be constructed by distributing path gains in different ways. Figure 4.20 shows how shifting of gains can result in scaling of some variables while leaving others unchanged. Specifically, dividing all gains by α in the branches entering a node divides the signal leaving the node by α. Thus, multiplying the branch gain by α for the branch leaving the node leaves all other signals unchanged.

This procedure can often be used to reduce the number of different parameters appearing in a simulation diagram. Such a reduction usually aids in the determination of the role of the various parameters from inspection of the diagram. Furthermore, fewer parameters would need adjustment in computer simulation studies. Finally, amplitude scaling is frequently necessary in analog computer simulation if amplifier saturation is to be avoided (see Example 4-8).

Example 4-8. Inspection of Figure 4.19 reveals that identical parameters appear in several places on the diagram. The vehicle moment of inertia J_v, for example, appears in four places. The term $(D + K_1^2/R_a)$ appears twice. A three-step modification of Figure 4.19 is shown in

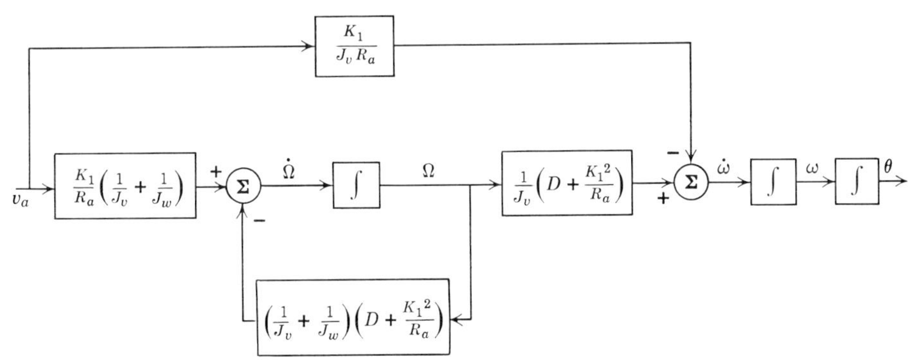

Figure 4.19 A simulation for a reaction-wheel attitude-control system.

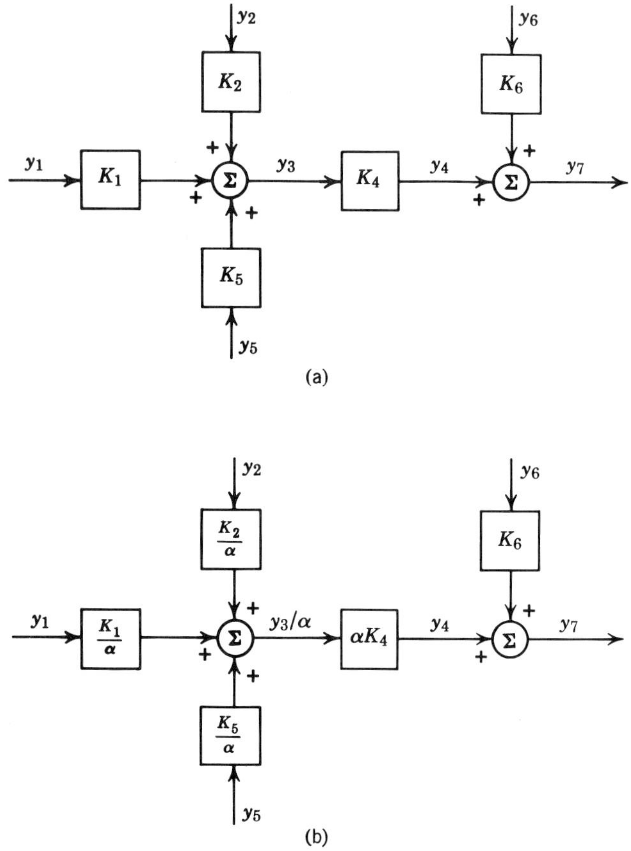

Figure 4.20 An example of scaling. (*a*) Original simulation diagram. (*b*) Scaled simulation diagram.

Figure 4.21. The modifications involve scaling as depicted in Figure 4.20. Notice that the original "state variables" θ, ω, and Ω are retained. (Alternative simulations can also be generated by choosing new variables, of course.) In Figure 4.21*c* fewer parameter-dependent gains appear than in Figure 4.19. ▲

4.3 Sensitivity Functions via Simulation Diagrams

The use of sensitivity functions in the estimation, to first order, of parameter variation effects on system responses was discussed in Chapter 3. A procedure was also given for drawing a network model from which sensitivity functions can be determined. In this section,

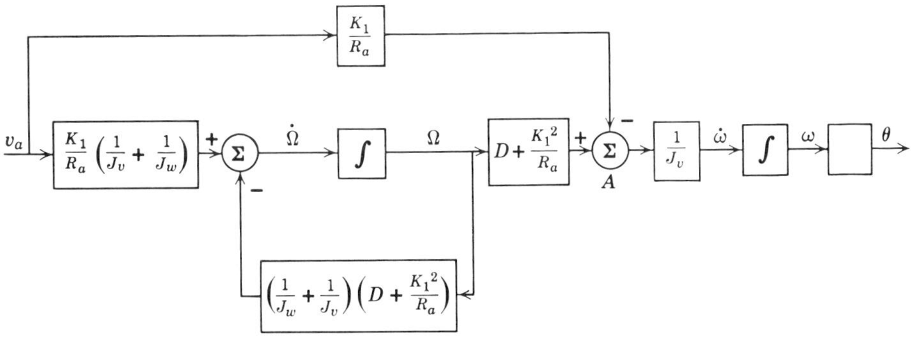

(a)

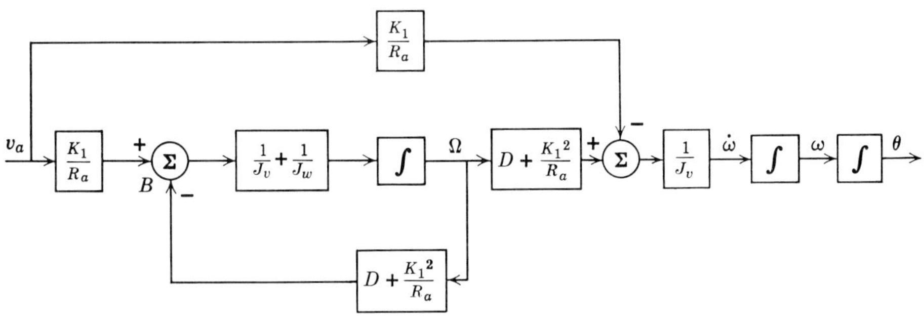

(b)

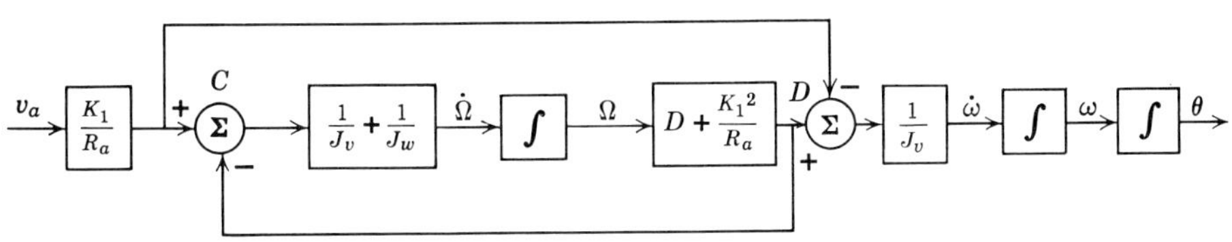

(c)

Figure 4.21 Amplitude-scaled simulation diagram of reaction-wheel attitude-control system. (*a*) Gains in branches entering summer *A* are multiplied by J_v; gains in branches leaving are divided by J_v. (*b*) Gains in branches entering summer *B* are divided by $(1/J_v + 1/J_w)$; gains in branches leaving are multiplied by $(1/J_v + 1/J_w)$. (*c*) Gains in branches entering point *C* are multiplied by K_1/R_a; gains in branches leaving point *C* are divided by K_1/R_a; gains in branches entering point *D* are multiplied by $(D + K_1^2/R_a)$; gains in branches leaving point *D* are divided by $(K_1^2/R_a + D)$.

we discuss the problem of determining sensitivity functions from simulation diagrams. We consider here only systems with linear elements. The nonlinear case is discussed in Chapter 9.

The method is illustrated by the following example.

Example 4-9. Suppose that we wish to investigate the effects of changes in the vehicle moment of inertia J_v on the vehicle angular rate Ω. We thus seek to obtain the sensitivity function

$$\sigma(t) = \frac{\partial \omega}{\partial J_v} \cdot \tag{4.41}$$

Taking the partial derivatives of (4.38 , (4.39), and (4.40) with respect to J_v, and assuming that the time derivative and the partial with respect to J_v can be interchanged, we obtain

$$\frac{d}{dt}\left(\frac{\partial \theta}{\partial J_v}\right) = \frac{\partial \omega}{\partial J_v} \tag{4.42}$$

$$\frac{d}{dt}\left(\frac{\partial \omega}{\partial J_v}\right) = \frac{1}{J_v}\left(D + \frac{K_1^2}{R_a}\right)\frac{\partial \Omega}{\partial J_v} - \frac{1}{J_v^2}\left(D + \frac{K_1^2}{R_a}\right)\Omega + \frac{K_1}{J_v^2 R_a} v_a \tag{4.43}$$

$$\frac{d}{dt}\left(\frac{\partial \Omega}{\partial J_v}\right) = -\left(\frac{1}{J_v} + \frac{1}{J_w}\right)\left(\frac{K_1^2}{R_a} + D\right)\frac{\partial \Omega}{\partial J_v}$$

$$+ \frac{1}{J_v^2}\left(\frac{K_1^2}{R_a} + D\right)\Omega - \frac{K_1}{J_v^2 R_a} v_a \tag{4.44}$$

Observe that these are *ordinary* (not partial) differential equations. These equations have a very interesting structure. They resemble the original equations (4.38, 4.39, and 4.40), but with the variables θ, ω, and Ω replaced by their partial derivatives with respect to J_v, and also different inputs are present. These new inputs involve both v_a and Ω. A simulation diagram for (4.42), (4.43), and (4.44) can be readily drawn, with inputs generated from a simulation of the original system as in Figure 4.21*c*. This diagram is given in Figure 4.22. In drawing the diagram, use is made of the fact that the signal at the output of the left-hand summing junction is

$$\frac{K_1}{R_a} v_a - \left(D + \frac{K_1^2}{R_a}\right)\Omega. \blacktriangle \tag{4.45}$$

Several other methods are available for obtaining simulation diagrams from which sensitivity functions can be determined. For example, equations can be written from the sensitivity network model as discussed in Chapter 3. Simulation of these equations will produce the desired sensitivity functions. A method of obtaining a sensitivity function simulation diagram by *inspection* of a system simulation diagram will be discussed in Chapters 6 and 7.

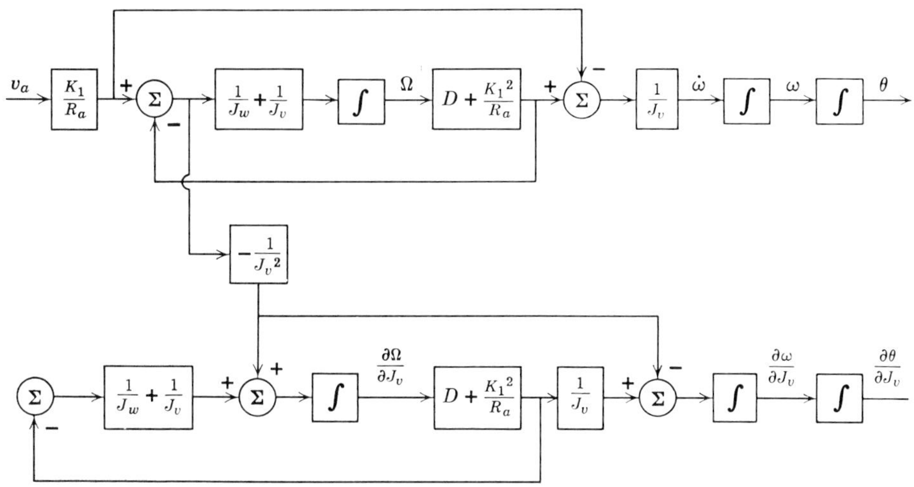

Figure 4.22 Simulation diagram for Example 4-9.

Problems

4-1. Draw an all-integrator simulation diagram for the network shown in Figure P4-1, where v is input and i is output. Label the input, the output, and state variables.

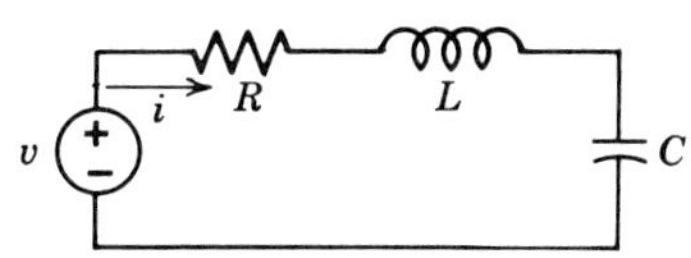

Figure P4-1

4-2. Repeat Problem 4-1 for the network shown in Figure P4-2, where i is the input, and v_4 and i_6 are the outputs.

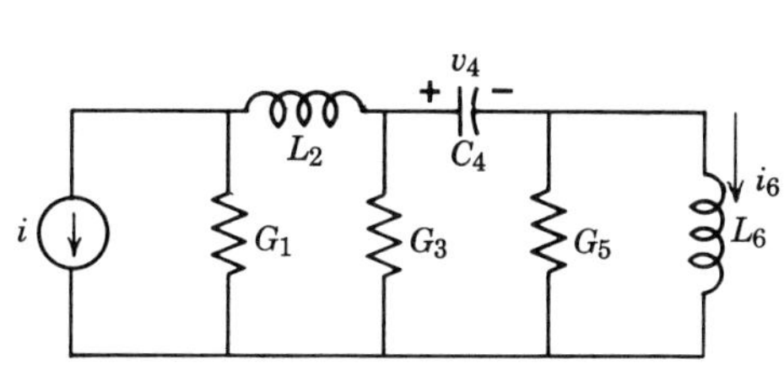

Figure P4-2

4-3. Consider the mechanical system shown in Figure P4-3, with external force $f(t)$ as input. Draw an all-integrator simulation diagram from which the *displacements* $x_1(t)$ and $x_2(t)$ can be found.

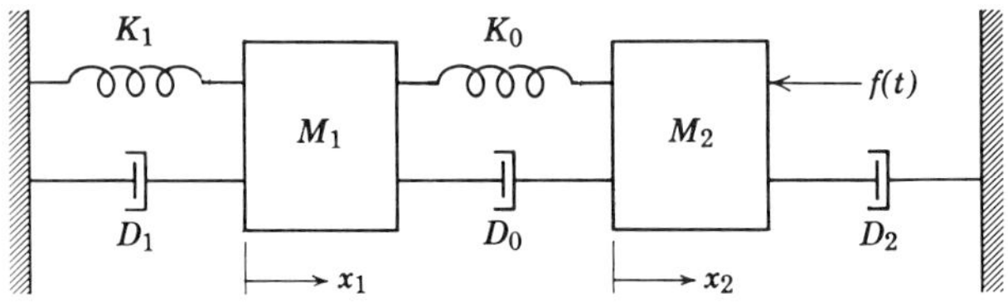

Figure P4-3

4-4. A second-order system is modeled by

$$\ddot{y} + 2\zeta\omega_n\dot{y} + \omega_n^2 y = r$$

where y is the output and r is the input. Draw an all-integrator simulation diagram for determining y. All initial conditions are zero.

4-5. Repeat Problem 4-4 for initial conditions $y(0) = y_0$ and $\dot{y}(0) = \dot{y}_0$.

4-6. Figure P4-6*a* shows a pictorial model of an automobile suspension system (one wheel only—coupling with other wheels is neglected).

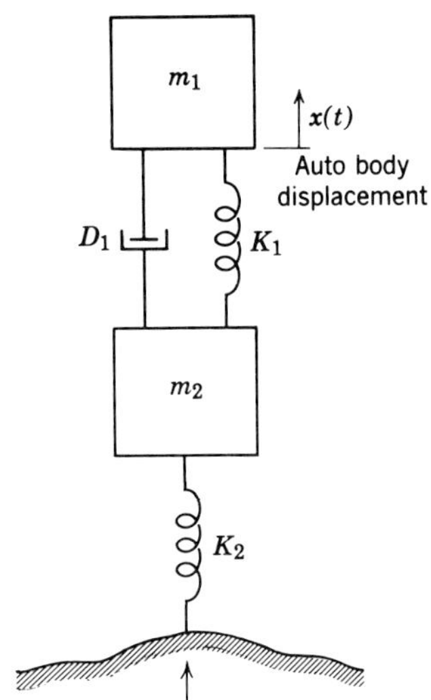

Figure P4-6 (*a*)

The parameters have the following physical identifications: $m_1 = \frac{1}{4}$ of mass of auto, $m_2 =$ mass of wheel, $K_1 =$ spring constant of auto spring, $K_2 =$ spring constant of tire, and $D_1 =$ shock absorber damping constant. Inputs and outputs are indicated on the functional diagram of Figure P4-6b. Develop an all-integrator simulation diagram for this system.

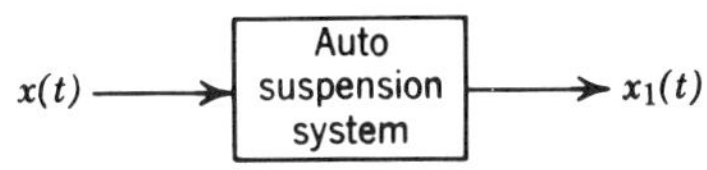

Figure P4-6 (b)

4-7. Develop a sensitivity simulation diagram for the suspension system of Problem 4-6 from which $\partial x_1/\partial D_1$ can be determined.

4-8. Repeat Problem 4-7 for determining $\partial x_1/\partial K_1$.

4-9. Repeat Problem 4-7 for determining $\partial x_1/\partial K_2$.

4-10. Develop a sensitivity simulation diagram for the system of Problem 4-3 from which $\partial x_1/\partial D_0$ and $\partial x_2/\partial D_0$ can be determined.

4-11. Draw an all-integrator simulation diagram for solving the differential equation

$$\frac{d^4y}{dt^4}+2.6131\frac{d^3y}{dt^3}+3.4142\frac{d^2y}{dt^2}+2.6131\frac{dy}{dt}+y=x$$

where y is the unknown function to be determined and x is a given function of time. Express the initial conditions in terms of $y(0)$, $\dot{y}(0)$, $\ddot{y}(\ )$, . . . , $x(0)$, $\dot{x}\ 0)$, ...

4-12. Repeat Problem 4-11 for

$$\frac{d^4y}{dt^4}+2.6131\frac{d^3y}{dt^3}+3.4142\frac{d^2y}{dt^2}+2.6131\frac{dy}{dt}+y=10\frac{d^2x}{dt^2}+20\frac{dx}{dt}+10x.$$

4-13. Repeat Problem 4-11 for

$$\frac{d^ny}{dt^n}+a_{n-1}\frac{d^{n-1}y}{dt^{n-1}}+\cdots+a_1\frac{dy}{dt}+a_0y=b_n\frac{d^nx}{dt^n}+b_{n-1}\frac{d^{n-1}x}{dt^{n-1}}+\cdots$$
$$+b_1\frac{dx}{dt}+b_0x.$$

4-14. Repeat for

$$\frac{d^2y}{dt^2}+a_1(t)\frac{dy}{dt}+a_0(t)y=x.$$

4-15. Draw an all-integrator simulation diagram for the accelerometer shown in Figure P4-15, where y is the position of the test mass M and $\ddot{x}$ is the acceleration of the accelerometer case. $\ddot{x}$ is input and y is output.

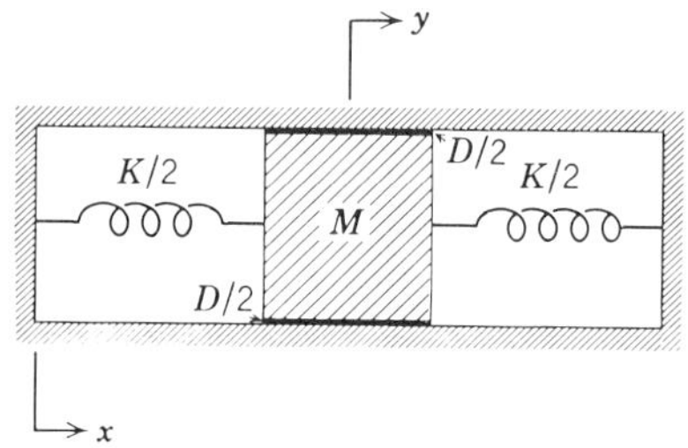

Figure P4-15

4-16. Draw an all-integrator simulation diagram for the network shown in Figure P4-16. Make the capacitor voltages correspond with integrator outputs.

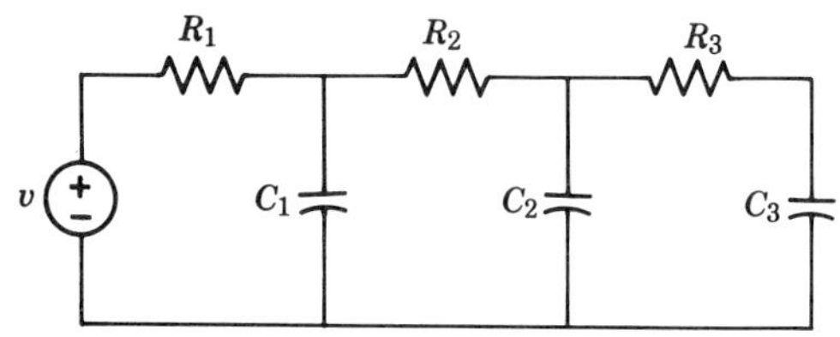

Figure P4-16

4-17. Figure P4-17 shows a motor drive for a load of moment of inertia J with rotational damping D, through a flexible shaft with rotational stiffness K, and a pair of gears. Consider the motor as a torque source. Draw an all-integrator simulation diagram.

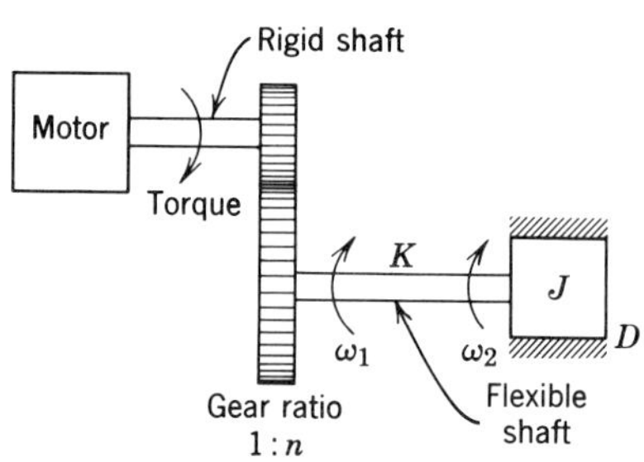

Figure P4-17

CHAPTER 5

Modeling Spacecraft Motion

The purpose of this chapter is to model the three-dimensional motion of a spacecraft. Two active actuators will be employed: gas jets and reaction wheels. A model for the gyro, a widely used sensor and controller, will also be developed.

The modeling of spacecraft motion may be decomposed into three steps

1. Choice of coordinate axes.
2. Selection of input and output variables.
3. Development of equations relating input and output variables.

Coordinate axes are discussed in Section 5.1. Euler's dynamical equations, which are Newton's laws in rotating coordinates, are also developed in Section 5.1. These equations are the basic equations for modeling rigid body motion. The spacecraft attitude and attitude rates are often selected as output variables for attitude-control systems. The representation of these quantities in terms of Euler angles is discussed in Section 5.2. The single-axis gyro is modeled in Section 5.3. Chapter 5 concludes with the modeling of a spacecraft with gas jet actuator (Section 5.4) and the modeling of a spacecraft with reaction-wheel actuators (Section 5.5).

5.1 Rigid Body Equations of Motion

Development of the equations of motion of a rotating spacecraft moving in three dimensions requires generalization of the mechanics discussed in Chapters 2 and 3. We begin with a discussion of rigid body motion.

The general motion of a rigid body can be described using Newton's law. Consider first a set of *inertial* coordinate axes with fixed orientation in space (that is, not rotating with the body) having its origin also fixed in space (that is, not translating with the body). For this case the translational and rotational aspects of the general motion may be described separately. Translational motion of the center of mass is governed by the vector equation

$$\vec{\mathbf{f}} = \frac{d}{dt}(m\vec{\mathbf{v}}), \tag{5.1}$$

where m is the mass of the body, $\vec{\mathbf{f}}$ the vector sum of the forces acting on the body, and $\vec{\mathbf{v}}$ is the velocity of the center of mass of the body with respect to the inertial axes. The quantity $m\vec{\mathbf{v}}$ is called the (linear) *momentum*. Rotational motion is governed by[1]

$$\vec{\boldsymbol{\tau}} = \frac{d\vec{\mathbf{L}}}{dt}, \tag{5.2}$$

where $\vec{\boldsymbol{\tau}}$ is the vector sum of the torques acting on the body, and $\vec{\mathbf{L}}$ is the total angular momentum of the body about its center of mass.

It is stressed that (5.1) and (5.2) are valid for *inertial* axes. For many situations that we shall consider, it will be more convenient to examine motions relative to axes *fixed in the body and rotating with it*. These axes are called *body axes*. In Figure 5.1, the unprimed axes

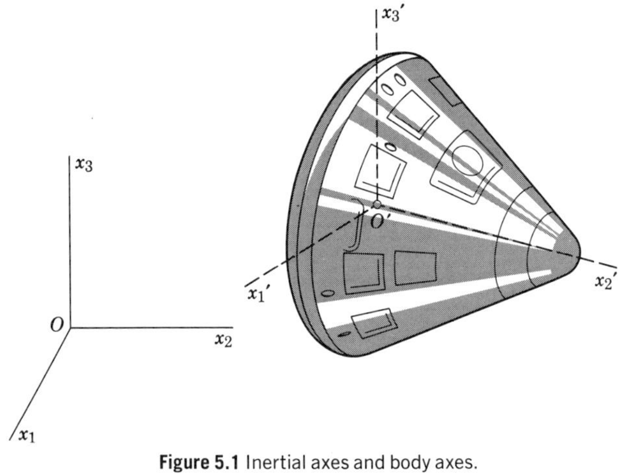

Figure 5.1 Inertial axes and body axes.

[1]It is illuminating to derive (5.2) for a rigid body by summing terms like (5.1) over all particles in the body. See, for example, J. L. Meriam, *Dynamics*, Wiley, New York, 1966, Chapter 4.

are inertial axes, and the primed axes are body axes. To do this, we express the angular momentum of (5.2) in terms of its components in body coordinates (the primed coordinates of Figure 5.1):

$$\vec{\mathbf{L}} = L'_1\hat{\mathbf{x}}'_1 + L'_2\hat{\mathbf{x}}'_2 + L'_3\hat{x}'_3 = \sum_{i=1}^{3} L'_i\hat{x}'_i, \tag{5.3}$$

where $\hat{x}'_i$ is the ith unit vector in the body coordinate system. Following this, we form the time derivative. Since (5.2) requires that the time derivative be computed in inertial coordinates, we shall obtain derivatives of both L'_i and $\hat{x}'_i$. The primed unit vectors are time-dependent because their directions are changing with time as the primed axes rotate with respect to the unprimed axes. The derivative of (5.3) becomes

$$\frac{d\vec{\mathbf{L}}}{dt} = \sum_{i=1}^{3} \frac{dL'_i}{dt}\hat{x}'_i + \sum_{i=1}^{3} L'_i\frac{d\hat{x}'_i}{dt}. \tag{5.4}$$

For clarity, we denote the left-hand side of (5.4) by

$$\left(\frac{d\vec{\mathbf{L}}}{dt}\right)_I, \tag{5.5}$$

since it is the time derivative of $\vec{\mathbf{L}}$ computed in inertial coordinates. The first summation on the right-hand side is just the time derivative of $\vec{\mathbf{L}}$ computed in body coordinates. Thus

$$\sum_{i=1}^{3} \frac{dL'_i}{dt}\hat{x}'_i = \left(\frac{d\vec{\mathbf{L}}}{dt}\right)_B. \tag{5.6}$$

Next, we examine the summation containing the rates of change of the unit vectors. The derivative of the unit vector $\hat{\mathbf{x}}'_1$ can be computed from

$$\frac{d\hat{\mathbf{x}}'_1}{dt} = \lim_{\Delta t \to 0} \frac{\hat{\mathbf{x}}'_1(t+\Delta t) - \hat{\mathbf{x}}'_1(t)}{\Delta t}. \tag{5.7}$$

The geometry of the situation, showing the *body* unit vectors at time t and at time $t+\Delta t$ is shown in Figure 5.2.

The angle $\Delta\alpha$ is in the $[\hat{\mathbf{x}}'_1(t), \hat{\mathbf{x}}'_2(t)]$ plane. $\Delta\beta$ is in the $[\hat{\mathbf{x}}'_1(t), \hat{\mathbf{x}}'_3(t)]$ plane. $\Delta\gamma$ is the elevation of $\hat{\mathbf{x}}'_1(t+\Delta t)$ from the $[\hat{\mathbf{x}}'_1(t), \hat{\mathbf{x}}'_2(t)]$ plane. Taking projections, we obtain

$$\hat{\mathbf{x}}'_1(t+\Delta t) = \cos\Delta\gamma\cos\Delta\alpha\hat{\mathbf{x}}'_1(t) + \cos\Delta\gamma\sin\Delta\alpha\hat{\mathbf{x}}'_2(t) + \sin\Delta\gamma\hat{\mathbf{x}}'_3(t). \tag{5.8}$$

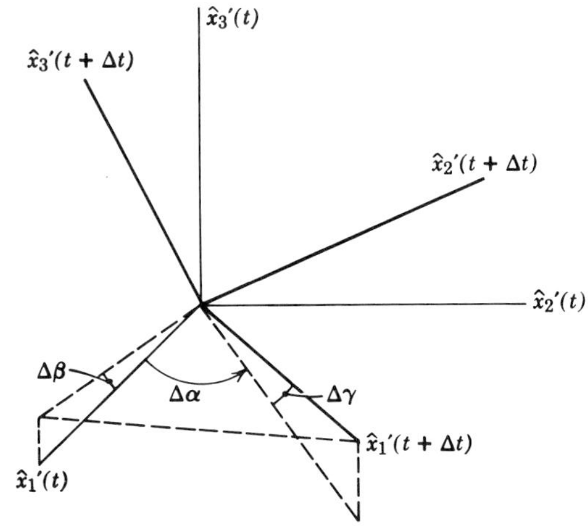

Figure 5.2 Geometry for calculation of dx_1'/dt.

It will be convenient to express $\sin \Delta\gamma$ in terms of β. Note that

$$\tan \Delta\beta = \frac{\sin \Delta\gamma}{\cos \Delta\gamma \cos \Delta\alpha}. \tag{5.9}$$

We subtract $\hat{x}_1'(t)$ from both sides of (5.8), divide by Δt, and take the limit as $\Delta t \to 0$. We obtain

$$\frac{d\hat{x}_1'}{dt} = \left(\lim_{\Delta t\to 0} \frac{\Delta\alpha}{\Delta t}\right)\hat{x}_2'(t) + \left(\lim_{\Delta t\to 0} \frac{\Delta\beta}{\Delta t}\right)\hat{x}_3'(t). \tag{5.10}$$

However,

$$\lim_{\Delta t\to 0} \frac{\Delta\alpha}{\Delta t} = \frac{d\alpha}{dt} \stackrel{\Delta}{=} \omega_3, \tag{5.11}$$

where ω_3 is the component of the vehicle angular velocity about the x_3' body axis.[2] Similarly,

$$\lim_{\Delta t\to 0} \frac{\Delta\beta}{\Delta t} = \frac{d\beta}{dt} \stackrel{\Delta}{=} -\omega_2, \tag{5.12}$$

the negative of the component of the vehicle angular velocity about

[2]This angular velocity component is expressed in *body* coordinates. However, we do not prime the component, as was done for angular momentum, because it is used extensively throughout the book, and the notation would become unwieldy.

x_2'. (Here, the "right-hand" rule is used to determine the sense of the rotations.) Thus

$$\frac{d\hat{\mathbf{x}}_1'}{dt} = \omega_3 \hat{\mathbf{x}}_2' - \omega_2 \hat{\mathbf{x}}_3'. \tag{5.13}$$

Note that if we *define* a vector

$$\vec{\boldsymbol{\omega}} = \omega_1 \hat{\mathbf{x}}_1' + \omega_2 \hat{\mathbf{x}}_2' + \omega_3 \hat{\mathbf{x}}_3', \tag{5.14}$$

then (5.13) is a vector cross product,

$$\frac{d\hat{\mathbf{x}}_1'}{dt} = \vec{\boldsymbol{\omega}} \times \hat{\mathbf{x}}_1'. \tag{5.15}$$

In a similar fashion, we may show that

$$\frac{d\hat{\mathbf{x}}_2'}{dt} = \vec{\boldsymbol{\omega}} \times \hat{\mathbf{x}}_2'.$$

$$\frac{d\hat{\mathbf{x}}_3'}{dt} = \vec{\boldsymbol{\omega}} \times \hat{\mathbf{x}}_3'. \tag{5.16}$$

Thus the second summation in (5.4) can be written as

$$(\vec{\boldsymbol{\omega}} \times \vec{\mathbf{L}})_B. \tag{5.17}$$

So (5.2) can be written as

$$\vec{\boldsymbol{\tau}}_B = \left[\frac{d\vec{\mathbf{L}}}{dt}\right]_B + (\vec{\boldsymbol{\omega}} \times \vec{\mathbf{L}})_B. \tag{5.18}$$

This set of equations is Newton's law for rotation written in a rotating coordinate system. The equations are called Euler's dynamical equations. They are the basic equations for studying rigid body rotations.

It will be convenient to rewrite (5.18) in matrix notation. Matrices will be extremely useful not only in the modeling discussion here but also in the control and optimization chapters later. A discussion of basic matrix algebra appears in Appendix A. Readers unfamiliar with this topic or desiring review of the fundamental ideas should consult Appendix A before continuing with this chapter.

We shall use the capital letters for matrices, with the exception of matrices having a single column. Such matrices are denoted by a small letter. Column matrices are sometimes called vectors. However, do not confuse $\vec{\mathbf{x}}$ and $\mathbf{x}$. $\vec{\mathbf{x}}$ is a space vector, and $\mathbf{x}$ is its matrix representation in a particular coordinate system, the elements of which are the components of $\vec{\mathbf{x}}$ in that particular coordinate system.

The components of (5.2) in inertial coordinates can be represented by matrices. Define

$$\boldsymbol{\tau} = \begin{bmatrix} \tau_1 \\ \tau_2 \\ \tau_3 \end{bmatrix}, \tag{5.19}$$

where τ_i is the component of $\vec{\tau}$ in the direction of the inertial axis x_i. Similarly, define

$$\mathbf{L} = \begin{bmatrix} L_1 \\ L_2 \\ L_3 \end{bmatrix}. \tag{5.20}$$

(Here we depart from our convention of a lower-case symbol for column matrices.) Defining the derivative of a matrix to mean a new matrix whose elements are the derivatives of the elements of the old matrix, we may represent the components of (5.2) in inertial coordinates by

$$\boldsymbol{\tau} = \frac{d}{dt}\mathbf{L}, \tag{5.21}$$

in the sense that both Equation 5.2 in inertial coordinates and Equation 5.21 stand for the same three equations,

$$\begin{aligned} \tau_1 &= \frac{dL_1}{dt} \\ \tau_2 &= \frac{dL_2}{dt} \\ \tau_3 &= \frac{dL_3}{dt}. \end{aligned} \tag{5.22}$$

Similarly, we can express Euler's dynamical equations (5.18) in matrix notation. Define

$$\mathbf{L}' = \begin{bmatrix} L_1' \\ L_2' \\ L_3' \end{bmatrix} \tag{5.23}$$

where L'_i is the component of the angular momentum $\vec{\mathbf{L}}$ in the direction of the body axis x'_i. In addition, define a torque matrix in terms of the torque vector body-axis components.

$$\boldsymbol{\tau}' = \begin{bmatrix} \tau'_1 \\ \tau'_2 \\ \tau'_3 \end{bmatrix} \tag{5.24}$$

It remains to represent the cross product $(\vec{\boldsymbol{\omega}} \times \vec{\mathbf{L}})_B$ in (5.18) as a matrix. The cross product has components given by the following determinant:

$$\begin{vmatrix} \hat{\mathbf{x}}'_1 & \hat{\mathbf{x}}'_2 & \hat{\mathbf{x}}'_3 \\ \omega_1 & \omega_2 & \omega_3 \\ L'_1 & L'_2 & L'_3 \end{vmatrix} = (\omega_2 L'_3 - \omega_3 L'_2)\hat{\mathbf{x}}'_1 + (\omega_3 L'_1 - \omega_1 L'_3)\hat{\mathbf{x}}'_2 + (\omega_1 L'_2 - \omega_2 L'_1)\hat{\mathbf{x}}'_3 \tag{5.25}$$

Thus the matrix representation of (5.25) is a column matrix

$$\begin{bmatrix} \omega_2 L'_3 - \omega_3 L'_2 \\ \omega_3 L'_1 - \omega_1 L'_3 \\ \omega_1 L'_2 - \omega_2 L'_1 \end{bmatrix} \tag{5.26}$$

This can be written in terms of $\mathbf{L}'$, defined previously, if we define another matrix

$$\mathbf{P} = \begin{bmatrix} 0 & -\omega_3 & \omega_2 \\ \omega_3 & 0 & -\omega_1 \\ -\omega_2 & \omega_1 & 0 \end{bmatrix}. \tag{5.27}$$

Equation 5.18 can then be written in matrix form as

$$\boldsymbol{\tau}' = \frac{d\mathbf{L}'}{dt} + \mathbf{P}\mathbf{L}'. \tag{5.28}$$

This equation is *the basic equation* for describing rigid body rotation. The angular momentum of the body, $\vec{\mathbf{L}}$ in (5.18), is related to $\vec{\boldsymbol{\omega}}$ by the following equations in component form in the primed system:

$$L'_1 = J_{11}\omega_1 + J_{12}\omega_2 + J_{13}\omega_3$$

$$L'_2 = J_{21}\omega_1 + J_{22}\omega_2 + J_{23}\omega_3$$

$$L'_3 = J_{31}\omega_1 + J_{32}\omega_2 + J_{33}\omega_3 \tag{5.29}$$

These equations can be conveniently written in matrix form. Using $\mathbf{L}'$ as given by (5.23), and defining

$$\boldsymbol{\omega} = \begin{bmatrix} \omega_1 \\ \omega_2 \\ \omega_3 \end{bmatrix} \tag{5.30}$$

and

$$\mathbf{J} = \begin{bmatrix} J_{11} & J_{12} & J_{13} \\ J_{21} & J_{22} & J_{23} \\ J_{31} & J_{32} & J_{33} \end{bmatrix}, \tag{5.31}$$

we have

$$\mathbf{L}' = \mathbf{J}\boldsymbol{\omega}. \tag{5.32}$$

The diagonal terms of the matrix $\mathbf{J}$ are called *moments of inertia*, while the off-diagonal terms are *products of inertia*. These quantities depend on the geometry of the vehicle in relation to the choice of body axes. In particular, one can always choose a set of body axes such that J is diagonal,

$$\mathbf{J} = \begin{bmatrix} J_1 & 0 & 0 \\ 0 & J_2 & 0 \\ 0 & 0 & J_3 \end{bmatrix}. \tag{5.33}$$

These body axes are called the *principal axes*. Such axes often can be identified by considering the symmetries of the vehicle.

Using (5.32), Euler's dynamical equations in matrix form, (5.28), become

$$\boldsymbol{\tau}' = \frac{d}{dt}(\mathbf{J}\boldsymbol{\omega}) + \mathbf{P}\mathbf{J}\boldsymbol{\omega}. \tag{5.34}$$

For most cases to be considered here, we can suppose that the moments and products of inertia do not change with time. Then (5.34) becomes

$$\boldsymbol{\tau}' = \mathbf{J}\dot{\boldsymbol{\omega}} + \mathbf{P}\mathbf{J}\boldsymbol{\omega}. \tag{5.35}$$

These equations have a relatively simple form where the body axes are also principal axes with origin at the center of mass. Then, using (5.33), (5.35) becomes, in component form,

$$\begin{aligned} \tau_1' &= J_1\dot{\omega}_1 - (J_2 - J_3)\omega_2\omega_3 \\ \tau_2' &= J_2\dot{\omega}_2 - (J_3 - J_1)\omega_3\omega_1 \\ \tau_3' &= J_3\dot{\omega}_3 - (J_1 - J_2)\omega_1\omega_2. \end{aligned} \tag{5.36}$$

We illustrate these notions by the following examples.

Example 5-1. Plane (Two-Dimensional) Motion. Consider a rigid body constrained to the (x_1, x_2) plane and rotating about the $x_3 = x_3'$ axis. The primed axes are not necessarily principal axes. Since there is no rotation about the x_1' and x_2' axes, $\omega_1 \equiv \omega_2 \equiv 0$. Then (5.35) becomes

$$\begin{bmatrix} \tau_1' \\ \tau_2' \\ \tau_3' \end{bmatrix} = \begin{bmatrix} J_{11} & J_{12} & J_{13} \\ J_{21} & J_{22} & J_{23} \\ J_{31} & J_{32} & J_{33} \end{bmatrix} \begin{bmatrix} 0 \\ 0 \\ \dot{\omega}_3 \end{bmatrix} + \begin{bmatrix} 0 & -\omega_3 & 0 \\ \omega_3 & 0 & 0 \\ 0 & 0 & 0 \end{bmatrix} \begin{bmatrix} J_{11} & J_{12} & J_{13} \\ J_{21} & J_{22} & J_{23} \\ J_{31} & J_{32} & J_{33} \end{bmatrix} \begin{bmatrix} 0 \\ 0 \\ \omega_3 \end{bmatrix}. \tag{5.37}$$

Performing the indicated matrix multiplications, we obtain

$$\begin{aligned} \tau_1' &= J_{13}\dot{\omega}_3 - J_{23}\omega_3^2 \\ \tau_2' &= J_{23}\dot{\omega}_3 + J_{13}\omega_3^2 \\ \tau_3' &= J_{33}\dot{\omega}_3. \end{aligned} \tag{5.38}$$

Furthermore, if the body axes are principal axes, then only one equation of interest remains:

$$\tau_3' = J_3\dot{\omega}_3. \tag{5.39}$$

This is the situation considered in Chapter 2 where the single-axis rotational inertia element was modeled. ▲

Example 5-2. Consider a spinning circular cylindrical satellite with body axes as shown in Figure 5.3. The axis x_3' is normal to the circular face

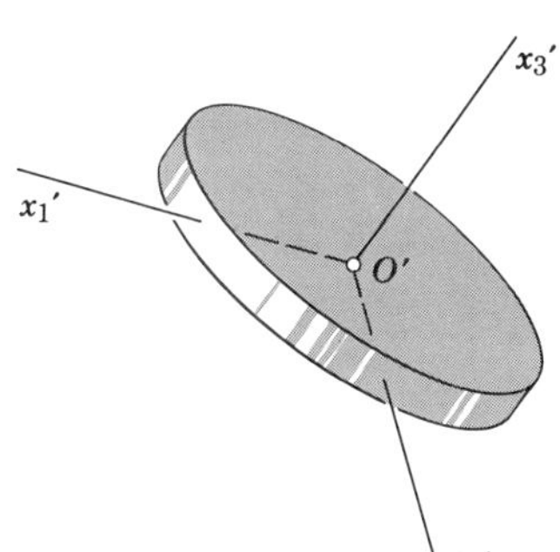

Figure 5.3 Body axes for spinning cylinder.

of the disk. The body axes are principal axes. No external torques are present, that is, $\boldsymbol{\tau}' \equiv \mathbf{0}$. Because of symmetry, $J_1 = J_2 = J$.

For this case, (5.36) becomes

$$\begin{aligned} 0 &= J\dot{\omega}_1 - (J - J_3)\omega_2\omega_3 \\ 0 &= J\dot{\omega}_2 - (J_3 - J)\omega_3\omega_1 \\ 0 &= J_3\dot{\omega}_3. \end{aligned} \tag{5.40}$$

The third equation reveals that

$$\omega_3 = \Omega = \text{constant}. \tag{5.41}$$

Defining

$$\Omega\left(\frac{J_3 - J}{J}\right) = K, \tag{5.42}$$

the first two equations in (5.40) become

$$\begin{aligned} \dot{\omega}_1 + K\omega_2 &= 0 \\ \dot{\omega}_2 - K\omega_1 &= 0. \end{aligned} \tag{5.43}$$

Thus

$$\ddot{\omega}_1 + K^2\omega_1 = 0. \tag{5.44}$$

This equation has the solution

$$\omega_1(t) = \omega_1(0)\cos Kt + \frac{\dot{\omega}_1(0)}{K}\sin Kt. \tag{5.45}$$

Moreover,

$$\omega_2(t) = \frac{-\dot{\omega}_1}{K} = \omega_1(0)\sin Kt - \frac{\dot{\omega}_1(0)}{K}\cos Kt. \tag{5.46}$$

Observe that $\omega_2(0) = -\dot{\omega}_1(0)/K$. Furthermore, note that, for all t,

$$\omega_1^2 + \omega_2^2 = \text{constant} = \omega_1^2(0) + \omega_2^2(0). \tag{5.47}$$

Thus the vector $\vec{\boldsymbol{\omega}}$ has *constant magnitude*,

$$\omega = \sqrt{\omega_1^2 + \omega_2^2 + \omega_3^2} = \sqrt{\omega_1^2(0) + \omega_2^2(0) + \Omega^2}. \blacktriangle \tag{5.48}$$

5.2 Representation of Spacecraft Attitude

The last section was devoted to the derivation of the basic equation of motion for rotating rigid bodies, (5.28). This equation involves the angular velocities ω_1, ω_2, and ω_3 of the spacecraft about the body axes. Next, these variables must be expressed in terms of the spacecraft attitude, that is, the spacecraft axis orientation with respect to the inertial reference axes. Thus we must investigate the orientation of one set of axes with respect to another.

One of the most widely used descriptions, involving three independent angles, is the method of Euler angles. The rigid body orientation is specified by means of *three ordered rotations* about a set of axes. Unfortunately, several alternative definitions of Euler angles are used. These involve several different conventions for measuring angles, as well as different types of axis rotations. We consider only one possibility here.

Consider the following sequence of rotations:

1. With the axes initially coincident, rotate about the inertial x_3 axis by an angle ψ. The new axis are denoted by ξ', η', and $x_3 = \zeta'$. See Figure 5.4.
2. Next rotate about ξ' by an angle θ. The new axes are $\xi = \xi'$, η, ζ. See Figure 5.5.
3. Finally, rotate about ζ by an angle φ. The final body axes are labeled x_1', x_2', x_3' in Figure 5.6.

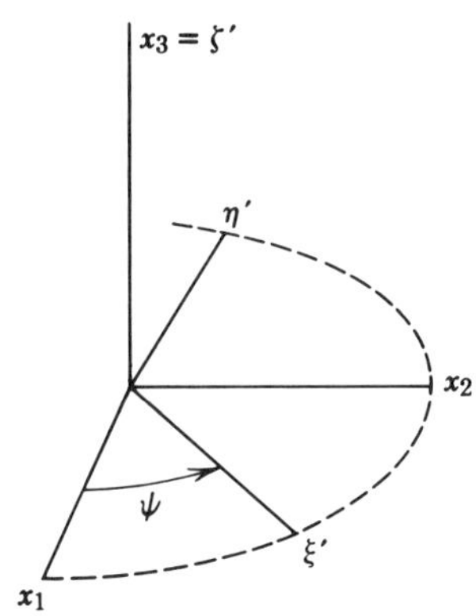

Figure 5.4 Definition of ψ. The coordinates are related by

$$\begin{bmatrix} \xi' \\ \eta' \\ \zeta' \end{bmatrix} = \begin{bmatrix} \cos\psi & \sin\psi & 0 \\ -\sin\psi & \cos\psi & 0 \\ 0 & 0 & 1 \end{bmatrix} \begin{bmatrix} x_1 \\ x_2 \\ x_3 \end{bmatrix} = A_1 \begin{bmatrix} x_1 \\ x_2 \\ x_3 \end{bmatrix}. \tag{5.49}$$

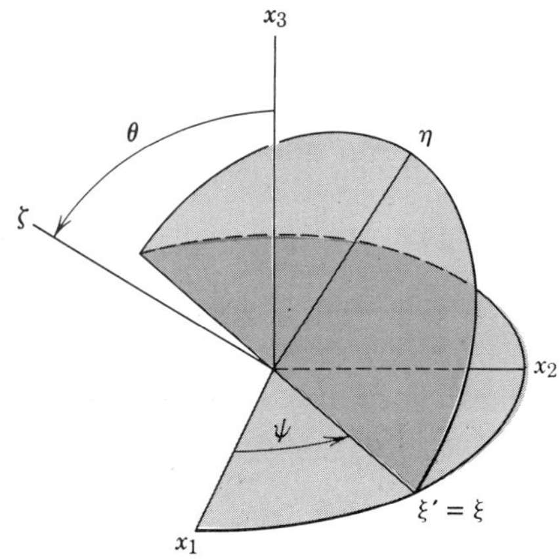

Figure 5.5 Definition of θ. This rotation is described by

$$\begin{bmatrix}\xi\\ \eta\\ \zeta\end{bmatrix}=\begin{bmatrix}1 & 0 & 0\\ 0 & \cos\theta & \sin\theta\\ 0 & -\sin\theta & \cos\theta\end{bmatrix}\begin{bmatrix}\xi'\\ \eta'\\ \zeta'\end{bmatrix}=A_2\begin{bmatrix}\xi'\\ \eta'\\ \zeta'\end{bmatrix}. \tag{5.50}$$

The variables ψ, θ, φ are the *Euler angles*. They describe the orientation of the body (primed) axes with respect to the inertial (unprimed) axes. The final relation between the primed and unprimed axes can be obtained by combining, via matrix multiplication,

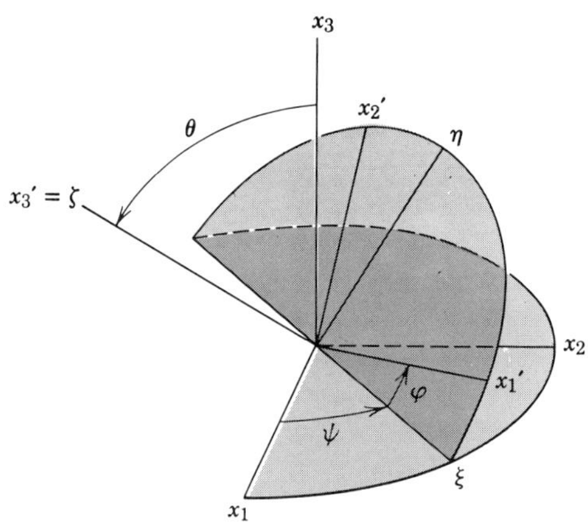

Figure 5.6 Euler angles ψ, θ, φ. This final rotation is described by

$$\begin{bmatrix}x_1'\\ x_2'\\ x_3'\end{bmatrix}=\begin{bmatrix}\cos\varphi & \sin\varphi & 0\\ -\sin\varphi & \cos\varphi & 0\\ 0 & 0 & 1\end{bmatrix}\begin{bmatrix}\xi\\ \eta\\ \zeta\end{bmatrix}=A_3\begin{bmatrix}\xi\\ \eta\\ \zeta\end{bmatrix}. \tag{5.51}$$

the matrix transformations of (5.49), (5.50), and (5.51). The result is

$$\mathbf{x}' = \begin{bmatrix} x_1' \\ x_2' \\ x_3' \end{bmatrix} = B\mathbf{x} = B\begin{bmatrix} x_1 \\ x_2 \\ x_3 \end{bmatrix}, \tag{5.52}$$

where

$$B = A_3 A_2 A_1 =$$

$$\begin{bmatrix} (\cos\varphi\cos\psi - \sin\varphi\cos\theta\sin\psi) & (\cos\varphi\sin\psi + \sin\varphi\cos\theta\cos\psi) & (\sin\varphi\sin\theta) \\ (-\sin\varphi\cos\psi - \cos\varphi\cos\theta\sin\psi) & (-\sin\varphi\sin\psi + \cos\varphi\cos\theta\cos\psi) & (\cos\varphi\sin\theta) \\ (\sin\theta\sin\psi) & (-\sin\theta\cos\psi) & (\cos\theta) \end{bmatrix} \tag{5.53}$$

To complete the description of rigid body motion, the Euler angles and the body angular velocities (omegas of Section 5.1) must be related. Observe that, as components of an angular velocity vector, $\dot{\psi}$ is directed along x_3, $\dot{\theta}$ along ξ, and $\dot{\varphi}$ along x_3'. (Refer to Figure 5.6 and recall the "right-hand screw" convention for defining the direction of angular velocity vectors.) The situation is depicted in Figure 5.7.

Geometric resolution of the components of $\dot{\psi}$, $\dot{\theta}$, and $\dot{\varphi}$ along the body axis x_1' will yield ω_1. Similarly, resolution of $\dot{\psi}$, $\dot{\theta}$, $\dot{\varphi}$ along x_2' and x_3' yield ω_2 and ω_3, respectively. The result is

$$\begin{aligned} \omega_1 &= \dot{\psi}\sin\theta\sin\varphi + \dot{\theta}\cos\varphi \\ \omega_2 &= \dot{\psi}\sin\theta\cos\varphi - \dot{\theta}\sin\varphi \\ \omega_3 &= \dot{\varphi} + \dot{\psi}\cos\theta \end{aligned} \tag{5.54}$$

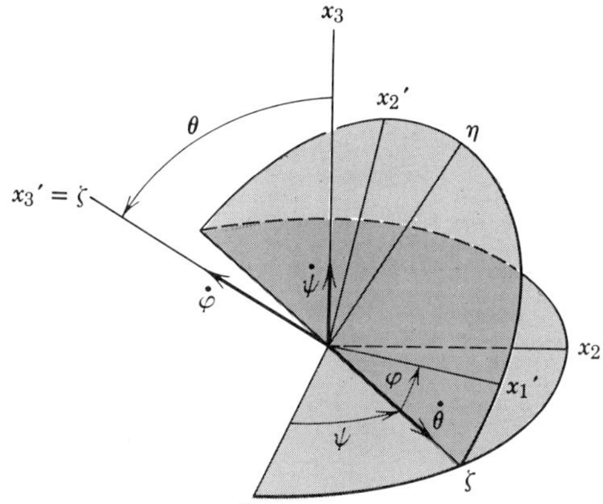

Figure 5.7 Resolution of velocities.

Equation 5.54 can be expressed in matrix form as

$$\begin{bmatrix}\omega_1\\ \omega_2\\ \omega_3\end{bmatrix}=\begin{bmatrix}\sin\theta\sin\varphi & 0 & \cos\varphi\\ \sin\theta\cos\varphi & 0 & -\sin\varphi\\ \cos\theta & 1 & 0\end{bmatrix}\begin{bmatrix}\dot{\psi}\\ \dot{\varphi}\\ \dot{\theta}\end{bmatrix}. \tag{5.55}$$

The following inverse relationship will also be useful.

$$\begin{bmatrix}\dot{\psi}\\ \dot{\varphi}\\ \dot{\theta}\end{bmatrix}=\frac{1}{\sin\theta}\begin{bmatrix}\sin\varphi & \cos\varphi & 0\\ -\sin\varphi\cos\theta & -\cos\varphi\cos\theta & \sin\theta\\ \cos\varphi\sin\theta & -\sin\varphi\sin\theta & 0\end{bmatrix}\begin{bmatrix}\omega_1\\ \omega_2\\ \omega_3\end{bmatrix}. \tag{5.56}$$

Euler's dynamical equations (5.35), together with (5.56), provide a complete description of the orientation (attitude) of a rigid body (such as a spacecraft) during rigid body rotational motion. They constitute a somewhat formidable set of six simultaneous nonlinear differential equations. To illustrate the meaning and use of these equations, we consider an example.

Example 5-3. Consider once again the "spinning disk" satellite of Example 5-2. Again no external torques are present. The body axes are the principal axis as before. We shall choose a convenient set of inertial axes. Since there are no external torques, the derivative of the angular momentum vector $\vec{\mathbf{L}}$ expressed in inertial coordinates is zero (see (5.2)). Thus $\vec{\mathbf{L}}$ is a constant, that is, $\vec{\mathbf{L}}$ is fixed in the inertial reference frame. We orient the inertial axis x_3 along $\vec{\mathbf{L}}$. Since θ is the Euler angle between x_3 and x_3', θ is the angle between $\vec{\mathbf{L}}$ and x_3'.

Expanding (5.56) in components,

$$\begin{aligned}\dot{\psi}&=\frac{\sin\varphi}{\sin\theta}\omega_1+\frac{\cos\varphi}{\sin\theta}\omega_2\\ \dot{\varphi}&=-\sin\varphi\cot\theta\,\omega_1-\cos\varphi\cot\theta\,\omega_2+\omega_3\\ \dot{\theta}&=\cos\varphi\,\omega_1-\sin\varphi\,\omega_2.\end{aligned} \tag{5.57}$$

These are nonlinear differential equations and cannot be solved analytically. However, useful information concerning the motion can be obtained from close study of these equations.

Since $\vec{\mathbf{L}}$ lies along the x_3 axis,

$$\begin{aligned}L_1'&=J_1\omega_1=L\sin\varphi\sin\theta=J\omega_1\\ L_2'&=J_2\omega_2=L\cos\varphi\sin\theta=J\omega_2\\ L_3'&=J_3\omega_3=L\cos\theta=J_3\Omega\end{aligned} \tag{5.58}$$

where L is the magnitude of $\vec{L}$. We shall combine (5.57) and (5.58) to express the Euler angle rates in terms of the Euler angles and the moments of inertia, that is, the omegas will be eliminated.

Consider the $\dot{\theta}$ equation first.

$$\dot{\theta} = \cos\varphi\left(\frac{L\sin\varphi\sin\theta}{J}\right) - \sin\varphi\left(\frac{L\cos\varphi\sin\theta}{J}\right) = 0. \tag{5.59}$$

Thus θ is a constant, that is, the motion is such that the angle between the inertial axis x_3 and the body axis x_3' is fixed.

Consider now the $\dot{\psi}$ equation in (5.57).

$$\dot{\psi} = \frac{\sin\varphi}{\sin\theta}\left(\frac{L\sin\varphi\sin\theta}{J}\right) + \frac{\cos\varphi}{\sin\theta}\left(\frac{L\cos\varphi\sin\theta}{J}\right) \tag{5.60}$$

$$\dot{\psi} = \frac{L}{J} = \text{constant}.$$

The quantity $\dot{\psi}$ is the rate at which the x_3' axis (the "body spin axis") *precesses* about the x_3 axis (the angular momentum vector) (see Figure 5.6). For this problem the precession rate is constant.

Finally, consider $\dot{\varphi}$ in (5.57). Notice that the "spin" $\dot{\varphi}$ is given by

$$\dot{\varphi} = -\dot{\psi}\cos\theta + \omega_3 = \text{constant}. \tag{5.61}$$

This can be rewritten as

$$\dot{\varphi} = L\left(\frac{1}{J_3} - \frac{1}{J}\right)\cos\theta. \ \blacktriangle \tag{5.62}$$

We consider next another example of rigid body motion—the gyro. Because of the considerable importance of this device as a sensing device in attitude-control systems, the gyro will be discussed extensively in the following section.

5.3 The Gyro

Measurements of attitude angles and angular velocities are required in spacecraft attitude-control systems, as is indicated in Figure 1.2. They are also needed in a multitude of other vehicle guidance and navigation systems. A basic sensor for performing these measurements is the gyroscope. The gyro is basically a spinning body with appropriate means for supporting it. From Newton's law, the rate of change of momentum with respect to inertial axes is equal to the torque applied to the body. Hence, if no torque is applied, momentum remains constant in magnitude as well as direction, so that the spin axis will remain fixed in inertial space. Various attitude angles associated with such a fixed spin axis may be measured or computed. The purpose of this section is to model a gyro, that is, to obtain its equations of motion. Simulation of these equations will be discussed also.

Figure 5.8 shows a schematic diagram of an elementary single-axis gyro. Its basic parts are the *rotor* or spinning wheel, a *gimbal* which is a supporting frame for the rotor and its axle, and the *case* which supports the gimbal. Angular momentum is imparted to the rotor by means of a motor (not shown) which drives the rotor at constant speed Ω about the x_3'' axis. The rotor is assumed to be perfectly centered. The gimbal is free to rotate about the x_2' axis only. The case is mounted firmly in a moving vehicle. The body axes of the case (and hence of the vehicle) are denoted by x_1', x_2', x_3'. The vehicle (and case) is moving with respect to an inertial frame.

The motion of the gimbal-rotor assembly with *respect to the case* can be described by a single angle α, the displacement of x_3'' with respect to x_3'. The angle α is considered to be the output for the gyro subsystem. This output may be measured conveniently in many applications with an electrical "pickoff" such as a potentiometer.

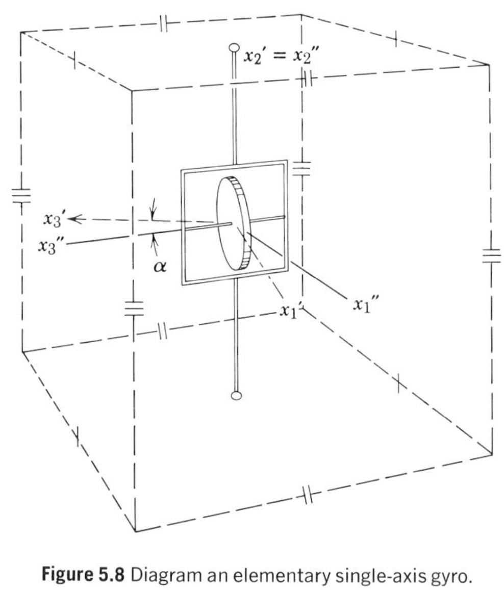

Figure 5.8 Diagram an elementary single-axis gyro.

The gyro inputs are torques about the x_2' axis exerted on the gimbal. These torques have two sources. One source is the movement of the case, due to the movement of the vehicle relative to the inertial reference frame. The second source of torques on the gimbal is due to springs or dampers placed between the gimbal and the case in many applications. The first torque source, due to case motion, will be expressed in terms of the variables characterizing the motion of the case relative to inertial axes, that is, in terms of the omegas and Euler angles (or their rates) of the case. The spring and damper torques will be represented by τ_2'. Notice that τ_2' may be selected by the designer to produce desirable gyro response characteristics.

Thus, for our purposes here, the gyro may be represented by the diagram of Figure 5.9.

We begin the modeling by applying Euler's dynamical equations (5.35) to the *gimbal-rotor assembly*.

$$\boldsymbol{\tau}'' = \frac{d\mathbf{L}''}{dt} + \mathbf{PL}'' \tag{5.35}$$

where $\boldsymbol{\tau}''$ is a column matrix whose components are the component torques applied to the gimbal assembly, *with respect to the gimbal coordinate axes*, $\mathbf{L}''$ is the angular momentum column matrix for the gimbal assembly in the gimbal coordinate axes, and $d\mathbf{L}''/dt$ is the time rate of change of $\mathbf{L}''$ with respect to gimbal coordinate axes. (We reserve single primes for case coordinates.) The matrix $\mathbf{P}$ is, from (5.27),

$$\mathbf{P} = \begin{bmatrix} 0 & -\omega_3'' & \omega_2'' \\ \omega_3'' & 0 & -\omega_1'' \\ -\omega_2'' & \omega_1'' & 0 \end{bmatrix} \tag{5.63}$$

where the double-primed omegas are the components of the gimbal angular velocity expressed in the gimbal coordinate system, x_1'', x_2'', x_3''. Notice that in these coordinates the spring-damper torque represents the only external input. The torques produced by movement of the case will be accounted for by relating the components of angular

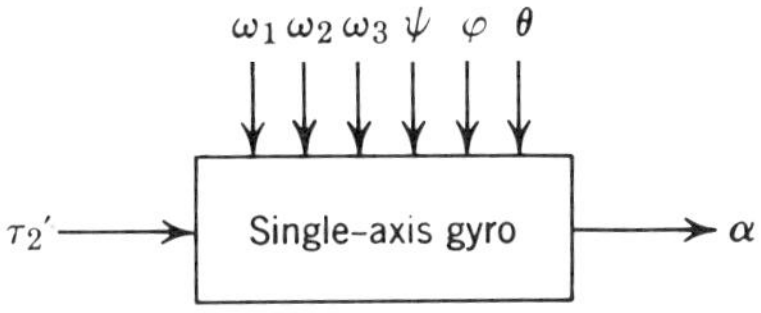

Figure 5.9 Single-axis gyro subsystem.

velocity of the gimbal-rotor assembly, written in the gimbal system (double-primed omegas), to the components of the angular velocity of the case expressed in the case system (primed omegas). An examination of the geometry of Figure 5.8 reveals that the column matrix representation $\mathbf{C}'$ in the case axes of any vector is related to the column matrix representation $\mathbf{C}''$ in the gimbal axes by the transformation

$$\mathbf{C}'' = \begin{bmatrix} \cos\alpha & 0 & \sin\alpha \\ 0 & 1 & 0 \\ -\sin\alpha & 0 & \cos\alpha \end{bmatrix} \mathbf{C}'. \tag{5.64}$$

In particular, if $\boldsymbol{\omega}'_{\text{case}}$ is the angular velocity of the case in case axes coordinates, then $\boldsymbol{\omega}''_{\text{case}}$, the angular velocity of the case in gimbal coordinate axes, is

$$\boldsymbol{\omega}''_{\text{case}} = \begin{bmatrix} \cos\alpha & 0 & \sin\alpha \\ 0 & 1 & 0 \\ -\sin\alpha & 0 & \cos\alpha \end{bmatrix} \begin{bmatrix} \omega_1 \\ \omega_2 \\ \omega_3 \end{bmatrix} \tag{5.65}$$

where $\boldsymbol{\omega}'_{\text{case}} = [\omega_1\ \omega_2\ \omega_3]^T$. Also, the angular velocity of the gimbal relative to the case axes has a matrix representation *in the gimbal coordinate axes* of

$$\boldsymbol{\omega}''_{\text{rel}} = \begin{bmatrix} 0 \\ \dot{\alpha} \\ 0 \end{bmatrix}. \tag{5.66}$$

The sum of (5.65) and (5.66) yields the gimbal angular velocity in gimbal coordinates:

$$\begin{bmatrix} \omega''_1 \\ \omega''_2 \\ \omega''_3 \end{bmatrix} = \boldsymbol{\omega}''_{\text{gimbal}} = \begin{bmatrix} \cos\alpha & 0 & \sin\alpha \\ 0 & 1 & 0 \\ -\sin\alpha & 0 & \cos\alpha \end{bmatrix} \begin{bmatrix} \omega_1 \\ \omega_2 \\ \omega_3 \end{bmatrix} + \begin{bmatrix} 0 \\ \dot{\alpha} \\ 0 \end{bmatrix}. \tag{5.67}$$

The angular momentum L'' of the gimbal-rotor assembly (in gimbal coordinates) is

$$\mathbf{L}'' = \begin{bmatrix} 0 \\ 0 \\ L''_r \end{bmatrix} + \begin{bmatrix} J_1 & 0 & 0 \\ 0 & J_2 & 0 \\ 0 & 0 & J_3 \end{bmatrix} \begin{bmatrix} \omega''_1 \\ \omega''_2 \\ \omega''_3 \end{bmatrix}, \tag{5.68}$$

where L''_r is the product of the spin axis rotor moment of inertia and the rotor speed relative to the gimbal, that is,

$$L''_r = J_3^r\ \Omega = \text{constant}. \tag{5.69}$$

The J's are the moments of inertia of the gimbal-rotor assembly about the principal axes x_1'', x_2'', x_3''. (Notice that if the mass of the gimbal is neglected, $J_1 = J_2$.)

Equations 5.36, 5.63, 5.67, and 5.68 can now be combined to yield the desired equations of motion. Usually only the x_2' component of (5.36) is of interest, since this is the axis for input torques producing an output deflection α. The x_2' component of (5.36) is

$$\tau_2' = J_2\dot{\omega}_2'' - (J_3 - J_1)\omega_3''\omega_1'' - L_r''\omega_1'', \tag{5.70}$$

where $\mathbf{L}''$ of (5.68) and $\mathbf{P}$ of (5.63) have been used. Equation 5.67 is used to replace the double-primed omegas in (5.70):

$$\begin{aligned}\tau' &= J_2(\dot{\omega}_2 + \ddot{\alpha}) - L_r''(\omega_1 \cos\alpha + \omega_3 \sin\alpha)\\ &\quad - (J_3 - J_1)(\omega_1 \cos\alpha + \omega_3 \sin\alpha)(-\omega_1 \sin\alpha + \omega_3 \cos\alpha).\end{aligned} \tag{5.71}$$

The omegas are related to the spacecraft Euler angles via (5.54):

$$\begin{aligned}\omega_1 &= \sin\theta \sin\varphi\, \dot{\psi} + \cos\varphi\, \dot{\theta}\\ \omega_2 &= \sin\theta \cos\varphi\, \dot{\psi} - \sin\varphi\, \dot{\theta}\\ \omega_3 &= \cos\theta\, \dot{\psi} + \dot{\varphi}.\end{aligned} \tag{5.72}$$

Finally, we suppose that the torque τ_2' arises because of a linear spring and linear damper mounted between the case and gimbal, producing a restoring toruqe given by

$$\tau_2' = -D\dot{\alpha} - K\alpha. \tag{5.73}$$

Equations 5.71, 5.72, and 5.73 constitute the desired model for the single-axis gyro.

It is illuminating to consider a linearized version of these equations. In many precision attitude-control problems, the spacecraft attitude variations will be quite small (milliradians). Consider the operating point

$$\begin{aligned}\psi^0 &= 0 \text{ radians}\\ \varphi^0 &= 0 \text{ radians}\\ \theta^0 &= \pi/2 \text{ radians}.\end{aligned} \tag{5.74}$$

[θ^0 is chosen as $\pi/2$ rather than 0 to avoid a singularity at $\theta = \pi/2$ in

(5.57)]. Observe, from (5.72), that a constant Euler angle operating point corresponds to

$$\begin{aligned} \omega_1^0 &= 0 \\ \omega_2^0 &= 0 \\ \omega_3^0 &= 0, \end{aligned} \tag{5.75}$$

as is clear physically. Linearizing (5.72) about these operating conditions, we obtain the small signal equations,

$$\begin{aligned} \omega_1 &= \dot{\theta} \\ \omega_2 &= \dot{\psi} \\ \omega_3 &= \dot{\varphi} \end{aligned} \tag{5.76}$$

where all quantities are now small-signal variations. Examination of (5.71) and (5.73) reveals that

$$\alpha^0 = 0 \text{ radians} \tag{5.77}$$

is the gyro output operating point corresponding to (5.75). Thus, linearization of (5.71) produces

$$-D\dot{\alpha} - K\alpha = J_2(\ddot{\psi} + \ddot{\alpha}) - L_r'\dot{\theta}, \tag{5.78}$$

or, rearranging,

$$J_2\ddot{\alpha} + D\dot{\alpha} + K\alpha = L_r''\dot{\theta} - J_2\ddot{\psi}. \tag{5.79}$$

Furthermore, in typical applications,

$$L_r''\dot{\theta} >> J_2\ddot{\psi},$$

because the rotor speed Ω is large, while the acceleration $\ddot{\psi}$ is generally quite small. This allows further simplification of the model:

$$J_2\ddot{\alpha} + D\dot{\alpha} + K\alpha = L_r''\dot{\theta}. \tag{5.80}$$

If $K \neq 0$, the single-axis gyro is called a *rate gyro*, because the steady-state (constant) value of α is proportional to $\dot{\theta}$. If $K = 0$, then the steady-state value of $\dot{\alpha}$ is proportional to $\dot{\theta}$, that is, α is proportional to θ. The gyro is then called a *rate-integrating* or *position* gyro.

A simulation of (5.80) is shown in Figure 5.10.

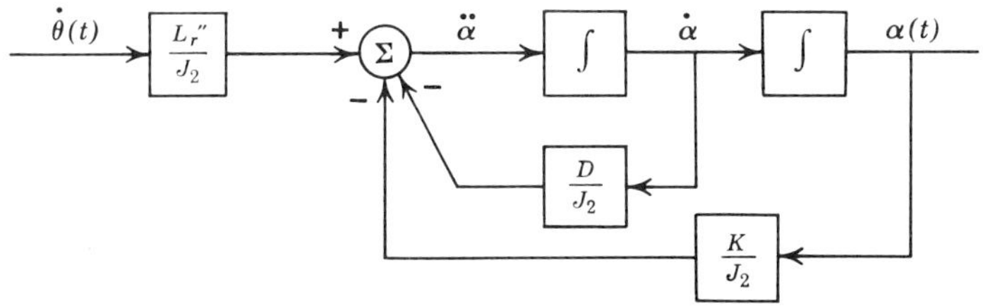

Figure 5.10 Simulation of linear model for single-axis rate gyro.

Example 5-4. Suppose that a spacecraft undertakes a roll with constant roll rate, $\dot{\theta} = \dot{\theta}_0$. The roll rate is measured with a single-axis gyro modeled by (5.80), with

$$\sqrt{\frac{K}{J_2}} = 600 \text{ sec}^{-1}$$

$$\frac{D}{J_2} = 960 \text{ sec}^{-1} \tag{5.81}$$

$$\frac{L_r''}{J_2} = 100 \text{ sec}^{-1}.$$

We wish to find $\alpha(t)$. With gyro parameters as above, (5.80) becomes

$$\ddot{\alpha} + 960\dot{\alpha} + (600)^2\alpha = 100\dot{\theta}_0. \tag{5.82}$$

We solve this equation by classical techniques, computing first the complementary solution and then the particular integral. The characteristic equation is

$$D^2 + 960D + (600)^2 = 0,$$

or

$$D = -480 \pm j360. \tag{5.83}$$

Thus the complementary solution may be written

$$\alpha_c(t) = A\epsilon^{-480t} \sin 360t + B\epsilon^{-480t} \cos 360t. \tag{5.84}$$

The particular integral of (5.82) is, by inspection,

$$\alpha_p = \frac{100}{(600)^2}\dot{\theta}_0 = \frac{1}{3600}\dot{\theta}_0. \tag{5.85}$$

Thus

$$\alpha(t) = \frac{1}{3600}\dot{\theta}_0 + \epsilon^{-480t}(A \sin 360t + B \cos 360t). \tag{5.86}$$

Assuming the gyro to be initially undeflected $[\alpha(0)=0]$, and with zero initial deflection velocity $[\dot{\alpha}(0)=0]$, (5.86) becomes

$$\alpha(t)=\frac{\dot{\theta}_0}{3600}[1-\epsilon^{-480t}(\tfrac{4}{3}\sin 360t+\cos 360t)]. \tag{5.87}$$

Ideally, $\alpha(t)$ would be proportional to $\dot{\theta}_0$ for all t (The factor 1/3600 is the "steady-state gain" of the gyro. It does not affect the form of the solution). We may define an *error* as

$$e(t)\stackrel{\Delta}{=}\dot{\theta}_0-3600\,\alpha(t)=\epsilon^{-480t}(\tfrac{4}{3}\sin 360t+\cos 360t). \tag{5.88}$$

Although the error is not zero and, theoretically, does not become zero for any finite time, it nevertheless becomes very small in comparison with unity [see (5.87)] in a short time. For example,

$$\epsilon^{-4.8}=0.008, \tag{5.89}$$

a number much less than unity. The error is less than this number for all time t greater than t_1, where t_1 is given by

$$480t_1=4.80,$$

$$t_1=0.01\text{ sec.} \tag{5.90}$$

Thus the error in the gyro reading is less than 1% after only 0.01 sec!! This is practically instantaneous in comparison with times involved in a spacecraft maneuver. ▲

We have developed general equations for rigid body motion and have applied them to several examples, including the gyroscope. We are now ready to model the motion of a spacecraft with two different actuators: gas jets and reaction wheels. Single-axis models of these situations were discussed in Chapters 2 and 3. We concentrate here on three-axis models.

5.4 Modeling a Spacecraft with Gas-Jet Actuator

The modeling of a spacecraft may be decomposed into several steps, as was mentioned in the introduction to this chapter.

1. Choice of coordinate axes.
2. Selection of input and output variables.
3. Development of equations of motion relating input and output variables.

In this section, the three-step procedure outlined above is followed to model a spacecraft with gas jets as actuators. In the following section, the three-step procedure is followed to model a spacecraft with reaction wheel actuation.

The choice of coordinate axes depends on the details of the space mission being considered. For example, axes appropriate for earth-centered satellite missions are probably not suitable for convenient description of interplanetary spacecraft missions. Generally speaking, we seek axes leading to the greatest analytical simplicity (avoidance of time-varying moments of inertia, for example), as well as ease of interpretation and physical significance of the variables involved. Clearly, the ability to select "optimum" axes can be something of an art, a skill developed through experience with such problems.

Generally, two sets of axes will be used in our work here. A set of reference axes will be chosen, with respect to which the spacecraft orientation (attitude) is to be controlled. Usually, for simplicity, we pick the reference axes to be inertial axes.[3] A second set of axes—body axes—will be used to express conveniently the motion of the spacecraft. Usually the origin of the body axes will be at the spacecraft center of mass so that the rotational motion can be separated from translational motion. Moreover, principal axes will be used for body axes. Finally, Euler angles will be used to express attitude, that is, to express the vehicle orientation with respect to the inertial reference axes.

In summary, then, we choose inertial reference axes, and we choose as body axes a set of spacecraft principal axes.

The system output variables consists of the Euler angles (attitudes) ψ, φ, θ, and the "body rates" ω_1, ω_2, and ω_3. The inputs consists of torques applied about the body axes, x_1', x_2', and x_3'. There are two kinds of torque inputs: *control torques* and *disturbance torques*. Control torques are at the designer's disposal and may be chosen to obtain desired system responses. For gas jets, the torques are "off-on" or "relay" type, as discussed in Chapter 2. Disturbance torques cannot be controlled by the designer. They occur due to the nature of the environment of the spacecraft. The major sources of disturbance torques include solar radiation pressure, magnetic field effects, meteor impact, gravity gradient effects, aerodynamic effects (during spacecraft atmospheric entry), and internal mass shifts. Thus we write

$$\boldsymbol{\tau}' = \boldsymbol{\tau}_c' + \boldsymbol{\tau}_d', \tag{5.91}$$

[3]In many missions, such as earth-orbit missions, the reference axes are not inertial. However, they may be considered to be nearly so as long as the angular velocity of the reference axes is small compared with the significant vehicle angular velocities.

where $\boldsymbol{\tau}_c'$ and $\boldsymbol{\tau}_d'$ are column matrices of control torques and disturbance torques, respectively.

The spacecraft motion is described by Euler's dynamic equations, (5.36), if we assume it to be a rigid body, and by the Euler angle relation of (5.56). Thus

$$\begin{aligned}
J_1\dot{\omega}_1 &= (J_2-J_3)\omega_2\omega_3+\tau_{c1}'+\tau_{d1}' \\
J_2\dot{\omega}_2 &= (J_3-J_1)\omega_3\omega_1+\tau_{c2}'+\tau_{d2}' \\
J_3\dot{\omega}_3 &= (J_1-J_2)\omega_1\omega_2+\tau_{c3}'+\tau_{d2}' \\
\dot{\psi} &= \frac{\sin\varphi}{\sin\theta}\omega_1+\frac{\cos\varphi}{\sin\theta}\omega_2 \\
\dot{\varphi} &= -\sin\varphi\cot\theta\,\omega_1-\cos\varphi\cot\theta\,\omega_2+\omega_3 \\
\dot{\theta} &= \cos\varphi\,\omega_1-\sin\varphi\,\omega_2.
\end{aligned} \tag{5.92}$$

These six simultaneous first-order equations, together with the observation that the control torques have only three permitted values, $+T$, $-T$, or 0, constitutes the desired model of a spacecraft with gas jet control.

Example 5-5. Linearization. If $\boldsymbol{\tau}_c'$ and $\boldsymbol{\tau}_d'$ are transient torques, that is, if

$$\lim_{t\to\infty}\boldsymbol{\tau}_c' = \mathbf{0}, \tag{5.93}$$

and if

$$\lim_{t\to\infty}\boldsymbol{\tau}_d' = \mathbf{0}, \tag{5.94}$$

then an operating point for (5.92) is given by

$$\begin{aligned}
\omega_1^0 = \omega_2^0 = \omega_3^0 &= 0 \\
\psi^0 = \varphi^0 = 0,\ \theta^0 &= \frac{\pi}{2}.
\end{aligned} \tag{5.95}$$

Linearization of (5.92) about the operating point of (5.95) yields

$$\begin{aligned}
J_1\dot{\omega}_1 &= \tau_{c1}'+\tau_{d1}' \\
J_2\dot{\omega}_2 &= \tau_{c2}'+\tau_{d2}' \\
J_3\dot{\omega}_3 &= \tau_{c3}'+\tau_{d3}' \\
\dot{\psi} &= \omega_2 \\
\dot{\varphi} &= \omega_3 \\
\dot{\theta} &= \omega_1.
\end{aligned} \tag{5.96}$$

This represents three *decoupled* single-axis gas-jet systems of the type discussed in Chapter 2, as is made explicit when the Euler angle rates are substituted in the rigid body equations:

$$
\begin{aligned}
J_1\ddot{\theta} &= \tau_1' \\
J_2\ddot{\psi} &= \tau_2' \\
J_3\ddot{\varphi} &= \tau_3'.
\end{aligned}
\tag{5.97}
$$

In typical applications, the control torques of (5.96) are switched between $+T$, $-T$, and 0, according to the values of the output variables. This constitutes *feedback control* of the gas jet-spacecraft system. Analysis and design of such gas jet-feedback systems will be discussed in later chapters. ▲

5.5 Modeling a Spacecraft with Reaction-Wheel Actuators

A single-axis model of a spacecraft with a reaction-wheel actuator was discussed in Chapter 3. As was discussed there, the reaction wheel exerts on the spacecraft a control torque that is equal in magnitude and opposite in direction to the torque produced by the motor on the wheel (motor rotor). In this chapter, a three-axis spacecraft motor is modeled. We shall suppose that there are three identical reaction wheels located at the vehicle center of mass, each having a spin axis directed along one of the vehicle principal axes. The wheels are driven by three identical d-c motors. This is the simplest three-axis case and is a natural generalization of the single-axis configurations discussed previously.

As was done for gas jet modeling, we choose inertial axes as reference axes, and spacecraft principal axes as body axes. Moreover, the Euler angles ψ, φ, θ and the body rates ω_1, ω_2, ω_3 are again chosen as outputs. There is a difference in selection of inputs, however. We shall apply Euler's dynamical equations to the *combined* spacecraft-reaction wheel system. Thus the torques produced by the wheels will be *internal* torques, leaving disturbance torques as the only external torques:

$$
\boldsymbol{\tau}' = \boldsymbol{\tau}_d'. \tag{5.98}
$$

The equations of motion will be written first with the wheel speeds appearing. These speeds are connected to the motor input voltages, as was discussed in Chapter 3. These voltages, then, represent the inputs to the spacecraft-reaction wheel system. The functional description is depicted in Figure 5.11.

Equations of motion are developed by applying (5.28),

$$\boldsymbol{\tau}' = \frac{d\mathbf{L}'}{dt} + \mathbf{PL}', \tag{5.99}$$

to the system composed of the spacecraft together with the three reaction wheels.

The total angular momentum $\mathbf{L}'$ can be thought of as composed of two components. Consider, for example, L_1', the first row in $\mathbf{L}'$, or the component of $\vec{\mathbf{L}}'$ along the x_1' axis. This momentum can be separated into two components, the momentum of the spacecraft together with wheels 2 and 3 whose spin axes are along x_2' and x_3', plus the momentum of the spinning wheel 1. Since spacecraft principal axes were chosen as body axes, the first of these components is just

$$L_{v1}' = J_{v1}\omega_1, \tag{5.100}$$

where J_{v1} is the moment of inertia about axis x_1' of the *spacecraft together with wheels 2 and 3*. The second momentum component about x_1' is

$$L_{w1}' = J_w(\omega_1 + \Omega_1), \tag{5.101}$$

where J_w is the wheel spin axis moment of inertia, and Ω is the wheel speed *relative to the spacecraft*. Similar arguments may be made for the other momenta, L_2' and L_3'. Thus

$$\mathbf{L}' = \mathbf{L}_v' + \mathbf{L}_w', \tag{5.102}$$

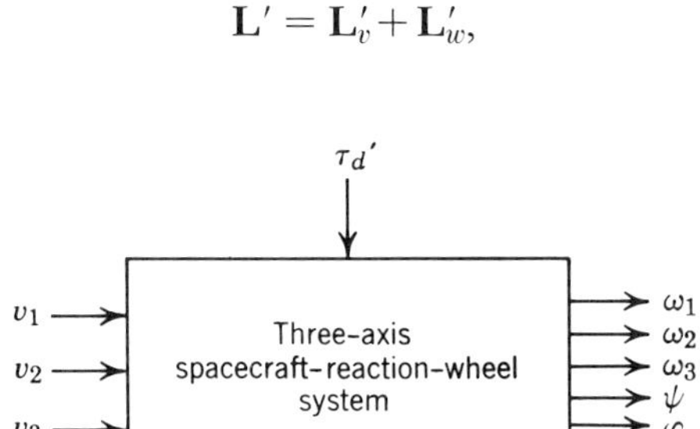

Figure 5.11 Functional representation of reaction-wheel control of spacecraft.

where

$$\mathbf{L}_v' = \begin{bmatrix} J_{v1} & 0 & 0 \\ 0 & J_{v2} & 0 \\ 0 & 0 & J_{v3} \end{bmatrix} \begin{bmatrix} \omega_1 \\ \omega_2 \\ \omega_3 \end{bmatrix}, \tag{5.103}$$

and where

$$\mathbf{L}_w' = J_w \begin{bmatrix} \omega_1 + \Omega_1 \\ \omega_2 + \Omega_2 \\ \omega_3 + \Omega_3 \end{bmatrix}. \tag{5.104}$$

In the above expressions, J_{vi} is the moment of inertia about axis x_i' of the spacecraft reaction-wheel system with wheel i removed, and J_w is the spin axis moment of inertia of the (identical) wheels.

Substitution of (5.98) and (5.102) into (5.99), with $\mathbf{L}_v'$ and $\mathbf{L}_w'$ as in (5.103) and (5.104), yields

$$\begin{aligned} J_{v1}\dot{\omega}_1 &= (J_{v2} - J_{v3})\omega_2\omega_3 + J_w(\omega_3\Omega_2 - \omega_2\Omega_3 - \dot{\omega}_1 - \dot{\Omega}_1) + \tau_{d1}' \\ J_{v2}\dot{\omega}_2 &= (J_{v3} - J_{v1})\omega_3\omega_2 + J_w(\omega_1\Omega_3 - \omega_3\Omega_1 - \dot{\omega}_2 - \dot{\Omega}_2) + \tau_{d2}' \\ J_{v3}\dot{\omega}_3 &= (J_{v1} - J_{v2})\omega_1\omega_2 + J_w(\omega_2\Omega_1 - \omega_1\Omega_2 - \dot{\omega}_3 - \dot{\Omega}_3) + \tau_{d3}'. \end{aligned} \tag{5.105}$$

The motor speed is related to the motor input voltage and the body rates by (3.102). Substitution of $\dot{\omega}_1$ from the first equation of (5.105) into (3.102) yields the following motor speed equation.

$$\begin{aligned} \dot{\Omega}_1 = &- \left[\frac{1}{J_{v1}} + \frac{1}{J_w}\right]\left[D + \frac{K^2 i_f^2}{R_a}\right]\Omega_1 - \frac{J_{v2} - J_{v3}}{J_{v1}}\omega_2\omega_3 - \frac{J_w}{J_{v1}}(\omega_3\Omega_2 - \omega_2\Omega_3) \\ &+ \frac{K i_f}{R_a}\left[\frac{1}{J_{v1}} + \frac{1}{J_w}\right]v_a - \frac{1}{J_{v1}}\tau_d. \end{aligned} \tag{5.106}$$

There is one such equation for each of the three wheels. Finally, Euler angles are used to describe the vehicle attitude. The Euler angles and the omegas are related by (5.56). Thus the three-dimensional motion of a spacecraft with reaction wheel actuators is modeled by (5.105), three equations of the form of (5.106), and (5.56). The model is a ninth-order coupled set of nonlinear differential equations. Clearly, analytic study of a fully general problem cannot be done. Usually simplifying assumptions are made for preliminary analysis and design. Computer simulation is then used to evaluate the effects of originally neglected effects.

These procedures are discussed in detail in the following chapters.

Problems

5-1. Consider the circular cylindrical satellite of Example 5.2. Suppose that, instead of zero external torques, the torque is

$$\boldsymbol{\tau}' = \begin{bmatrix} T \\ 0 \\ 0 \end{bmatrix}$$

where T is a constant. Find $\omega_1(t)$, $\omega_2(t)$, and $\omega_3(t)$.

5-2. Consider the axially symmetric spacecraft rotating about its center of mass, as shown in Figure P5-2. The moments of inertia are $J_1 = J_2 =$

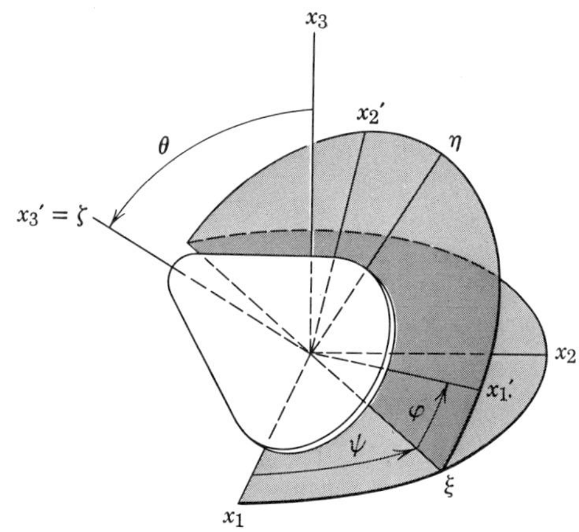

Figure P5-2

J, and J_3. Suppose it is desired that the spacecraft spin at a constant rate ($\dot{\varphi}$ = constant) with constant values of θ and ψ. Find the necessary applied torque $\boldsymbol{\tau}'$. Discuss the special case $\theta = \pi/2$.

5-3. Methods other than Euler angles are sometimes used to specify the orientation of one set of axes with respect to another. In this problem we consider *direction cosines*. The components x_1', x_2', x_3' of a vector $\vec{x}$ in body axes may be related to the components x_1, x_2, x_3 of $\mathbf{x}$ in inertial axes by ordinary vector resolution:

$$x_1' = a_{11}x_1 + a_{12}x_2 + a_{13}x_3$$
$$x_2' = a_{21}x_1 + a_{22}x_2 + a_{23}x_3$$
$$x_3' = a_{31}x_1 + a_{32}x_2 + a_{33}x_3,$$

where the direction cosines are

$$a_{ij} = \cos\theta_{ij},$$

θ_{ij} being the plane angle between x_i' and x_j. Observe that there are nine direction cosines. Show that

$$\sum_{j=1}^{3} a_{ij}a_{kj} = \delta_{ik}, \qquad i, k = 1, 2, 3,$$

where $\delta_{ik} = 1$ for $i = k$, and zero if $i \neq k$ (use orthogonality of unit vectors).

5-4. For the direction cosines of Problem 5-3, show that

$$\sum_{i=1}^{3} a_{ij}a_{ik} = \delta_{jk} \qquad (k, j = 1, 2, 3).$$

5-5. Find expressions relating the time derivatives of the direction cosines to the ω's of Euler's dynamical equations. The first of these is

$$\dot{a}_{11} = \omega_3 a_{21} - \omega_2 a_{31}.$$

(*Hint.* For any vector $\vec{\mathbf{x}}$,

$$\left.\frac{d\vec{\mathbf{x}}}{dt}\right|_I = \left.\frac{d\vec{\mathbf{x}}}{dt}\right|_B + (\vec{\boldsymbol{\omega}} \times \vec{\mathbf{x}}).$$

5-6. Interpret the results of Example 5-3 geometrically by a sketch showing $\vec{\mathbf{L}}$ and $\vec{\boldsymbol{\omega}}$, as well as the Euler angles. What happens to these vectors and angles as the rigid body motion occurs?

5-7. Consider a rate gyro with parameters as in Example 5-4. Suppose

$$\dot{\theta}(t) = \dot{\theta}_0(1 - \epsilon^{-t/T})u(t).$$

Find $\alpha(t)$ and the error

$$e(t) = \dot{\theta}_0 - 3600\,\alpha(t).$$

Discuss how well the rate gyro follows the input rate $\dot{\theta}(t)$.

5-8. Suppose the spring is omitted ($K = 0$) in the gyro of (5.80), making it a *rate-integrating* gyro. Suppose

$$\frac{D}{L_r''} = 2, \qquad \frac{J_2}{D} = 0.1 \text{ sec}.$$

Find and sketch $\alpha(t)$ when $\dot{\theta}(t) = u(t)$, the unit step. Does this gyro deserve the name *rate-integrating*?

5-9. Draw a simulation diagram for the rate gyro of Example 5.4 from which the sensitivity function $\partial\alpha/\partial D$ can be determined. Sketch $\partial\alpha/\partial D$ vs t for the parameter values and input given in Example 5-4.

5-10. Repeat Problem 5-9 for the gyro parameters and input of Problem 5-8.

5-11. Show that the three-axis reaction-wheel attitude-control system equations of Section 5.5 reduce to the single-axis equations of Chapter 3 by linearization about an appropriate operating point.

PART II

System Analysis

Chapter 6. System Response via Transfer Functions
Chapter 7. System Response via State Variables
Chapter 8. Stability of Linear Time-Invariant Systems
Chapter 9. Response of Nonlinear Systems
Chapter 10. Analysis of Discrete-Time Systems

The purpose of system analysis is to predict system behavior from the system model, system input, and system initial conditions. Dynamic systems that are described by ordinary linear differential equations with constant coefficients are treated first. In Chapter 6, such systems are analyzed using transfer functions that relate the Laplace transforms of inputs to Laplace transforms of outputs. Linear systems are also analyzed directly in the time domain in Chapter 7, using the state formulation whereby the system is described by means of simultaneous first-order differential equations. The connections between the transfer function representation and the state representation of systems is pointed out.

One of the essential ingredients of system analysis is stability analysis. Again linear time-invariant systems are treated first. Various methods for checking stability are described in Chapter 8. The response and the stability of nonlinear systems are discussed in Chapter 9.

The dynamic systems that have been treated are characterized by differential equations that describe the evolution of system variables continuously in time. The modeling and analysis concepts that have been developed thus far apply equally well to dynamic systems that are described by difference equations. Difference equations describe the evolution of system variables in discrete time. Digital computers are examples of discrete-time systems. When a system is composed of discrete-time subsystems and continuous-time subsystems, it is called a sampled-data system. In Chapter 10, the analysis of discrete-time systems is presented.

CHAPTER 6

System Response via Transfer Functions

In the modeling of lumped continuous-time dynamic systems, ordinary differential equations are used to describe the time histories of the various physical variables of the model. Such equations are nonlinear and time-varying, in general. If the signal inputs to the system are small, a less accurate but simpler model may be used in the preliminary stages of the analysis or design. A useful simplification results when the differential equations are linearized at the equilibrium point or points, as discussed in Chapter 3. If the elements of the model are time-invariant, the linear differential equations have constant coefficients. System models that are described by ordinary linear differential equations with constant coefficients are much easier to analyze and synthesize than those described by nonlinear equations. As a consequence, linearization is employed to obtain a first approximation to the behavior of the nonlinear system.

There are two general methods of analysis and synthesis of linear time-invariant models. One is based on the Laplace transform[1] and the transfer function. This method is referred to as the frequency domain method, and it is the subject of the present chapter. The other method is called the time-domain method. The state space concepts that are discussed in the next chapter are associated with the time-domain method.

6.1 The Transfer Function Concept

Suppose that the output and input of a system are related by a linear ordinary differential equation with constant coefficients. Then, if the Laplace transform applied to the equation, the transform of the output

[1]Readers unfamiliar with the Laplace transform should consult Appendix B where simple properties of the transform needed for this chapter are developed.

is linearly related to the transform of the input and linearly related to initial conditions also.

Let $X(s)$ be the transform of the input to a linear time-invariant system and let $Y(s)$ be the transform of the output of the system. Then $Y(s)$ is

$$Y(s) = H(s)X(s) + (\text{initial condition terms}). \tag{6.1}$$

The initial condition terms are absent if the initial conditions are zero. The coefficient $H(s)$ is called the *transfer function* of the system. Thus the output transform is equal to the product of the transfer function and the input transform plus initial condition terms if initial conditions are not zero.

Example 6-1. For single axis attitude motion of a spacecraft the equations of motion are

$$J\dot{\omega} = \tau \tag{6.2}$$

$$\dot{\theta} = \omega \tag{6.3}$$

where τ is the input torque, ω is the angular rate of motion, θ is the output attitude position, and J is the moment of inertia of the vehicle. The angular velocity ω may be eliminated from (6.2) and (6.3) yielding

$$J\ddot{\theta} = \tau \tag{6.4}$$

which is a second-order ordinary linear differential equation with constant coefficients. The Laplace transform of (6.4) is

$$J[s^2\theta(s) - s\theta(0-) - \dot{\theta}(0-)] = T(s) \tag{6.5}$$

or

$$\theta(s) = \frac{1}{Js^2}T(s) + \frac{s\theta(0-) + \dot{\theta}(0-)}{s^2}. \tag{6.6}$$

Thus the transfer function relating $T(s)$ to $\theta(s)$ for initial conditions equal to zero is

$$H(s) = \frac{1}{Js^2}. \quad \blacktriangle \tag{6.7}$$

If the linear time-invariant system consists of lumped components, $H(s)$ is rational in s. Suppose that a linear time-invariant system is represented by the differential equation

$$\frac{d^ny}{dt^n} + a_{n-1}\frac{d^{n-1}y}{dt^{n-1}} + \cdots + a_1\frac{dy}{dt} + a_0y = b_m\frac{d^mx}{dt^m} + b_{m-1}\frac{d^{m-1}x}{dt^{m-1}} + \cdots + b_0x \tag{6.8}$$

where $y(t)$ is the output and $x(t)$ is the input. Then, if we take the Laplace transform of (6.8), we obtain

$$(s^n + a_{n-1}s^{n-1} + \cdots + a_0)Y(s) = (b_m s^m + b_{m-1}s^{m-1} + \cdots + b_0)X(s) + \text{initial condition terms.} \quad (6.9)$$

Even if we start with a system of integrodifferential equations describing the system, take the transform, and then apply Cramer's rule, the resulting equation is the same as (6.9). This means that the transfer function is

$$H(s) = \frac{b_m s^m + b_{m-1}s^{m-1} + \cdots + b_0}{s^n + a_{n-1}s^{n-1} + \cdots + a_0}. \quad (6.10)$$

Conversely, given that the transfer function for a linear time-invariant system is as in (6.10), then the input and output transforms must be related as in (6.9), and the time domain input and output signals must be related by the differential equation in (6.8).

The numerator and denominator polynomials of the rational transfer function may be written in factored form:

$$H(s) = K\frac{(s - z_1)(s - z_2)\cdots(s - z_m)}{(s - p_1)(s - p_2)\cdots(s - p_n)} \quad (6.11)$$

where the z_i are called the zeros of $H(s)$ and the p_i are called the poles of $H(s)$. The constant K in (6.11) is equal to b_m in (6.10). The zero locations are indicated by O and the poles by X on the complex s plane. The rational function $H(s)$ is completely determined by specifying the value of K and the values of all the poles and zeros.

For electrical systems, if the output variable is voltage and the input variable is voltage also, the transfer function is called voltage transfer function. If input and output are both current, the transfer function is called current transfer function. If output is voltage and input is current, the transfer function is called an *impedance* function. Finally, if output is current and input is voltage, the transfer function is called an *admittance* function. Similarly, for mechanical systems, if the input and output variables are both force, the transfer function is called a force transfer function. If both variables are torque, velocity, or angular velocity, the transfer function is called torque, velocity, and angular velocity transfer function, respectively. *Mechanical impedance* refers to the transfer function when the output is force (or torque) and the input is velocity (or angular velocity). If output is velocity (or angular velocity) and input is force (or torque), the transfer function is called *mechanical mobility*.

The transfer-function representation places emphasis on the relationship between the input transform and the output transform. In fact, the output transform $Y(s)$ is equal to the transfer function $H(s)$ times the input transform $X(s)$ when all initial conditions are zero. The effects of initial conditions are studied separately.

Example 6-2 The accelerometer of Example 3-5 in Chapter 3 is described by the input-output differential equation

$$\ddot{y}+\frac{D}{M}\dot{y}+\frac{K}{M}y=\ddot{x} \tag{6.12}$$

where the vehicle acceleration $\ddot{x}$ is the input and the test mass displacement y is the output. The transfer function is

$$H(s)=\frac{1}{s^2+(D/M)s+(K/M)}. \tag{6.13}$$

For

$$\frac{K}{M}-\left(\frac{D}{2M}\right)^2>0, \tag{6.14}$$

the poles are complex and the system is underdamped (see Appendix B). Since the transfer function is second order, we may relate the parameters K, M, and D to the standard parameters ω_n and ζ of Appendix B. In terms of the natural frequency ω_n and damping ratio ζ, the second-degree polynomial is written as

$$s^2+2\zeta\omega_n s+\omega_n{}^2.$$

Comparing with (6.13) we have

$$\omega_n=\sqrt{\frac{K}{M}} \tag{6.15}$$

and

$$2\zeta\omega_n=\frac{D}{M} \tag{6.16}$$

or

$$\zeta=\frac{D}{2\sqrt{KM}}. \tag{6.17}$$

The transfer function has no finite zeros and it has two finite complex conjugate poles. The constant K is unity. ▲

6.2 Transfer Function Block Diagrams

In the functional block diagram of a system, input and output quantities associated with subsystems are identified. However, the relationship between the input and the output of a subsystem is not indicated. In Chapter 4, simulation diagrams were discussed. The input-output relationship for blocks in a simulation diagram are indicated, but the relationships, such as multiplication by a constant or integration, are elementary.

For a transfer function block, the input and output are transforms of signals on the corresponding functional block or simulation block, and the transfer function is indicated inside the block. Summers are also used in transfer function block diagrams, as in simulation diagrams. The definition of a summer is the same as in a simulation diagram except that transforms of signals are involved instead of time domain signals.

One of the advantages of the transfer function representation for linear time-invariant systems is the simplicity of the algebraic relations between the subsystem or component transfer functions and the transfer function of the whole system. For example, suppose we have a cascade of two linear time-invariant systems as represented by the block diagram in Figure 6.1. One system has a transfer function $H_1(s)$. By definition, the transfer function block for $H_1(s)$ relates $X_2(s)$ to $X_1(s)$ by

$$H_1(s) = \frac{X_2(s)}{X_1(s)}. \tag{6.18}$$

The other system has a transfer function $H_2(s)$. By definition, the transfer function block for $H_2(s)$ relates $X_3(s)$ to $X_2(s)$ by

$$H_2(s) = \frac{X_3(s)}{X_1(s)}. \tag{6.19}$$

The output of the block described by (6.18), $X_2(s)$, is the input of the block described by (6.19). Thus the transfer function of the cascaded

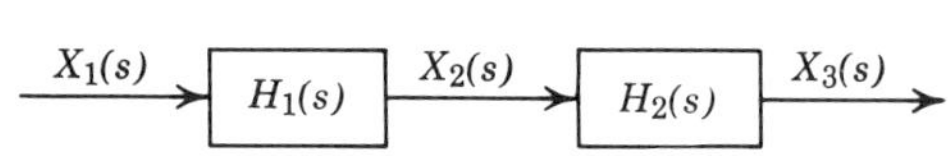

Figure 6.1 Block diagram for two linear time-invariant systems in cascade.

system, which is defined as the ratio of the output $X_3(s)$ and the input $X_1(s)$,

$$H(s) = \frac{X_3(s)}{X_1(s)}, \tag{6.20}$$

is

$$H(s) = \frac{X_3(s)}{X_1(s)} = \frac{X_3(s)}{X_2(s)}\frac{X_2(s)}{X_1(s)} = H_1(s)H_2(s). \tag{6.21}$$

Thus the transfer function of two systems in cascade, as in Figure 6.1, is equal to the product of the individual transfer functions. By combining transfer functions two at a time, it is clear that the transfer function of a cascade of systems is equal to the product of the individual transfer functions in the cascade.

Example 6-3. Suppose that in Figure 6.1, H_1 is described by

$$\frac{dx_2}{dt} + 3x_2(t) = 2\frac{dx_1}{dt} + 5x_1 \tag{6.22}$$

and H_2 described by

$$5\frac{dx_3}{dt} + 4x_3 = 3\frac{dx_2}{dt} + 2x_2. \tag{6.23}$$

Then $H_1(s)$ is

$$H_1(s) = \frac{X_2(s)}{X_1(s)} = \frac{2s+5}{s+3} \tag{6.24}$$

and $H_2(s)$ is

$$H_2(s) = \frac{X_3(s)}{X_2(s)} = \frac{3s+2}{5s+4}. \tag{6.25}$$

The transfer function of the cascaded system is

$$H(s) = \frac{X_3(s)}{X_1(s)} = H_1(s)H_2(s) = \frac{(2s+5)(3s+2)}{(s+3)(5s+4)} = \frac{6s^2+19s+10}{5s^2+19s+12}. \tag{6.26}$$

From (6.26), the input-output differential equation relating $x_3(t)$ and $x_1(t)$ of the cascaded system is easily obtained by inspection. From (6.26)

$$(5s^2+19s+12)X_3(s) = (6s^2+19s+10)X_1(s) \tag{6.27}$$

and

$$5\frac{d^2x_3}{dt^2} + 19\frac{dx_3}{dt} + 12x_3 = 6\frac{d^2x_1}{dt^2} + 19\frac{dx_1}{dt} + 10x_1. \blacktriangle \tag{6.28}$$

Now consider the simple combination of blocks shown in Figure 6.2. The two blocks have a common input $X_0(s)$; block H_1 has $X_1(s)$ as output, and block H_2 has $X_2(s)$ as output. The system output $X_3(s)$ is the sum of $X_1(s)$ and $X_2(s)$. This defines the summer. Since

$$X_1(s) = H_1(s)X_0(s), \tag{6.29}$$

and

$$X_2(s) = H_2(s)X_0(s), \tag{6.30}$$

then

$$X_3(s) = X_1(s) + X_2(s) = [H_1(s) + H_2(s)]X_0(s). \tag{6.31}$$

Thus the transfer function $H(s)$, which relates $X_0(s)$ to $X_3(s)$, is

$$H(s) = H_1(s) + H_2(s). \tag{6.32}$$

For two systems in parallel whose inputs are common and whose outputs add, the transfer function of the combined system is equal to the sum of the transfer functions of the individual systems.

The transfer-function block diagram corresponding to a simulation diagram is easily determined. First, all the indicated signals are replaced by their Laplace transforms. The transfer function for a scalar block whose output is equal to the input times a constant is simply equal to the constant. This follows immediately by taking the Laplace transform of the input-output equation $y(t) = Kx(t)$ for a scalor block, where $y(t)$ is the output and $x(t)$ is the input. The scalor block is replaced by a transfer function block with K indicated inside the block. For an integrator block with an input-output equation

$$y(t) = \int_0^t x(\tau)\, d\tau$$

for zero initial conditions, the Laplace transform of the input-output equation is

$$Y(s) = \frac{1}{s}X(s).$$

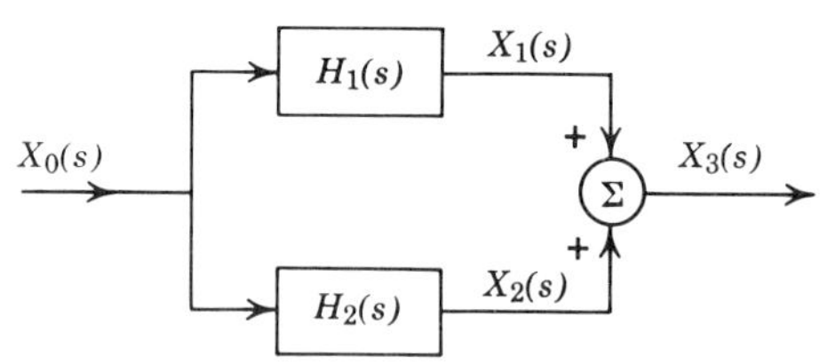

Figure 6.2 Block diagram of systems in parallel.

Thus the transfer function of an integrator block is $1/s$. The integrator block is replaced by a transfer function block with $1/s$ indicated inside the block. The summers are unchanged. The resulting diagram is called a transfer function block diagram. Similarly, a functional block diagram may be connected to a transfer function block diagram by replacing the functional blocks by transfer function blocks and replacing the indicated signals by their Laplace transforms.

When a summer has only one input, and if the input feeds the summer with a plus sign, then the output equals the input, and the transfer function of the summer is unity. If the single input feeds the summer with a minus sign, then the output is equal to the negative of the input, so that in this case the transfer function of the summer is -1. Thus a single input summer has a transfer function equal to $+1$ or -1, depending on whether the input feeds the summer with a plus sign or a minus sign. The single input summer is treated like a transfer function block with a transfer function of $+1$ or -1.

From the cascade and parallel operations described above, it is possible to write the system of equations for any block diagram no matter how complex it is. Consider the simple feedback structure shown in Figure 6.3 where $X_1(s)$ is the input and $X_4(s)$ is the output. The equations for the block diagram are

$$X_2(s) = aX_1(s) + cX_3(s) \tag{6.33}$$

$$X_3(s) = bX_2(s) \tag{6.34}$$

and

$$X_4(s) = dX_3(s). \tag{6.35}$$

Substituting (6.34) in (6.33), we have

$$X_2(s) = aX_1(s) + cbX_2(s)$$

or

$$X_2(s) = \frac{a}{1-cb}X_1(s). \tag{6.36}$$

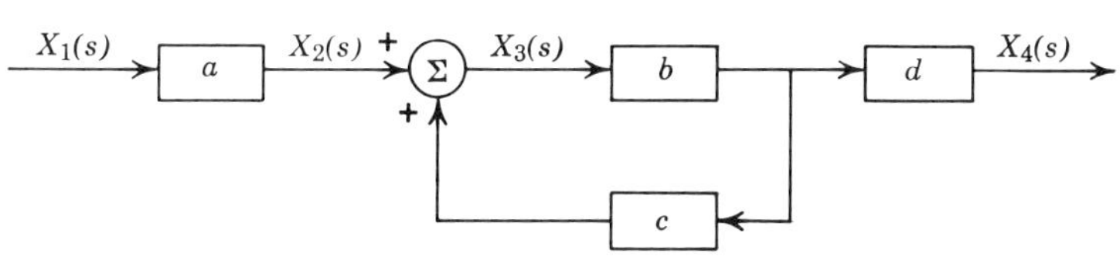

Figure 6.3 Simple feedback structure.

Hence, from (6.34), (6.35), and (6.36),

$$\frac{X_4(s)}{X_1(s)} = \frac{abd}{1-cb}. \tag{6.37}$$

This is seen to be the transfer function for the block diagram in Figure 6.4 where $X_2(s)$ has been eliminated.

Given an arbitrary interconnection of block diagrams where each block is characterized by a transfer function and the signals are combined by means of summers, a system of linear algebraic equations in the transforms of the signals can be written. The system of equations may be solved using any algebraic variable elimination method, such as Cramer's rule. The algebraic elimination of variables is equivalent to the application of the three basic results above on cascade, parallel, and feedback connections.

6.3 Mason's Formula for Transfer Functions

For a specified pair of input and output of a transfer function block diagram, it is possible to write the transfer function expression in one step using *Mason's formula*.[2] In this section, Mason's formula will be stated without proof. The terms in the formula are obtained from the transfer function block diagram *by inspection*. Before stating the formula, we must define the symbols and terminology used. We shall define forward path, forward-path transfer function, loop, loop transfer function, nontouching loops, and determinant of a block diagram. A *forward path* is a connected portion of the block diagram that connects the *input* of the block diagram to the *output* of the block diagram so that no summer has more than one input. For example, for the transfer function block diagram in Figure 6.5, the portion shown in Figure 6.6 is a forward path. The portion shown in Figure 6.7*a* is not a forward path because one summer has two inputs. The portion in Figure 6.7*b* is not a forward path because it is not connected. The protion in Figure 6.7*c* is not a forward path because it does not connect the block diagram input to the block diagram output. In this

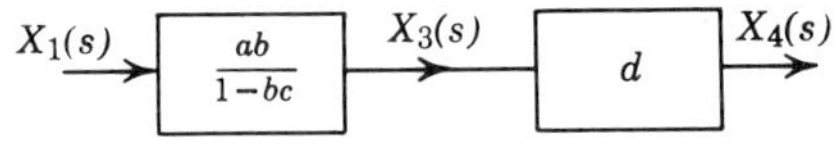

Figure 6.4 Structure with the same transfer function as that for Figure 6.3.

[2]S. J. Mason, "Feedback Theory—Further Properties of Signal Flow Graphs," *Proc. IRE*, **44**, 920–926 (July 1956).

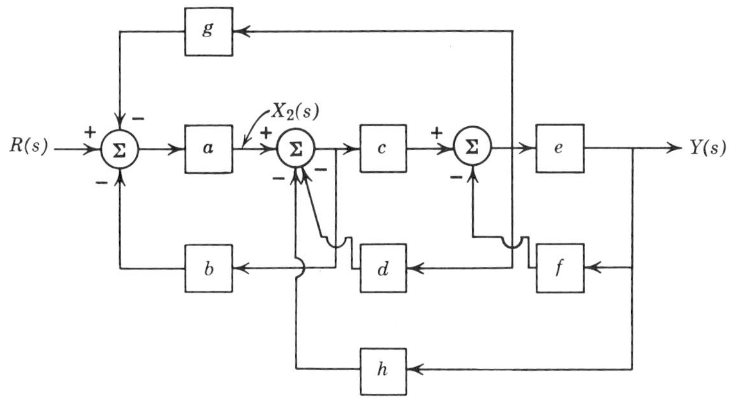

Figure 6.5 Example for explaining Mason's rule.

example, there is only one forward path. In general, there may be more than one forward path. Notice that a forward path can always be traced from input to output along the arrow directions of the blocks and summe s. The transfer function of a forward path is called a *forward-path transfer function.* By a repeated application of the result for cascaded systems the forward-path transfer function for the forward path in Figure 6.6 is *ace*.

A *loop* is defined as a connected portion of a block diagram so that no summer has more than one input, and the output of *every* block or summer of the portion is an input of a block or summer of the portion. Figure 6.8 shows some of the loops associated with the block diagram in Figure 6.5. Starting from any signal on the loop, a loop can always be traced along the arrow directions of the blocks and summers of the loop and eventually reach the same signal. Figure 6.9 shows some portions that are not loops. The portion in Figure 6.9*a* is not a loop because one summer has two inputs. The portion in Figure 6.9*b* is not a loop because the output of Figure 6.9*c* is not an input of a block or a summer of the portion. The portion in Figure 6.9*c* is not a loop because one summer has two inputs. The *loop transfer function* is defined as the product of the transfer functions of the blocks and summers of the loop. Recall that a summer with a single input has a transfer function that is equal to +1 or −1, depending on the sign at the input of the summer. For the loop in Figure 6.8*a*, the loop transfer function is $-ab$; for Figure 6.8*b*, it is $-ef$; for Figure 6.8*c* it is $-ceh$; and for Figure 6.8*d*, it is $-cd$.

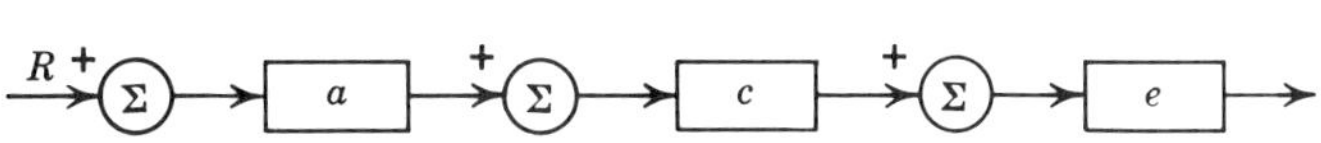

Figure 6.6 A connected portion of the block diagram in Figure 6.5 that is a forward path.

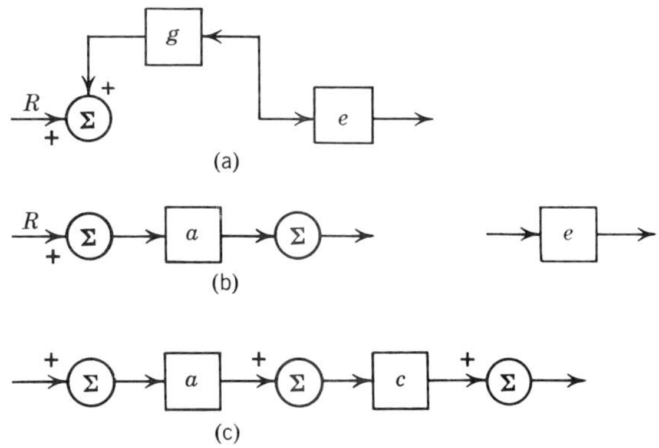

Figure 6.7 Portions that are not forward paths of the block diagram in Figure 6.5.

Two loops are said to be a *nontouching loop pair* if there is no common summer present in both loops. For example, the loops in Figure 6.8*a* and Figure 6.8*b* constitute a nontouching loop pair. However, the loops in Figure 6.8*a* and Figure 6.8*d* are touching. The loops in Figure 6.8*c* and Figure 6.8*d* are touching because they have a summer in common. Notice that if two loops have a block in common, they must also have a summer in common. A set of N loops is said to be a *nontouching loop N-tuple* if no summer in any loop is in any of the other N-1 loops.

We shall now define the terms directly involved in the determinants. Denote the loop transfer functions by l_1, l_2, ... Denote the sum of *all* the loop transfer functions by Σl_i. If the ith and jth loops constitute a nontouching loop pair, form the product $l_i l_j$. Form similar products for all other nontouching loop pairs. Denote the sum of all

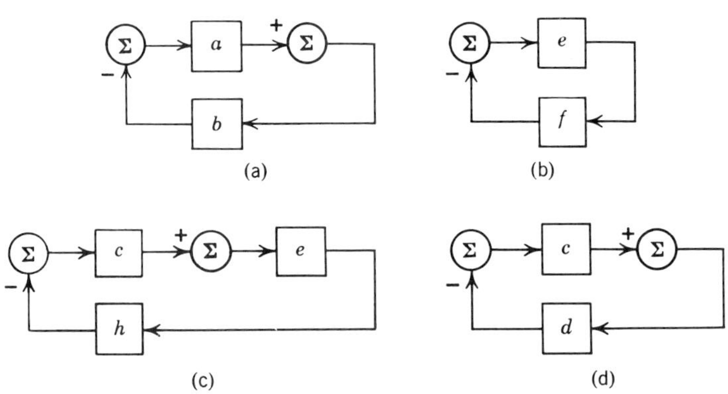

Figure 6.8 Some loops for the block diagram of Figure 6.5.

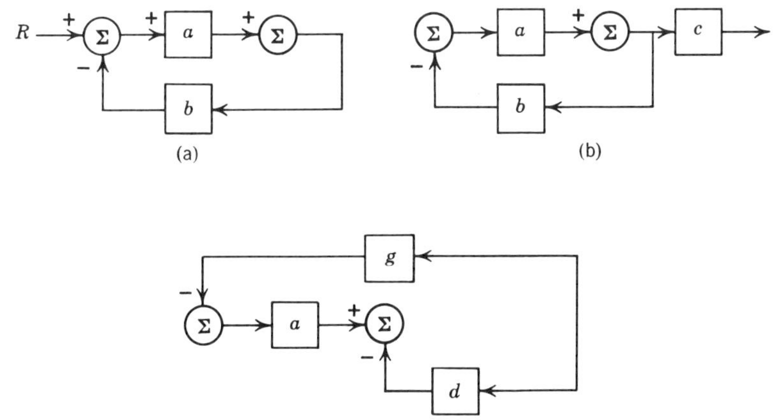

Figure 6.9 Portions that are not loops.

such products by $\Sigma l_i l_j$. Similarly, if the ith, jth, and kth loops constitute a nontouching loop triple, form $l_i l_j l_k$. Form similar triple products for all other nontouching loop triples. Denote the sum of all such triple products by $\Sigma l_i l_j l_k$. Continue the process of forming k-tuple products up to N-tuples, so that there exist no nontouching loop N-tuples and there exist at least one nontouching loop $(N-1)$-tuples. The determinant Δ of a transfer-function block diagram is defined as

$$\Delta = 1 - \sum l_i + \sum l_i l_j - \sum l_i l_j l_k + \cdots \tag{6.38}$$

for all nonzero sums of k-tuple products of nontouching loop k-tuples. Note the alternation of signs. For the block diagram in Figure 6.5, the determinant is

$$\begin{aligned} \Delta &= 1 - (-ab - cd - ef - ceh - acg) + [(-ab)(-ef)] \\ &= 1 + (ab + cd + ef + ceh + acg) + abef. \end{aligned} \tag{6.39}$$

Denote the forward transfer function of the kth forward path by P_k. Let Δ_k denote the determinant of the transfer function block diagram that remains after deleting the blocks and summers of the kth forward path.

Mason's rule for the transfer function from $R(s)$ to $Y(s)$ is

$$\frac{Y(s)}{R(s)} = H(s) = \frac{\Sigma P_k \Delta_k}{\Delta} \tag{6.40}$$

where the summation is taken over all forward paths.

For example, for Figure 6.5, there is only one forward path. Its transfer function is *ace*. The block diagram that remains after the forward path is deleted consists of the blocks g, b, d, f, and h, which are disconnected from each other. Hence the determinant Δ associated with the forward path is equal to 1. The transfer function from $R(s)$ to $Y(s)$ using Mason's rule is

$$\frac{Y(s)}{R(s)} = \frac{ace}{1+(ab+cd+ef+ceh+acg)+abef}. \tag{6.41}$$

Example 6-5. Suppose that in Figure 6.5, $X_2(s)$ is regarded as the output and it is desired to obtain the transfer function $X_2(s)/R(s)$. With $X_2(s)$ as output and $R(s)$ as input, it is seen from Figure 6.5 that there is only one forward path. The transfer function of this forward path is equal to a. The determinant of the block diagram is the same as before. It is given in (6.39). This is because the calculation of Δ has nothing to do with the choice of output in the same block diagram. The block diagram that remains after deleting the forward path is shown in Figure 6.10. The determinant Δ_1 for this block diagram is $\Delta_1 = 1 - (-ef)$. Hence, by Mason's rule,

$$\frac{X_2(s)}{R(s)} = \frac{a(1+ef)}{1+(ab+cd+ef+ceh+acg)+abef}. \quad \blacktriangle \tag{6.42}$$

In the application of Mason's rule one of the key steps is to find *all* the loops. Once the complete set of loops is determined, it is a simple matter to determine the sets of nonthouching loops taken two at a time, three at a time, etc. Another key step is to find *all* the forward paths from the specified input to the specified output. When a forward path is deleted, the remaining block diagram does not have any loop

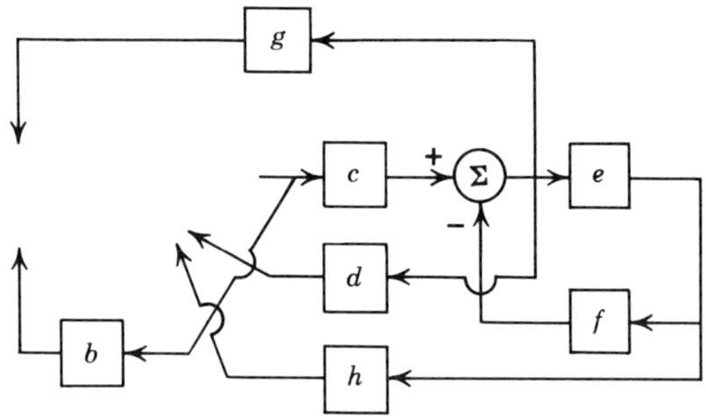

Figure 6.10 Block diagram of Figure 6.5 with X_2 as output after the forward path is deleted.

that is not also a loop of the complete diagram. In general, the reduced block diagram has fewer loops. Similarly, every nontouching loop pair in the reduced block diagram is also a nontouching loop pair in the complete block diagram. Higher order nontouching loop N-tuples of the reduced block diagram are also nontouching loop N-tuples of the complete block diagram. Hence, every term in Δ_k is contained in Δ for all forward paths. Thus Δ_k can be obtained from Δ by replacing by zero every l_i corresponding to a missing loop i.

Example 6-6. Consider the simulation diagram for a single-axis reaction-wheel attitude system in the upper portion of Figure 4.20. The input voltage v_a is the output of an amplifier whose input is comprised of a reference voltage θ_r minus the amplified output of an attitude position sensor minus the amplified output of an attitude rate sensor. From Chapter 5, the step response of a typical gyroscope sensor is very close to the final value in a relatively short time. As an approximation, we shall assume that the gyroscope is modeled by a scalor, whose transfer function is a constant. As discussed earlier, a simulation diagram for a linear time-invariant system is easily converted to a transfer function diagram. Each integrator is replaced by a transfer function $1/s$, and all the signals are replaced by their Laplace transforms. The transfer function for each scalor is equal to the corresponding multiplication constant. The transfer function block diagram is shown in Figure 6.11. The motor and vehicle parameters K_1, R_a, D, J_v, and J_w are usually given constants. The amplifier gain associated with the attitude position gyro is usually chosen so that $K_a = 1$. This

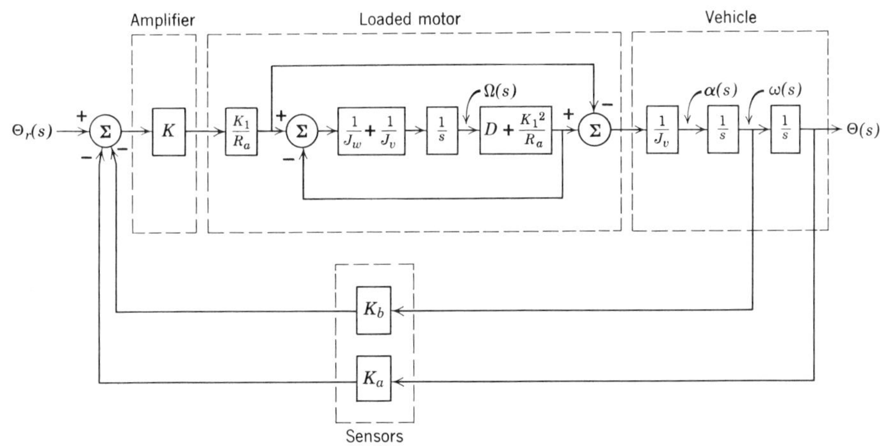

Figure 6.11 Transfer-function block diagram of single-axis reaction-wheel attitude-control system with attitude position and attitude rate feedback.

makes the output of the sensor numerically equal to $\Theta(s)$, which is directly compared with the desired $\Theta_r(s)$. Let K_a be equal to 1. The parameters K_b and K are control parameters to be chosen by the designer.

The object of this example is to illustrate the use of Mason's rule for determining the transfer function from $\Theta_r(s)$ to $\Theta(s)$. By inspection of the block diagram of Figure 6.11, it is seen that there are two forward paths and five loops. There are two nontouching loop pairs. There is no nontouching loop triple.

To simplify the writing of equations, denote by K_c the constant

$$\left(\frac{1}{J_v}+\frac{1}{J_w}\right)\left(D+\frac{K_1{}^2}{R_a}\right) = K_c \tag{6.43}$$

and by K_d the constant

$$K\left(\frac{K_1}{R_a}\right)\left(\frac{1}{J_v}\right) = K_d. \tag{6.44}$$

With $K_a = 1$, the determinant Δ of the block diagram is

$$\begin{aligned}\Delta = 1 - &\left[-\frac{K_c}{s}-\frac{K_cK_dK_b}{s^2}-\frac{K_cK_d}{s^3}+\frac{K_dK_b}{s}+\frac{K_d}{s^2}\right]\\ &+\left[\left(-\frac{K_c}{s}\right)\left(\frac{K_dK_b}{s}\right)+\left(-\frac{K_c}{s}\right)\left(\frac{K_d}{s^2}\right)\right]\end{aligned} \tag{6.45}$$

Denote by P_1 the forward-path transfer function

$$P_1 = \frac{K_dK_c}{s^3}. \tag{6.46}$$

Since the deletion of forward path 1 breaks all the loops,

$$\Delta_1 = 1. \tag{6.47}$$

Denote by P_2 the forward-path transfer function

$$P_2 = -\frac{K_d}{s^2}. \tag{6.48}$$

The determinant of the block diagram that remains after deleting path 2 is

$$\Delta_2 = 1-\left(-\frac{K_c}{s}\right). \tag{6.49}$$

Thus the transfer function $\theta(s)/\theta_r(s)$ is

$$\frac{\Theta(s)}{\Theta_1(s)} = \frac{P_1\Delta_1 + P_2\Delta_2}{\Delta} \tag{6.50}$$

where Δ, P_1, Δ_1, P_2, and Δ_2 are given by (6.45)–(6.49). Substituting (6.45) to (6.49) in (6.50), we have

$$\frac{\Theta(s)}{\Theta_r(s)} = \frac{\dfrac{K_dK_c}{s^3} - \dfrac{K_d}{s^2}\left(1 + \dfrac{K_c}{s}\right)}{1 + \left(\dfrac{K_c}{s} + \dfrac{K_cK_dK_b}{s^2} + \dfrac{K_cK_d}{s^3} - \dfrac{K_dK_b}{s} - \dfrac{K_d}{s^2}\right) - \dfrac{K_cK_dK_b}{s^2} - \dfrac{K_cK_d}{s^3}}$$

$$= \frac{-K_d}{s^2 + (K_c - K_dK_b)s - K_d}. \tag{6.51}$$

For K_d positive, the system will have poles with positive real part and for a finite input such as a step input, the variable $\theta(t)$ will grow without bounds as $t \to \infty$. So let us choose K_d to be negative. That is since K_1, R_a, and J_v are positive, then from (6.44), the amplifier gain K must be negative. From (6.43), since J_v, J_w, D, K_1, and R_a are positive, K_c is positive. Hence if K_b is also positive, $K_c - K_dK_b$ will be positive. In fact, $K_c - K_dK_b$ remains positive provided $K_b > K_c/K_d$, where K_c/K_d is a negative quantity.

Comparing (6.51) with the standard second-order transfer function of Appendix B, we identify

$$\omega_n^2 = -K_d \tag{6.52}$$

and

$$2\zeta\omega_n = K_c - K_dK_b. \tag{6.53}$$

For $0 < \zeta < 1$, the system is underdamped. For a unit step input $\theta_r(t) = u(t)$, $\Theta_r(s) = 1/s$ and the final value theorem yields

$$\lim_{t\to\infty} \theta(t) = \lim_{s\to 0} s\frac{1}{s}\frac{-K_d}{s^2 + (K_c - K_dK_b)s - K_d} = 1. \tag{6.54}$$

Thus, after some transient, the attitude angle θ becomes approximately equal to the desired attitude θ_r. Typical response curves for an underdamped second-order system are shown in Figure B.4 of Appendix B. ▲

In Mason's formula in (6.40), multiplying both sides by $R(s)$ yields

$$Y(s) = R(s)\frac{\sum P_k\Delta_k}{\Delta} = \frac{\sum(P_kR(s))\Delta_k}{\Delta}. \tag{6.55}$$

But $P_k R(s)$ is the output signal of forward path k. Notice that the output value of forward path k is not the output value of the original block diagram because the two diagrams are different. Denote the output signal of forward path k by $Y_k(s)$. Hence (6.55) may be written as

$$Y(s) = \sum Y_k(s)\frac{\Delta_k}{\Delta}. \tag{6.56}$$

The path transfer function P_k is expressed in terms of $Y_k(s)$ by

$$P_k = \frac{Y_k(s)}{R(s)}. \tag{6.57}$$

6.4 Transfer Function Block Diagrams with Multiple Inputs and Multiple Outputs

Transfer functions can also be defined for linear time-invariant systems with several inputs and several outputs. If a system has two inputs $x_1(t)$ and $x_2(t)$, and two outputs $y_1(t)$ and $y_2(t)$, then for zero initial conditions, the Laplace transforms are related by

$$Y_1(s) = H_{11}(s)X_1(s) + H_{12}(s)X_2(s) \tag{6.58}$$

and

$$Y_2(s) = H_{21}(s)X_1(s) + H_{22}(s)X_2(s). \tag{6.59}$$

Using superposition, the output $Y_1(s)$ can be computed by first setting $X_2(s) = 0$ and determining the output due to $X_1(s)$. Denoting this output by $Y_{1a}(s)$, it is given by

$$Y_{1a}(s) = H_{11}(s)X\ (s). \tag{6.60}$$

Similarly, $X_1(s)$ is set to zero, and the output due to $X_2(s)$ is computed. Denoting the output by $Y_{1b}(s)$, it is

$$Y_{1b}(s) = H_{12}(s)X_2(s). \tag{6.61}$$

Hence

$$Y_1(s) = Y_{1a}(s) + Y_{1b}(s). \tag{6.62}$$

Thus

$$H_{11}(s) = \frac{Y_{1a}(s)}{X_1(s)} = \left.\frac{Y_1(s)}{X_1(s)}\right|_{X_2(s)=0} \tag{6.63}$$

and

$$H_{12}(s) = \frac{Y_{1b}(s)}{X_2(s)} = \left.\frac{Y_1(s)}{X_2(s)}\right|_{X_1(s)=0} \tag{6.64}$$

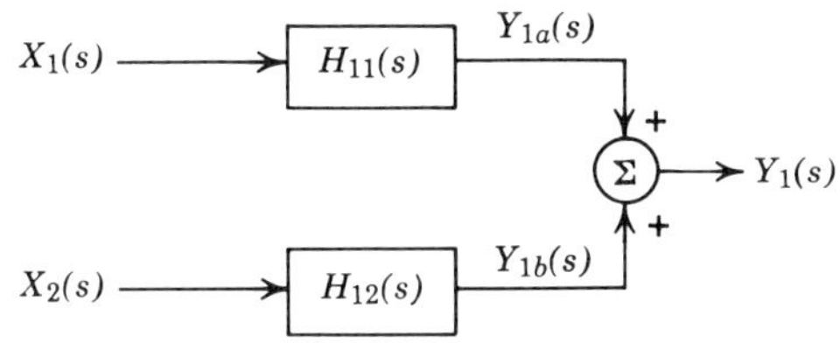

Figure 6.12 Block-diagram representation of (6.58).

Figure 6.12 shows a representation of (6.58). Similarly, the relation in (6.59) may be represented by the block diagram of Figure 6.13.

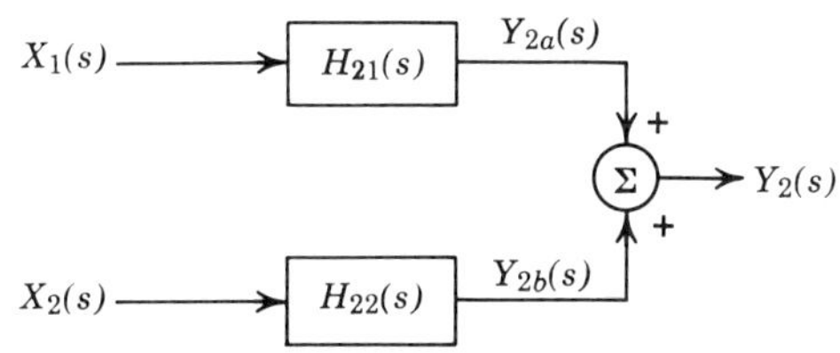

Figure 6.13 Block-diagram representation of (6.59).

Thus

$$H_{21}(s) = \frac{Y_{2a}(s)}{X_1(s)} = \left.\frac{Y_2(s)}{X_1(s)}\right|_{X_2(s)=0} \tag{6.65}$$

$$H_{22}(s) = \frac{Y_{2b}(s)}{X_2(s)} = \left.\frac{Y_2(s)}{X_2(s)}\right|_{X_1(s)=0} \tag{6.66}$$

Mason's rule can be employed for determining these transfer functions by simply setting some inputs to zero, and a single input system is considered.

Example 6-7. Figure 6.14 is a transfer-function block diagram where a and b are the transfer functions of the boxes, $X_1(s)$ and $X_2(s)$ are inputs, and $Y_1(s)$ and $Y_2(s)$ are outputs. Assuming zero initial conditions, $Y_1(s)$ and $Y_2(s)$ must be linearly related to $X_1(s)$ and $X_2(s)$ by (6.58)

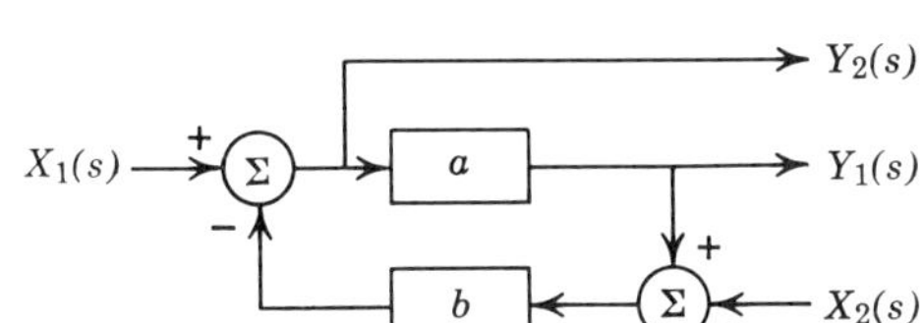

Figure 6.14 Multi-input multi-output system for Example 6-7.

and (6.59). To compute $H_{11}(s)$, set $X_2(s)$ to zero and denote the new output $Y_1(s)$ by $Y_{1a}(s)$. By Mason's formula

$$\frac{Y_{1a}(s)}{X_1(s)} = H_{11}(s) = \frac{a}{1+ab}. \tag{6.67}$$

To compute $H_{12}(s)$, set $X_1(s)$ to zero and denote the new output $Y_1(s)$ by $Y_{1b}(s)$. By Mason's formula

$$\frac{Y_{1b}(s)}{X_2(s)} = H_{12}(s) = \frac{-ab}{1+ab}. \tag{6.68}$$

Similarly, set $X_2(s)$ to zero and denote the new output $Y_2(s)$ by $Y_{2a}(s)$. By Mason's formula,

$$\frac{Y_{2a}(s)}{X_1(s)} = H_{21}(s) = \frac{1}{1+ab}. \tag{6.69}$$

To compute $H_{22}(s)$, set $X_1(s)$ to zero and denote the new $Y_2(s)$ by $Y_{2b}(s)$. By Mason's formula

$$\frac{Y_{2b}(s)}{X_2(s)} = \frac{-b}{1+ab}. \tag{6.70}$$

Hence

$$\begin{bmatrix} Y_1(s) \\ Y_2(s) \end{bmatrix} = \begin{bmatrix} H_{11}(s) & H_{12}(s) \\ H_{21}(s) & H_{22}(s) \end{bmatrix} \begin{bmatrix} X_1(s) \\ X_2(s) \end{bmatrix} = \begin{bmatrix} \dfrac{a}{1+ab} & \dfrac{-ab}{1+ab} \\ \dfrac{1}{1+ab} & \dfrac{-b}{1+ab} \end{bmatrix} \begin{bmatrix} X_1(s) \\ X_2(s) \end{bmatrix}. \ \blacktriangle \tag{6.71}$$

Initial conditions have been ignored thus far. In view of the multi-input multi-output analysis above, the initial conditions may be treated as additional inputs. Consider the integrator simulation diagram shown in Figure 6.15, which relates the input $x(t)$ to the output $y(t)$ by

$$y(t) = y_0 + \int_0^t x(\tau)\, d\tau. \tag{6.72}$$

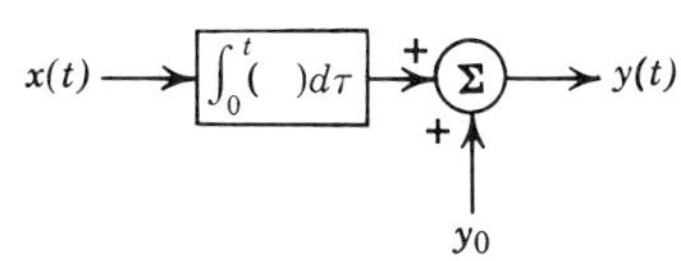

Figure 6.15 Simulation diagram of integrator.

The Laplace transform of (6.72) is

$$Y(s) = \frac{y_0}{s} + \frac{X(s)}{s}. \tag{6.73}$$

Notice that $Y(s)$ is the output of a two-input system in Figure 6.16. Figure 6.17 is another transfer-function representation of (6.73).

Thus a transfer-function block diagram corresponding to a simulation diagram of a linear time-invariant system is obtained as before for the case of zero initial conditions except that, when initial conditions are not zero, the integrator transfer-function block diagram is as shown in Figure 6.16 or Figure 6.17. The resulting transfer-function block diagram is multi-input in general. The transfer function from any input to any output is equal to the output contribution due to that input alone divided by the input with all other inputs equal to zero.

Example 6-8. Figure 6.18 shows a transfer-function block diagram corresponding to a simulation diagram that has two integrators with initial conditions x_{10} and x_{20}. There are two inputs $r_1(t)$ and $r_2(t)$, and one output $y(t)$. The transfer-function block diagram is obtained by replacing the integrator block by a transfer function $1/s$ and replacing the initial conditions constants x_{10} and x_{20} by their Laplace transforms x_{10}/s and x_{20}/s, respectively. The signals x_{10}/s and x_{20}/s may be regarded as inputs, so that the system has one output and four inputs. The output $Y(s)$ can be expressed in terms of the inputs as

$$Y(s) = H_1(s)R_1(s) + H_2(s)R_2(s) + H_3(s)\frac{x_{10}}{s} + H_4(s)\frac{x_{20}}{s}. \tag{6.74}$$

$H_1(s)$ is the transfer function from $R_1(s)$ to the component of $Y(s)$, $Y_1(s)$ when $R_2(s)$, x_{10}, and x_{20} are zero. From Figure 6.18, Mason's formula yields

$$\frac{Y_1(s)}{R_1(s)} = H_1(s) = \frac{K/s^2}{1 + (KK_2/s) + (KK_1/s^2)} = \frac{K}{s^2 + KK_2 s + KK_1}. \tag{6.75}$$

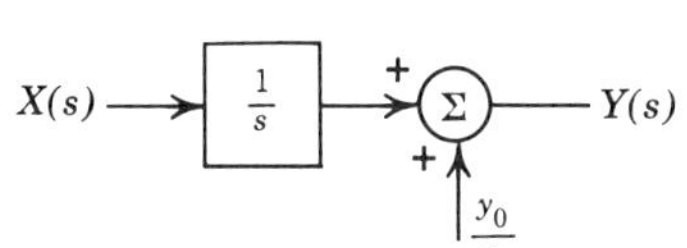

Figure 6.16 Transfer-function block diagram of integrator.

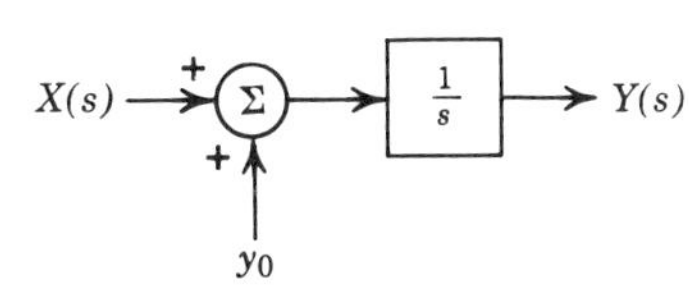

Figure 6.17 Alternative transfer-function block diagram of an integrator.

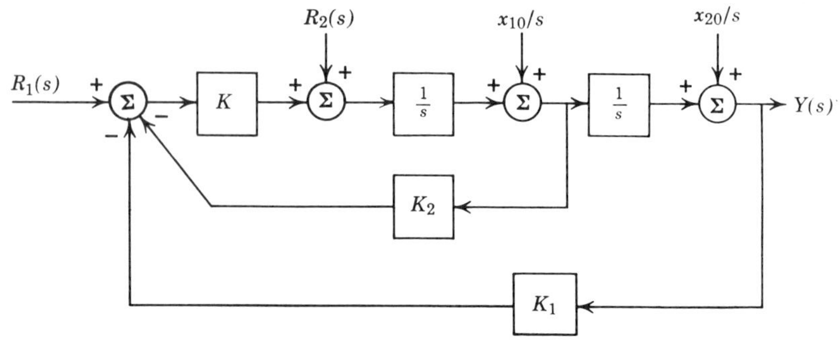

Figure 6.18 Transfer-function block diagram corresponding to a simulation diagram with two inputs, one output, and two integrators with initial conditions.

$H_2(s)$ is the transfer function from $R_2(s)$ to the component of $Y(s)$, $Y_2(s)$ when $R_1(s)$, x_{10}, and x_{20} are zero. From Figure 6.18 and Mason's formula,

$$\frac{Y_2(s)}{R_2(s)} = H_2(s) = \frac{1/s^2}{1 + (KK_2/s) + (KK_1/s^2)} = \frac{1}{s^2 + KK_2 s + KK_1}. \tag{6.76}$$

$H_3(s)$ is the transfer function from x_{10}/s to the component of $Y(s)$, $Y_3(s)$ when $R_1(s)$, $R_2(s)$, and x_{20} are zero. From Figure 6.18,

$$\frac{Y_3(s)}{x_{10}/s} = H_3(s) = \frac{1/s}{1 + KK_2/s + KK_1/s^2} = \frac{s}{s^2 + KK_2 s + KK_1} \tag{6.77}$$

Finally, the transfer function $H_4(s)$ is the ratio of the component of $Y(s)$, $Y_4(s)$ when $R_1(s)$, $R_2(s)$, and x_{10} are zero. From Figure 6.18 and Mason's formula,

$$\frac{Y_4(s)}{x_{20}/s} = H_4(s) = \frac{1(1 + KK_2/s)}{1 + KK_2/s + KK_1/s^2} = \frac{s^2 + KK_2 s}{s^2 + KK_2 s + KK_1}. \tag{6.78}$$

Hence, from (6.74)–(6.78),

$$Y(s) = \frac{KR_1(s) + R_2(s) + x_{10} + (s + KK_2)x_{20}}{s^2 + KK_2 s + KK_1}. \blacktriangle \tag{6.79}$$

In summary, a linear time-invariant simulation diagram can be converted to a transfer-function block diagram by replacing the integrator blocks by the transfer function $1/s$, replacing the scalor blocks by their transfer functions, which are equal to their multiplication constants, leaving the summers undisturbed and replacing all the signals, *including initial conditions*, by their Laplace transforms. The transforms of the original time domain inputs and the transforms of initial conditions constitute the set of inputs for the transfer-function block diagram. Any output may be determined by using the superposition property of linear systems and Mason's formula for computing a transfer function from any input to any output.

6.5 Frequency Response via Sinusoidal Steady-State Response

In many applications, input signals are either sinusoids or sums of sinusoids. If a sinusoidal input to a linear time-invariant system produces a response asymptotic for large t to a sinusoidal function of the same frequency, a sinusoidal steady-state solution is said to exist. The steady-state response is closely connected with the transfer function with s replaced by $j\omega$ as we shall demonstrate in this section. By observing the sinusoidal steady-state response to a sinusoidal input, the function $H(j\omega)$ may be determined. But $H(j\omega)$ and the transfer function $H(s)$ are very closely related, being equal for $s = j\omega$. As discussed earlier, transfer functions, transforms of input signals, and transforms of initial conditions combine to yield transforms of output signals. Thus the response to a nonsinusoidal input is related to the response to a sinusoidal input via $H(j\omega)$.

If a cosine function $x(t) = \cos \omega_0 t$ is applied for all $t > 0$ to a system whose transfer function is $H(s)$, then the Laplace transform of the output $y(t)$ is

$$\begin{aligned} Y(s) &= [\mathscr{L}\{\cos \omega_0 t\}]H(s) + \text{sum of initial condition terms} \\ &= \frac{sH(s)}{(s-j\omega_0)(s+j\omega_0)} + \text{sum of initial condition terms.} \end{aligned} \tag{6.80}$$

Using Figure 6.17 as the transfer function block diagram for an integrator and the results on multi-input block diagrams of the previous section, it is seen that the sum of initial condition terms in (6.80) is a sum of the products of initial conditions and the corresponding transfer functions. Thus (6.80) may be written as

$$Y(s) = \frac{sH(s)}{(s+j\omega_0)(s+j\omega_0)} + x_{10}H_1(s) + x_{20}H_2(s) + \ldots + x_{n0}H_n(s) \tag{6.81}$$

where $x_{10}, x_{20}, \ldots, x_{n0}$ are the initial conditions and $H_1(s), H_2(s), \ldots,$

$H_n(s)$ are the transfer functions from the input of each integrator to the specified output.

The partial fraction expansion of $Y(s)$ in (6.81) is

$$Y(s) = \frac{A}{s-j\omega_0} + \frac{B}{s+j\omega_0} + \{\text{terms corresponding to the poles of } H(s), H_1(s), \ldots, \text{and } H_n(s)\}. \tag{6.82}$$

We shall show that if all the poles of $H(s), H_1(s), \ldots,$ and $H_n(s)$ are in the left half of the s-plane, then $y(t)$ has a sinusoidal steady-state component $y_{ss}(t)$, which is equal to the inverse transform of the first two terms in (6.82),

$$y_{ss}(t) = \mathscr{L}^{-1}\left(\frac{A}{s-j\omega_0} + \frac{B}{s+j\omega_0}\right). \tag{6.83}$$

After some time, $y(t)$ will be approximately equal to $y_{ss}(t)$. Furthermore, if $H(s)$, $H_1(s), \ldots,$ or $H_n(s)$ has a pole on the $j\omega$ axis or in the right half of the s-plane, there is no sinusoidal steady state.

If all the poles of $H(s)$, $H_1(s), \ldots,$ and $H_n(s)$ are in the left half of the s-plane, then every real pole must be negative and every complex pole must have a negative real part. Denote a typical pole of $H(s)$, $H_1(s), \ldots,$ and $H_n(s)$ by $s_i = -\alpha_i + j\omega_i$ where α_i is real and positive and ω_i is real. Then a typical term corresponding to this pole in the partial fraction expansion of (6.82) is

$$\frac{K_i}{(s+\alpha_i-j\omega_i)^k}.$$

If the pole is simple, k equals one only. If the pole is multiple, with multiplicity m, then a term with $k=m$ is present. In addition, other terms with $k<m$ may be present. From Tables B.1 and B.2, the inverse Laplace transform of such an expression is

$$\frac{K_i t^{(k-1)}\epsilon^{-\alpha_i t}\epsilon^{j\omega_i t}}{(k-1)!}.$$

If ω_i is not zero, the complex conjugate of the above term corresponding to the pole at $\bar{s}_i = -\alpha_i - j\omega_i$ is present also, and their sum is a real quantity. Because of the factor $\epsilon^{-\alpha_i t}$, the terms approach zero exponentially. For any specified degree of accuracy, there is a finite waiting time T so that, for all $t>T$, the inverse transform of $Y(s)$ in (6.82) is approximately equal to $y_{ss}(t)$ in (6.83).

Let us now compute A and B of the partial fraction expansion in (6.83). From Appendix B, A is computed from

$$A = \left.\frac{(s-j\omega_0)sH(s)}{(s-j\omega_0)(s+j\omega_0)}\right|_{s=j\omega_0} = \tfrac{1}{2}H(j\omega_0). \tag{6.84}$$

Similarly,

$$B = \left.\frac{(s+j\omega_0)sH(s)}{(s-j\omega_0)(s+j\omega_0)}\right|_{s=-j\omega_0} = \tfrac{1}{2}H(-j\omega_0). \tag{6.85}$$

Hence

$$y_{ss}(t) = \mathscr{L}^{-1}\left[\frac{A}{s-j\omega_0}+\frac{B}{s+j\omega_0}\right] = \tfrac{1}{2}H(j\omega_0)\,\epsilon^{j\omega_0 t}+\tfrac{1}{2}H(-j\omega_0)\,\epsilon^{-j\omega_0 t}. \tag{6.86}$$

But $H(s)$ is a ratio of polynomials with real coefficients, so that $H(-j\omega_0)$ is the complex conjugate of $H(j\omega_0)$. Since $\epsilon^{-j\omega_0 t}$ is the complex conjugate of $\epsilon^{j\omega_0 t}$, then the two terms in (6.86) are complex conjugates. The sum of a complex quantity and its conjugate is equal to twice the real part of the complex quantity. Hence

$$\begin{aligned}\tfrac{1}{2}H(j\omega_0)\,\epsilon^{j\omega_0 t}+\tfrac{1}{2}H(-j\omega_0)\,\epsilon^{-j\omega_0 t} &= 2\,Re[\tfrac{1}{2}H(j\omega_0)\,\epsilon^{j\omega_0 t}]\\ &= Re[H(j\omega_0)\,\epsilon^{j\omega_0 t}].\end{aligned} \tag{6.87}$$

Writing $H(j\omega_0)$ in exponential form

$$H(j\omega_0) = |H(j\omega_0)|\epsilon^{j\phi} \tag{6.88}$$

where $|H(j\omega_0)|$ is the magnitude of $H(j\omega_0)$ and ϕ is the angle of $H(j\omega_0)$, (6.86), (6.87), and (6.88) yield

$$y_{ss}(t) = Re\{|H(j\omega_0)|\epsilon^{(j\omega t+\phi)}\} = |H(j\omega_0)|\cos(\omega_0 t+\phi). \tag{6.89}$$

Thus, after waiting for some finite time T, the response $y(t)$ to an input $x(t) = \cos\omega_0 t$ for all $t \geqslant 0$ is approximately equal to the sinusoidal steady-state component $y_{ss}(t)$ in (6.89). For most practical systems, the waiting time T is relatively short. If the input $x(t)$ is equal to $K\cos\omega_0 t$ for all $t \geqslant 0$, the steady-state response is $y_{ss}(t) = K|H(j\omega_0)|\cos(\omega_0 t+\phi)$ because of linearity. Thus the magnitude $|H(j\omega_0)|$ can be measured as the ratio of the peak values of the output sinusoid and the input sinusoid for the given frequency ω_0. The angle ϕ can be measured as the difference between the phase angles of the output sinusoid and the input sinusoid. The experiment may be conducted for several values of ω. The function $|H(j\omega)|$ [or $20\log|H(j\omega)|$

in decibels (db)], and the function $\phi(\omega)$ may be plotted versus ω. 20 log $|H(j\omega)|$ and $\phi(\omega)$ are commonly plotted on a logarithmic scale for ω, and the pair of plots is called the *frequency response* of the system. From the frequency response, $H(j\omega)$ is readily obtained using (6.88). Because of the experimental as well as conceptual simplicity of the sinusoidal steady-state interpretation for the frequency response and the simple connection between frequency response and transfer-function concepts, sinusoidal steady-state analysis is widely used for linear time-invariant systems.

If $H(s)$ or any of the transfer functions $H_1(s), H_2(s), \ldots,$ or $H_n(s)$ has a pole on the $j\omega$ axis or in the right half of the s-plane, then in the inverse transform of $Y(s)$, at least one term in addition to $y_{ss}(t)$ will not approach zero for $t \to \infty$. Either $y(t)$ eventually becomes unbounded for $t \to \infty$ or it contains sinusoidal terms at frequencies other than those of the input. In either case, by definition, there is no sinusoidal steady state.

Example 6-9. Suppose that for a given frequency, the steady-state input and output sinusoidal signals are as shown in Figure 6.19. The period or time duration per cycle is T. Thus the frequency ω_0 for the wave forms is

$$\omega_0 = \frac{2\pi}{T}. \tag{6.90}$$

If the wave forms are displayed on an oscilloscope, the period T can be computed from the time per unit length scale used and the actual distance corresponding to one period.

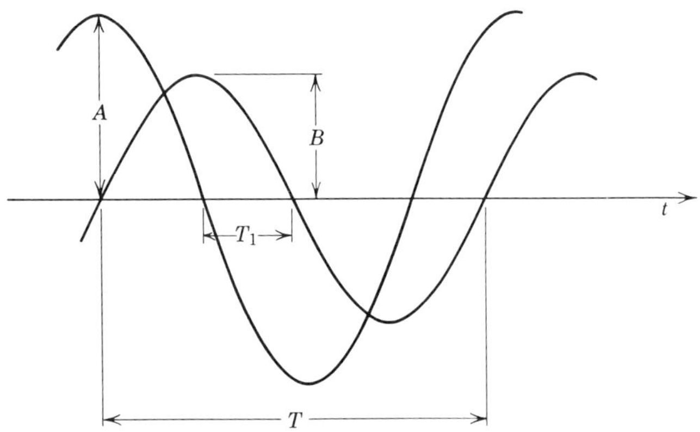

Figure 6.19 Input and output sinusoidal waveforms for computing magnitude and phase of $H(j\omega_0)$ for a fixed ω_0.

The peak value of the input wave form is A, and the peak value of the output wave form is B. Hence the magnitude of $H(j\omega_0)$ is

$$|H(j\omega_0)| = \frac{B}{A}. \tag{6.91}$$

From Figure 6.19, it is seen that the phase angle displacement between the two curves is

$$\theta_1 = 2\pi \frac{T_1}{T} \text{ radians} \tag{6.92}$$

because the actual displacement in time is T_1, and T is the period Figure 6.19 also shows that the output occurs later in time by T_1 compared to the input. That is, if the input wave form is represented by the equation

$$x(t) = A \cos(\omega_0 t + \alpha) \tag{6.93}$$

then the output representation is

$$y(t) = B \cos(\omega_0 t + \alpha - \theta_1). \tag{6.94}$$

We say that the output lags the input by θ_1 in this example. The phase angle of $H(j\omega_0)$ is equal to the phase angle of the output minus the phase angle of the input. Hence

$$\phi = (\alpha - \theta_1) - \alpha = -\theta_1 = -\frac{2\pi T_1}{T}. \tag{6.95}$$

From (6.88), (6.91), and (6.95), the complex value of $H(j\omega_0)$ for the given fixed ω_0 is

$$H(j\omega_0) = \frac{B}{A}\epsilon^{-j\theta_1} = \frac{B}{A}\epsilon^{-j(2\pi T_1/T)}. \tag{6.96}$$

In general, for another value of ω, $H(j\omega)$ and ϕ will be different from B/A and $-\theta$, respectively. Similar experiments can be conducted for other values of ω, and the resulting $|H(j\omega)|$ and $\phi(\omega)$ may be plotted versus ω. Instead of plotting the frequency response, the complex function $|H(j\omega)\epsilon^{j\phi}$ may be plotted in a complex plane with ω as a parameter. As ω varies, a locus is traced in the complex plane. ▲

6.6 Frequency Response of Rational Transfer Functions

In this section, the frequency response of a rational transfer function will be studied. Based on a knowledge of the properties of such a frequency response, a rational transfer function may be constructed so that its associated frequency response approximates a given set of experimentally obtained frequency-response data.

Since a rational function consists of a product of factors divided by a product of factors, as in (6.11), the logarithm of the magnitude of $H(j\omega)$ is a sum of the logarithms of magnitudes of the individual numerator factors minus the sum of logarithms of magnitudes of the individual denominator factors. Conventionally, twenty times the logarithm of the magnitude is called "the magnitude in decibels (db)." Thus the frequency response magnitudes (in db) is expressed as a linear combination of frequency response magnitudes (in db) of simple first-order functions. Similarly if the individual factors of $H(j\omega)$ are expressed in exponential form, it is seen that the angle of $H(j\omega)$ is equal to the sum of the angles of the individual numerator factors minus the sum of the angles of the individual denominator angles. For these reasons, the frequency response of a first-order transfer function is basic to the study of the frequency response of a high-order transfer function.

Let us consider a simple transfer function

$$H(s) = \frac{1}{1 + s/\omega_0} \tag{6.97}$$

where ω_0 is a real constant. For $s = j\omega$, the magnitude function in decibels (db) is

$$\begin{aligned} 20 \log |H(j\omega)| &= 20 \log \left| \frac{1}{1 + j(\omega/\omega_0)} \right| \\ &= -20 \log \left| 1 + j\frac{\omega}{\omega_0} \right| \\ &= -20 \log \sqrt{1 + \left(\frac{\omega}{\omega_0}\right)^2}. \end{aligned} \tag{6.98}$$

Notice that for $(\omega/\omega_0) << 1$,

$$20 \log |H(j\omega)| \approx -20 \log 1 = 0 \tag{6.99}$$

which is a horizontal line. For $(\omega/\omega_0) >> 1$,

$$20 \log |H(j\omega)| \approx -20 \log \left(\frac{\omega}{\omega_0}\right) = 20 \log \omega_0 - 20 \log \omega \tag{6.100}$$

which is a straight line with slope -20 and zero ordinate at $\omega = \omega_0$ if $20 \log |H(j\omega)|$ is plotted as a function of $\log \omega$. Equations 6.99 and 6.100 specify the asymptotes for values of ω away from ω_0. Figure 6.20

shows the two straight line asymptotes. The actual plot of 20 log $|H(j\omega)|$ is shown as a dotted curve. For $\omega = \omega_0$, the value of 20 log $|H(j\omega)|$ is $-20 \log \sqrt{2} = -3$, but the intersection of the two asymptotes has a zero ordinate. If the asymptotes are taken as an approximation to the curve, there is an error of 3 db at the corner or break frequency ω_0. The errors for ω away from ω_0 are less. For example, for $\omega = 2\omega_0$ and $\omega = 1/2\,\omega_0$, the error is 1 db. For better accuracy, it is a simple matter to first plot the asymptotes and the three correct points at $1/2\,\omega_0$, ω_0, and $2\omega_0$. A smooth curve passing through the three points and asymptotic to the two straight lines provides a quick procedure for plotting 20 log $|H(j\omega)|$ against log ω for the $H(s)$ in (6.97). Instead of plotting against log ω in linear coordinates, it is more convenient to plot against ω in logarithmic coordinates, resulting in the same shape for the plot. In logarithmic coordinates, the distance between $\omega = 1$ and $\omega = 10$ is the same as that between $\omega = 10$ and $\omega = 100$. This is the same result if log ω is plotted on linear scale, since $\log 10 - \log 1 = \log 100 - \log 10$. The slope of the asymptote is 20 db per decade.

For $s = j\omega$, the angle of $H(j\omega)$ for $H(s)$ in (6.97) is

$$\phi = -\tan^{-1}\left(\frac{\omega}{\omega_0}\right). \tag{6.101}$$

The plot of ϕ versus ω in logarithmic coordinates is shown in Figure 6.21.

The basic shapes in Figures 6.20 and 6.21 are useful for determining

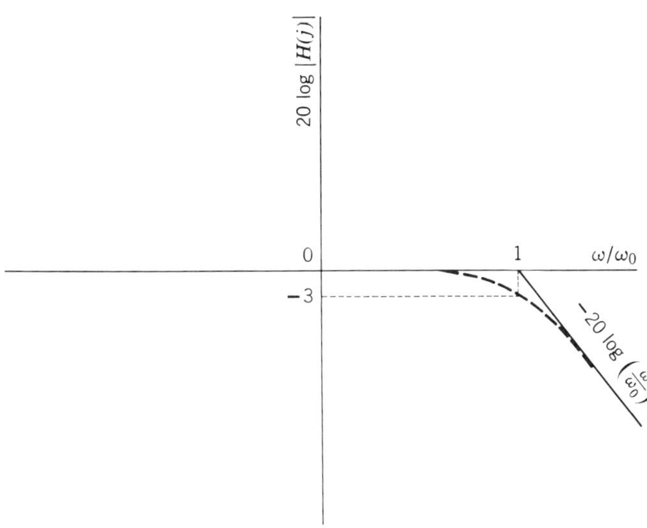

Figure 6.20 Magnitude frequency response for $H(s)$ in (6.97).

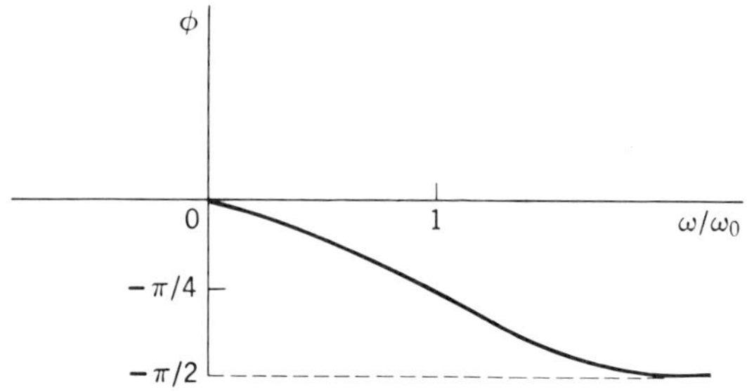

Figure 6.21 Phase-angle frequency response for $H(s)$ in (6.97).

the magnitude and phase-angle frequency response for a rational function with real poles and real zeros. The first step is to factor a constant from the usual factors in (6.11) so that the factor is of the form $(1+s/\omega_0)$. This factor is now the same as that in (6.97). Then s is replaced by $j\omega$. Finally, 20 log $|H(j\omega)|$ is formed, and the angle of $H(j\omega)$ is formed.

Example 6-10. Let us consider a transfer function

$$H(s)=5\frac{(s+2)}{s+20}. \tag{6.102}$$

First we factor 2 from $(s+2)$ and 20 from $(s+20)$ so that the factors will be of the form $1+s/\omega_0$. Equation 6.102 becomes

$$H(s)=\frac{5(2)(1+s/2)}{20(1+s/20)}=\frac{1}{2}\frac{(1+s/2)}{(1+s/20)}. \tag{6.103}$$

Then 20 log $|H(j\omega)|$ is

$$20\log|H(j\omega)|=20\left(\log\tfrac{1}{2}+\log\left|1+\frac{j\omega}{2}\right|-\log\left|1+\frac{j\omega}{20}\right|\right). \tag{6.104}$$

The first term on the right-hand side of (6.104) is simply a constant. The two other factors are of the form considered previously. Thus the whole expression has a piecewise linear approximation because 20 log $|H(j\omega)|$ is a sum of expressions that have piecewise linear approximations. The corners or break frequencies are at $\omega=2$ and $\omega=20$. The asymptotes for the individual terms are shown in Figure 6.22, and the composite approximation, which is obtained by adding the ordinates of the individual asymptotes, is shown in Figure 6.23. The actual magnitude curve in db is shown in dotted lines.

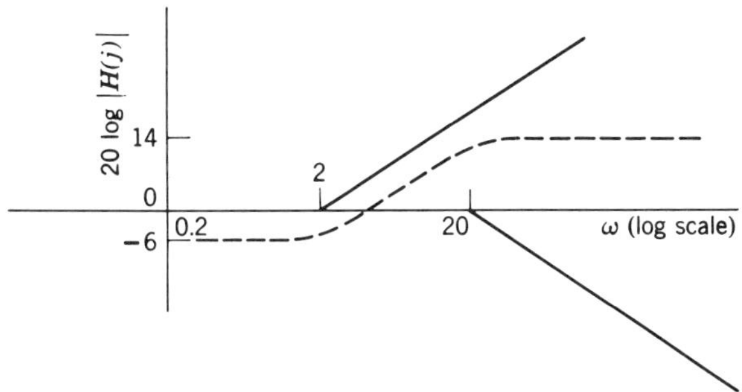

Figure 6.22 Asymptotes for individual terms in (6.104).

For the angle ϕ, first we substitute $s = j\omega$ in (6.103),

$$H(j\omega) = \frac{1}{2}\frac{(1 + j\omega/2)}{(1 + j\omega/20)}. \tag{6.105}$$

The angle of $H(j\omega)$ is equal to the angle of $(1 + j\omega/2)$ minus the angle of $(1 + j\omega/20)$. Hence

$$\phi = \tan^{-1}\frac{\omega}{2} - \tan^{-1}\frac{\omega}{20}. \tag{6.106}$$

Figure 6.24 shows the plots for the individual terms, and Figure 6.25 shows the sum of the terms for ϕ in (6.106). ▲

If a zero or pole is at the origin, then the corner break frequency is at $\omega = 0$. But on a logarithmic coordinate, $\omega = 0$ corresponds to $-\infty$. Hence the asymptote for the magnitude frequency response is one straight line only. Furthermore, the asymptote is the exact value for the curve ($20 \log |H(j\omega)| = 20 \log \omega$ is a straight line with a slope

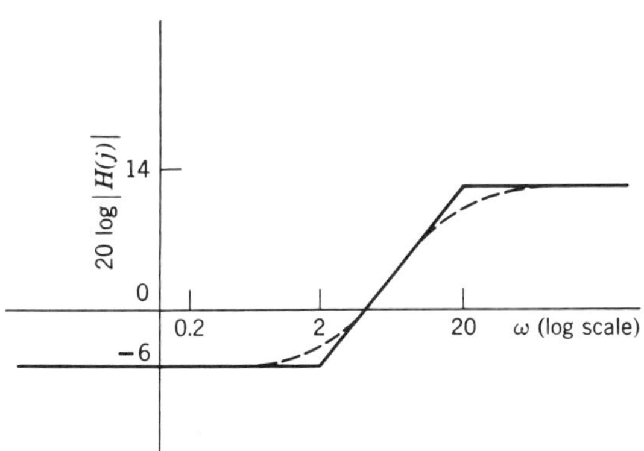

Figure 6.23 Piecewise linear and actual magnitude for 20 log $|H(j\omega)|$ in (6.104).

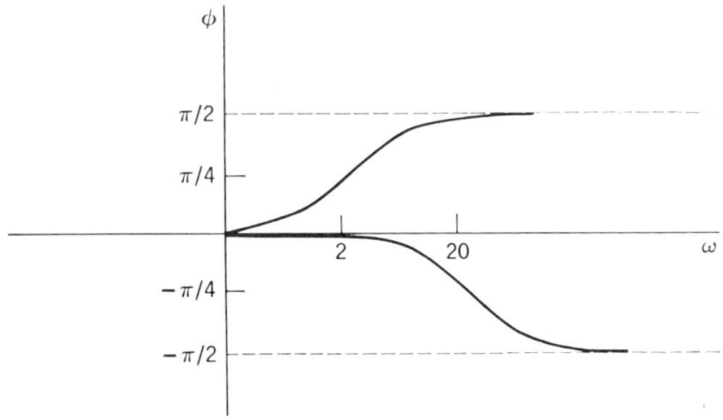

Figure 6.24 Phase angles for individual terms in (6.106).

of 20 db per decade and a zero ordinate at $\omega = 1$). The phase angle, corresponding to a zero at the origin, is a constant $\pi/2$ for all ω and, for a pole at the origin, the phase angle is a constant $-\pi/2$ for all ω.

Example 6-11. Let us consider the frequency response for

$$H(s) = \frac{10s}{s+5}. \tag{6.107}$$

First, factor 5 from the denominator and rewrite (6.107) as

$$H(s) = \frac{10s}{5(s/5+1)} = \frac{2s}{(1+s/5)}. \tag{6.108}$$

Hence, for $20 \log |H(j\omega)|$, we add the three plots of the individual magnitude frequency responses shown in Figure 6.26. The composite magnitude frequency response and its asymptote are also shown. ▲

For complex poles or zeros, it is more convenient to combine the pairs of complex conjugate first-order factors. A typical quadratic

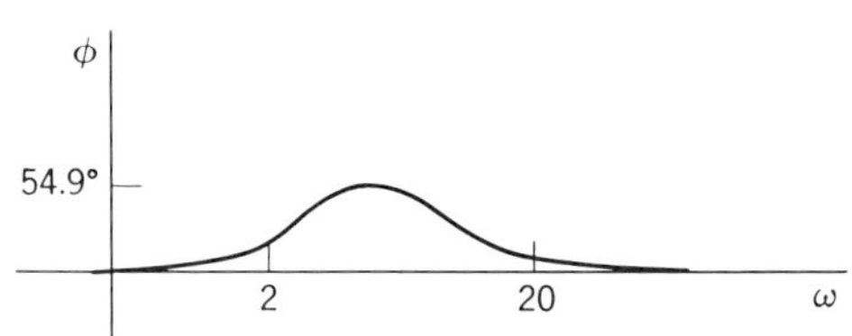

Figure 6.25 Total phase-angle ϕ for (6.106).

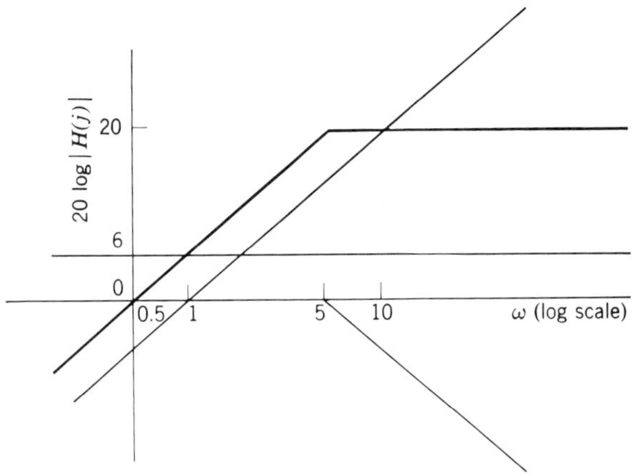

Figure 6.26 Individual magnitude frequency response for determining 20 log $|H(J\omega)|$ for Example 6-11.

factor is $(s^2+2\zeta\omega_n s+\omega_n{}^2)$ with $0<\zeta<1$ for complex zeros of the factor. As in the linear case, it is also convenient to factor the constant term so that the constant term in the quadratic factor is 1. Thus the quadratic factor we wish to consider is $[1+2\zeta(s/\omega_n)+(s/\omega_n)^2]$. For example, for a transfer function

$$H(s)=\frac{\omega_n{}^2}{s^2+2\zeta\omega_n s+\omega_n{}^2} \tag{6.109}$$

considered in Appendix B, if we factor $\omega_n{}^2$ from the denominator, (6.109) becomes

$$H(s)=\frac{1}{(s/\omega_n)^2+2\zeta(s/\omega_n)+1}. \tag{6.110}$$

Hence

$$20\log|H(j\omega)|=-20\log\sqrt{\left[1-\left(\frac{\omega}{\omega_n}\right)^2\right]^2+\left(2\zeta\frac{\omega}{\omega_n}\right)^2}. \tag{6.111}$$

Thus for $(\omega/\omega_n) << 1$, $20\log|H(j\omega)|\approx 0$ and for $(\omega/\omega_n) >> 1$, 20 log $|H(j\omega)|\approx -40\log(\omega/\omega_n)$, which are the asymptotes. The actual magnitude frequency response depends on the value of ζ. Figure 6.27 shows the magnitude frequency-response plots for several values of ζ. Observe that for $(\omega/\omega_n) << 1$, the curves are asymptotic to zero and for $(\omega/\omega_n) >> 1$, the curves are asymptotic to -40 db/decade. Also, for $\zeta=1$, the curve is like that in Figure 6.20 for $\omega_n=\omega_0$ except

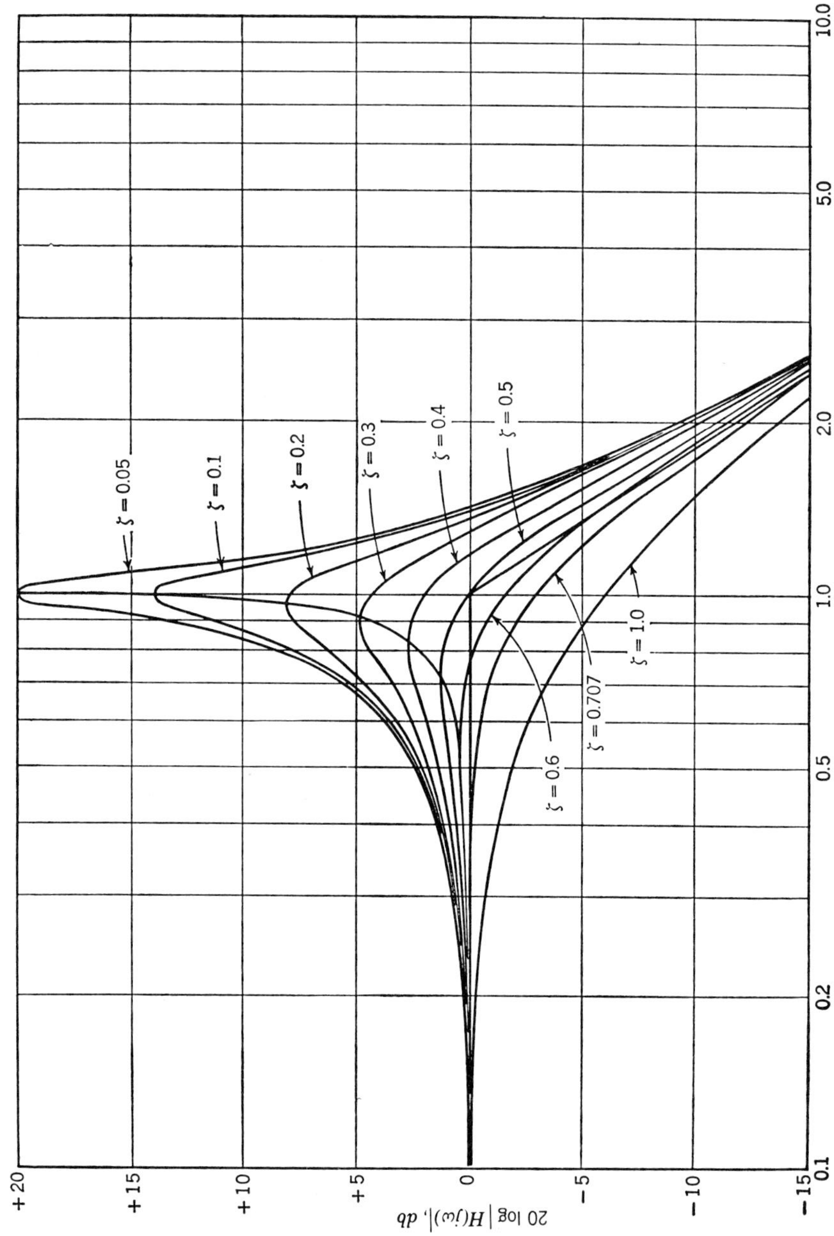

Figure 6.27 Magnitude frequency response, 20 log $|H(j\omega)|$ for $H(s) = \dfrac{1}{(s/\omega_n)^2 + 2\zeta(s/\omega_n) + 1}$ for several values of ζ, $0 < \zeta \leq 1$.

that the magnitude in db is doubled. This is because for $\zeta = 1$, the poles are real and coincident so that the response should be double that in Figure 6.20.

The phase function of $H(j\omega)$ corresponding to $H(s)$ in (6.110) is

$$\phi = -\tan^{-1}\frac{2\zeta(\omega/\omega_n)}{1-(\omega/\omega_n)^2}. \tag{6.112}$$

It is shown in Figure 6.28 for various values of ζ. For $\zeta = 1$, the angle is exactly twice that for a first-order factor, as in Figure 6.21 for $\omega_0 = \omega_n$, because $\zeta = 1$ corresponds to a double pole at $s = -\omega_n$.

Example 6-12. For the accelerometer discussed in Chapter 3 and Example 6-2, the transfer function from the transform of the vehicle acceleration to the transform of the test mass displacement is given by (6.13). Using the standard notation for second-order all-pole transfer functions, (6.13) is rewritten as

$$H(s) = \frac{1}{s^2 + 2\zeta\omega_n s + \omega_n^2} \tag{6.113}$$

where ω_n and ζ are related to D, K, and M by (6.15) and (6.17). Equation (6.113) is rewritten as

$$H(s) = \frac{1}{\omega_n^2}\frac{1}{1+2\zeta(s/\omega_n)+(s/\omega_n)^2}. \tag{6.114}$$

Thus

$$20\log|H(j\omega)| = -20\log\omega_n^2 - 20\log\left|1-\left(\frac{\omega}{\omega_n}\right)^2 + j2\zeta\frac{\omega}{\omega_n}\right| \tag{6.115}$$

which is a composition of a horizontal line and a typical second-order curve as in Figure 6.27. Figure 6.29 shows a plot of the two terms for $\omega_n > 1$ and $\zeta = .7$ together with their asymptotes. The complete magnitude frequency response is shown in Figure 6.30.

The magnitude response may be obtained experimentally by mounting the accelerometer on a movable platform and shaking the platform horizontally in sinusoidal fashion. That is, if the position of the platform $x(t)$ is varied as a sine function

$$x(t) = X\sin\omega_0 t \tag{6.116}$$

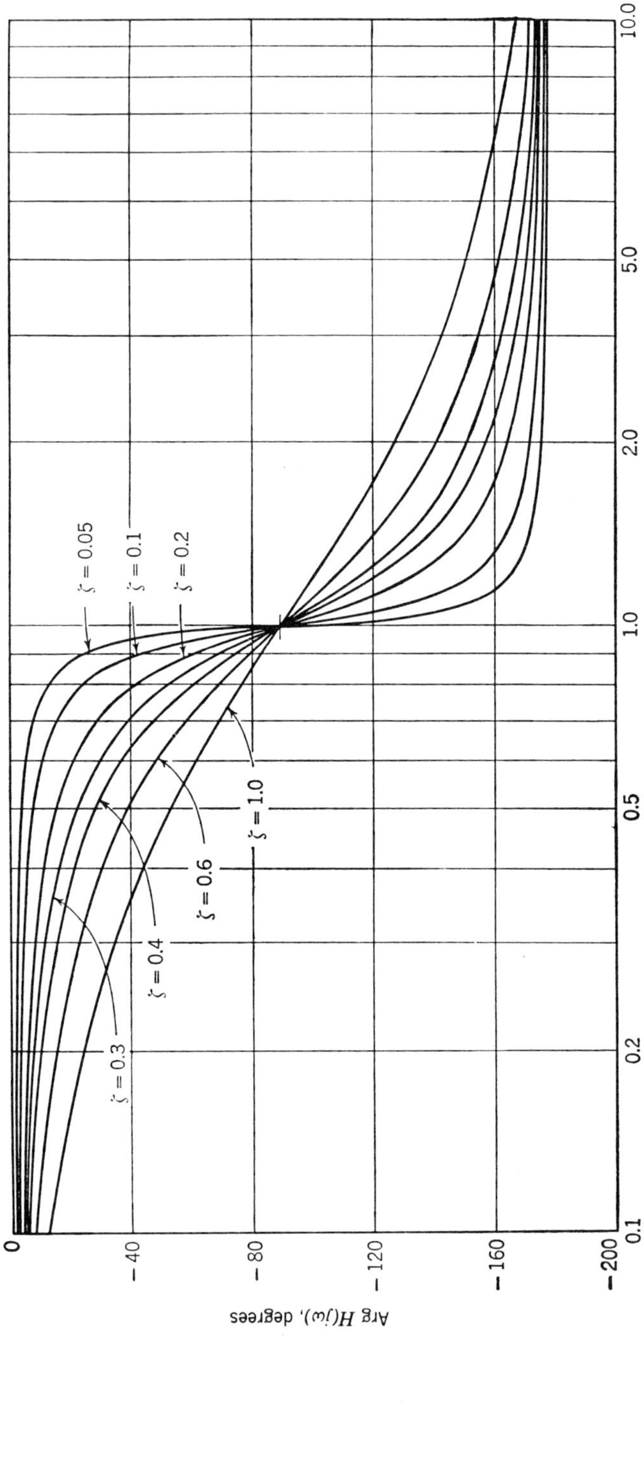

Figure 6-28 Phase function $\phi = \text{Arg}\, H(j\omega)$ for

$$H(s) = \frac{1}{(s/\omega_n)^2 + 2\zeta(s/\omega_n) + 1}$$

for several values of ζ, $0 < \zeta \leq 1$.

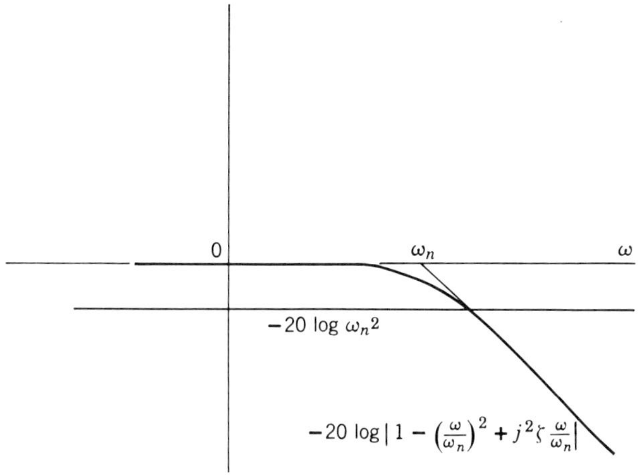

Figure 6.29 Magnitude frequency response for individual terms in (6.115).

where X and ω_0 are positive real constants, then

$$\dot{x}(t) = \omega_0 X \cos \omega_0 t \tag{6.117}$$

and

$$\ddot{x} = -\omega_0^2 X \sin \omega_0 t = \omega_0^2 X \sin (\omega_0 - \pi). \tag{6.118}$$

The test mass will move sinusoidally in the steady state. It is of the form

$$y(t) = Y \sin (\omega_0 t + \alpha) \tag{6.119}$$

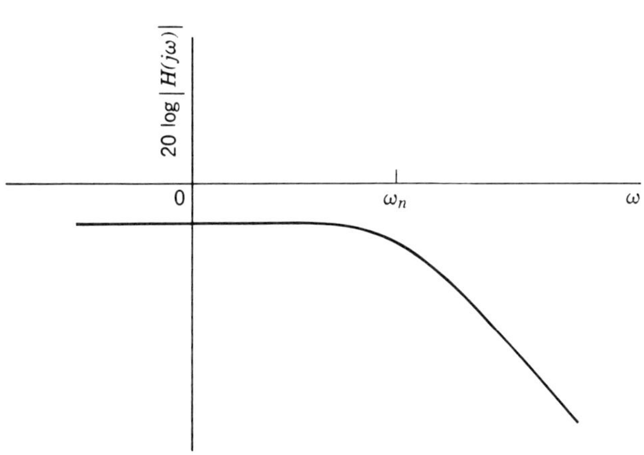

Figure 6.30 Magnitude frequency response of accelerometer in Example 6-12.

where Y is a real constant. Hence the magnitude of the transfer function $|H(j\omega_0)|$ can be computed from

$$|H(j\omega_0)| = \frac{|Y|}{\omega_0^2|X|}. \tag{6.120}$$

This experiment can be repeated for several values of ω and a plot like Figure 6.30 obtained.

The parameters ω_n and ζ may be predicted as follows. Draw the asymptotes for the curve as in Figure 6.30. The value of ω at which the asymptotes intersect is ω_n. Since the height of the horizontal part is $-20 \log \omega_n^2$, this can be used as a check for the accuracy of the measured ω_n. The value of ζ can be predicted by measuring the height of the peak with respect to the horizontal part of the frequency response. This peak can be compared with the standard curves in Figure 6.27, and the value of ζ can be read off directly. From the predicted values of ω_n and ζ, the ratios K/M and D/M can be computed if desired. ▲

Given an experimental plot of $20 \log |H(j\omega)|$ for a linear time-invariant system, it may be desired to obtain an analytic expression for $H(s)$. From the discussions above, it is noted that if $H(s)$ is a rational function, the composite asymptote of $20 \log |H(j\omega)$ is piecewise linear, and the slopes of the portions of the asymptote are multiples of 20 db per decade. Keeping in mind that the actual curve is smooth and slightly off at the corners, one can anticipate these discrepancies and arrive at a piecewise linear asymptotic approximation to the curve. As a general guide, it is desirable to arrive at a transfer-function representation $H(s)$ with as few zeros and poles as possible and, at the same time, meet the accuracy or tolerance limit for the approximation. This means that the number of corners in the piecewise linear approximation should be as few as can be allowed without sacrificing accuracy.

Example 6-13. It is desired to obtain a rational transfer function $H(s)$ whose magnitude frequency response approximates the curve given in Figure 6.31. The asymptote for the desired magnitude frequency response must be piecewise linear with slopes of multiples of 20 db per decade. After a few trials, the piecewise linear curve in Figure 6.31 is chosen. The corners are at $\omega = 1.5$, 6, and 10. Based on these corners and the low-frequency asymptote, we choose a zero at the origin, a pole at $s = -1.5$, a pole at $s = -6$, and a pole at $s = -10$. $H(s)$ has the form

$$H(s) = \frac{Ks}{(1 + s/1.5)(1 + s/6)(1 + s/10)}. \tag{6.121}$$

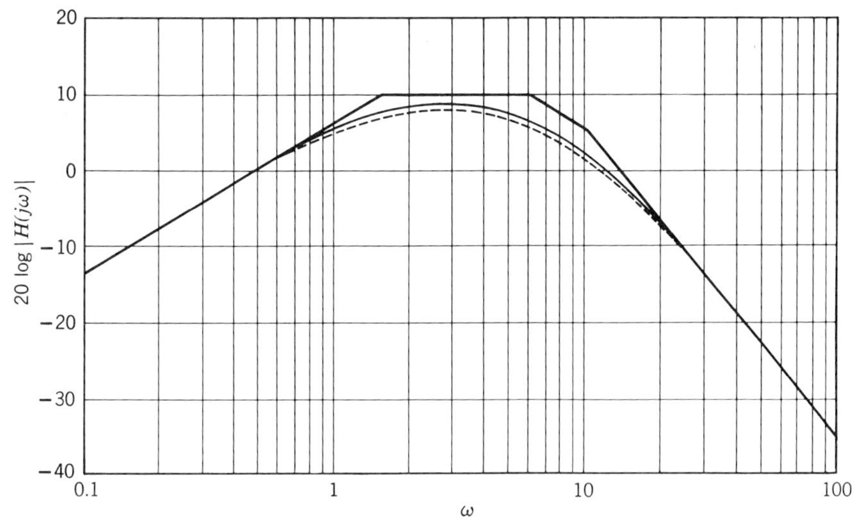

Figure 6.31 Magnitude frequency response for Example 6-13.

To adjust the horizontal part of the asymptote to be 10 db, 20 log $[(K)(1.15)]$ must have a value of 10 db or $K = \sqrt{10}/1.5$. So

$$H(s) = \frac{(60)(\sqrt{10})s}{(s+1.5)(s+6)(s+10)}. \tag{6.122}$$

The actual magnitude response $20 \log |H(j\omega)|$ for $H(s)$ in (6.122) is shown dotted in Figure 6.31. The logarithmic frequency-response plots as well as the piecewise linear asymptotes are known as *Bode plots*. ▲

6.7 Frequency Response from Pole-Zero Diagrams

As described in the previous sections, the function $H(j\omega)$ is obtainable from the transfer function by replacing s by $j\omega$. In this section, we shall give a geometric interpretation to the calculation of $H(j\omega)$ from $H(s)$.

Suppose that the transfer function $H(s)$ is expressed in factored form, as in (6.11), in terms of the gain factor K, the zeros $z_1, z_2, \ldots,$ and the poles $p_1, p_2, \ldots$. The zeros and poles may be complex but complex zeros or poles are present in conjugate pairs if the coefficients of the polynomials in $H(s)$ are real. Consider a typical zero factor $(s-z_i)$. For any s and any z_i, the difference $(s-z_i)$ is represented by an arrow in the complex plane from the point z_i to the point s, as shown in Figure 6.32. The point s could be anywhere in the s-plane. In particular, it could be on the $j\omega$ axis. For a fixed value of ω, ω_0, draw arrows from all the zero locations to the point $s=j\omega_0$ in the complex s-plane. Draw arrows from all the pole locations to the point $s=j\omega_0$. Let the arrow from z_1 to $s=j\omega_0$ represent $l_1\epsilon^{j\alpha_1}$, let the arrow from z_2 to $s=j\omega_0$ represent $l_2\epsilon^{j\alpha_2}, \ldots$, and let the arrow from z_m to $s=j\omega_0$ represent $l_m\epsilon^{j\alpha_m}$. Similarly, let the arrow from the pole p_1 to $s=j\omega_0$ represent $L_1\epsilon^{j\beta_1}$, let the arrow from the pole p_2 to $s=j\omega_0$ represent $L_2\epsilon^{j\beta_2}, \ldots$, and let the arrow from the pole p_n to $s=j\omega_0$ represent $L_n\epsilon^{j\beta_n}$. Then from (6.11), $H(j\omega_0)$ is

$$H(j\omega_0)=\frac{K(l_1\epsilon^{j\alpha_1})(l_2\epsilon^{j\alpha_2})\ldots(l_m\epsilon^{j\alpha_m})}{(L_1\epsilon^{j\beta_1})(L_2\epsilon^{j\beta_2})\ldots(L_n\epsilon^{j\beta_n})} \tag{6.123}$$

or

$$H(j\omega_0)=\frac{Kl_1l_2\ldots l_m}{L_1L_2\ldots L_n}\epsilon^{j(\alpha_1+\alpha_2\cdots+\alpha_m-\beta_1-\beta_2-\cdots-\beta_n)}. \tag{6.124}$$

Hence

$$|H(j\omega_0)|=\frac{Kl_1l_2\ldots l_m}{L_1L_2\ldots L_n} \tag{6.125}$$

and the angle ϕ of $H(j\omega_0)$ is

$$\phi=(\alpha_1+\alpha_2+\cdots+\alpha_m)-(\beta_1+\beta_2+\cdots+\beta_n). \tag{6.126}$$

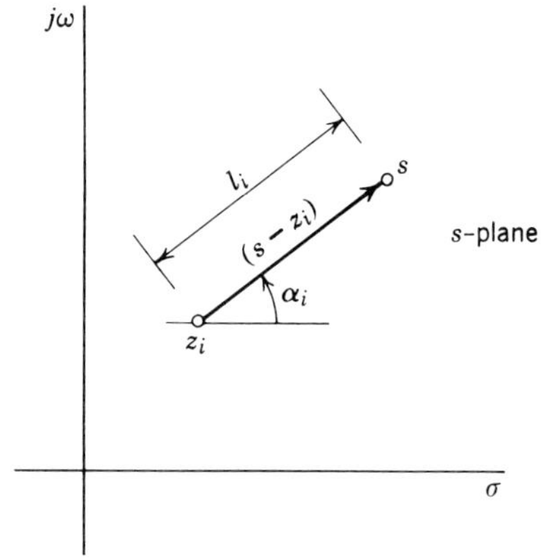

Figure 6.32 Geometrical representation of $(s-z_1)$.

The magnitude and phase angle may be computed using a protractor and ruler for the measurement of $l_1, l_2, \ldots l_m$, $L_1, \ldots, L_n$, $\alpha_1, \ldots \alpha_m$, $\beta_1, \ldots \beta_n$, provided that the *same scale* is used for both a real and an imaginary s-plane axis. The process is repeated for several values of ω, and thus $|H(j\omega)|$ and $\phi(j\omega)$ may be sketched versus ω. This graphical procedure is useful for a rough sketch. Its main advantage is that a general shape of the curves could be visualized by inspection of the pole-zero diagram and imagining a point $s = j\omega_0$ moving on the $j\omega$ axis. For example, as $j\omega_0$ is moved along the $j\omega$ axis, there may be a pole close to the $j\omega$ axis so that, as the $j\omega_0$ point approaches the vicinity of the pole, the length L_i decreases. Since L_i is in the denominator, this tends to increase the magnitude $|H(j\omega)|$ for $j\omega_0$ close to the pole location.

Example 6-14. The pole-zero diagram for a system is given in Figure 6.33. The scale factor K is equal to 1. The arrows from the poles and the zero are drawn to a point $s = j\omega_0$. Denote the lengths and angles as shown. For this value of $s = j\omega_0$,

$$H(j\omega_0) = \frac{l_1 \epsilon^{j\pi/2}}{L_1 \epsilon^{j\theta_1} L_2 \epsilon^{j\theta_2}}$$

$$= \frac{l_1}{L_1 L_2} \epsilon^{j(\pi/2 - \theta_2 - \theta_1)}. \tag{6.127}$$

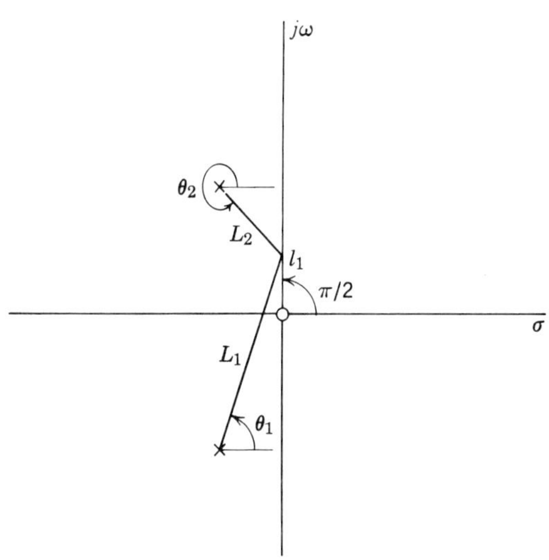

Figure 6.33 Pole-zero diagram and calculation of $H(j\omega)$ for Example 6-14.

It is also seen that for $s = 0$, $H(0) = 0$. As $s = j\omega$ moves upward on the $j\omega$ axis, lengths l_1 and L_1 increase, but length L_2 decreases. L_2 will have its greatest effect in the vicinity of the upper pole. The increases in l_1 and L_1 will approximately maintain their ratio to be constant, but the decrease in L_2 will tend to increase $|H(j\omega)|$. As $j\omega$ passes the pole, l_1, L_1, and L_2 all will increase, tending to decrease $|H(j\omega)|$ for very large values of ω. The lengths l_1, L_1, and L_2 are large and approximately equal, so that $l_1/(L_1L_2)$ tends to zero for $\omega \to \infty$. Thus, by inspection of the pole-zero diagram, we sketch the magnitude response as in Figure 6.34. ▲

6.8 Sensitivity Functions from Block Diagrams

A study of the effects of parameter variations on system behavior and performance is an important part of systems engineering analysis and design. In the evaluation of a preliminary design with respect to tolerances on system parameters, as well as in optimization with respect to control parameters at the disposal of the designer, influence coefficients or partial derivatives with respect to parameters are useful.

In Chapter 3, we described a procedure for obtaining a sensitivity function, associated with a signal and a parameter in a linear time-invariant network, as a signal from another linear time-invariant network. A sensitivity function, associated with a signal and a parameter in a linear time-invariant system simulation diagram, is also obtainable as a signal from another simulation diagram of a linear

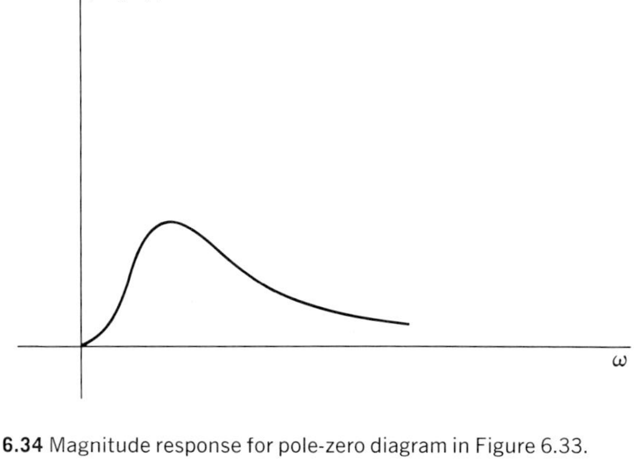

Figure 6.34 Magnitude response for pole-zero diagram in Figure 6.33.

time-invariant system. This was done in Chapter 4 by differentiating the state equations with respect to the perturbed parameter, thus obtaining the state equations for the sensitivity functions. A simulation diagram was then drawn on the basis of the sensitivity state equations In this section, we shall derive a rule for directly drawing a sensitivity-simulation diagram from the given simulation diagram of the linear time-invariant system without writing any equations. Before discussing the general case, let us consider a simple example.

Example 6-15. In Figure 6.35, R, Z, Y, X, and W are Laplace transforms of signals, x_0 is an initial condition constant, and K is a constant gain. It is desired to study the sensitivity of the signal transforms with respect to the gain K. By inspection of the structural connections, it is evident that the signal transforms are related by

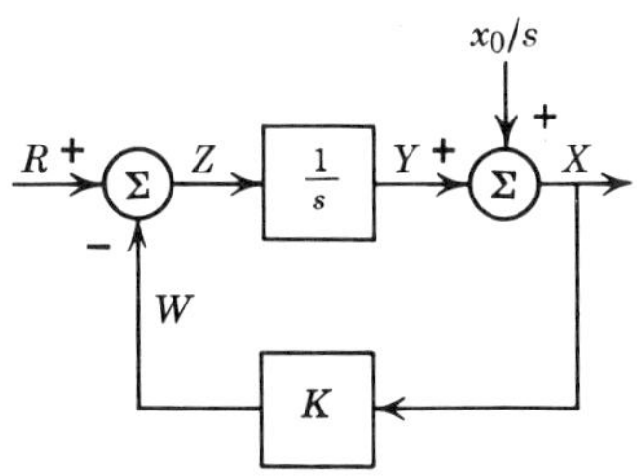

Figure 6.35 Block diagram for Example 6-15.

$$Z = R - W \tag{6.128}$$

and

$$X = Y + \frac{x_0}{s}. \tag{6.129}$$

By definition of the individual block transfer functions, the signal transforms are further related by

$$Y = \frac{1}{s}Z \tag{6.130}$$

and

$$W = KX. \tag{6.131}$$

Suppose that the input R and the initial condition x_0 do not depend on K. Differentiating (6.128) and (6.129), we have

$$\frac{\partial Z}{\partial K} = -\frac{\partial W}{\partial K} \tag{6.132}$$

and

$$\frac{\partial X}{\partial K} = -\frac{\partial Y}{\partial K}. \tag{6.133}$$

Similarly, differentiating (6.130) and (6.131) we obtain

$$\frac{\partial Y}{\partial K} = -\frac{1}{s}\frac{\partial Z}{\partial K} \tag{6.134}$$

and

$$\frac{\partial W}{\partial K} = K\frac{\partial X}{\partial K} + X \tag{6.135}$$

The above four equations in comparison with (6.128) to (6.131) suggest that the sensitivity functions can be obtained as signals in a system that is structurally the duplicate of the original system except for a simple modification, as shown in Figure 6.36.

In the duplicate system, instead of R and x_0 we have zero input signals. Furthermore, the input signal of the K block in the original system is added to the output of the K block of the duplicate system. The partial derivative of every signal in the original system appears at corresponding points in the duplicate system. ▲

Now consider a general simulation diagram of a linear system that consists of integrators, scalors, and summers. By considering the summers, it is seen that the signals are related by linear algebraic equations. Differentiating these equations with respect to a parameter yields the same structural relationships among the partial derivatives as those among the original signals. Differentiating the input-output equations for each block yields the same input-output equations for the partial derivatives except for those blocks that depend on the parameter. For the latter blocks, additional terms appear, similar to (6.135). The same argument applies when we consider general block transfer functions rather than integrators and scalors only. Also, the parameter may appear in more than one block.

Let us now state the rule for drawing a sensitivity block diagram for a transfer-function block diagram, as in Figure 6.37. Let there be p blocks which depend on the parameter q. These blocks are represented by the transfer functions $H_1(s, q)$, $H_2(s, q)$, . . . , $H_p(s, q)$. All other blocks are independent of q. The objective is to draw a sensitivity block diagram whose signals are partial derivatives, with respect to q, of signals in the original transfer-function block diagram, evaluated at $q = q_0$, a nominal parameter value.

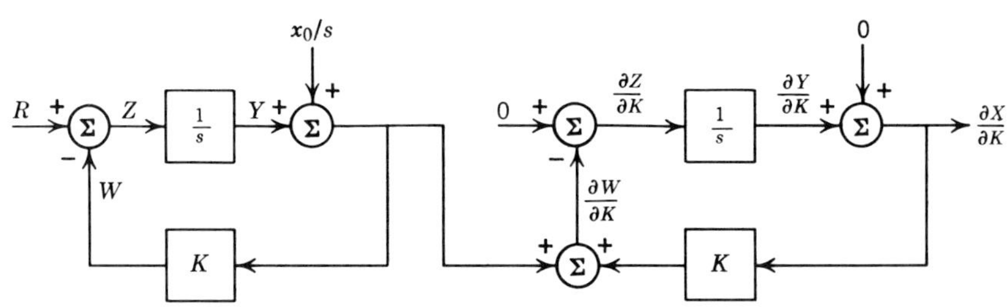

Figure 6.36 Sensitivity block diagram for Example 6-15.

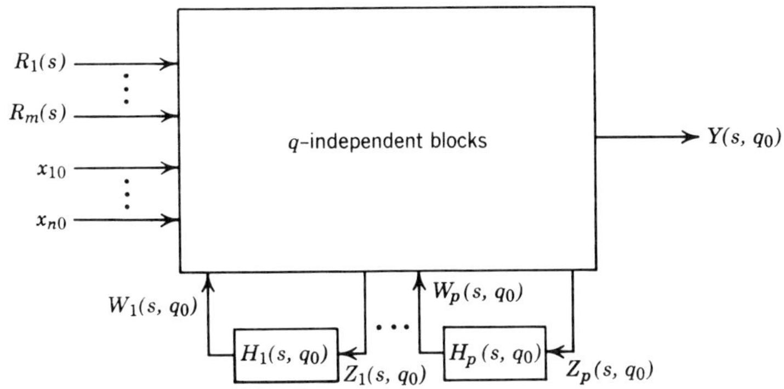

Figure 6.37 Block diagram with p blocks depending on the parameter q. No other blocks are dependent on q.

Rule for Drawing Sensitivity Block Diagram.[3] To draw the sensitivity block diagram, we start with the original block diagram (Figure 6.37) whose sensitivity with respect to q is to be analyzed. Identify all the blocks $H_1, \ldots, H_p$ containing q. Draw a duplicate of the block diagram, with all external inputs $R_1(s), \ldots, R_m(s)$ and all initial conditions $x_{10}, \ldots, x_{n0}$ set to zero. From the *input* of H_i of the original diagram connect a block whose transfer function is $\partial H_i(s, q_0)/\partial q$. Add the output of this block to the *output* of H_i of the duplicate diagram. Repeat this construction for $i = 1, 2, \ldots, p$. Every signal of the duplicate diagram is the sensitivity function of the corresponding signal in the original diagram. The resulting diagram is shown in Figure 6.38.

The notation $\partial H_i(s, q_0)/\partial q$ means

$$\frac{\partial H_i(s, q_0)}{\partial q} = \left.\frac{\partial H_i(s, q)}{\partial q}\right|_{q=q_0}.$$

It is assumed that the inputs $R_1(s), \ldots, R_m(s)$ and the initial conditions $x_{10}, \ldots, x_{n0}$ do not depend on q. The signal $Y(s, q_0)$ is *any* signal in the original transfer-function block diagram. Hence the scheme in Figure 6.38 yields the sensitivity functions of all the signals at the corresponding points of the duplicate block diagram.

Example 6-16. Let us draw a sensitivity block diagram for obtaining $\partial Y/\partial \tau$ for the system in Figure 6.39. From the pattern depicted in

[3]M. L. Bihovski, "Sensitivity and Dynamic Accuracy of Control Systems," *Tekhnicheskaya Kibernetika*, No. 6 (1964); see also, P. V. Kokotovic and R. S. Rutman, "Sensitivity of Automatic Control Systems (Survey), *Automation and Remote Control*, **26** (**4**), 727–749 (April 1965).

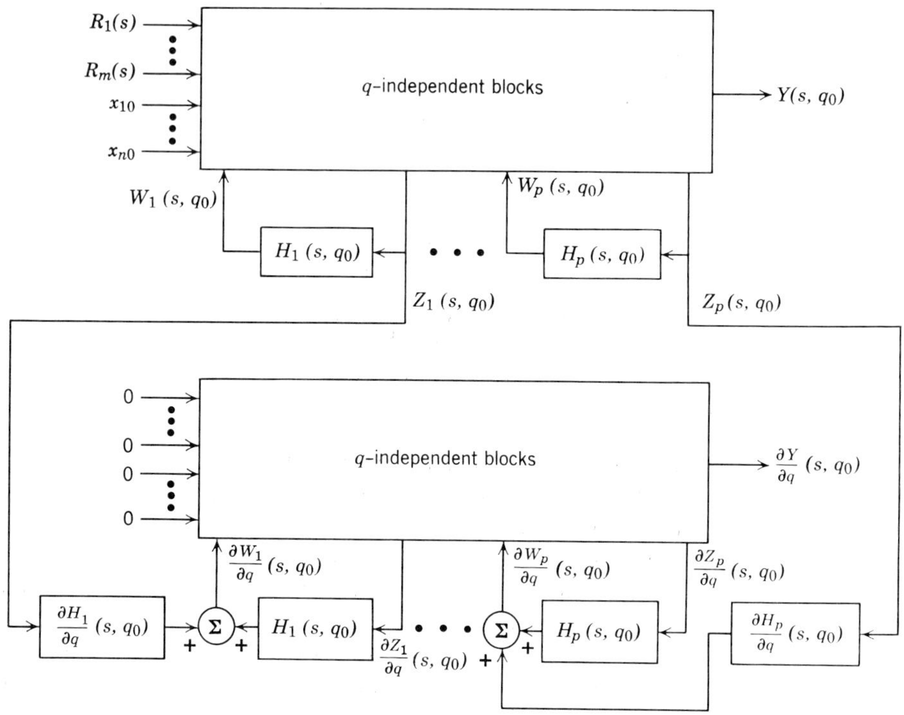

Figure 6.38 Block diagram for obtaining sensitivity functions.

Figure 6.38, we need $\partial H/\partial\tau$ where $H = K_0/[s(\tau s+1)]$:

$$\frac{\partial}{\partial\tau}\left[\frac{K_0}{s(\tau s+1)}\right] = \frac{-K_0}{(\tau s+1)^2}.$$

Following the rule, we obtain the block diagram of Figure 6.40. ▲

Example 6-17. Consider the simulation diagram of the single-axis reaction-wheel attitude control system shown in the upper portion of Figure 4.20. The initial conditions are zero. Since the transfer function of an integrator is $1/s$, the transfer-function block diagram is as shown in Figure 6.41. It is desired to draw a block diagram for a sensitivity analysis with respect to J_v. The first step is to draw the original diagram with its input. Then a duplicate diagram is drawn with inputs

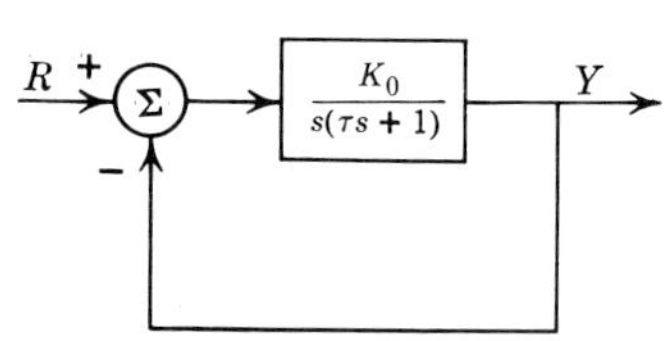

Figure 6.39 Block diagram for Example 6-16.

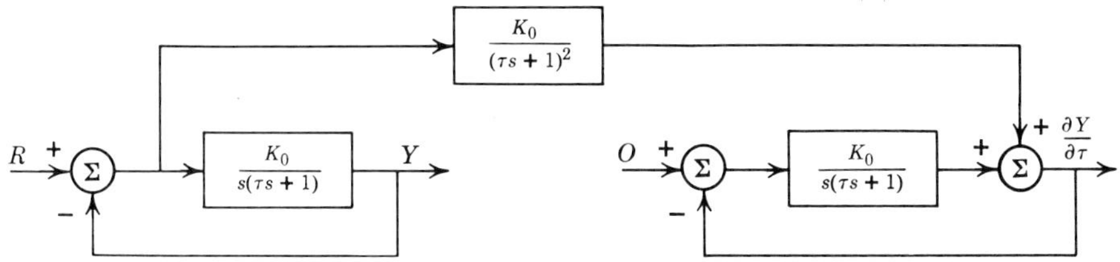

Figure 6.40 Sensitivity with respect to τ for Example 6-16.

set to zero. The third step is to draw the connections between the original diagram and the duplicate diagram. J_v appears in two blocks of the original block diagrams so that two connections will be involved. The first is obtained by using the input of the block in the original diagram whose transfer function is $(1/J_v + 1/J_w)$ as input to a block whose transfer function is

$$\frac{\partial}{\partial J_v}\left(\frac{1}{J_v} + \frac{1}{J_w}\right) = -\frac{1}{J_v^2}. \tag{6.136}$$

The output of this block is added to the output of the block in the duplicate diagram whose transfer function is $(1/J_v + 1/J_w)$. This is shown in Figure 6.42. The next step is to use the input of the block in the original diagram whose transfer function is $1/J_v$ as the input to a block whose transfer function is

$$\frac{\partial}{\partial J_v}\left(\frac{1}{J_v}\right) = -\frac{1}{J_v^2}. \tag{6.137}$$

The output of this block is added to the output of the block in the duplicate diagram whose transfer function is $1/J_v$. The connection is shown in Figure 6.42. The signals in the duplicate diagram are the sensitivity functions of the corresponding signals of the original diagram. The transfer-function block diagram in Figure 6.42 is easily converted to a simulation diagram by replacing the $1/s$ blocks by integrator blocks. Likewise, the simulation diagram of Figure 4.20 is easily converted to a transfer-function block diagram by replacing each integrator block by a transfer-function block $1/s$.

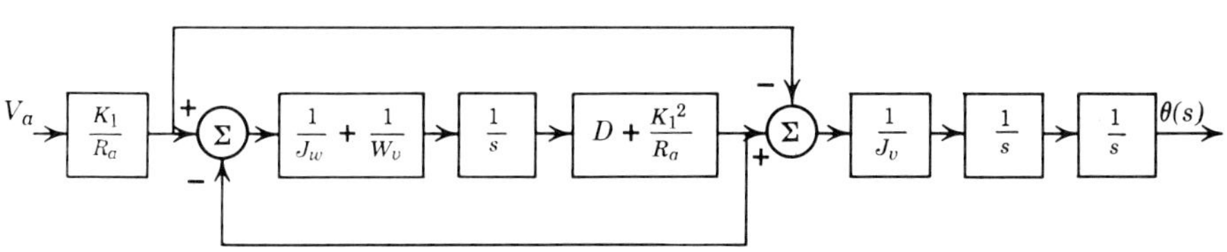

Figure 6.41 Transfer-function block diagram for Example 6-17.

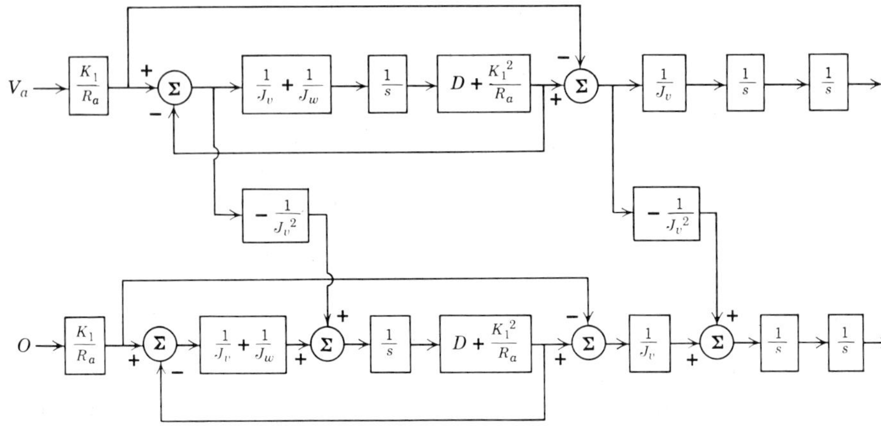

Figure 6.42 Block diagram for sensitivity analysis of block diagram of Figure 6.41 for variations in J_v.

Aside from the minor change of converting $1/s$ to an integrator block, the composite block diagram in Figure 6.42 is not quite the same as the composite diagram of Figure 4.20. However, they yield the same result. By an examination of the two diagrams, it is seen that the difference is in the input signal to the summer following the $1/J_v$ block of the duplicate diagram. In Figure 4.20, the input from the original diagram to the summer is

$$\frac{1}{J_v^2}\left[\frac{K_1}{R_a}V_a - \left(D + \frac{K_1^2}{R_a}\right)\Omega\right].$$

In Figure 6.42, the corresponding input is

$$-\frac{1}{J_v^2}(J_v)(\dot{\omega}).$$

But from (4.34),

$$-\frac{1}{J_v^2}(J_v)\dot{\omega} = -\frac{1}{J_v^2}\left[\left(D + \frac{K_1^2}{R_a}\right)\Omega - \frac{K_1}{R_a}V_a\right].$$

Hence the two inputs are equal, and the sensitivity block diagrams in Figure 4.20 and Figure 6.42 are equivalent. ▲

Sensitivity functions with respect to initial conditions are also obtainable from block diagrams. The scheme is very simple. Let us derive the structure for determining $\partial Y(s, x_{i0})/\partial x_{i0}$ where $Y(s, x_{i0})$ is *any* signal in a transfer function block diagram and x_{i0} is an initial condition that is subject to a small perturbation. Figure 6.43 shows a

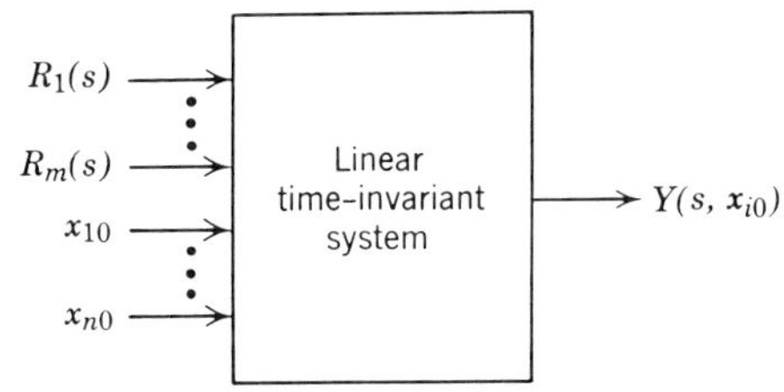

Figure 6.43 Block diagram for sensitivity with respect to x_{i0}.

block diagram with inputs $R_1(s), R_2(s), \ldots, R_m(s)$, and initial conditions $x_{10}, x_{20}, \ldots, x_{n0}$. By differentiating all the structural connection equations and the input-output equations for each block, with respect to x_{i0}, it is seen that the partial derivatives are related by the same equations as those relating the signals in the original system. The only modification is that the inputs are all zero, and the initial conditions are all zero except that which corresponds to x_{i0} which becomes unity. This is shown in Figure 6.44. Since $Y(s, x_{i0})$ is any signal in the original transfer-function block diagram, the scheme in Figure 6.44 applies to each signal of the block diagram. Hence, each signal in the block diagram of Figure 6.44 is a sensitivity function of the corresponding signal in the block diagram of Figure 6.43.

Example 6-18. Let us obtain the sensitivity block diagram for the block diagram in Figure 6.18, with respect to the parameter x_{10}. It should be noted that the signal x_{10}/s may be regarded as the output of a block whose transfer function is $1/s$ and whose input is x_{10}. Similarly, x_{20}/s may be regarded as the output of a block whose transfer function is $1/s$ and whose input is x_{20}. Hence x_{10} and x_{20} can be considered as input signals. The scheme of Figure 6.44 is then applied. The inputs $R_1(s)$, $R_2(s)$, and x_{20} are set to zero, and x_{10} is replaced by 1. The structure is shown in Figure 6.45. Each signal in the block diagram of Figure 6.45

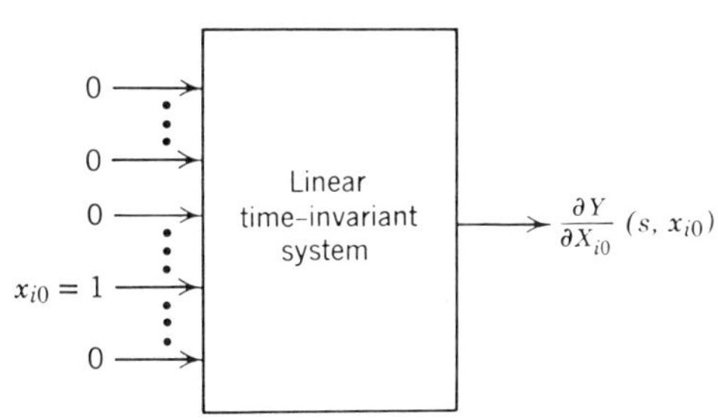

Figure 6.44 Sensitivity block diagram for the block diagram in Figure 6.43 with respect to x_{i0}.

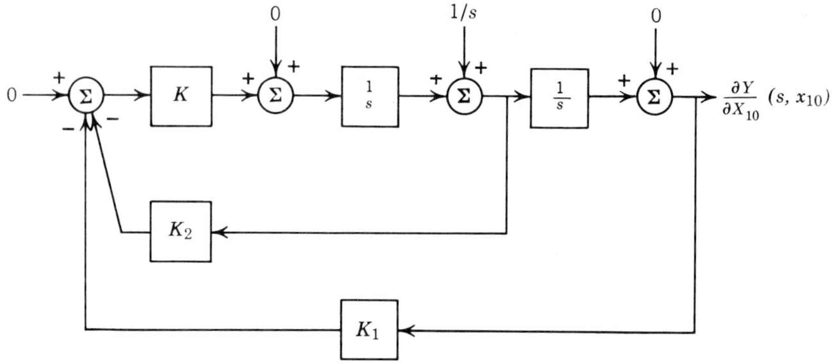

Figure 6.45 Sensitivity block diagram for the block diagram in Figure 6.18 with respect to x_{10}.

is the sensitivity function of the corresponding signal in the block diagram of Figure 6.18. In terms of a simulation diagram, the sensitivity simulation diagram is the same as the original simulation diagram except that the inputs $r_1(t)$ and $r_2(t)$ are set to zero, the initial condition x_{10} is set to 1.

Problems

6-1. A simplified single-axis attitude control system is shown in Figure P6-1. Suppose $K = 4J\sigma^2$.

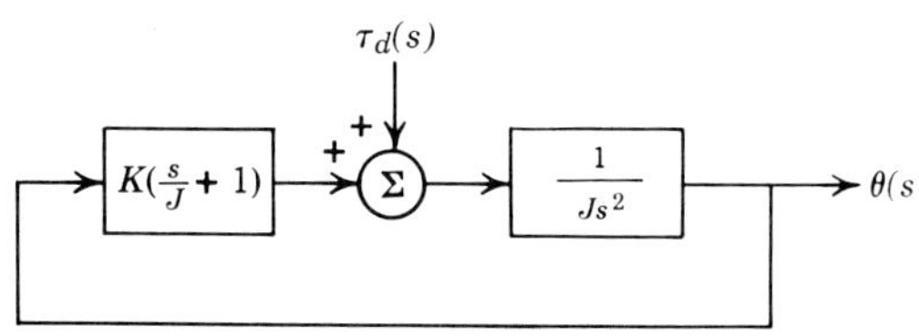

Figure P6-1

(a) Show that the system is critically damped.

(b) Suppose the system is subjected to an external torque

$$\tau_d(t) = T_d \cos \omega_d t.$$

The steady-state attitude response can be written as

$$\theta(t) = \theta_m \cos(\omega_d t + \psi).$$

Find θ_m and ψ in terms of the system parameters.

(c) Sketch the piecewise linear approximation for

$$20 \log \left| \frac{\theta(j\omega)}{\tau_d(j\omega)} \right|.$$

(d) If it is desired that the effects of τ_d be "small," what parameter combinations do you suggest? Why?

6-2. Modify the transfer-function block diagram of the single-axis attitude control system of Figure 6.11 to include the disturbance input torque τ_d and the initial conditions θ_0, $\dot{\theta}_0$, and Ω_0 in addition to θ_r. Use the convention in Figure 6.17 for the initial condition inputs.

6-3. Show from the block diagram of Problem 6-2 that the initial attitude rate $\dot{\theta}_0$ may be represented equivalently by an *impulse* disturbance torque input.

6-4. As discussed in Section 6.4, a multiple-input multiple-output system can be represented in transfer-function matrix form by

$$\mathbf{Y}(s) = \mathbf{H}(s)\mathbf{X}(s).$$

Consider the reaction-wheel system of Problem 6-2. Let

$$\mathbf{X}(s) = \begin{bmatrix} \theta_r(s) \\ \tau_d(s) \\ \theta_0 \\ \Omega_0 \end{bmatrix}, \qquad Y(s) = \begin{bmatrix} \theta(s) \\ \Omega(s) \end{bmatrix}.$$

Find $\mathbf{H}(s)$ by applying Mason's formula to the diagram in Problem 6-2. Use the notation of (6.43) and (6.44) when convenient.

6-5. Suppose that for the reaction-wheel system in Problem 6-2, the parameters of $\theta(s)/\theta_r(s)$ are selected for *critical damping*, $\zeta = 1$. That is, from (6.52) and (6.53), $-K_d = \omega_n^2$ and $K_c - K_d K_b = 2\omega_n$, so that $s^2 + (K_c - K_d K_b)s - K_d = (s + \omega_n)^2$. Find $\theta(t)$ when $\theta(0) = \theta_0$, and $\theta_r(t) = \tau_d(t) \equiv 0$. Sketch $\theta(t)/\theta_0$ vs. $\omega_n t$. Label initial and final values carefully.

6-6. Consider again the critically damped system in Problem 6-5. Suppose the system is subjected to a meteorite impact, represented by an impulse disturbance torque, $\tau_d(t) = T_d \delta(t)$. $\theta_r = \theta_0 = \dot{\theta}_0 = \Omega_0 = 0$. Find $\theta(t)$, and sketch $\theta(t)$ vs. t. Determine the maximum attitude angle $\theta_{\max}$.

6-7. Find and sketch $\Omega(t)$ vs t for the impulsive disturbance of Problem 6-6.

6-8. Using Mason's formula, verify that the transfer function for the system shown in Figure P6-8 is

$$H(s) = \frac{Y(s)}{X(s)} = \frac{b_n s^n + b_{n-1} s^{n-1} + \cdots + b_1 s + b_0}{s^n + a_{n-1} s^{n-1} + \cdots + a_1 s + a_0}.$$

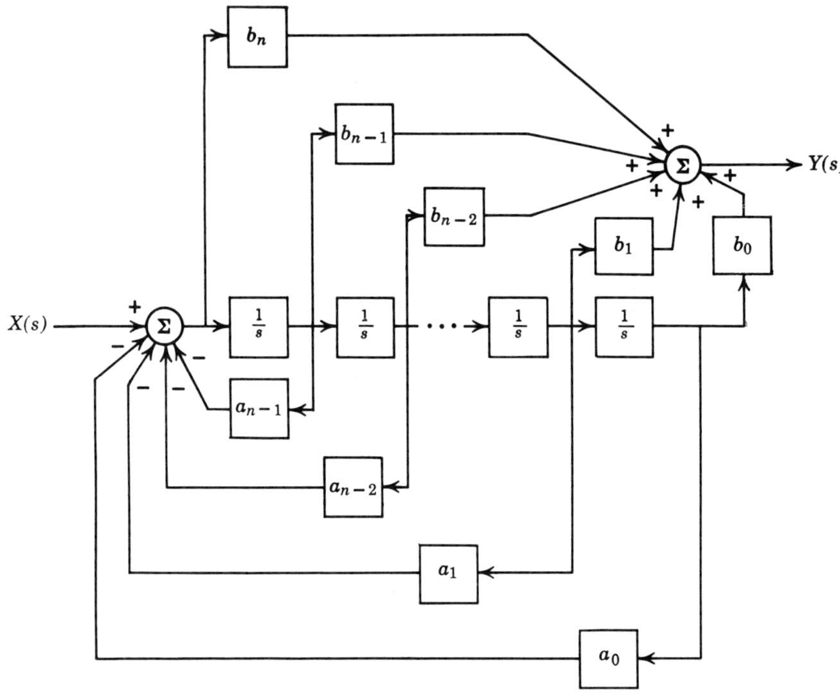

Figure P6-8

Thus the system represents an all-integrator simulation for

$$\frac{d^n y}{dt^n} + a_{n-1}\frac{d^{n-1}y}{dt^{n-1}} + \cdots + a_1\frac{dy_1}{dt} + a_0 y = b_n\frac{d^n x}{dt^n} + b_{n-1}\frac{d^{n-1}x}{dt^{n-1}} + \cdots + b_1\frac{dx}{dt} + b_0 x.$$

6-9. Using Mason's formula, verify that the system shown in Figure P6-9 has a transfer function

$$H(s) = \frac{Y(s)}{X(s)} = \frac{b_n s^n + b_{n-1}s^{n-1} + \cdots + b_1 s + b_0}{s^n + a_{n-1}s^{n-1} + \cdots + a_1 s + a_0}$$

where

$$b_n = \beta_0$$

$$b_{n-1} = a_{n-1}\beta_0 + \beta_1$$

$$b_{n-2} = a_{n-2}\beta_0 + a_{n-1}\beta_1 + \beta_2$$

$$\vdots$$

$$b_0 = a_0\beta_0 + a_1\beta_1 + \cdots + a_{n-1}\beta_{n-1} + \beta_n.$$

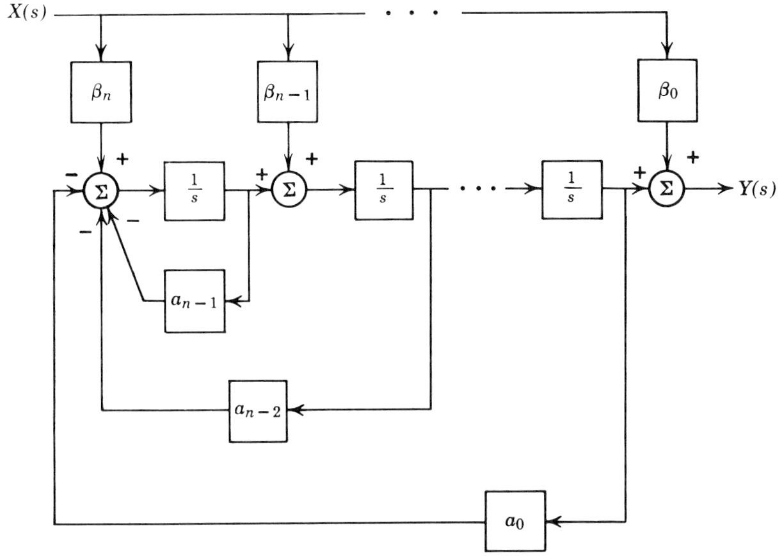

Figure P6-9

6-10. Determine the transfer function $V_0(s)/V(s)$ for the network in Figure P6-10.

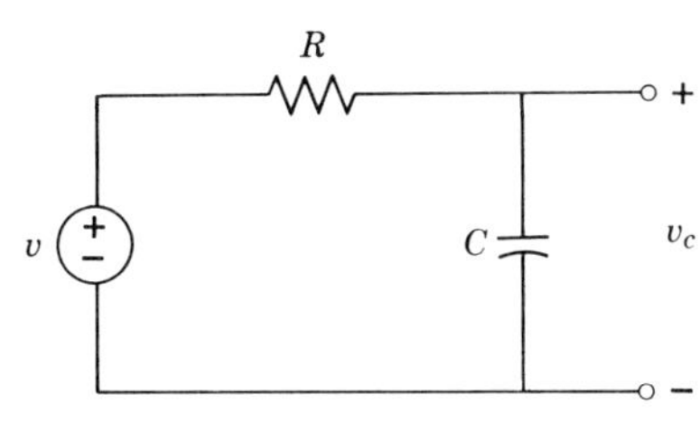

Figure P6-10

6-11. Repeat Problem 6-10 for the network in Figure P6-11.

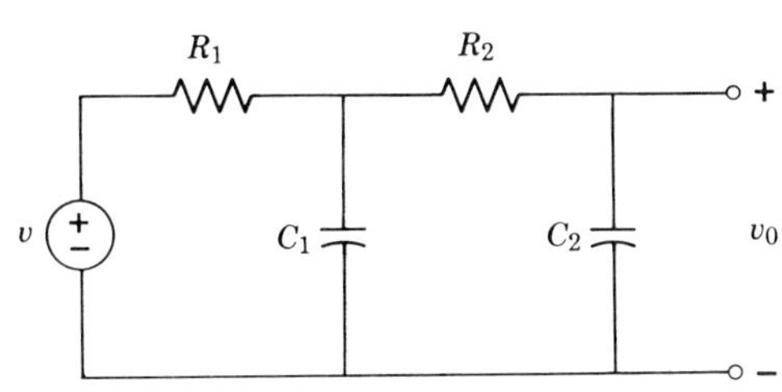

Figure P6-11

6-12. Repeat Problem 6-10 for the network in Figure P6-12.

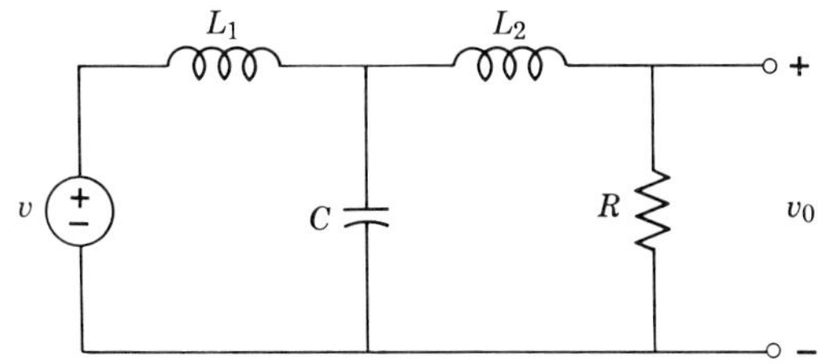

Figure P6-12

6-13. Figure P6-13 shows a pictorial model for an automobile suspension system using one wheel only, where M_1 is $\frac{1}{4}$ of the automobile mass, K_1 and D_1 are the spring and shock absorber for one wheel, M_2 is the mass of one wheel, and K_2 is the stiffness coefficient of the tire. $x(t)$ represents the road terrain elevation, x_1 is the vertical position of the automobile, and x_2 is the vertical position of the tire. Determine the transfer function $X_1(s)/X(s)$.

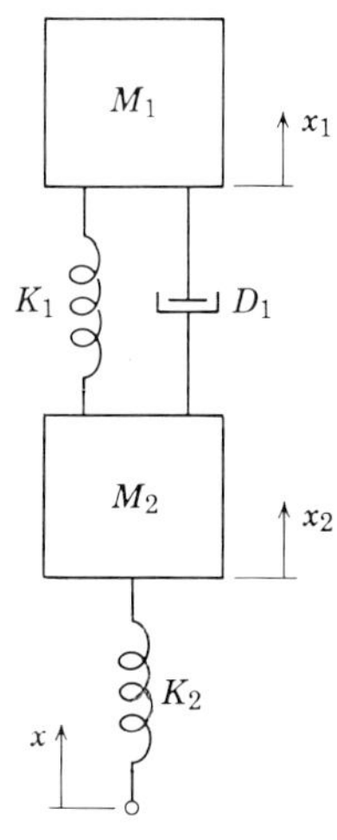

Figure P6-13

6-14. Determine the transfer function $X_2(s)/X(s)$ for the system in Problem 6-13.

6-15. Using Mason's formula, determine the transfer function Y/X for the block diagram in Figure P6-15.

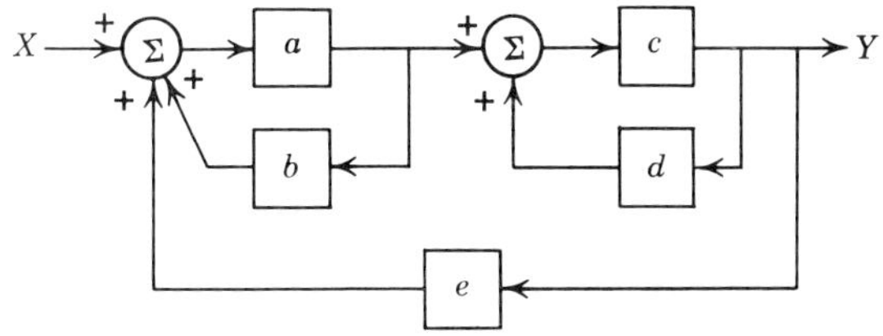

Figure P6-15

6-16. Repeat Problem 6-15 for the block diagram in Figure P6-16.

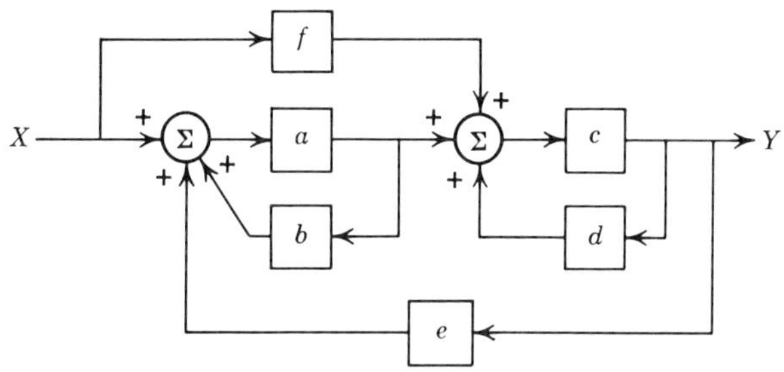

Figure P6-16

6-17. Expand the following functions in partial fractions:

$$\text{(a)}\ H(s) = \frac{(s+1)(s+2)}{s(s+3)(s+4)}$$

$$\text{(b)}\ H(s) = \frac{(s+1)^2}{(s+2)(s+3)}$$

$$\text{(c)}\ H(s) = \frac{3(s+1)}{(s+2)^2(s+3)}$$

$$\text{(d)}\ H(s) = \frac{5}{(s+1)^2(s+2)^3}$$

$$\text{(e)}\ H(s) = \frac{8}{[(s+1)^2+1](s+2)}$$

6-18. Sketch $20 \log |H(j\omega)|$ and the argument (angle) of $H(j\omega)$ vs. ω on logarithmic scale, for $H(j\omega)$ in Problem 6-17(a). Show the asymptotes.

6-19. Repeat Problem 6-18 for $H(s)$ in Problem 6-17(b).

6-20. Repeat Problem 6-18 for $H(s)$ in Problem 6-17(c).

6-21. Repeat Problem 6-18 for $H(s)$ in Problem 6-17(d).

6-22. Repeat Problem 6-18 for $H(s)$ in Problem 6-17(e).

6-23. The input to a linear time-invariant system is $v(t) = 10 \sin 5t$. If the transfer function $H(s)$ of the system is given by Problem 6-17(b), determine the steady-state output.

6-24. Repeat Problem 6-23 for $H(s)$ given by Problem 6-17(c).

6-25. Repeat Problem 6-23 for $H(s)$ given by Problem 6-17(d).

6-26. Repeat Problem 6-23 for $H(s)$ given by Problem 6-17(e).

6-27. For the network in Figure P6-27, the parameters are chosen so that $R_C = R_L = \sqrt{L/C}$.

(a) Determine the transfer function $V_0(s)/V(s)$.

(b) Sketch the actual frequency response diagrams for the transfer function.

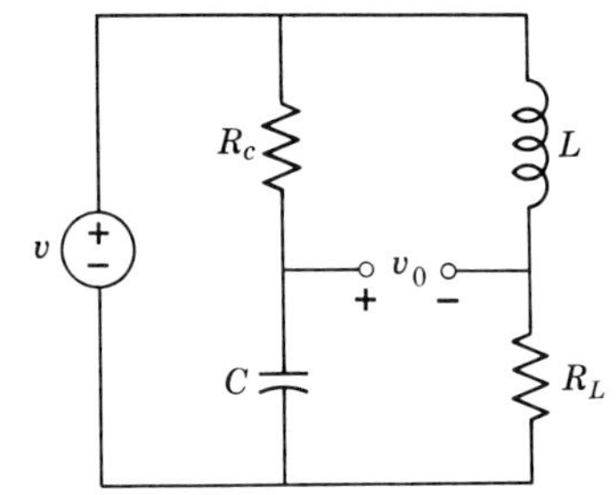

Figure P6-27

6-28. The input to a linear time-invariant system is a unit step function $u(t)$. If the initial conditions are zero and if the transfer function is given by $H(s)$ in Problem 6-17(a), determine the output response for $t \geqslant 0$.

6-29. Repeat Problem 6-28 for $H(s)$ given in Problem 6-17(b).

6-30. Repeat Problem 6-28 for $H(s)$ given in Problem 6-17(c).

6-31. Repeat Problem 6-28 for $H(s)$ given in Problem 6-17(d).

6-32. Repeat Problem 6-28 for $H(s)$ given in Problem 6-17(e).

6-33. The piecewise linear approximation for the magnitude response in db of a rational function is shown in Figure P6-33. Determine a corresponding function $H(s)$.

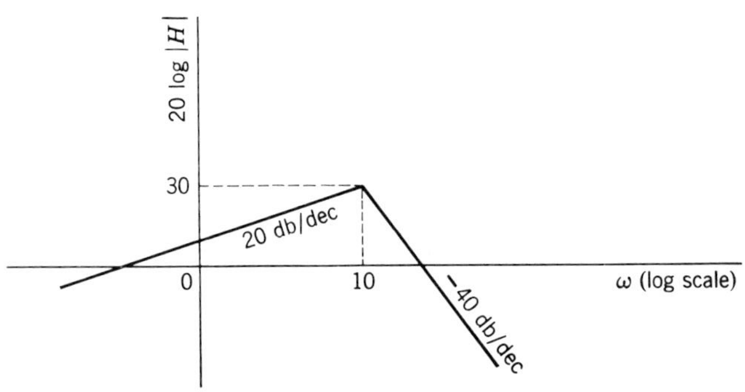

Figure P6-33

6-34. Repeat Problem 6-33 for the piecewise linear approximation in Figure P6-34.

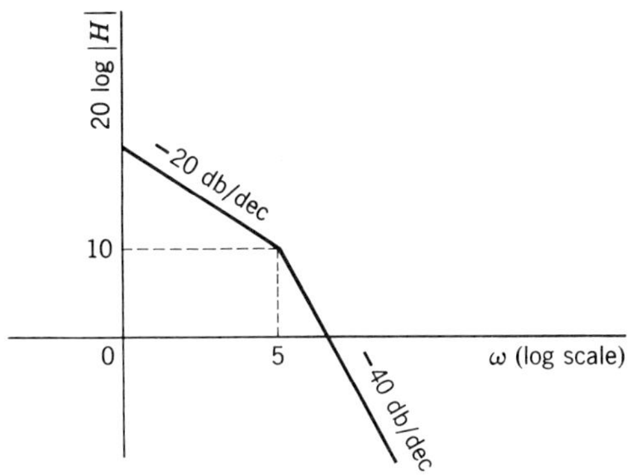

Figure P6-34

6-35. The pole-zero diagram of a rational function $H(s)$ is shown in Figure P6-35. It is known that $|H(j5)| = 10$. Determine $H(s)$.

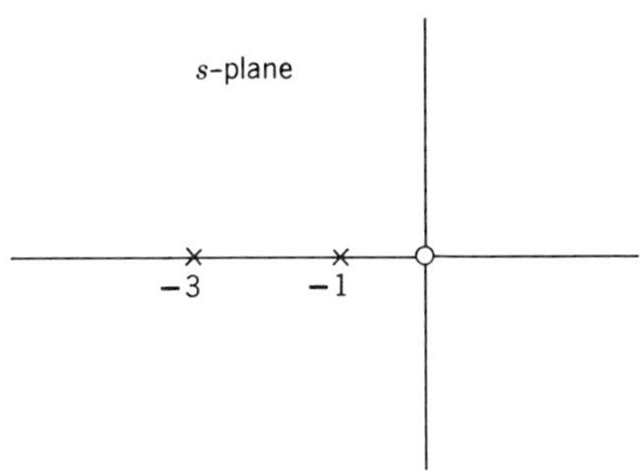

Figure P6-35

6-36. Repeat Problem 6-35 for the pole-zero diagram of Figure P6-36 and $|H(j5)| = 4$.

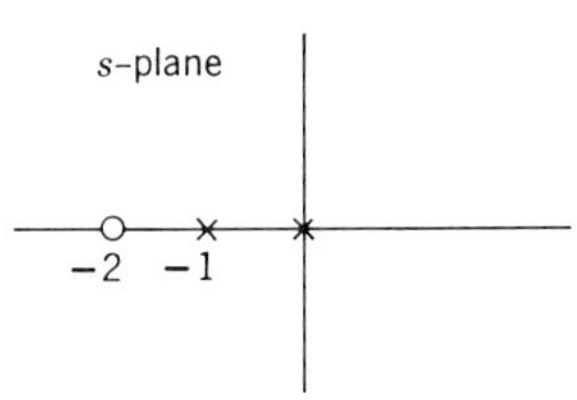

Figure P6-36

6-37. Repeat Problem 6-35 for the pole-zero diagram of Figure P6-37 and $H(0) = 1$.

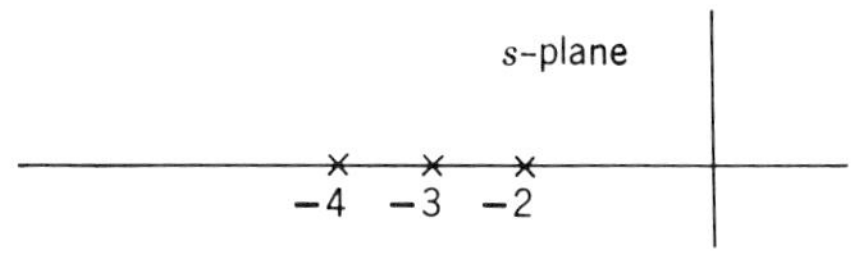

Figure P6-37

6-38. Figure P6-38 shows a simplified model for a rotating machine with an unbalanced rotor causing vibration on the floor. The machine is

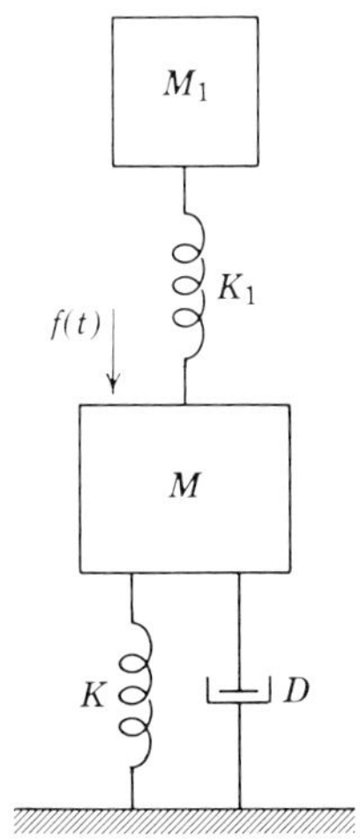

Figure P6-38

represented by mass M. The floor support is represented by the spring constant K and damping D. The unbalance in the rotor shift is represented by a vertical force $f(t)$

$$f(t) = F_0 \sin \omega_0 t.$$

The vibration force $f_2(t)$ transmitted to the floor is the sum of the forces transmitted through K and D. This force excludes the force due to gravity. M_1 and K_1 are mounted on the machine to reduce the force transmitted to the floor.

(a) With M_1 and K_1 removed, determine the transfer function $F_2(s)/F(s)$. Determine the peak value of the vibration force transmitted to the floor in the steady state, in terms of F_0, ω_0, M, K, and D.

(b) The vibration absorber (M_1 and K_1) is to be designed so that the steady-state vibration force transmitted to the floor is zero. Determine the expression for the transfer function $F_2(s)/F(s)$. Determine the relationship among M_1, K_1, and ω_0 in order for the steady-state vibration force to be zero.

(c) Draw an electrical network analog with force analogous to current for the mechanical system with the vibration absorber mounted. From the electrical network diagram, determine the condition to be satisfied for the analogs of M_1 and K_1 so that the current analogous to f_2 is zero. (*Hint.* Recall that a series *LC* circuit in series resonance behaves as a short circuit.)

6-39. For the system in Problem 6-38, compute the peak value of the steady-state vibration force f_2 with $f(t) = F_0 \sin(\frac{1}{2}\omega_0 t)$ for:

(a) The vibration absorber omitted.

(b) The vibration absorber added where M_1 and K_1 are designed to eliminate floor vibration at the frequency ω_0.

6-40. Repeat Problem 6-39 for $f(t) = F_0 \sin(2\omega_0 t)$.

6-41. Draw a sensitivity block diagram for a sensitivity analysis of the system in Figure P6-1 with respect to K.

6-42. Repeat Problem 6-41 for a sensitivity analysis with respect to σ.

6-43. Repeat Problem 6-41 for a sensitivity analysis with respect to J.

6-44. In Figure P6-15, it is given that the transfer functions a and c depend on a parameter q and that the other transfer functions are independent of q. Draw a sensitivity block diagram for the sensitivity block diagram for the sensitivity analysis of the system with respect to q.

6-45. Repeat Problem 6-44 for the system in Figure P6-16.

6-46. Draw a sensitivity block diagram for a sensitivity analysis with respect to x_{20} of the system in Figure 6.18.

6-47. For the system in Figure 6.11, verify that

$$\frac{\partial\theta}{\partial K_b} = \frac{\partial\dot{\theta}}{\partial K_a}$$

and

$$\frac{\partial\dot{\theta}}{\partial K_b} = \frac{\partial\ddot{\theta}}{\partial K_a}$$

Hence $\partial\theta/\partial K_a$, $\partial\dot{\theta}/\partial K_a$, $\partial\theta/\partial K_b$ and $\partial\dot{\theta}/\partial K_b$ can be obtained from a single sensitivity block diagram. Draw the sensitivity block diagram.

CHAPTER 7

System Response via State Variables

One of the advantages of a system description in state form is that an all-integrator simulation diagram may be drawn directly from an inspection of the equations. This leads almost immediately to an analog computer plugging diagram for simulation or computation. Even for digital or hybrid simulation and computation, the vector first-order form is more convenient to program than a single *n*th-order form of differential equation. A second advantage of the state form is that most of the available mathematical theory for handling differential equations, especially nonlinear differential equations, is in terms of a set of first-order differential equations.

In previous chapters, we have written equations using conservation of energy, D'Alembert's principle, Kirchhoff's laws, and element equations based on postulates such as Ohm's law and Hooke's law. The differential equations describing the complete behavior of a system may be expressed in terms of a set of first-order differential equations together with a set of algebraic equations. The complete solution of the differential equations is obtained if certain initial conditions are given. For example, in an electrical network, if all the capacitor voltages and all the inductor currents are given at the initial time, then the complete solution for any of the voltages or currents can be obtained. In a mechanical translational system, if all the spring displacements and mass velocities are given at the initial time, then the complete solution for any of the forces or velocities can be obtained. If initial mass positions are also given, then the complete solution for mass positions can be obtained. The initial conditions usually pertain to initial stored energies in elements. For example a capacitor voltage determines the stored energy in the capacitor's electric field. An inductor current determines the stored energy in the inductors magnetic field. A mass velocity determines the kinetic energy of the mass. A spring displacement determines the potential energy of the spring.

It may not be possible to arbitrarily specify all the initial conditions. For example, in an electrical network which contains three capacitors which are in a loop with no other element in the loop, then by Kirchhoff's voltage law, $v_1 + v_2 + v_3 = 0$. In this case, once v_1 and v_2 are specified at the initial instant, v_3 must equal $-(v_1 + v_2)$ at the initial instant. Thus v_3 cannot be specified independently of v_1 and v_2 at the initial instant.

A minimum set of initial conditions, sufficient for determining the complete system behavior (such as all currents and all voltages in a network) when the input and system parameters are known is called an *initial state* of the system. An initial state may be represented by a vector whose components are the initial conditions in the minimum set of initial conditions described above. Denote this vector by $\mathbf{x}(t_0)$ where t_0 is the initial time. The components of an initial state vector are called *initial state variables*. These components are initial values of certain signals (such as currents or voltages) in a system. At an arbitrary time t later than t_0, these same currents and voltages will generally have different values. Thus if $x_i(t_0)$ is the ith component of $\mathbf{x}(t_0)$ at time t_0, the value at time t of the signal whose initial value is $x_i(t_0)$, denoted by $x_i(t)$, will be different from $x_i(t_0)$ in general. The vector $\mathbf{x}(t)$ whose components are $x_i(t)$ is called a *state vector* at time t. The components of $\mathbf{x}(t)$ are called *state variables* at time t. A state vector at time t serves as an initial state vector for determining the complete system behavior for all time greater than t.

For an electrical network, if there are no loops containing capacitors and voltage sources only, or nodes of inductors and current sources only, then inductor currents and capacitor voltages may be chosen as state variables. The dimension of $\mathbf{x}(t)$ is equal to the total number of capacitors and inductors minus the number of loops of capacitors and voltage sources, and minus the number of nodes of inductors and current sources.

Example 7-1. As an example of a state variable description, consider the electrical network of Figure 7.1.

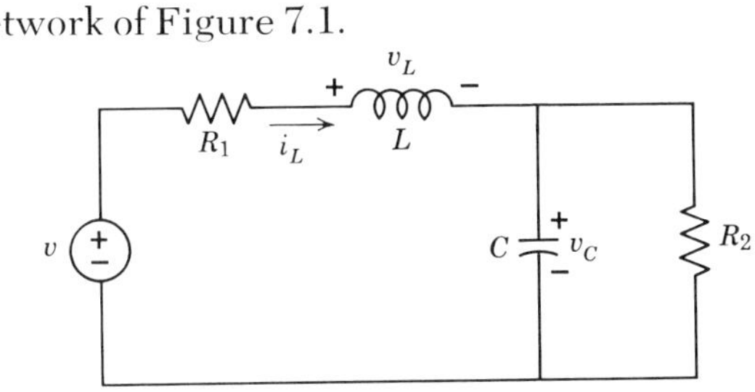

Figure 7.1 Simple electrical network described by two first-order differential equations.

Applying KVL to the loop containing v, R_1, L, and C, and using the branch characteristics

$$-v+R_1 i_L+L\frac{di_L}{dt}+v_C=0 \tag{7.1}$$

or

$$\frac{di_L}{dt}=-\frac{R_1}{L}i_L-\frac{1}{L}v_C+\frac{1}{L}v. \tag{7.2}$$

Applying KCL to the node connecting R_2, C, and L, and using the branch characteristics,

$$C\frac{dv_C}{dt}+\frac{1}{R_2}v_C-i_L=0 \tag{7.3}$$

or

$$\frac{dv_C}{dt}=\frac{1}{C}i_L-\frac{1}{R_2C}v_C. \tag{7.4}$$

Equations (7.2) and (7.4) may be written in matrix form as

$$\begin{bmatrix}\dfrac{di_L}{dt}\\ \dfrac{dv_C}{dt}\end{bmatrix}=\begin{bmatrix}-\dfrac{R_1}{L} & -\dfrac{1}{L}\\ \dfrac{1}{C} & -\dfrac{1}{R_2C}\end{bmatrix}\begin{bmatrix}i_L\\ v_C\end{bmatrix}+\begin{bmatrix}\dfrac{1}{L}v\\ 0\end{bmatrix} \tag{7.5}$$

The two-dimensional vector $[i_L\, v_C]^T$ qualifies as a state vector because by knowning i_L and v_C for any time t_1 and the input voltage $v(t)$ for $t_1 < t < t_2$, then (7.5) may be solved uniquely for $i_L(t)$ and $v_C(t)$ for $t_1 < t < t_2$. Once i_L and v_C are determined, all currents and voltages in the network may be expressed in terms of i_C, v_C, and the network parameters. That is, if we know the initial conditions $v_C(t_1)$ and $i_L(t_1)$, the differential equations in (7.5) have a unique solution. Furthermore, since the second-order system of equations require two independent initial conditions for a unique solution, a state vector must be two-dimensional. $[i_L\, v_C]^T$ may be chosen as a state vector for the network in Figure 7.1. If v_L is the output, it may be expressed in terms of the state variables by applying KVL to the loop containing v, R_1, L, and C,

$$-v+R_1 i_L+v_L+v_C=0 \tag{7.6}$$

or

$$v_L=-R_1 i_L-v_C+v. \tag{7.7}$$

Thus with v_L as output, (7.5) and (7.7) describe the network. Instead of i_L and v_C, it is readily verified that v_{R_1} and i_{R_2} may be chosen as state variables because v_{R_1} is R_1 times i_L and i_{R_2} is v_C divided by R_2. Hence knowledge of v_{R_1} and i_{R_2} immediately yield values for i_L and v_C. There are many other possible state vectors for this network. ▲

For any choice of a state vector such as $\mathbf{x}_1(t)$ and if $\mathbf{T}$ is any nonsingular matrix $n \times n$ where the dimension of $\mathbf{x}_1(t)$ is n, then $\mathbf{x}_2(t) = \mathbf{T}\mathbf{x}_1(t)$ is also a state vector. This is because for a given $\mathbf{x}_2(t)$, $\mathbf{x}_1(t)$ is determined from $\mathbf{x}_1(t) = \mathbf{T}^{-1}\mathbf{x}_2(t)$. Since $\mathbf{x}_2(t)$ has dimension n, it satisfies the requirements for a state vector if $\mathbf{x}_1(t)$ does. Thus, there are infinitely many choices for a state vector of a system. However, once a choice is made, it is customary to refer to the chosen vector as the state vector of the system.

The state equations for a *linear system* are written as

$$\frac{d\mathbf{x}}{dt} = \mathbf{A}\mathbf{x} + \mathbf{B}\mathbf{u} \tag{7.8}$$

$$\mathbf{y} = \mathbf{C}\mathbf{x} + \mathbf{D}\mathbf{u} \tag{7.9}$$

where $\mathbf{x}(t)$ is an n-dimensional state vector, $\mathbf{y}(t)$ is the p-dimensional output vector, and $\mathbf{u}(t)$ is the m-dimensional input vector. The dimension of $\mathbf{x}$ is defined as the order of the systems. The matrices $\mathbf{A}$, $\mathbf{B}$, $\mathbf{C}$, and $\mathbf{D}$ are either constants or functions of time. If $\mathbf{A}$, $\mathbf{B}$, $\mathbf{C}$, and $\mathbf{D}$ do not depend on t, that is, the matrices are constants, we say that the system is *linear time-invariant*. For instance, the system in Figure 7.1 is linear time-invariant because the equations describing it, Equations 7.5 and 7.7, involve constant matrices only. A system for which the matrices explicitly depend on t is said to be *time-varying*. More general definitions of linearity and time-invariance are given in Section 7.7.

A system described by nonlinear differential equations is called a nonlinear system.

Example 7-2. Recall the equations for angular motion of a spacecraft discussed in Chapter 5. Assuming that the body-fixed axes coincide with the principal axes of inertia, the Euler dynamic equations are reproduced below [see (5.92)]:

$$J_1\dot{\omega}_1 = \tau'_{c_1} + \tau'_{d_1} + (J_2 - J_3)\omega_2\omega_3 \tag{7.10}$$

$$J_2\dot{\omega}_2 = \tau'_{c_2} + \tau'_{d_2} + (J_3 - J_1)\omega_3\omega_1 \tag{7.11}$$

$$J_3\dot{\omega}_3 = \tau'_{c_3} + \tau'_{d_3} + (J_1 - J_2)\omega_1\omega_2. \tag{7.12}$$

Dividing by J_1, J_2, and J_3, respectively, we obtain

$$\dot{\omega}_1 = \frac{J_2 - J_3}{J_1}\omega_2\omega_3 + \frac{1}{J_1}\tau'_{c_1} + \frac{1}{J_1}\tau'_{d_1} \tag{7.13}$$

$$\dot{\omega}_2 = \frac{J_3 - J_1}{J_2}\omega_3\omega_1 + \frac{1}{J_2}\tau'_{c_2} + \frac{1}{J_2}\tau'_{d_2} \tag{7.14}$$

$$\dot{\omega}_3 = \frac{J_1 - J_2}{J_3}\omega_1\omega_2 + \frac{1}{J_3}\tau'_{c_3} + \frac{1}{J_3}\tau'_{d_3} \tag{7.15}$$

The variables ω_1, ω_2, and ω_3 are the angular velocities with respect to the body fixed axes. These angular velocities are related to the fixed inertial reference by means of the Euler angles θ_1, θ_2, and θ_3 as discussed in Chapter 5 also. These equations [see (5.42)] are reproduced below:

$$\dot{\psi} = \frac{\sin\varphi}{\sin\theta}\omega_1 + \frac{\cos\varphi}{\sin\theta}\omega_2 \tag{7.16}$$

$$\dot{\varphi} = -\sin\varphi\cot\theta\,\omega_1 - \cos\varphi\cot\theta\,\omega_2 + \omega_3 \tag{7.17}$$

$$\dot{\theta} = \cos\varphi\,\omega_1 - \sin\varphi\,\omega_2 \tag{7.18}$$

From a study of mechanics, it is known that six initial conditions are required to determine the complete behavior of the system. The variables ω_1, ω_2, ω_3, θ_1, θ_2, and θ_3 may be regarded as the state variables of the system and the six equations (7.13) to (7.18), constitute the state description of the angular motion of the spacecraft. The state variables are assumed to be the outputs in this case. The first-order differential equations are obviously nonlinear and the spacecraft attitude-control system is said to be nonlinear. Notice, however, that the differential equations do not contain time t explicitly, so that the nonlinear spacecraft attitude system is time-invariant. ▲

In general, we may write the state equations of a nonlinear physical system in the form

$$\frac{dx_1}{dt} = f_1(x_1, x_2, \ldots, x_n, u_1, u_2, \ldots, u_m, t)$$

$$\frac{dx_2}{dt} = f_2(x_1, x_2, \ldots, x_n, u_1, u_2, \ldots, u_m, t)$$

$$\vdots$$

$$\frac{dx_n}{dt} = f_n(x_1, x_2, \ldots, x_n, u_1, u_2, \ldots, u_m, t)$$

where x_i are the state variables, u_i are the input variables and f_i are functions of the state variables, the input variables, and time t. For the output variables, the component equations are

$$
\begin{aligned}
y_1 &= h_1(x_1, x_2, \ldots, x_n, u_1, \ldots, u_m, t)\\
y_2 &= h_2(x_1, x_2, \ldots, x_n, u_1, \ldots, u_m, t)\\
&\bullet\\
&\bullet\\
&\bullet\\
y_p &= h_p(x_1, \ldots, x_n, u_1, \ldots, u_m, t)
\end{aligned}
$$

these $n+p$ equations together with n initial conditions $x_1(t_0), x_2(t_0), \ldots, x_n(t_0)$ describe the behavior of the system. The above equations may be rewritten in compact form as

$$\frac{d\mathbf{x}}{dt} = \mathbf{f}(\mathbf{x}, \mathbf{u}, t), \; t > t_0 \tag{7.19}$$

$$\mathbf{y} = \mathbf{h}(\mathbf{x}, \mathbf{u}, t), \; t > t_0 \tag{7.20}$$

$$\mathbf{x}(t_0) = \mathbf{x}_0 \tag{7.21}$$

where $\mathbf{x}(t)$ is the state n-vector at time t, $\mathbf{y}$ is the output p-vector, $\mathbf{u}$ is the input m-vector, $\mathbf{f}$ is an n-vector valued function (column matrix of functions), $\mathbf{h}$ is a vector valued function and $\mathbf{x}(t_0)$ is the initial state. The vector differential equation in (7.19) together with the initial state condition in (7.21) determines the state vector $\mathbf{x}(t)$ for all t in any interval $[t_0, T]$ for which the input vector $\mathbf{u}(t)$ is specified. Equation 7.20 then relates the state vector and the input vector to the output vector at the time instant t.

In this chapter, concepts and methods for the analysis of linear systems will be developed in terms of state variables. It will be shown that the response of a linear system may be expressed as a superposition of the response due to the input with zero initial state and the response due to the initial state with zero input. State and transfer function representations will be related.

For clarity, the basic ideas are presented for first- and second-order systems in the first six sections. These ideas are generalized for nth-order systems in Sections 7.7 and 7.8. Starred Sections 7.9, 7.10, and 7.11 concern advanced topics and may be omitted on first reading.

7.1 State Variable Response of First-Order Systems

The ideas of state variable analysis will be presented first in the context of first-order and second-order systems for simplicity. General nth-order systems will be considered later in the chapter.

Consider a linear time-invariant system whose behavior is completely described by the scalar first-order differential equation

$$\frac{dx}{dt} = ax + bu \tag{7.22}$$

where a and b are constants, x is the state, and u is the input. The output is equal to x. The initial state is specified as

$$x(t_0) = x_0 \tag{7.23}$$

For $u = 0$, it is easily verified that

$$x(t) = c\epsilon^{at} \tag{7.24}$$

satisfies (7.22), where c is an arbitrary constant. To solve for the complete solution of (7.22) the variation of parameters method may be employed. The procedure is to assume that c in (7.24) is time varying. Then $x(t)$ in (7.24) is substituted in (7.22) for $u \neq 0$, yielding an equation for c. If we follow this, we have

$$ca\epsilon^{at} + \dot{c}\epsilon^{at} = ac\epsilon^{at} + bu. \tag{7.25}$$

Thus

$$\frac{dc(t)}{dt} = b\epsilon^{-at}u(t). \tag{7.26}$$

Integrating (7.26),

$$c(t) - c(t_0) = \int_{t_0}^{t} b\epsilon^{-a\tau} u(\tau)\, d\tau \tag{7.27}$$

or

$$c(t) = c(t_0) + \int_{t_0}^{t} b\epsilon^{-a\tau} u(\tau)\, d\tau. \tag{7.28}$$

From (7.24) and (7.28),

$$x(t) = \epsilon^{at} c(t_0) + \epsilon^{at} \int_{t_0}^{t} b\epsilon^{-a\tau} u(\tau)\, d\tau. \tag{7.29}$$

For $t = t_0$, (7.29) yields

$$x(t_0) = \epsilon^{at_0} c(t_0). \tag{7.30}$$

Equations 7.23 and 7.30 together yield

$$c(t_0) = \epsilon^{-at_0} x_0 \tag{7.31}$$

Thus (7.29) may be written as

$$x(t) = \epsilon^{a(t-t_0)} x_0 + \int_{t_0}^{t} b\epsilon^{a(t-\tau)} u(\tau)\, d\tau. \tag{7.32}$$

Observe that ϵ^{at} has been shifted inside the integral, but this is allowable since it is independent of τ.

Equation (7.32) gives the state $x(t)$, which is also the output, in terms of the initial state x_0 and the input function $u(\tau)$ for $t_0 \leqslant \tau \leqslant t$. When $u(\tau) \equiv 0$, the output is called the *zero-input response*. In this case, it is simply an exponential times the initial state x_0. When $x_0 = 0$, the output is called the *zero-state response*. It is the second term of (7.32). The zero-state response for an input that is a unit impulse located at $t = t_0$,

$$u(t) = \delta(t - t_0) \tag{7.33}$$

is called the *unit impulse response*. For the first-order case, it is

$$x(t) = \int_{t_0}^{t} b\epsilon^{a(t-\tau)} \delta(\tau - t_0)\, d\tau. \tag{7.34}$$

Thus

$$x(t) = \begin{cases} b\epsilon^{a(t-t_0)} & \text{for} \quad t \geqslant t_0 \\ 0 \quad \text{for} & t < t_0 \end{cases} \tag{7.35}$$

The unit impulse response which is the zero-state response due to an input $\delta(t - t_0)$ is denoted by $h(t - t_0)$. Thus

$$h(t - t_0) = \begin{cases} b\epsilon^{a(t-t_0)} & \text{for} \quad t \geqslant t_0 \\ 0 \quad \text{for} & t < t_0 \end{cases} \tag{7.36}$$

Thus, using (7.36), (7.32) may be written as

$$x(t) = \epsilon^{a(t-t_0)} x_0 + \int_{t_0}^{t} h(t - \tau) u(\tau)\, d\tau. \tag{7.37}$$

The integral in (7.27) is defined as the *convolution* of $h(t)$ and $u(t)$.

The simple result of (7.37) is extended to second-order systems and to higher-order systems in later sections. However, the basic result for linear systems is that the zero-state response and zero-input response add to yield the total response. Furthermore, the zero-state response is a convolution of the input and the unit impulse response, and the zero-input response is a time function times the initial state.

7.2 The State Plane

For a second-order system, there are two state variables $x_1(t)$ and $x_2(t)$. The value of $x_2(t)$ may be plotted versus $x_1(t)$ in an $x_1 - x_2$ plane called the *state plane* or the *phase plane*, with time as a parameter. The locus of points traced as t increases from the initial time t_0 for a given input and initial state is called a *trajectory* on the phase plane. A family of trajectories for various initial states is called a *phase portrait*. Although the trajectory representation may be extended to higher-order systems, geometrical visualization is limited to systems of at most third order. It is most useful for second-order systems.

The state or phase plane is commonly used to portray the family of trajectories when the input is zero, but the initial state is not zero. If the state variables are the outputs, the trajectory represents the *zero-input response* of the system.

Consider a second-order linear time-invariant system modeled by

$$\begin{aligned} \dot{x}_1 &= a_{11}x_1 + a_{12}x_2 \\ \dot{x}_2 &= a_{21}x_1 + a_{22}x_2, \end{aligned} \tag{7.38}$$

where the a_{ij} are constants. The equilibrium points as defined in Chapter 3 are the points for which $\dot{x}_1$ and $\dot{x}_2$ are identically zero. From (7.22), the equilibrium points are the solutions of

$$\begin{aligned} a_{11}x_1 + a_{12}x_2 &= 0 \\ a_{21}x_1 + a_{22}x_2 &= 0. \end{aligned} \tag{7.39}$$

If $a_{11}a_{22} - a_{12}a_{21} \neq 0$, the only solution of (7.39) is

$$\mathbf{x}_e = \begin{bmatrix} 0 \\ 0 \end{bmatrix} \tag{7.40}$$

However, if $a_{11}a_{22} - a_{12}a_{21} = 0$, there would be infinitely many solutions of (7.39) on a straight line, called the *equilibrium line*.

To find the trajectories in the state plane, we need to relate x_1 directly to x_2. From (7.38) we can obtain the desired relation by dividing the second equation by the first:

$$\frac{dx_2}{dx_1} = \frac{a_{21}x_1 + a_{22}x_2}{a_{11}x_1 + a_{12}x_2}. \tag{7.41}$$

This equation defines the *slope* of the desired state plane trajectories. It is possible to integrate (7.41). However, many properties of the trajectories may be deduced by investigating their slopes. The locus of equal slopes called *isoclines* can be sketched based on (7.41) for different values of slopes. Short line segments of equal slope may be marked on the loci. The slope of the lines is the *slope* of the trajectories at each point of the corresponding isocline. The drawing can be made arbitrarily accurate by employing a large number of isoclines.

For any given constant slope m, dx_2/dx_1 in (7.41) may be replaced by m. The resulting equation specifies the locus of all points of the phase portrait so that all the trajectories passing through those points have slope m. Since the resulting equation is linear and homogeneous, the isocline for any m is a straight line passing through the origin.

Example 7-3. Suppose a system is modeled by the equations

$$\dot{x}_1 = x_2$$

$$\dot{x}_2 = -5x_1 - 6x_2.$$

It is easily seen that

$$\frac{dx_2}{dx_1} = \frac{-5x_1 - 6x_2}{x_2} \tag{7.43}$$

is the slope equation. The phase portrait can be sketched by drawing lines of equal slope on the isoclines, then sketching in the trajectory from the slopes. For example, the trajectories cross the line

$$-5x_1 - 6x_2 = 0$$

with zero slope. The line

$$x_2 = 0$$

is crossed with infinite slope by all trajectories. In general, the line

$$-5x_1 - 6x_2 = mx_2$$

is the locus of trajectories of slope m. The use of isoclines to sketch trajectories is shown by Figure 7.2. The isoclines give the slopes of the trajectories at each point. The drawing can be made arbitrarily accurate by employing a large number of isoclines. ▲

For any starting state on a trajectory, the arrows on the loci indicate that as time increases, the locus of points (x_1, x_2) which satisfy the state equations would trace the loci in the directions indicated. In Example 7-3, $\dot{x}_1 = x_2$. For this situation, the arrow directions are easily determined. If x_2 is positive, $\dot{x}_1$ is positive and x_1 tends to increase with time. Thus, in the upper half plane, the arrow directions are to the right. Similarly, if x_2 is negative, $\dot{x}_1$ is negative so that x_1 tends to decrease with time. Figure 7.2 illustrates these directions.

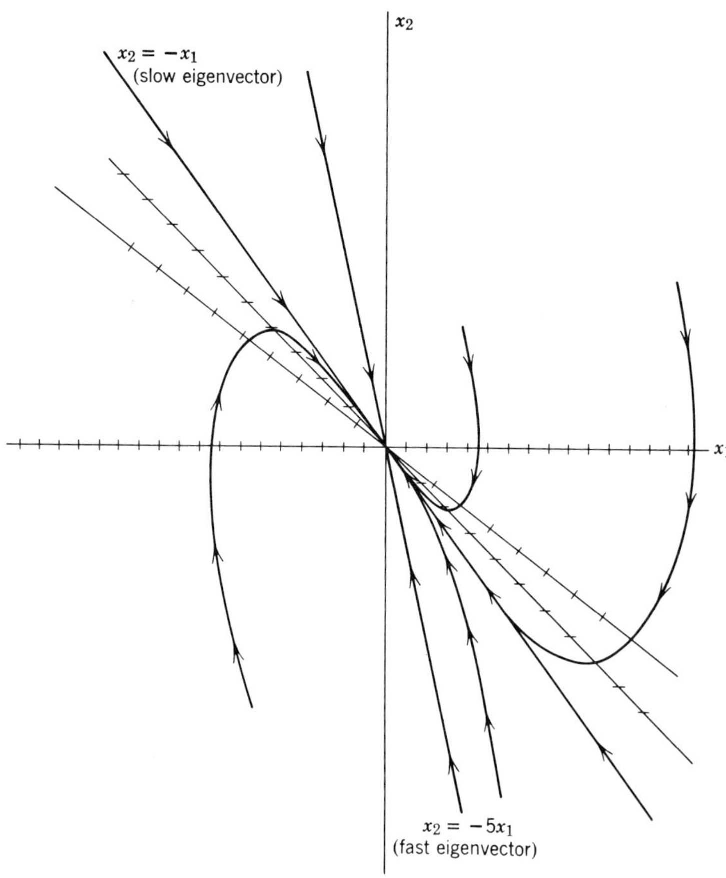

Figure 7.2 Phase-portrait plot from isoclines for Example 7-3.

7.3 State-Transition Matrix for Second-Order Systems

Let us continue our investigation of the second-order system (7.38). We shall demonstrate that the zero-input response is of the form

$$\begin{bmatrix} x_1(t) \\ x_2(t) \end{bmatrix} = \begin{bmatrix} \phi_{11}(t) & \phi_{12}(t) \\ \phi_{21}(t) & \phi_{22}(t) \end{bmatrix} \begin{bmatrix} x_1(0) \\ x_2(0) \end{bmatrix}.$$

Taking the Laplace transform of (7.38), we have

$$sX_1(s) - x_1(0) = a_{11}X_1(s) + a_{12}X_2(s) \tag{7.44}$$

$$sX_2(s) - x_2(0) = a_{21}X_1(s) + a_{22}X_2(s) \tag{7.45}$$

or

$$(s - a_{11})X_1(s) - a_{12}X_2(s) = x_1(0) \tag{7.46}$$

$$-a_{21}X_1(s) + (s - a_{22})X_2(s) = x_2(0). \tag{7.47}$$

In matrix form,

$$\begin{bmatrix} s - a_{11} & -a_{12} \\ -a_{21} & s - a_{22} \end{bmatrix} \begin{bmatrix} X_1(s) \\ X_2(s) \end{bmatrix} = \begin{bmatrix} x_1(0) \\ x_2(0) \end{bmatrix} \tag{7.48}$$

Provided that

$$[s\mathbf{I} - \mathbf{A}] = \begin{bmatrix} s - a_{11} & -a_{12} \\ -a_{21} & s - a_{22} \end{bmatrix} \tag{7.49}$$

is nonsingular, we may solve for $X_1(s)$ and $X_2(s)$ from (7.48):

$$\begin{bmatrix} X_1(s) \\ X_2(s) \end{bmatrix} = \begin{bmatrix} s - a_{11} & -a_{12} \\ -a_{21} & s - a_{22} \end{bmatrix}^{-1} \begin{bmatrix} x_1(0) \\ x_2(0) \end{bmatrix} \tag{7.50}$$

The matrix $sI - A$ is singular when

$$\det[s\mathbf{I} - \mathbf{A}] = (s - a_{11})(s - a_{22}) - a_{12}a_{21} = 0. \tag{7.51}$$

Since (7.51) is a quadratic equation in s, there are exactly two values of s for which the determinant is zero. These two values of s are called the *eigenvalues* of the matrix $\mathbf{A}$. Thus the solution in (7.50) is valid except for s equal to these eigenvalues. Thus

$$\begin{bmatrix} X_1(s) \\ X_2(s) \end{bmatrix} = \frac{1}{s^2 - (a_{11} + a_{22})s + (a_{11}a_{22} - a_{12}a_{21})} \begin{bmatrix} s - a_{22} & a_{12} \\ a_{21} & s - a_{11} \end{bmatrix} \begin{bmatrix} x_1(0) \\ x_2(0) \end{bmatrix} \tag{7.52}$$

First, suppose that the eigenvalues are distinct, denoted by λ_1 and λ_2. Thus

$$s^2-(a_{11}+a_{22})s+(a_{11}a_{22}-a_{12}a_{21})=(s-\lambda_1)(s-\lambda_2) \tag{7.53}$$

Then, expanding in partial fractions, we have

$$\begin{bmatrix} \dfrac{s-a_{22}}{(s-\lambda_1)(s-\lambda_2)} & \dfrac{a_{12}}{(s-\lambda_1)(s-\lambda_2)} \\ \\ \dfrac{a_{21}}{(s-\lambda_1)(s-\lambda_2)} & \dfrac{s-a_{11}}{(s-\lambda_1)(s-\lambda_2)} \end{bmatrix}$$

$$=\frac{1}{\lambda_1-\lambda_2}\begin{bmatrix} \dfrac{\lambda_1-a_{22}}{s-\lambda_1}-\dfrac{\lambda_2-a_{22}}{s-\lambda_2} & \dfrac{a_{12}}{s-\lambda_1}-\dfrac{a_{12}}{s-\lambda_2} \\ \\ \dfrac{a_{21}}{s-\lambda_1}-\dfrac{a_{21}}{s-\lambda_2} & \dfrac{\lambda_1-a_{11}}{s-\lambda_1}-\dfrac{\lambda_2-a_{11}}{s-\lambda_2} \end{bmatrix} \tag{7.54}$$

Substituting (7.54) in (7.52), and taking the inverse Laplace transform,

$$\begin{bmatrix} x_1(t) \\ x_2(t) \end{bmatrix}=\begin{bmatrix} \dfrac{\lambda_1-a_{22}}{\lambda_1-\lambda_2}\epsilon^{\lambda_1 t}-\dfrac{\lambda_2-a_{22}}{\lambda_1-\lambda_2}\epsilon^{\lambda_2 t} & \dfrac{a_{12}}{\lambda_1-\lambda_2}\epsilon^{\lambda_1 t}-\dfrac{a_{12}}{\lambda_1-\lambda_2}\epsilon^{\lambda_2 t} \\ \\ \dfrac{a_{21}}{\lambda_1-\lambda_2}\epsilon^{\lambda_1 t}-\dfrac{a_{21}}{\lambda_1-\lambda_2}\epsilon^{\lambda_2 t} & \dfrac{\lambda_1-a_{11}}{\lambda_1-\lambda_2}\epsilon^{\lambda_1 t}-\dfrac{\lambda_2-a_{11}}{\lambda_1-\lambda_2}\epsilon^{\lambda_2 t} \end{bmatrix}\begin{bmatrix} x_1(0) \\ x_2(0) \end{bmatrix} \tag{7.55}$$

for $t \geqslant 0$.

Analogous to the first-order case, the zero-input response is equal to a matrix time function times the initial state. The matrix is called the *state-transition matrix* of the system. Although the inverse Laplace transform yields a time function specified only for $t \geqslant 0$, the state-transition matrix is defined for all t. The expression in (7.55) satisfies (7.38) for all time t. The state-transition matrix will be denoted by $\boldsymbol{\phi}(t)$:

$$\boldsymbol{\phi}(t)=\begin{bmatrix} \phi_{11}(t) & \phi_{12}(t) \\ \phi_{21}(t) & \phi_{22}(t) \end{bmatrix}=\begin{bmatrix} \dfrac{(\lambda_1-a_{22})\epsilon^{\lambda_1 t}-(\lambda_2-a_{22})\epsilon^{\lambda_2 t}}{\lambda_1-\lambda_2} & \dfrac{a_{12}(\epsilon^{\lambda_1 t}-\epsilon^{\lambda_2 t})}{\lambda_1-\lambda_2} \\ \\ \dfrac{a_{21}(\epsilon^{\lambda_1 t}-\epsilon^{\lambda_2 t})}{\lambda_1-\lambda_2} & \dfrac{(\lambda_1-a_{11})\epsilon^{\lambda_1 t}-(\lambda_2-a_{11})\epsilon^{\lambda_2 t})}{\lambda_1-\lambda_2} \end{bmatrix} \tag{7.56}$$

If $\lambda_1 = \lambda_2$, instead of (7.54) the partial fraction expansion of $[s\mathbf{I}-\mathbf{A}]^{-1}$ is

$$\begin{bmatrix} \dfrac{s-a_{22}}{(s-\lambda_1)^2} & \dfrac{a_{12}}{(s-\lambda_1)^2} \\ \\ \dfrac{a_{21}}{(s-\lambda_1)^2} & \dfrac{s-a_{11}}{(s-\lambda_1)^2} \end{bmatrix} = \begin{bmatrix} \dfrac{1}{s-\lambda_1}+\dfrac{\lambda_1-a_{22}}{(s-\lambda_1)^2} & \dfrac{a_{12}}{(s-\lambda_1)^2} \\ \\ \dfrac{a_{21}}{(s-\lambda_1)^2} & \dfrac{1}{s-\lambda_1}+\dfrac{\lambda_1-a_{11}}{(s-\lambda_1)^2} \end{bmatrix}$$

Hence the inverse transform of (7.52) becomes

$$\begin{bmatrix} x_1(t) \\ x_2(t) \end{bmatrix} = \begin{bmatrix} \epsilon^{\lambda_1 t} + (\lambda_1 - a_{22})t\epsilon^{\lambda_1 t} & a_{12}t\epsilon^{\lambda_1 t} \\ a_{21}t\epsilon^{\lambda_1 t} & \epsilon^{\lambda_1 t} + (\lambda_1 - a_{11})t\epsilon^{\lambda_1 t} \end{bmatrix} \begin{bmatrix} x_1(0) \\ x_2(0) \end{bmatrix} \tag{7.57}$$

for $t \geqslant 0$. Thus the state at time t is linearly related to the state at time zero regardless of whether $\lambda_1 = \lambda_2$.

The 2×2 matrix in (7.57) is the state-transition matrix for the system described by (7.38) when $\lambda_1 = \lambda_2$. The vector $[x_1(t)\ x_2(t)]^T$ of (7.57) satisfies (7.38) for all t including negative t, and the state-transition matrix is defined for all t. It is denoted by $\boldsymbol{\phi}(t)$:

$$\boldsymbol{\phi}(t) = \begin{bmatrix} \phi_{11}(t) & \phi_{12}(t) \\ \phi_{21}(t) & \phi_{22}(t) \end{bmatrix} = \begin{bmatrix} \epsilon^{\lambda_1 t} + (\lambda_1 - a_{22})t\epsilon^{\lambda_1 t} & a_{12}t\epsilon^{\lambda_1 t} \\ a_{21}t\epsilon^{\lambda_1 t} & \epsilon^{\lambda_1 t} + (\lambda_1 - a_{11})t\epsilon^{\lambda_1 t} \end{bmatrix} \tag{7.58}$$

The transition matrix for the second-order system (7.38) is given by (7.56) when $\lambda_1 \neq \lambda_2$ and by (7.58) when $\lambda_1 = \lambda_2$. In either case, the state at time t is linearly related to the state at $t=0$, as given by (7.55) and (7.57).

7.4 Phase-Plane Trajectories in Terms of Eigenvectors

In this section, we shall demonstrate that if the two eigenvalues of $\mathbf{A}$ are real and distinct, the corresponding eigenvectors are possible trajectories. In particular, if an initial state lies on one of the eigenvectors, the trajectory remains on the eigenvector. Otherwise, the trajectory can be expressed as a sum of the two eigenvectors.

Let us recall that for $\lambda_1 \neq \lambda_2$,

$$x_1(t) = \left[\frac{\lambda_1 - a_{22}}{\lambda_1-\lambda_2}x_1(0) + \frac{a_{12}}{\lambda_1-\lambda_2}x_2(0)\right]\epsilon^{\lambda_1 t} - \left[\frac{\lambda_2 - a_{22}}{\lambda_1-\lambda_2}x_1(0) + \frac{a_{12}}{\lambda_1-\lambda_2}x_2(0)\right]\epsilon^{\lambda_2 t} \tag{7.59}$$

$$x_2(t) = \left[\frac{a_{21}}{\lambda_1-\lambda_2}x_1(0) + \frac{\lambda_1-a_{11}}{\lambda_1-\lambda_2}x_2(0)\right]\epsilon^{\lambda_1 t}$$

$$-\left[\frac{a_{21}}{\lambda_1-\lambda_2}x_1(0) + \frac{\lambda_2-a_{11}}{\lambda_1-\lambda_2}x_2(0)\right]\epsilon^{\lambda_2 t} \tag{7.60}$$

Suppose that $x_1(0)$ and $x_2(0)$ are chosen so that the coefficients of $\epsilon^{\lambda_2 t}$ vanish. That is,

$$(\lambda_2 - a_{22})x_1(0) + a_{12}x_2(0) = 0 \tag{7.61}$$

$$a_{21}x_1(0) + (\lambda_2 - a_{11})x_2(0) = 0 \tag{7.62}$$

Equations 7.61 and 7.62 have nontrivial solutions provided that

$$\det\begin{bmatrix}\lambda_2 - a_{22} & a_{12}\\ a_{21} & \lambda_2 - a_{11}\end{bmatrix} = 0 \tag{7.63}$$

But

$$\det\begin{bmatrix}\lambda_2 - a_{22} & a_{12}\\ a_{21} & \lambda_2 - a_{11}\end{bmatrix} = (\lambda_2 - a_{22})(\lambda_2 - a_{11}) - a_{12}a_{21}. \tag{7.64}$$

Since λ_2 is an eigenvalue that satisfies (7.51), then the polynomial in (7.64) is equal to zero and (7.63) is satisfied.

If the coefficients a_{11}, a_{12}, a_{21}, and a_{22} are real and if λ_1 and λ_2 are assumed to be real, then the solution for $x_1(0)$ and $x_2(0)$ are real. For this case, (7.61) and (7.62) can be geometrically interpreted in the phase plane as straight lines passing through the origin. A nontrivial solution [$x_1(0)$ and $x_2(0)$ not both zero] satisfying both (7.61) and (7.62) is obtained only if the lines coincide, that is, if the equations are dependent. Thus, either (7.61) or (7.62) defines the line on which $x_1(0)$ and $x_2(0)$ must lie. If either $\lambda_2 - a_{22}$ or a_{12} is nonzero, then (7.61) defines the line. If both $\lambda_2 - a_{22}$ and a_{12} are zero then (7.62) defines the line. Not all the coefficients in (7.61) and (7.62) can be zero if $\lambda_1 \neq \lambda_2$. With this choice for $x_1(0)$ and $x_2(0)$, (7.59) and (7.60) become

$$x_1(t) = k_1\epsilon^{\lambda_1 t} \tag{7.65}$$

$$x_2(t) = k_2\epsilon^{\lambda_1 t}, \tag{7.66}$$

where k_1 and k_2 are real constants. Thus

$$(k_2\epsilon^{\lambda_1 t})x_1(t) = (k_1\epsilon^{\lambda_1 t})x_2(t)$$

or

$$k_2x_1(t) = k_1x_2(t). \tag{7.67}$$

Equation 7.67 gives the equation for a trajectory that starts from an initial state $x_1(0)$ and $x_2(0)$, which satisfies (7.61) and (7.62). Since $x_1(0)$ and $x_2(0)$ must also satisfy (7.67), then k_2 and k_1 must be in the same proportion as $(\lambda_2 - a_{22})$ and $-a_{12}$, or a_{21} and $-(\lambda_2 - a_{11})$. Thus if a trajectory starts on the line

$$(\lambda_2 - a_{22})x_1(t) = -a_{12}x_2(t) \tag{7.68}$$

or

$$a_{21}x_1(t) = -(\lambda_2 - a_{11})x_2(t) \tag{7.69}$$

it stays on the line so that the line is a trajectory. Alternatively, the line can be described by the vector

$$\begin{bmatrix} c_{11} \\ c_{21} \end{bmatrix} = k \begin{bmatrix} -a_{12} \\ \lambda_2 - a_{22} \end{bmatrix} \tag{7.70}$$

or

$$\begin{bmatrix} c_{11} \\ c_{21} \end{bmatrix} = k \begin{bmatrix} -(\lambda_2 - a_{11}) \\ a_{21} \end{bmatrix} \tag{7.71}$$

where k is an arbitrary real number. The vector $[c_{11}\ c_{21}]^T$ in (7.70) or (7.71) satisfies

$$[\lambda_1 \mathbf{I} - \mathbf{A}] \begin{bmatrix} c_{11} \\ c_{21} \end{bmatrix} = \begin{bmatrix} \lambda_1 - a_{11} & -a_{12} \\ -a_{21} & \lambda_1 - a_{22} \end{bmatrix} \begin{bmatrix} c_{11} \\ c_{21} \end{bmatrix} = 0 \tag{7.72}$$

This is verified by substituting (7.70) or (7.71) in (7.72) and noting that λ_1 and λ_2 are related to a_{11}, a_{12}, a_{21}, and a_{22} from (7.53). A vector $[c_{11}\ c_{21}]^T$, which satisfies (7.72), is called an *eigenvector* of $\mathbf{A}$ corresponding to the eigenvalue λ_1 of $\mathbf{A}$. Thus the trajectory in (7.68) or (7.69) lies on the eigenvector corresponding to λ_1 for real λ_1.

Similarly, for real λ_1 and λ_2, real initial conditions $x_1(0)$ and $x_2(0)$ may be chosen so that the coefficients of $\epsilon^{\lambda_1 t}$ in (7.59) and (7.60) vanish. These initial conditions are solutions of

$$(\lambda_1 - a_{22})x_1(0) + a_{12}x_2(0) = 0 \tag{7.73}$$

$$a_{21}x_1(0) + (\lambda_1 - a_{11})x_2(0) = 0. \tag{7.74}$$

The equations have nontrivial solutions because

$$\det \begin{bmatrix} \lambda_1 - a_{22} & a_{12} \\ a_{21} & \lambda_1 - a_{11} \end{bmatrix} = 0$$

by virtue of λ_1 being an eigenvalue of **A** satisfying (7.51). For these choices of $x_1(0)$ and $x_2(0)$, the terms in $\epsilon^{\lambda_1 t}$ of (7.59) and (7.60) vanish so that they are of the form

$$x_1(t) = k_3\epsilon^{\lambda_2 t} \tag{7.75}$$

$$x_2(t) = k_4\epsilon^{\lambda_2 t} \tag{7.76}$$

where k_3 and k_4 are constants. From (7.75) and (7.76), we have

$$k_4\epsilon^{\lambda_2 t}x_1(t) = k_3\epsilon^{\lambda_2 t}x_2(t)$$

or

$$k_4x_1(t) = k_3x_2(t). \tag{7.77}$$

Equation (7.77) defines a trajectory which is a straight line passing through the origin. Since $x_1(0)$ and $x_2(0)$ which satisfy (7.73) and (7.74) must also satisfy (7.77), k_4 and k_3 must be in the same proportion as $\lambda_1 - a_{22}$ and $-a_{12}$, or a_{21} and $-(\lambda_1 - a_{11})$. Thus the equation of the trajectory in (7.77) may be written as

$$(\lambda_1 - a_{22})x_1(t) = -a_{12}x_2(t) \tag{7.78}$$

or

$$a_{21}x_1(t) = -(\lambda_1 - a_{11})x_2(t). \tag{7.79}$$

It is easily verified that

$$\begin{bmatrix} c_{12} \\ c_{22} \end{bmatrix} = k\begin{bmatrix} -a_{12} \\ \lambda_1 - a_{22} \end{bmatrix} \tag{7.80}$$

or

$$\begin{bmatrix} c_{12} \\ c_{22} \end{bmatrix} = k\begin{bmatrix} -(\lambda_1 - a_{11}) \\ a_{21} \end{bmatrix} \tag{7.81}$$

is an *eigenvector* of A corresponding to the eigenvalue λ_2 where k is an arbitrary number. That is, $[c_{12}\ c_{22}]^T$ satisfies

$$\begin{bmatrix} \lambda_2 - a_{11} & -a_{12} \\ -a_{21} & \lambda_2 - a_{22} \end{bmatrix}\begin{bmatrix} c_{12} \\ c_{22} \end{bmatrix} = 0. \tag{7.82}$$

Hence the trajectory of (7.78) or (7.79) lies on the eigenvector of **A** corresponding to the eigenvalue λ_2 of **A**.

The two straight lines of (7.68) or (7.69) and (7.78) or (7.79) are trajectories when λ_1 and λ_2 are real and $\lambda_1 \neq \lambda_2$. These two lines lie on the two eigenvectors corresponding to λ_1 and λ_2. If an initial condition lies on one of the eigenvectors, the trajectory remains on that eigenvector. If an initial condition is on neither one of the eigenvectors, then neither of the eigenvectors is a trajectory for that initial state. However, the trajectory can be expressed as a sum of the eigenvectors.

From (7.55), $\mathbf{x}(t)$ is linear in $\mathbf{x}(0)$. But any $\mathbf{x}(0)$ may be resolved as a sum of two vectors, one along one eigenvector and another along the other eigenvector, since the eigenvectors are not coincident. That is, any $\mathbf{x}(0)$ can be written as $\mathbf{x}(0)=\mathbf{x}_1(0)+\mathbf{x}_2(0)$ where $\mathbf{x}_1(0)$ has the same direction as the eigenvector $[c_{11}\,c_{21}]^T$ and $\mathbf{x}_2(0)$ has the same direction as the eigenvector $[c_{12}\,c_{22}]^T$. Denote the transition matrix in (7.55) by $\boldsymbol{\phi}(t)$. Then

$$\begin{aligned}\mathbf{x}(t) &= \boldsymbol{\phi}(t)\mathbf{x}(0) = \boldsymbol{\phi}(t)[\mathbf{x}_1(0)+\mathbf{x}_2(0)]\\ &= \boldsymbol{\phi}(t)\mathbf{x}_1(0)+\boldsymbol{\phi}(t)\mathbf{x}_2(0)\end{aligned} \tag{7.83}$$

But from (7.65) and (7.66), if $\mathbf{x}_1(0)$ is on the eigenvector corresponding to λ_1,

$$\boldsymbol{\phi}(t)\mathbf{x}_1(0) = k_5\begin{bmatrix}c_{11}\\ c_{21}\end{bmatrix}\epsilon^{\lambda_1 t} \tag{7.84}$$

and from (7.75) and (7.76), if $x_2(0)$ is on the eigenvector corresponding to λ_2

$$\boldsymbol{\phi}(t)\mathbf{x}_2(0) = k_6\begin{bmatrix}c_{12}\\ c_{22}\end{bmatrix}\epsilon^{\lambda_2 t} \tag{7.85}$$

Hence

$$\mathbf{x}(t) = \begin{bmatrix}x_1(t)\\ x_2(t)\end{bmatrix} = k_5\begin{bmatrix}c_{11}\\ c_{21}\end{bmatrix}\epsilon^{\lambda_1 t}+k_6\begin{bmatrix}c_{12}\\ c_{22}\end{bmatrix}\epsilon^{\lambda_2 t} \tag{7.86}$$

Equation 7.86 states that the solution $\mathbf{x}$ is a linear combination of the eigenvectors.

Example 7-4. In Example 7-3, the $\mathbf{A}$ matrix corresponding to (7.42) is

$$\mathbf{A} = \begin{bmatrix}0 & 1\\ -5 & -6\end{bmatrix} \tag{7.87}$$

The eigenvalues are the solutions of

$$\lambda(\lambda+6)+5=0 \tag{7.88}$$

which are $\lambda_1 = -5$ and $\lambda_2 = -1$. For $\lambda_1 = -5$, the eigenvector is given by the solutions of

$$\begin{bmatrix} 0 & 1 \\ -5 & -6 \end{bmatrix}\begin{bmatrix} c_{11} \\ c_{21} \end{bmatrix} = -5\begin{bmatrix} c_{11} \\ c_{21} \end{bmatrix}. \tag{7.89}$$

From (7.89),

$$c_{21} = -5c_{11}. \tag{7.90}$$

Thus the eigenvector corresponding to λ_1 is

$$\begin{bmatrix} c_{11} \\ c_{21} \end{bmatrix} = \begin{bmatrix} k_1 \\ -5k_1 \end{bmatrix} \tag{7.91}$$

where k_1 is arbitrary. Note that a *line* is determined by the eigenvector equation, not a *point*, even though we have two equations in two unknowns. The two equations are redundant, since the determinant of $[\lambda_1\mathbf{I} - \mathbf{A}]$ is zero.

The function

$$\mathbf{x} = \begin{bmatrix} x_1(t) \\ x_2(t) \end{bmatrix} = \begin{bmatrix} k_1 \\ -5k_1 \end{bmatrix}\epsilon^{-5t} \tag{7.92}$$

satisfies (7.42). From (7.92),

$$x_2 = -5x_1. \tag{7.93}$$

Equation 7.93 is a straight line on the phase plane. It is also a representation of the eigenvector in (7.93) on the phase plane. The line is shown in Figure 7.2. For positive x_2, x_1 is increasing and for negative x_2, x_1 is decreasing, as indicated by the arrows on the eigenvector line. Similarly the eigenvector corresponding to $\lambda_2 = -1$ is found from

$$\begin{bmatrix} 0 & 1 \\ -5 & -6 \end{bmatrix}\begin{bmatrix} c_{12} \\ c_{22} \end{bmatrix} = -1\begin{bmatrix} c_{12} \\ c_{22} \end{bmatrix}, \tag{7.94}$$

which yields

$$c_{22} = -c_{12} \tag{7.95}$$

Hence

$$\begin{bmatrix} c_{12} \\ c_{22} \end{bmatrix} = \begin{bmatrix} k_2 \\ -k_2 \end{bmatrix} \tag{7.96}$$

where k_2 is arbitrary. The function

$$\mathbf{x}(t) = \begin{bmatrix} x_1(t) \\ x_2(t) \end{bmatrix} = \begin{bmatrix} k_2 \\ -k_2 \end{bmatrix}\epsilon^{-t} \tag{7.97}$$

satisfies (7.42) and

$$x_2 = -x_1 \tag{7.98}$$

is a possible trajectory. This trajectory is indicated on Figure 7.2. For any initial condition on the line $x_2 = -x_1$, the trajectory stays on the line $x_2 = -x_1$. Similarly, for any initial condition on the line $x_2 = -5x_1$, the trajectory stays on the line $x_2 = -5x_1$. The eigenvector represented by (7.93) may be called the fast eigenvector and that represented by (7.96) may be called the slow eigenvector because the trajectory in (7.93) approaches the equilibrium point faster than that in (7.98). For an initial condition that is not on either line, the solution is given by

$$\mathbf{x}(t) = k_1 \begin{bmatrix} 1 \\ -5 \end{bmatrix} \epsilon^{-5t} + k_2 \begin{bmatrix} 1 \\ -1 \end{bmatrix} \epsilon^{-t} \tag{7.99}$$

where k_1 and k_2 are determined from the specified initial state $\mathbf{x}_0$. Thus for $t = 0$,

$$\mathbf{x}_0 = k_1 \begin{bmatrix} 1 \\ -5 \end{bmatrix} + k_2 \begin{bmatrix} 1 \\ -1 \end{bmatrix} = \begin{bmatrix} 1 & 1 \\ -5 & -1 \end{bmatrix} \begin{bmatrix} k_1 \\ k_2 \end{bmatrix} \tag{7.100}$$

and

$$\begin{bmatrix} k_1 \\ k_2 \end{bmatrix} = \begin{bmatrix} 1 & 1 \\ -5 & -1 \end{bmatrix}^{-1} \mathbf{x}_0 = \begin{bmatrix} -\frac{1}{4} & -\frac{1}{4} \\ \frac{5}{4} & \frac{1}{4} \end{bmatrix} \mathbf{x}_0 \tag{7.101}$$

From (7.99), it is seen that the first term approaches zero faster than the second term does, so that for t large

$$\mathbf{x}(t) \approx k_2 \begin{bmatrix} 1 \\ -1 \end{bmatrix} \epsilon^{-t} \tag{7.102}$$

and

$$\frac{x_2}{x_1} \approx -1. \tag{7.103}$$

That is, the trajectories become asymptotic to $x_2 = -x_1$ as they approach the equilibrium point. This is seen in Figure 7.2. This property is useful in sketching the phase portrait in Figure 7.2. ▲

The equilibrium point of the phase portrait in Figure 7.2 corresponding to two negative real eigenvalues is defined as a *stable node*. There are six types of the equilibrium point $\mathbf{x} = \mathbf{0}$ as shown in Figure 7.3. In Figure 7.3*a* the stable node is shown once more. When the eigenvalues are real and both positive, the equilibrium point is called an *unstable node*, as illustrated in Figure 7.3*b*. When one eigenvalue is negative real and the other eigenvalue is positive real, the equilibrium point is called a *saddle point*. Figure 7.3*c* shows a typical

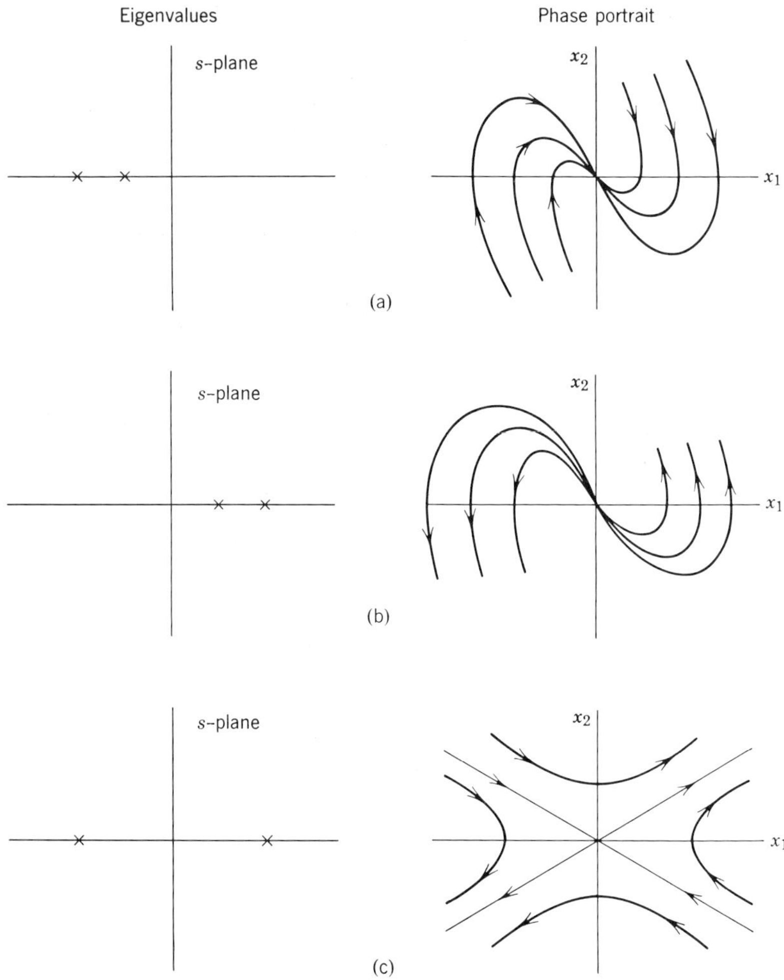

Figure 7.3 Six types of equilibrium point $\mathbf{x} = \mathbf{0}$. (*a*) Stable node. (*b*) Unstable node. (*c*) Saddle point.

phase portrait for a saddle point. Complex eigenvalues appear in conjugate pairs. When the complex eigenvalues have negative real parts, the equilibrium point is called a *stable focus*. If the complex eigenvalues have positive real parts, the equilibrium point is called an *unstable focus*. Finally, if the complex eigenvalues have zero real part, the equilibrium is called a *center*. These latter three types of equilibrium point are illustrated in Figure 7.3*d*, 7.3*e*, and 7.3*f*. For these three, the eigenvectors are complex. Hence the geometric interpretation of the trajectory as a sum of the eigenvectors does not lead to a simple visualization.

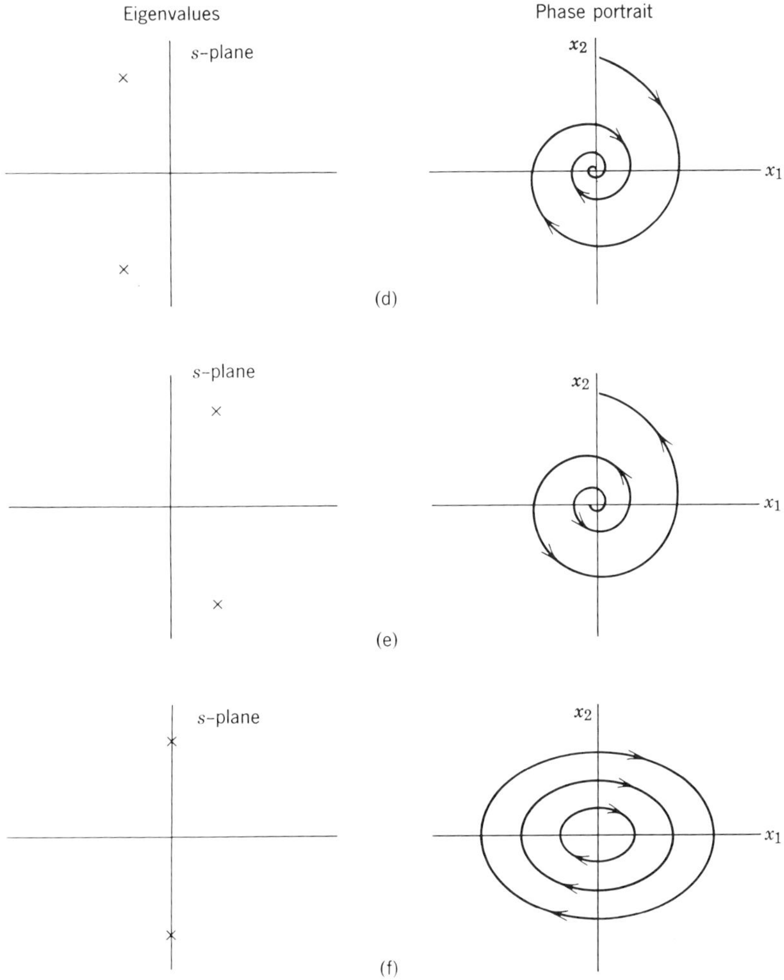

Figure 7.3 (*d*) Stable focus. (*e*) Unstable focus. (*f*) Center.

7.5 Properties of the State-Transition Matrix—Second-Order Systems

In this section, we derive some properties of the transition matrix. For clarity, the discussion is restricted to second-order systems. Higher-order systems are treated in later sections.

Property 1. From an examination of (7.55) and (7.57), the transition matrix $\phi(t)$ at $t = 0$ is the identity matrix:

$$\boldsymbol{\phi}(0) = \begin{bmatrix} 1 & 0 \\ 0 & 1 \end{bmatrix} \tag{7.104}$$

Property 2. Det $\boldsymbol{\phi}(t) \neq 0$ for any finite t. First, we note that for the $\phi(t)$ in (7.56) for $\lambda_1 \neq \lambda_2$,

$$\det \boldsymbol{\phi}(t) = \frac{1}{(\lambda_1-\lambda_2)^2}\{[(\lambda_1-a_{22})\epsilon^{\lambda_1 t}-(\lambda_2-a_{22})\epsilon^{\lambda_2 t}][(\lambda_1-a_{11})\epsilon^{\lambda_1 t}$$
$$-(\lambda_2-a_{11})\epsilon^{\lambda_2 t}]-[a_{21}(\epsilon^{\lambda_1 t}-\epsilon^{\lambda_2 t})a_{12}(\epsilon^{\lambda_1 t}-\epsilon^{\lambda_2 t})]\}$$
$$=\frac{1}{(\lambda_1-\lambda_2)^2}\{(\lambda_1-a_{22})(\lambda_1-a_{11})\epsilon^{2\lambda_1 t}+(\lambda_2-a_{22})(\lambda_2-a_{11})\epsilon^{2\lambda_2 t}$$
$$-[(\lambda_2-a_{22})(\lambda_1-a_{11})+(\lambda_1-a_{22})(\lambda_2-a_{11})]\epsilon^{(\lambda_1+\lambda_2)t}$$
$$-a_{12}a_{21}(\epsilon^{\lambda_1 t}-\epsilon^{\lambda_2 t})^2\} \tag{7.105}$$

Using the fact that λ_1 and λ_2 satisfy (7.51), (7.105) becomes

$$\det \boldsymbol{\phi}(t) = \epsilon^{(\lambda_1+\lambda_2)t} \neq 0 \tag{7.106}$$

for any finite t.

For the $\boldsymbol{\phi}(t)$ in (7.58) when $\lambda_1 = \lambda_2$,

$$\det \boldsymbol{\phi}(t) = [\epsilon^{\lambda_1 t}+(\lambda_1-a_{22})t\epsilon^{\lambda_1 t}][\epsilon^{\lambda_1 t}+(\lambda_1-a_{11})t\epsilon^{\lambda_1 t}]-a_{12}a_{21}t^2\epsilon^{2\lambda_1 t} \tag{7.107}$$

Since λ_1 satisfies (7.51), (7.107) can be simplified to

$$\det \boldsymbol{\phi}(t) = \epsilon^{2\lambda_1 t} \neq 0 \tag{7.108}$$

for all finite t.

From (7.106) and (7.108), $\det \boldsymbol{\phi}(t) \neq 0$ regardless of whether λ_1 and λ_2 are equal or not. Hence, we have proved that $\boldsymbol{\phi}(t)$ is nonsingular for all finite t.

Property 3. $\boldsymbol{\phi}(t)$ satisfies

$$\dot{\boldsymbol{\phi}}(t) = \begin{bmatrix} \dot{\phi}_{11}(t) & \dot{\phi}_{12}(t) \\ \dot{\phi}_{21}(t) & \dot{\phi}_{22}(t) \end{bmatrix} = \begin{bmatrix} a_{11} & a_{12} \\ a_{21} & a_{22} \end{bmatrix}\begin{bmatrix} \phi_{11}(t) & \phi_{12}(t) \\ \phi_{21}(t) & \phi_{22}(t) \end{bmatrix} \tag{7.109}$$

In fact, it is common practice to define $\boldsymbol{\phi}(t)$ as the matrix that satisfies (7.109) and

$$\begin{bmatrix} \phi_{11}(0) & \phi_{12}(0) \\ \phi_{21}(0) & \phi_{22}(0) \end{bmatrix} = \begin{bmatrix} 1 & 0 \\ 0 & 1 \end{bmatrix}. \tag{7.110}$$

With this as definition for $\boldsymbol{\phi}(t)$, it is then shown that $\mathbf{x}(t) = \boldsymbol{\phi}(t)\mathbf{x}(0)$ satisfies $\dot{\mathbf{x}}(t) = \mathbf{A}\mathbf{x}(t)$ and when $t=0$ in $\mathbf{x}(t)$, the correct initial state is obtained. We show that the $\boldsymbol{\phi}(t)$ already obtained in (7.56) and

(7.58) satisfies (7.109). This is done by forming $\dot{\boldsymbol{\phi}}(t)$ from (7.56) and forming $\mathbf{A}\boldsymbol{\phi}(t)$ from (7.56). Then the two expressions are compared. For the $\boldsymbol{\phi}(t)$ in (7.56), $\dot{\boldsymbol{\phi}}(t)$ is

$$\begin{bmatrix} \dot{\phi}_{11}(t) & \dot{\phi}_{12}(t) \\ \dot{\phi}_{21}(t) & \dot{\phi}_{22}(t) \end{bmatrix}$$

$$= \begin{bmatrix} \dfrac{\lambda_1(\lambda_1 - a_{22})\epsilon^{\lambda_1 t} - \lambda_2(\lambda_2 - a_{22})\epsilon^{\lambda_2 t}}{\lambda_1 - \lambda_2} & \dfrac{a_{12}(\lambda_1\epsilon^{\lambda_1 t} - \lambda_2\epsilon^{\lambda_2 t})}{\lambda_1 - \lambda_2} \\ \\ \dfrac{a_{21}(\lambda_1\epsilon^{\lambda_1 t} - \lambda_2\epsilon^{\lambda_2 t})}{\lambda_1 - \lambda_2} & \dfrac{\lambda_1(\lambda_1 - a_{11})\epsilon^{\lambda_1 t} - \lambda_2(\lambda_2 - a_{11})}{\lambda_1 - \lambda_2} \end{bmatrix} \tag{7.111}$$

Next, form

$$\begin{bmatrix} a_{11} & a_{12} \\ a_{21} & a_{22} \end{bmatrix} \begin{bmatrix} \phi_{11}(t) & \phi_{12}(t) \\ \phi_{21}(t) & \phi_{22}(t) \end{bmatrix}$$

$$= \begin{bmatrix} a_{11} & a_{12} \\ a_{21} & a_{22} \end{bmatrix} \begin{bmatrix} \dfrac{(\lambda_1 - a_{22})\epsilon^{\lambda_1 t} - (\lambda_2 - a_{22})\epsilon^{\lambda_2 t}}{\lambda_1 - \lambda_2} & \dfrac{a_{12}(\epsilon^{\lambda_1 t} - \epsilon^{\lambda_2 t})}{\lambda_1 - \lambda_2} \\ \dfrac{a_{21}(\epsilon^{\lambda_1 t} - \epsilon^{\lambda_2 t})}{\lambda_1 - \lambda_2} & \dfrac{(\lambda_1 - a_{11})\epsilon^{\lambda_1 t} - (\lambda_2 - a_{11})\epsilon^{\lambda_2 t}}{\lambda_1 - \lambda_2} \end{bmatrix} \tag{7.112}$$

Carrying out the indicated multiplication in (7.112), we have

$$\begin{bmatrix} a_{11} & a_{12} \\ a_{21} & a_{22} \end{bmatrix} \begin{bmatrix} \phi_{11}(t) & \phi_{12}(t) \\ \phi_{21}(t) & \phi_{22}(t) \end{bmatrix}$$

$$= \begin{bmatrix} \dfrac{-\Delta(\epsilon^{\lambda_1 t} - \epsilon^{\lambda_2 t}) + a_{11}(\lambda_1\epsilon^{\lambda_1 t} - \lambda_2\epsilon^{\lambda_2 t})}{\lambda_1 - \lambda_2} & \dfrac{a_{12}(\lambda_1\epsilon^{\lambda_1 t} - \lambda_2\epsilon^{\lambda_2 t})}{\lambda_1 - \lambda_2} \\ \\ \dfrac{a_{21}(\lambda_1\epsilon^{\lambda_1 t} - \lambda_2\epsilon^{\lambda_2 t})}{\lambda_1 - \lambda_2} & \dfrac{-\Delta(\epsilon^{\lambda_1 t} - \epsilon^{\lambda_2 t}) + a_{22}(\lambda_1\epsilon^{\lambda_1 t} - \lambda_2\epsilon^{\lambda_2 t})}{\lambda_1 - \lambda_2} \end{bmatrix} \tag{7.113}$$

where $\Delta = \det \mathbf{A}$. From the fact that λ_1 and λ_2 satisfy (7.51),

$$(\lambda_1 - a_{11})(\lambda_1 - a_{22}) - a_{12}a_{21} = 0 \tag{7.114}$$

and

$$(\lambda_2 - a_{11})(\lambda_2 - a_{22}) - a_{12}a_{21} = 0. \tag{7.115}$$

By using (7.114) and (7.115), (7.113) can be written as

$$\begin{bmatrix} a_{11} & a_{12} \\ a_{21} & a_{22} \end{bmatrix}\begin{bmatrix} \phi_{11}(t) & \phi_{12}(t) \\ \phi_{21}(t) & \phi_{22}(t) \end{bmatrix}$$

$$= \begin{bmatrix} \dfrac{\lambda_1(\lambda_1 - a_{22})\epsilon^{\lambda_1 t} - \lambda_2(\lambda_2 - a_{22})\epsilon^{\lambda_2 t}}{\lambda_1 - \lambda_2} & \dfrac{a_{12}(\lambda_1\epsilon^{\lambda_1 t} - \lambda_2\epsilon^{\lambda_2 t})}{\lambda_1 - \lambda_2} \\ \dfrac{a_{21}(\lambda_1\epsilon^{\lambda_1 t} - \lambda_2\epsilon^{\lambda_2 t})}{\lambda_1 - \lambda_2} & \dfrac{\lambda_1(\lambda_1 - a_{11})\epsilon^{\lambda_1 t} - \lambda_2(\lambda_2 - a_{11})\epsilon^{\lambda_2 t}}{\lambda_1 - \lambda_2} \end{bmatrix}. \tag{7.116}$$

Comparing (7.111) and (7.116), it is seen that $\boldsymbol{\phi}(t)$ in (7.56) satisfies (7.109).

Similarly, from $\boldsymbol{\phi}(t)$ in (7.58), form

$$\dot{\phi}(t) = \begin{bmatrix} \dot{\phi}_{11}(t) & \dot{\phi}_{12}(t) \\ \dot{\phi}_{21}(t) & \dot{\phi}_{22}(t) \end{bmatrix}$$

$$= \begin{bmatrix} (2\lambda_1 - a_{22})\epsilon^{\lambda_1 t} + \lambda_1(\lambda_2 - a_{22})t\epsilon^{\lambda_1 t} & a_{12}\epsilon^{\lambda_1 t} + \lambda_1 a_{12} t\epsilon^{\lambda_1 t} \\ a_{21}\epsilon^{\lambda_1 t} + \lambda_1 a_{21} t\epsilon^{\lambda_1 t} & (2\lambda_1 - a_{11})\epsilon^{\lambda_1 t} + \lambda_1(\lambda_1 - a_{11})t\epsilon^{\lambda_1 t} \end{bmatrix}. \tag{7.117}$$

Then form

$$\begin{bmatrix} a_{11} & a_{12} \\ a_{21} & a_{22} \end{bmatrix}\begin{bmatrix} \phi_{11}(t) & \phi_{12}(t) \\ \phi_{21}(t) & \phi_{22}(t) \end{bmatrix} = \begin{bmatrix} a_{11} & a_{12} \\ a_{21} & a_{22} \end{bmatrix}\begin{bmatrix} \epsilon^{\lambda_1 t} + (\lambda_1 - a_{22})t\epsilon^{\lambda_1 t} & a_{12}t\epsilon^{\lambda_1 t} \\ a_{21}t\epsilon^{\lambda_1 t} & \epsilon^{\lambda_1 t} + (\lambda_1 - a_{11})t\epsilon^{\lambda_1 t} \end{bmatrix}$$

$$= \begin{bmatrix} [a_{11} + a_{11}(\lambda_1 - a_{22})t + a_{12}a_{21}t]\epsilon^{\lambda_1 t} & [a_{11}a_{12}t + a_{12} + a_1(\lambda_1 - a_{11})t]\epsilon^{\lambda_1 t} \\ [a_{21} + a_{21}(\lambda_1 - a_{22})t + a_{22}a_{21}t]\epsilon^{\lambda_1 t} & [a_{12}a_{21}t + a_{22} + a_{22}(\lambda_1 - a_{11})t]\epsilon^{\lambda_1 t} \end{bmatrix}. \tag{7.118}$$

Noting that λ_1 satisfies (7.114) and

$$2\lambda_1 = a_{11} + a_{22} \tag{7.119}$$

from (7.53) because $\lambda_1 = \lambda_2$, then (7.118) becomes

$$\begin{bmatrix} a_{11} & a_{12} \\ a_{21} & a_{22} \end{bmatrix}\begin{bmatrix} \phi_{11}(t) & \phi_{12}(t) \\ \phi_{21}(t) & \phi_{22}(t) \end{bmatrix}$$

$$= \begin{bmatrix} (2\lambda_1 - a_{22})\epsilon^{\lambda_1 t} + \lambda_1(\lambda_1 - a_{22})t\epsilon^{\lambda_1 t} & a_{12}\epsilon^{\lambda_1 t} + \lambda_1 a_{12} t\epsilon^{\lambda_1 t} \\ a_{21}\epsilon^{\lambda_1 t} + \lambda_1 a_{21} t\epsilon^{\lambda_1 t} & (2\lambda_1 - a_{11})\epsilon^{\lambda_1 t} + \lambda_1(\lambda_1 - a_{11})t\epsilon^{\lambda_1 t} \end{bmatrix} \tag{7.120}$$

Comparing (7.117) and (7.120), it is seen that the $\boldsymbol{\phi}(t)$ in (7.58) satisfies (7.109).

The transition matrix also satisfies

$$\begin{bmatrix} \frac{d}{dt}\phi_{11}(t-t_0) & \frac{d}{dt}\phi_{12}(t-t_0) \\ \frac{d}{dt}\phi_{21}(t-t_0) & \frac{d}{dt}\phi_{22}(t-t_0) \end{bmatrix} = \begin{bmatrix} a_{11} & a_{12} \\ a_{21} & a_{22} \end{bmatrix} \begin{bmatrix} \phi_{11}(t-t_0) & \phi_{12}(t-t_0) \\ \phi_{21}(t-t_0) & \phi_{22}(t-t_0) \end{bmatrix} \tag{7.121}$$

or

$$\frac{d}{dt}\boldsymbol{\phi}(t-t_0) = \mathbf{A}\boldsymbol{\phi}(t-t_0) \tag{7.122}$$

where t_0 is an arbitrary constant. To show this, let $t-t_0$ be denoted by τ,

$$\tau = t - t_0 \tag{7.123}$$

Since t_0 is a constant

$$dt = d\tau. \tag{7.124}$$

Hence

$$\begin{bmatrix} \frac{d}{dt}\phi_{11}(t-t_0) & \frac{d}{dt}\phi_{12}(t-t_0) \\ \frac{d}{dt}\phi_{21}(t-t_0) & \frac{d}{dt}\phi_{22}(t-t_0) \end{bmatrix} = \begin{bmatrix} \frac{d}{d\tau}\phi_{11}(\tau) & \frac{d}{d\tau}\phi_{12}(\tau) \\ \frac{d}{d\tau}\phi_{21}(\tau) & \frac{d}{d\tau}\phi_{22}(\tau) \end{bmatrix} \tag{7.125}$$

and

$$\begin{bmatrix} a_{11} & a_{12} \\ a_{21} & a_{22} \end{bmatrix} \begin{bmatrix} \phi_{11}(t-t_0) & \phi_{12}(t-t_0) \\ \phi_{21}(t-t_0) & \phi_{22}(t-t_0) \end{bmatrix} = \begin{bmatrix} a_{11} & a_{12} \\ a_{21} & a_{22} \end{bmatrix} \begin{bmatrix} \phi_{11}(\tau) & \phi_{12}(\tau) \\ \phi_{21}(\tau) & \phi_{22}(\tau) \end{bmatrix} \tag{7.126}$$

Since (7.109) can be written for any independent variable, such as τ, then

$$\begin{bmatrix} \frac{d}{d\tau}\phi_{11}(\tau) & \frac{d}{d\tau}\phi_{12}(\tau) \\ \frac{d}{d\tau}\phi_{21}(\tau) & \frac{d}{d\tau}\phi_{22}(\tau) \end{bmatrix} = \begin{bmatrix} a_{11} & a_{12} \\ a_{21} & a_{22} \end{bmatrix} \begin{bmatrix} \phi_{11}(\tau) & \phi_{12}(\tau) \\ \phi_{21}(\tau) & \phi_{22}(\tau) \end{bmatrix} \tag{7.127}$$

Comparing (7.125) and (7.126) and, in view of (7.127), (7.121) is proved.

Using (7.121), it is seen that

$$\begin{bmatrix} x_1(t) \\ x_2(t) \end{bmatrix} = \begin{bmatrix} \phi_{11}(t-t_0) & \phi_{12}(t-t_0) \\ \phi_{21}(t-t_0) & \phi_{22}(t-t_0) \end{bmatrix} \begin{bmatrix} x_1(t_0) \\ x_2(t_0) \end{bmatrix} \tag{7.128}$$

satisfies
$$\begin{bmatrix} \dot{x}_1(t) \\ \dot{x}_2(t) \end{bmatrix} = \begin{bmatrix} a_{11} & a_{12} \\ a_{21} & a_{22} \end{bmatrix} \begin{bmatrix} x_1(t) \\ x_2(t) \end{bmatrix}, \tag{7 129}$$

and for $t = t_0$, (7.128) reduces to

$$\begin{bmatrix} x_1(t_0) \\ x_2(t_0) \end{bmatrix} = \begin{bmatrix} 1 & 0 \\ 0 & 1 \end{bmatrix} \begin{bmatrix} x_1(t_0) \\ x_2(t_0) \end{bmatrix} = \begin{bmatrix} x_1(t_0) \\ x_2(t_0) \end{bmatrix} \tag{7.130}$$

That is, (7.128) gives the solution for (7.129), satisfying the initial condition in (7.130) at $t = t_0$.

7.6 Complete Solution for Second-Order Systems

When the input **u** is not zero, the state equation is

$$\dot{\mathbf{x}} = \begin{bmatrix} \dot{x}_1 \\ \dot{x}_2 \end{bmatrix} = \mathbf{Ax} + \mathbf{Bu} = \begin{bmatrix} a_{11} & a_{12} \\ a_{21} & a_{22} \end{bmatrix} \begin{bmatrix} x_1 \\ x_2 \end{bmatrix} + \begin{bmatrix} b_{11} & b_{12} \\ b_{21} & b_{22} \end{bmatrix} \begin{bmatrix} u_1 \\ u_2 \end{bmatrix} \tag{7.131}$$

The solution is given by

$$\mathbf{x}(t) = \boldsymbol{\phi}(t-t_0)\mathbf{x}_0 + \int_{t_0}^{t} [\boldsymbol{\phi}(t-\tau)\mathbf{Bu}(\tau)]\, d\tau \tag{7 132}$$

where $\boldsymbol{\phi}(t)$ is the transition matrix and $\mathbf{x}_0$ is the initial state at $t = t_0$. The integral of a matrix means a matrix of integrals of the elements of the matrix integrand. Let us show that $\mathbf{x}(t)$ in (7.132) satisfies (7.131) and $\mathbf{x}(t_0) = \mathbf{x}_0$. For $t = t_0$, (7.132) becomes

$$\mathbf{x}(t_0) = \boldsymbol{\phi}(0)\mathbf{x}_0 + \mathbf{0} = \mathbf{x}_0. \tag{7.133}$$

Hence the initial condition is satisfied. Next, we form $\mathbf{Ax} + \mathbf{Bu}$ from $\mathbf{x}$ in (7.132):

$$\begin{aligned} \mathbf{Ax} + \mathbf{Bu} = {} & \begin{bmatrix} a_{11} & a_{12} \\ a_{21} & a_{22} \end{bmatrix} \begin{bmatrix} \phi_{11}(t-t_0) & \phi_{12}(t-t_0) \\ \phi_{21}(t-t_0) & \phi_{22}(t-t_0) \end{bmatrix} \begin{bmatrix} x_1(t_0) \\ x_2(t_0) \end{bmatrix} \\ & + \begin{bmatrix} a_{11} & a_{12} \\ a_{21} & a_{22} \end{bmatrix} \int_{t_0}^{t} \begin{bmatrix} \phi_{11}(t-\tau) & \phi_{12}(t-\tau) \\ \phi_{21}(t-\tau) & \phi_{22}(t-\tau) \end{bmatrix} \begin{bmatrix} b_{11} & b_{12} \\ b_{21} & b_{22} \end{bmatrix} \begin{bmatrix} u_1(t) \\ u_2(\tau) \end{bmatrix} d\tau \\ & + \begin{bmatrix} b_{11} & b_{12} \\ b_{21} & b_{22} \end{bmatrix} \begin{bmatrix} u_1(t) \\ u_2(t) \end{bmatrix} \end{aligned} \tag{7.134}$$

Since **A** is a constant matrix, it may be shifted inside the integral:

$$\mathbf{Ax} + \mathbf{Bu} = \mathbf{A}\boldsymbol{\phi}(t-t_0)\mathbf{x}(t_0) + \int_{t_0}^{t} \mathbf{A}\boldsymbol{\phi}(t-t_0)\mathbf{Bu}(\tau)\, d\tau + \mathbf{Bu}(t). \tag{7.135}$$

Next, form $\dot{\mathbf{x}}(t)$ from $\mathbf{x}(t)$ in (7.132):

$$\begin{bmatrix} \dot{x}_1(t) \\ \dot{x}_2(t) \end{bmatrix} = \begin{bmatrix} \dot{\phi}_{11}(t-t_0) & \dot{\phi}_{12}(t-t_0) \\ \dot{\phi}_{21}(t-t_0) & \dot{\phi}_{22}(t-t_0) \end{bmatrix} \begin{bmatrix} x_1(t_0) \\ x_2(t_0) \end{bmatrix}$$

$$+ \int_{t_0}^{t} \begin{bmatrix} \dot{\phi}_{11}(t-\tau) & \dot{\phi}_{12}(t-\tau) \\ \dot{\phi}_{21}(t-\tau) & \dot{\phi}_{22}(t-\tau) \end{bmatrix} \begin{bmatrix} b_{11} & b_{12} \\ b_{21} & b_{22} \end{bmatrix} \begin{bmatrix} u_1(\tau) \\ u_2(\tau) \end{bmatrix} d\tau$$

$$+ \begin{bmatrix} \phi_{11}(t-t) & \phi_{12}(t-t) \\ \phi_{21}(t-t) & \phi_{22}(t-t) \end{bmatrix} \begin{bmatrix} b_{11} & b_{12} \\ b_{21} & b_{22} \end{bmatrix} \begin{bmatrix} u_1(t) \\ u_2(t) \end{bmatrix} \tag{7.136}$$

As a consequence of (7.110) and (7.122), (7.136) becomes

$$\dot{\mathbf{x}}(t) = \begin{bmatrix} a_{11} & a_{12} \\ a_{21} & a_{22} \end{bmatrix} \begin{bmatrix} \phi_{11}(t-t_0) & \phi_{12}(t-t_0) \\ \phi_{21}(t-t_0) & \phi_{22}(t-t_0) \end{bmatrix} \begin{bmatrix} x_1(t_0) \\ x_2(t_0) \end{bmatrix} + \begin{bmatrix} b_{11} & b_{12} \\ b_{21} & b_{22} \end{bmatrix} \begin{bmatrix} u_1(t) \\ u_2(t) \end{bmatrix}$$

$$+ \int_{t_0}^{t} \begin{bmatrix} a_{11} & a_{12} \\ a_{21} & a_{22} \end{bmatrix} \begin{bmatrix} \phi_{11}(t-\tau) & \phi_{12}(t-\tau) \\ \phi_{21}(t-\tau) & \phi_{22}(t-\tau) \end{bmatrix} \begin{bmatrix} b_{11} & b_{12} \\ b_{21} & b_{22} \end{bmatrix} \begin{bmatrix} u_1(\tau) \\ u_2(\tau) \end{bmatrix} d\tau \tag{7.137}$$

or

$$\dot{x}(t) = \mathbf{A}\boldsymbol{\phi}(t-t_0)\mathbf{x}_0 + \mathbf{B}\mathbf{u}(t) + \int_{t_0}^{t} \mathbf{A}\boldsymbol{\phi}(t-\tau)\mathbf{B}\mathbf{u}(\tau)\, d\tau. \tag{7.138}$$

Comparing (7.135) and (7.138), it is seen that $\mathbf{x}(t)$ in (7.132) satisfies (7.131). If the system has two outputs represented by the vector

$$\mathbf{y}(t) = \begin{bmatrix} y_1(t) \\ y_2(t) \end{bmatrix} \tag{7.139}$$

related to $\mathbf{x}(t)$ by

$$\begin{bmatrix} y_1(t) \\ y_2(t) \end{bmatrix} = \begin{bmatrix} c_{11} & c_{12} \\ c_{21} & c_{22} \end{bmatrix} \begin{bmatrix} x_1(t) \\ x_2(t) \end{bmatrix} + \begin{bmatrix} d_{11} & d_{12} \\ d_{21} & d_{22} \end{bmatrix} \begin{bmatrix} u_1(t) \\ u_2(t) \end{bmatrix} \tag{7.140}$$

Then

$$\begin{bmatrix} y_1(t) \\ y_2(t) \end{bmatrix} = \begin{bmatrix} c_{11} & c_{12} \\ c_{21} & c_{22} \end{bmatrix} \begin{bmatrix} \phi_{11}(t-t_0) & \phi_{12}(t-t_0) \\ \phi_{21}(t-t_0) & \phi_{22}(t-t_0) \end{bmatrix} \begin{bmatrix} x_1(t_0) \\ x_2(t_0) \end{bmatrix}$$

$$+ \int_{t_0}^{t} \begin{bmatrix} c_{11} & c_{12} \\ c_{21} & c_{22} \end{bmatrix} \begin{bmatrix} \phi_{11}(t-\tau) & \phi_{12}(t-\tau) \\ \phi_{21}(t-\tau) & \phi_{22}(t-\tau) \end{bmatrix} \begin{bmatrix} b_{11} & b_{12} \\ b_{21} & b_{22} \end{bmatrix} \begin{bmatrix} u_1(\tau) \\ u_2(\tau) \end{bmatrix} d\tau$$

$$+ \begin{bmatrix} d_{11} & d_{12} \\ d_{21} & d_{22} \end{bmatrix} \begin{bmatrix} u_1(t) \\ u_2(t) \end{bmatrix} \tag{7.141}$$

The first term in (7.141) is called the *zero-input response*:

$$\text{Zero-input response} = \begin{bmatrix} c_{11} & c_{12} \\ c_{21} & c_{22} \end{bmatrix} \begin{bmatrix} \phi_{11}(t-t_0) & \phi_{12}(t-t_0) \\ \phi_{21}(t-t_0) & \phi_{22}(t-t_0) \end{bmatrix} \begin{bmatrix} x_1(t_0) \\ x_2(t_0) \end{bmatrix} \tag{7.142}$$

It is the output when the input is zero. The sum of the remaining terms in (7.141) is called the *zero-state response*:

$$\begin{aligned} &\text{Zero-state response} \\ &= \int_{t_0}^{t} \begin{bmatrix} c_{11} & c_{12} \\ c_{21} & c_{22} \end{bmatrix} \begin{bmatrix} \phi_{11}(t-\tau) & \phi_{12}(t-\tau) \\ \phi_{21}(t-\tau) & \phi_{22}(t-\tau) \end{bmatrix} \begin{bmatrix} b_{11} & b_{12} \\ b_{21} & b_{22} \end{bmatrix} \begin{bmatrix} u_1(\tau) \\ u_2(\tau) \end{bmatrix} d\tau \\ &\quad + \begin{bmatrix} d_{11} & d_{12} \\ d_{21} & d_{22} \end{bmatrix} \begin{bmatrix} u_1(t) \\ u_2(t) \end{bmatrix} \end{aligned} \tag{7.143}$$

It is the output when the initial state is zero. The total output is equal to the zero-input response plus the zero-state response.

The number of outputs does not have to be two. It may be one or more than two. If there is only one output, such as $y_1(t)$, then only one row from each of

$$\begin{bmatrix} c_{11} & c_{12} \\ c_{21} & c_{22} \end{bmatrix} \quad \text{and} \quad \begin{bmatrix} d_{11} & d_{12} \\ d_{21} & d_{22} \end{bmatrix}$$

is needed.

The last term in (7.143) may be placed inside the integral by using a delta function multiplier:

$$\begin{aligned} \text{zero-state response} &= \int_{t_0}^{t} \left\{ \begin{bmatrix} c_{11} & c_{12} \\ c_{21} & c_{22} \end{bmatrix} \begin{bmatrix} \phi_{11}(t-\tau) & \phi_{12}(t-\tau) \\ \phi_{21}(t-\tau) & \phi_{22}(t-\tau) \end{bmatrix} \begin{bmatrix} b_{11} & b_{12} \\ b_{21} & b_{22} \end{bmatrix} \right. \\ &\quad \left. + \begin{bmatrix} d_{11} & d_{12} \\ d_{21} & d_{22} \end{bmatrix} \delta(t-\tau) \right\} \begin{bmatrix} u_1(\tau) \\ u_2(\tau) \end{bmatrix} d\tau \end{aligned} \tag{7.144}$$

This is valid because

$$\int_{t_0}^{t} \delta(t-\tau) u(\tau)\, d\tau = u(t)$$

for $t \geq t_0$. The zero-state response expressed as a single integral in (7.144) is the two-dimensional generalization of the convolution in (7.37). It is said to be the *matrix convolution* of

$$\mathbf{h}(t) = \begin{bmatrix} c_{11} & c_{12} \\ c_{21} & c_{22} \end{bmatrix} \begin{bmatrix} \phi_{11}(t) & \phi_{12}(t) \\ \phi_{21}(t) & \phi_{22}(t) \end{bmatrix} \begin{bmatrix} b_{11} & b_{12} \\ b_{21} & b_{22} \end{bmatrix} + \begin{bmatrix} d_{11} & d_{12} \\ d_{21} & d_{22} \end{bmatrix} \delta(t) \tag{7.145}$$

and

$$\mathbf{u}(t) = \begin{bmatrix} u_1(t) \\ u_2(t) \end{bmatrix}.$$

The matrix $\mathbf{h}(t)$ is called the *unit impulse response matrix* of the system The (i,j)th element of $\mathbf{h}(t)$, $h_{ij}(t)$, is the unit impulse response $y_i(t)$ for $u_j(t)$ equal to a unit impulse, with the other input equal to zero and initial state equal to zero.

Example 7-5. The accelerometer considered in previous chapters where vehicle acceleration is the input and seismic mass position is the output has a transfer function of

$$H(s) = \frac{K}{s^2 + 2\zeta\omega_n s + \omega_n{}^2} \tag{7.146}$$

where K, ζ, and ω_n are positive constant parameters of the accelerometer. Recall that for such a second-order transfer function, ζ is called damping ratio and ω_n is called undamped natural frequency. The corresponding output-input differential equation is

$$\frac{d^2y}{dt^2} + 2\zeta\omega_n \frac{dy}{dt} + \omega_n{}^2 y = Kr \tag{7.147}$$

where $y(t)$ denotes the output time function and $r(t)$ denotes the input time function. The accelerometer is second order, and it may be characterized by two state variables. For this example, the mass position $y(t)$ and mass velocity $\dot{y}(t)$ are suitable state variables. Denote the state variables $y(t)$ and $\dot{y}(t)$ by $x_1(t)$ and $x_2(t)$, respectively. Then

$$\dot{x}_1(t) = x_2(t) \tag{7.148}$$

$$\dot{x}_2(t) = -\omega_n{}^2 x_1(t) - 2\zeta\omega_n(t) + Kr(t) \tag{7.149}$$

and

$$y(t) = x_1(t). \tag{7.150}$$

Thus the state equations may be written in matrix form as

$$\begin{bmatrix} \dot{x}_1(t) \\ \dot{x}_2(t) \end{bmatrix} = \begin{bmatrix} 0 & 1 \\ -\omega_n{}^2 & -2\zeta\omega_n \end{bmatrix} \begin{bmatrix} x_1(t) \\ x_2(t) \end{bmatrix} + \begin{bmatrix} 0 \\ K \end{bmatrix} r(t) \tag{7.151}$$

There is only one output, so that (7.150) is written as

$$y(t) = [1 \quad 0] \begin{bmatrix} x_1(t) \\ x_2(t) \end{bmatrix} \tag{7.152}$$

Thus

$$\mathbf{A} = \begin{bmatrix} 0 & 1 \\ -\omega_n^2 & -2\zeta\omega_n \end{bmatrix} \tag{7 153}$$

$$\mathbf{B} = \begin{bmatrix} 0 \\ K \end{bmatrix} \tag{7.154}$$

$$\mathbf{C} = [1 \quad 0] \tag{7.155}$$

and

$$\mathbf{D} = 0. \tag{7.156}$$

Note that for $a_{11} = 0$ and $a_{12} = 1$, $-\omega_n^2$ becomes a_{21} and $-2\zeta\omega_n$ becomes a_{22}. For ω_n equal to a normalized value of 1 and $\zeta = 1$ (critically damped),

$$\boldsymbol{\Phi}(s) = [s\mathbf{I} - \mathbf{A}]^{-1} = \begin{bmatrix} s & -1 \\ 1 & s+2 \end{bmatrix}^{-1}$$

$$= \begin{bmatrix} \dfrac{s+2}{s(s+2)+1} & \dfrac{1}{s(s+2)+1} \\[2ex] \dfrac{-1}{s(s+2)+1} & \dfrac{s}{s(s+2)+1} \end{bmatrix}$$

$$= \begin{bmatrix} \dfrac{(s+1)+1}{(s+1)^2} & \dfrac{1}{(s+1)^2} \\[2ex] \dfrac{-1}{(s+1)^2} & \dfrac{(s+1)-1}{(s+1)^2} \end{bmatrix} = \begin{bmatrix} \dfrac{1}{s+1} + \dfrac{1}{(s+1)^2} & \dfrac{1}{(s+1)^2} \\[2ex] \dfrac{-1}{(s+1)^2} & \dfrac{1}{(s+1)} - \dfrac{1}{(s+1)^2} \end{bmatrix} \tag{7.157}$$

Thus.

$$\mathcal{L}^{-1}\boldsymbol{\Phi}(s) = \begin{bmatrix} e^{-t} + te^{-t} & te^{-t} \\ -te^{-t} & \epsilon^{-t} - te^{-t} \end{bmatrix} u(t) \tag{7.158}$$

where $u(t)$ is the unit step function. Hence the transition matrix is

$$\boldsymbol{\phi}(t) = \begin{bmatrix} \epsilon^{-t} + t\epsilon^{-t} & t\epsilon^{-t} \\ -t\epsilon^{-t} & \epsilon^{-t} - t\epsilon^{-t} \end{bmatrix} \tag{7.159}$$

for all t. Note that

$$\boldsymbol{\phi}(0) = \begin{bmatrix} 1 & 0 \\ 0 & 1 \end{bmatrix} \tag{7.160}$$

Also, $\boldsymbol{\phi}(t)$ in (7.159) satisfies (7.109) for the **A** matrix of this example, and it is nonsingular for all t.

The zero-input response is

$$\text{zero-input response} = [1 \quad 0]\begin{bmatrix} \epsilon^{-t} + t\epsilon^{-t} & t\epsilon^{-t} \\ -t\epsilon^{-t} & \epsilon^{-t} - t\epsilon^{-t} \end{bmatrix}\begin{bmatrix} x_1(0) \\ x_2(0) \end{bmatrix}$$

$$= [(\epsilon^{-t} + t\epsilon^{-t})t\epsilon^{-t}]\begin{bmatrix} x_1(0) \\ x_2(0) \end{bmatrix} = (\epsilon^{-t} + t\epsilon^{-t})x_1(0) + t\epsilon^{-t}x_2(0) \quad (7.161)$$

The zero-state response is

$$\text{zero-state response} = \int_{t_1}^{t} [1 \quad 0]\begin{bmatrix} \epsilon^{-(t-\tau)} + (t-\tau)\epsilon^{-(t-\tau)} & (t-\tau)\epsilon^{-(t-\tau)} \\ -(t-\tau)\epsilon^{-(t-\tau)} & \epsilon^{-(t-\tau)} - (t-\tau)\epsilon^{-(t-\tau)} \end{bmatrix}$$

$$\cdot \begin{bmatrix} 0 \\ K \end{bmatrix} r(\tau)\, d\tau = \int_0^t (t-\tau)\epsilon^{-(t-\tau)} Kr(\tau)\, d\tau \quad (7.162)$$

From (7.162), the unit impulse response is

$$h(t) = Kt\epsilon^{-t}u(t) \quad (7.163)$$

where $u(t)$ is the unit step function. This impulse response is also equal to the inverse Laplace transform of

$$H(s) = \frac{K}{s^2 + 2\zeta\omega_n s + \omega_n{}^2} = \frac{K}{(s+1)^2} \quad (7.164)$$

Thus

$$h(t) = \mathscr{L}^{-1}H(s) = Kt\epsilon^{-t}u(t) \quad (7.165)$$

For a unit step input,

$$\text{zero-state response} = K\int_0^t (t-\tau)\epsilon^{-(t-\tau)}\, d\tau = K[1 - \epsilon^{-t} - t\epsilon^{-t}]u(t). \quad (7.166)$$

This may be verified by finding the inverse transform of

$$Y(s) = \frac{K}{s(s+1)^2} \quad (7.167)$$

The unit step response in (7.166) has the form shown in Figure B.4 (see Appendix B). ▲

7.7 The State-Transition Matrix—Nth-Order System

Let us now consider an nth-order linear time-invariant dynamic system described by state equations as in (7.8) and (7.9) where **A**, **B**, **C**, and **D** are constant matrices. We shall find it useful to treat the effect of initial state $\mathbf{x}(t_0)$ and input $\mathbf{u}(t)$ separately. The output when the input $\mathbf{u}(t)$ is identically zero for $t \geqslant t_0$ is called the *zero-input response*. The zero-input response is due to the initial state $\mathbf{x}(t_0)$ entirely. When **u** is identically zero, (7.8) becomes

$$\dot{\mathbf{x}} = \mathbf{A}\mathbf{x}. \tag{7.168}$$

Equation (7.168) has a *unique* solution[1] for $\mathbf{x}(t)$ satisfying the initial state at t_0. This means that no matter how we arrive at a function $\mathbf{x}(t)$, provided that it satisfies (7.168) and the initial conditions, the function we have is the only function that will satisfy (7.168) and the initial conditions. Analogous to the second-order case, we shall show that the solution is of the form

$$\mathbf{x}(t) = \boldsymbol{\phi}(t-t_0)x(t_0)$$

where $\boldsymbol{\phi}(t-t_0)$ is the transition matrix. Also, we shall describe several methods of determining the transition matrix.

One method of determining $\boldsymbol{\phi}(t-t_0)$ is by computing a matrix infinite series. Define

$$\epsilon^{\mathbf{A}(t-t_0)} \triangleq \mathbf{I} + (t-t_0)\mathbf{A} + \tfrac{1}{2}(t-t_0)^2\mathbf{A}^2 + \cdots + \frac{1}{n!}(t-t_0)^n\mathbf{A}^n + \cdots = \sum_{k=0}^{\infty} \frac{1}{k!}(t-t_0)^k\mathbf{A}^k \tag{7.169}$$

where **I** is the identity matrix of order n, A is a square matrix of order n, $\mathbf{A}^2 = (\mathbf{A})(\mathbf{A})$, and $\mathbf{A}^k = \mathbf{A}(\mathbf{A}^{k-1})$ for all integer values of k. This matrix series converges uniformly for all finite t, that is, each of its elements converges uniformly for all finite t.[2] Assuming this uniform convergence, the series may be differentiated term by term.

Let us verify that $\epsilon^{\mathbf{A}(t-t_0)}$ satisfies

$$\frac{d}{dt}\epsilon^{\mathbf{A}(t-t_0)} = \mathbf{A}\epsilon^{\mathbf{A}(t-t_0)} \tag{7.170}$$

and

$$\epsilon^{\mathbf{A}(t-t_0)} = \mathbf{I}, \tag{7.171}$$

[1] L. S. Pontriagin, "Ordinary Differential Equations," Addison-Wesley, Reading, Mass., 1962, pp. 18–23.

for $t = t_0$. By inspection of (7.169), it is seen that (7.171) follows immediately. Next, (7.170) is verified by computing the derivative of $\epsilon^{\mathbf{A}(t-t_0)}$ in (7.169). Thus

$$\begin{aligned}\frac{d}{dt}[\epsilon^{\mathbf{A}(t-t_0)}] &= \left[\frac{d}{dt}\sum_{k=0}^{\infty}\frac{1}{k!}(t-t_0)^k\mathbf{A}^k\right]\\ &= \left[\mathbf{A} + (t-t_0)\mathbf{A}^2 + \cdots \frac{1}{n!}(t-t_0)^n\mathbf{A}^{n+1} + \cdots\right]\\ &= \mathbf{A}\left[\mathbf{I} + (t-t_0)\mathbf{A} + \cdots + \frac{1}{n!}(t-t_0)^n\mathbf{A}^n + \cdots\right]\\ &= \mathbf{A}\epsilon^{\mathbf{A}(t-t_0)}. \end{aligned} \tag{7.172}$$

Using the properties of $\epsilon^{\mathbf{A}(t-t_0)}$ in (7.170) and (7.171), it is readily verified that

$$\mathbf{x}(t) = \epsilon^{\mathbf{A}(t-t_0)}\mathbf{x}_0 \tag{7.173}$$

is a solution of (7.168) satisfying the initial conditions

$$\mathbf{x}(t_0) = \mathbf{x}_0. \tag{7.174}$$

Thus, by setting $t = t_0$, (7.173) becomes

$$\mathbf{x}(t_0) = \epsilon^{\mathbf{A}(t_0-t_0)}\mathbf{x}_0 = \mathbf{I}\mathbf{x}_0 = \mathbf{x}_0. \tag{7.175}$$

Next, form $\dot{\mathbf{x}}(t)$ from $\mathbf{x}(t)$ in (7.173):

$$\frac{d}{dt}\mathbf{x}(t) = \frac{d}{dt}\epsilon^{\mathbf{A}(t-t_0)}\mathbf{x}_0. \tag{7.176}$$

Using the result of (7.170), (7.176) may be written as

$$\dot{\mathbf{x}}(t) = \mathbf{A}\epsilon^{\mathbf{A}(t-t_0)}\mathbf{x}_0 = \mathbf{A}\mathbf{x}(t) \tag{7.177}$$

which verifies that $\mathbf{x}(t)$ in (7.173) satisfies (7.168) and (7.174). Because the differential equation of (7.168) has a unique solution[2] satisfying the initial condition in (7.174), the solution of (7.173) must be the unique solution.

The matrix $\epsilon^{\mathbf{A}(t-t_0)}$ premultiplies $\mathbf{x}(t_0)$ to yield $\mathbf{x}(t)$ when there is no input. It is called the *state-transition matrix* for the vector differential

[2]See L. A. Zadeh and C. A. Desoer, *Linear System Theory*, McGraw-Hill, New York, 1963, pp. 294–295.

equation of (7.168). We shall denote the transition matrix by $\boldsymbol{\phi}(t-t_0)$:

$$\boldsymbol{\phi}(t-t_0) = \epsilon^{\mathbf{A}(t-t_0)}. \tag{7.178}$$

Thus, from (7.170), $\boldsymbol{\phi}(t-t_0)$ satisfies

$$\frac{d}{dt}\boldsymbol{\phi}(t-t_0) = \mathbf{A}\boldsymbol{\phi}(t-t_0). \tag{7.179}$$

For t close to t_0, $\boldsymbol{\phi}(t)$ may be approximated by taking only a few terms in the series of (7.169). Such a computational procedure is feasible on a digital computer. For a high-order system, such a computation may be just as practical as other available methods.

Let us obtain the state-transition matrix for (7.168) by another method. Suppose that we take the Laplace transform of (7.179) for $t_0 = 0$. Then we obtain

$$s\boldsymbol{\Phi}(s) - \boldsymbol{\phi}(0) = \mathbf{A}\boldsymbol{\Phi}(s). \tag{7.180}$$

Since $\boldsymbol{\phi}(0) = \mathbf{I}$, (7.180) can be written as

$$(s\mathbf{I}-\mathbf{A})\boldsymbol{\Phi}(s) = \mathbf{I}. \tag{7.181}$$

The matrix $[s\mathbf{I}-\mathbf{A}]$ is nonsingular except at the eigenvalues of $\mathbf{A}$. Hence

$$\boldsymbol{\Phi}(s) = [s\mathbf{I}-\mathbf{A}]^{-1} \tag{7.182}$$

and

$$\boldsymbol{\phi}(t) = \mathscr{L}^{-1}\{[s\mathbf{I}-\mathbf{A}]^{-1}\} \tag{7.183}$$

for $t \geqslant 0$. The inverse Laplace transform of $[s\mathbf{I}-\mathbf{A}]^{-1}$ is a matrix whose elements are the inverse transforms of the elements of the matrix $[s\mathbf{I}-\mathbf{A}]^{-1}$.

Example 7-6. Suppose

$$\dot{\mathbf{x}} = \begin{bmatrix} -1 & -1 \\ 1 & -1 \end{bmatrix}\mathbf{x}. \tag{7.184}$$

Form

$$s\mathbf{I}-\mathbf{A} = \begin{bmatrix} s & 0 \\ 0 & s \end{bmatrix} - \begin{bmatrix} -1 & -1 \\ 1 & -1 \end{bmatrix} = \begin{bmatrix} s+1 & +1 \\ -1 & s+1 \end{bmatrix}. \tag{7.185}$$

Next form

$$(s\mathbf{I}-\mathbf{A})^{-1}=\begin{bmatrix} s+1 & 1 \\ -1 & s+1 \end{bmatrix}^{-1}=\frac{1}{(s+1)^2+1}\begin{bmatrix} s+1 & -1 \\ +1 & s+1 \end{bmatrix}$$

$$=\begin{bmatrix} \dfrac{s+1}{(s+1)^2+1} & \dfrac{-1}{(s+1)^2+1} \\ \dfrac{+1}{(s+1)^2+1} & \dfrac{s+1}{(s+1)^2+1} \end{bmatrix}. \tag{7.186}$$

From Tables B.1 and B.2 (see Appendix B), we obtain the inverse transform

$$\mathscr{L}^{-1}\{[s\mathbf{I}-\mathbf{A}]^{-1}\}=\begin{bmatrix} \epsilon^{-t}\cos t & -\epsilon^{-t}\sin t \\ \epsilon^{-t}\sin t & \epsilon^{-t}\cos t \end{bmatrix} \tag{7.187}$$

for $t \geqslant 0$. The transition matrix is equal to this inverse transform for $t \geqslant 0$. The transition matrix for all t is obtained by assuming the form for $t < 0$ to be the same as that for $t \geqslant 0$. This assumption is verified by noting that the $\boldsymbol{\phi}(t)$ obtained satisfies $\dot{\boldsymbol{\phi}}(t)=\mathbf{A}\boldsymbol{\phi}(t)$ for all t. Thus, for this example, $\epsilon^{\mathbf{A}(t-t_0)}$ is

$$\epsilon^{\mathbf{A}(t-t_0)}=\begin{bmatrix} \epsilon^{-(t-t_0)}\cos(t-t_0) & -\epsilon^{-(t-t_0)}\sin(t-t_0) \\ \epsilon^{-(t-t_0)}\sin(t-t_0) & \epsilon^{-(t-t_0)}\cos(t-t_0) \end{bmatrix} \tag{7.188}$$

where t_0 is the instant at which the initial state is specified. Notice that the transition matrix for a linear time-invariant system depends on the difference between t and t_0. Thus, if the transition matrix for $t_0=0$ is known, the transition matrix for an arbitrary t_0 is simply obtained by replacing t by $t-t_0$. ▲

Next, consider an all-integrator simulation diagram of the system described by (7.168), where the initial condition at $t=0$ may be specified at our convenience. Our objective is to measure the transition matrix. Since the transition matrix does not depend on the values of the initial state, the initial conditions may be chosen at our convenience without changing ϵ^{At}. Suppose that all the components of $\mathbf{x}(0)$ are equal to zero except the jth:

$$\begin{aligned} x_i(0) &= 0, \quad \text{for all} \quad i \neq j, i=1,\ldots,n \\ &= x_j(0) \neq 0, \quad \text{for} \quad i=j. \end{aligned} \tag{7.189}$$

Then the solution

$$\mathbf{x}(t)=\boldsymbol{\epsilon}^{\mathbf{A}t}\mathbf{x}(0) \tag{7.190}$$

reduces to the jth column of ϵ^{At} times $x_j(0)$. Let us denote the ith row and jth column element of ϵ^{At} by $\phi_{ij}(t)$. Then the ith row of (7.190) becomes

$$x_i(t) = \phi_{ij}(t)x_j(0) \tag{7.191}$$

whenever (7.189) is satisfied. Equation 7.191 is valid for all rows from $i=1$ to $i=n$. This suggests an experimental procedure for determining $\phi_{ij}(t)$. First, set all initial conditions to zero except $x_1(0)$. Then, measure $x_1(t), x_2(t), \ldots, x_n(t)$, which are the integrator outputs. These signals are equal to $x_1(0)$ times $\phi_{11}(t), \phi_{21}(t), \ldots, \phi_{n1}(t)$, respectively. Repeat for all initial conditions equal to zero except $x_2(0)$. This time n simultaneous measurements of the state variables gives $x_2(0)$ times the second column of the $\boldsymbol{\phi}(t)$ matrix. We continue this process up to $x_n(0) \neq 0$ and all other initial conditions set to zero. Thus n sets of experiments with n measurements in each yield the n^2 elements of the matrix ϵ^{At}. Therefore,

$$\phi_{ij}(t) = \left.\frac{x_i(t)}{x_j(0)}\right|_{\substack{x_i(0)=0 \quad \text{for all} \quad i \neq j. \\ x_j(0) \neq 0}} \tag{7.192}$$

Another way of obtaining the transition matrix is to compute the elements of $[s\mathbf{I}-\mathbf{A}]^{-1}$ as appropriate transfer functions from various initial conditions to the various Laplace transforms of the state variables. Recall from Chapters 4 and 6 that for an integrator whose input is $x(t)$ and whose output is $y(t)$, the signals are related by

$$y(t) = y(0) + \int_0^t x(\tau)\, d\tau. \tag{7.193}$$

The transform of (7.193) is

$$Y(s) = \frac{y(0)}{s} + \frac{1}{s}X(s). \tag{7.194}$$

The block diagram for (7.194) is shown in Figure 7.4.

Thus the (i,j)th element of $[s\mathbf{I}-\mathbf{A}]^{-1}$ is the transfer function from the jth initial condition $x_j(0)$ to the ith integrator output $X_i(s)$.

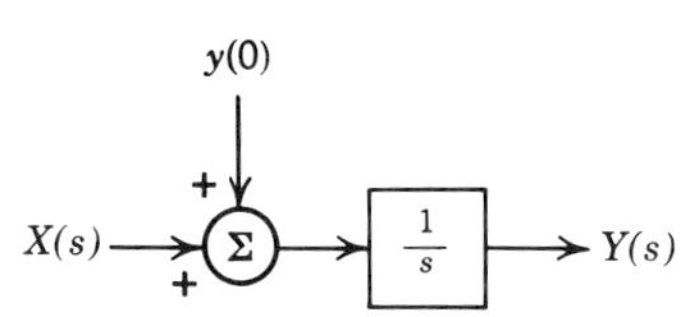

Figure 7.4 Transfer-function block diagram of an integrator.

Example 7-7. Consider the second-order system

$$\dot{x}_1 = a_{11}x_1 + a_{12}x_2 \tag{7.195}$$

$$\dot{x}_2 = a_{21}x_1 + a_{22}x_2 \tag{7.196}$$

whose transfer-function block diagram is shown in Figure 7.5. Let the elements of the matrix $[s\mathbf{I}-\mathbf{A}]^{-1}$ be represented by $\Phi_{ij}(s)$. From the vector Laplace transform of $\mathbf{x}(t) = \phi(t)\mathbf{x}(t)$, we have the component equations

$$X_1(s) = \Phi_{11}(s)x_1(0) + \Phi_{12}(s)x_2(0) \tag{7.197}$$

$$X_2(s) = \Phi_{21}(s)x_1(0) + \Phi_{22}(s)x_2(0). \tag{7.198}$$

If we set $x_2(0) = 0$, then $\Phi_{11}(s)$ is the transfer function from $x_1(0)$ to $X_1(s)$

$$\left.\frac{X_1(s)}{x_1(0)}\right|_{x_2(0)=0} = \Phi_{11}(s) \tag{7.199}$$

and

$$\left.\frac{X_2(s)}{x_1(0)}\right|_{x_2(0)=0} = \Phi_{21}(s). \tag{7.200}$$

The other components $\Phi_{12}(s)$ and $\Phi_{22}(s)$ are obtained from

$$\left.\frac{X_1(s)}{x_2(0)}\right|_{x_1(0)=0} = \Phi_{12}(s) \tag{7.201}$$

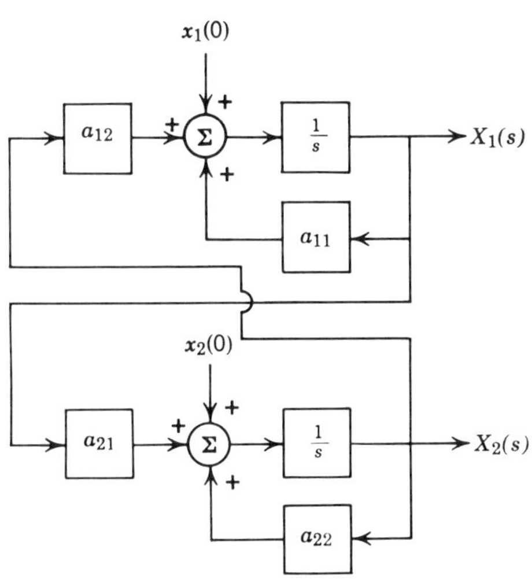

Figure 7.5 Transfer-function block diagram for a second-order system with no input.

and

$$\left.\frac{X_2(s)}{x_2(0)}\right|_{x_1(0)=0} = \Phi_{22}(s). \tag{7.202}$$

These transfer functions may be obtained from the transfer-function block diagram using Mason's rule. By inspection of Figure 7.5, we have

$$\Phi_{11}(s) = \frac{\frac{1}{s}\left(1 - \frac{a_{22}}{s}\right)}{1 - \left(\frac{a_{11}}{s} + \frac{a_{22}}{s} + \frac{a_{12}a_{21}}{s^2}\right) + \frac{a_{11}a_{22}}{s^2}} \tag{7.203}$$

or

$$\Phi_{11}(s) = \frac{(s - a_{22})}{s^2 - (a_{11} + a_{22})s + (a_{11}a_{22} - a_{12}a_{21})}. \tag{7.204}$$

The other elements $\Phi_{12}(s)$ and $\Phi_{22}(s)$ may be obtained easily using Mason's rule also. Finally, the transition matrix may be obtained as the inverse transform of the matrix of $\Phi_{ij}(s)$. For the system represented by (7.184) where $a_{11} = -1$, $a_{12} = -1$, $a_{21} = 1$ and $a_{22} = -1$, then

$$\Phi_{11} = \frac{s+1}{s^2+2s+2} = \frac{s+1}{(s+1)^2+1}, \tag{7.205}$$

which checks with $\Phi_{11}(s)$ in (7.186). Similarly, by Mason's rule,

$$\Phi_{12}(s) = \left.\frac{X_1(s)}{x_2(0)}\right|_{x_1(0)=0} = \frac{\frac{1}{s}(a_{12})\left(\frac{1}{s}\right)}{1 - \left(\frac{a_{11}}{s} + \frac{a_{22}}{s} + \frac{a_{12}a_{21}}{s^2}\right) + \frac{a_{11}a_{22}}{s^2}}$$

$$= \frac{a_{12}}{s^2 - (a_{11} + a_{22})s + (a_{11}a_{22} - a_{12}a_{21})}. \tag{7.206}$$

Again, for the example in (7.184), substitution of the various values of a_{ij} yields

$$\Phi_{12}(s) = \frac{-1}{s^2+2s+2} = \frac{-1}{(s+1)^2+1} \tag{7.207}$$

which checks with $\Phi_{12}(s)$ in (7.186). $\Phi_{21}(s)$ and $\Phi_{22}(s)$ may be computed using Mason's rule for Figure 7.5. Thus

$$\Phi_{21}(s) = \left.\frac{X_2(s)}{x_1(0)}\right|_{x_2(0)=0} = \frac{\frac{1}{s}(a_{21})\left(\frac{1}{s}\right)}{1-\left(\frac{a_{11}}{s}+\frac{a_{22}}{s}+\frac{a_{12}a_{21}}{s^2}\right)+\frac{a_{11}a_{22}}{s^2}}$$

$$= \frac{a_{21}}{s^2-(a_{11}+a_{22})s+(a_{11}a_{22}-a_{12}a_{21})} \tag{7.208}$$

and

$$\Phi_{22}(s) = \left.\frac{X_2(s)}{x_2(0)}\right|_{x_1(0)=0} = \frac{\frac{1}{s}\left(1-\frac{a_{11}}{s}\right)}{1-\left(\frac{a_{11}}{s}+\frac{a_{22}}{s}+\frac{a_{12}a_{21}}{s^2}\right)+\frac{a_{11}a_{22}}{s^2}}$$

$$= \frac{s-a_{11}}{s^2-(a_{11}+a_{22})s+(a_{11}a_{22}-a_{12}a_{21})}. \tag{7.209}$$

For the values of a_{11}, a_{12}, a_{21}, and a_{22} in (7.184), it is seen that the resulting $\Phi_{21}(s)$ and $\Phi_{22}(s)$ match those in (7.186). ▲

Example 7-8. For the **A** matrix of Example 7-3, $a_{11}=0$, $a_{12}=1$, $a_{21}=-5$, and $a_{22}=-6$, as obtained from (7.26). Using (7.204), (7.206), (7.208), and (7.209), we have

$$\mathbf{\Phi}(s) = \begin{bmatrix} \frac{s+6}{s^2+6s+5} & \frac{1}{s^2+6s+5} \\ \frac{-5}{s^2+6s+5} & \frac{s}{s^2+6s+5} \end{bmatrix}. \tag{7.210}$$

The poles are the roots of

$$s^2+6s+5=0 \tag{7.211}$$

which are $s_1=-5$ and $s_2=-1$. These are the eigenvalues of the **A** matrix as obtained from (7.42). Expanding $\Phi(s)$ in partial fractions we have

$$\mathbf{\Phi}(s) = \begin{bmatrix} \frac{\frac{5}{4}}{s+1}-\frac{\frac{1}{4}}{s+5} & \frac{-1}{s+5}+\frac{1}{s+1} \\ \frac{5}{s+5}+\frac{5}{s+1} & \frac{5}{s+5}-\frac{1}{s+1} \end{bmatrix}. \tag{7.212}$$

The inverse transform is

$$\mathcal{L}^{-1}[\mathbf{\Phi}(s)] = \begin{bmatrix} \frac{5}{4}\epsilon^{-t} - \frac{1}{4}\epsilon^{-5t} & \epsilon^{-t} - \epsilon^{-5t} \\ 5\epsilon^{-5t} - 5\epsilon^{-t} & 5\epsilon^{-5t} - \epsilon^{-t} \end{bmatrix} \tag{7.213}$$

for $t \geqslant 0$. Hence

$$\boldsymbol{\phi}(t) = \begin{bmatrix} \frac{5}{4}\epsilon^{-t} - \frac{1}{4}\epsilon^{-5t} & \epsilon^{-t} - \epsilon^{-5t} \\ 5\epsilon^{-5t} - 5\epsilon^{-t} & 5\epsilon^{-5t} - \epsilon^{-t} \end{bmatrix} \tag{7.214}$$

for all t. For an initial time other than zero, t is replaced by $(t - t_0)$ in (7.214). ▲

7.8 The Complete Solution

Let us turn our attention now to the complete solution for the linear time-invariant system when the input is applied. The complete solution for

$$\dot{\mathbf{x}} + \mathbf{A}\mathbf{x} + \mathbf{B}\mathbf{u} \tag{7.215}$$

and

$$\mathbf{x}(t_0) = \mathbf{x}_0 \tag{7.216}$$

is

$$\mathbf{x}(t) = \boldsymbol{\phi}(t - t_0)\mathbf{x}_0 + \int_{t_0}^{t} \boldsymbol{\phi}(t - \tau)\,\mathbf{B}\mathbf{u}(\tau)\,d\tau \tag{7.217}$$

where $\boldsymbol{\phi}(t)$ is the transition matrix $\epsilon^{\mathbf{A}t}$. This is verified by substitution of (7.217) in (7.215) and (7.216), and it is left to the reader as an exercise.

If the output $\mathbf{y}(t)$ is related to $\mathbf{x}(t)$ and $\mathbf{u}(t)$ by

$$\mathbf{y}(t) = \mathbf{C}\mathbf{x}(t) + \mathbf{D}\mathbf{u}(t), \tag{7.218}$$

then

$$\mathbf{y}(t) = \mathbf{C}\boldsymbol{\phi}(t - t_0)\mathbf{x}_0 + \int_{t_0}^{t} \mathbf{C}\boldsymbol{\phi}(t - \tau)\mathbf{B}\mathbf{u}(\tau)\,d\tau + \mathbf{D}\mathbf{u}(\tau). \tag{7.219}$$

The output due to $\mathbf{u}(t)$ alone, with $\mathbf{x}(t_0) = \mathbf{0}$ is called the *zero-state response* and, from (7.219), this is

$$\mathbf{y}(t) = \int_{t_0}^{t} \mathbf{C}\boldsymbol{\phi}(t - \tau)\mathbf{B}\mathbf{u}(\tau)\,d\tau + \mathbf{D}\mathbf{u}(\tau). \tag{7.220}$$

The input $\mathbf{u}(t)$ is applied starting at $t = t_0$. As in the second-order case, (7.220) may be rewritten as

$$\mathbf{y}(t) = \int_{t_0}^{t} [\mathbf{C}\boldsymbol{\phi}(t-\tau)\mathbf{B} + \mathbf{D}\delta(t-\tau)]\mathbf{u}(\tau)\, d\tau. \tag{7.221}$$

The integral of (7.221) is said to be the matrix convolution of

$$\mathbf{h}(t) = [\mathbf{C}\boldsymbol{\phi}(t)\mathbf{B} + \mathbf{D}\delta(t)] \tag{7.222}$$

and $\mathbf{u}(t)$. The matrix $\mathbf{h}(t)$ is called the *unit impulse response matrix* of the system. We see that the zero-input response and the zero-state response may be computed separately, and their sum is the total response. Observe also that the transition matrix $\boldsymbol{\phi}$ which, together with $\mathbf{x}(t_0)$, characterizes the zero input response, is used in the determination of the zero-state response.

As in the zero-input case, we may use Laplace transforms to solve for the complete response. The Laplace transform of (7.215) is

$$s\mathbf{X}(s) = \mathbf{x}(0) = \mathbf{AX}(s) + \mathbf{BU}(s) \tag{7.223}$$

which is rewritten as

$$[s\mathbf{I} - \mathbf{A}]\mathbf{X}(s) = \mathbf{x}(0) + \mathbf{BU}(s). \tag{7.224}$$

Hence

$$\mathbf{X}(s) = [s\mathbf{I} - \mathbf{A}]^{-1}\mathbf{x}(0) + [s\mathbf{I} - \mathbf{A}]^{-1}\mathbf{BU}(s) \tag{7.225}$$

or

$$\mathbf{X}(s) = \boldsymbol{\Phi}(s)\mathbf{x}(0) + \boldsymbol{\Phi}(s)\mathbf{BU}(s) \tag{7.226}$$

where $\boldsymbol{\Phi}(s)$ is $[s\mathbf{I} - \mathbf{A}]^{-1}$, the Laplace transform of the transition matrix. Using (7.226), the Laplace transform of (7.218) is

$$\begin{aligned}\mathbf{Y}(s) &= \mathbf{CX}(s) + \mathbf{DU}(s)\\ &= \mathbf{C}\boldsymbol{\Phi}(s)\mathbf{x}(0) + \mathbf{C}\boldsymbol{\Phi}(s)\mathbf{BU}(s) + \mathbf{DU}(s).\end{aligned} \tag{7.227}$$

The inverse transforms of (7.226) and (7.227) give $\mathbf{x}(t)$ and $\mathbf{y}(t)$ when $t_0 = 0$. In other words, for $t_0 = 0$, the Laplace transform of (7.217) is (7.226) and that for (7.219) is (7.227). Again, notice that the transform of the state-transition matrix $\boldsymbol{\Phi}(s)$ is involved not only in determining the transform of the zero-input response $\mathbf{C}\boldsymbol{\Phi}(s)\mathbf{x}(0)$ but also in determining the transform of the zero-state response $\mathbf{C}\boldsymbol{\Phi}(s)\mathbf{BU}(s) + \mathbf{DU}(s)$.

The zero-state response may be determined by computing the transfer function matrix from $\mathbf{U}(s)$ to $\mathbf{Y}(s)$:

$$\text{Zero-state } \mathbf{Y}(s) = [\mathbf{C}\boldsymbol{\Phi}(s)\mathbf{B}+\mathbf{D}]\mathbf{U}(s). \tag{7.228}$$

Thus the (i, j)th element of the transfer function matrix $\mathbf{C}\boldsymbol{\Phi}(s)\mathbf{B}+\mathbf{D}$ is the transfer function from $U_j(s)$ to $Y_i(s)$. Mason's rule may be used if the transfer function block diagram is given.

Notice that the transfer function matrix $[\mathbf{C}\boldsymbol{\Phi}(s)\mathbf{B}+\mathbf{D}]$ is the Laplace transform of the impulse response matrix $\mathbf{h}(t)$ in (7.222) and $\mathbf{U}(s)$ is the Laplace transform of the input $\mathbf{u}(t)$. Hence the Laplace transform of the matrix convolution of $\mathbf{h}(t)$ and $\mathbf{u}(t)$ in (7.221) is the product of the transform of $\mathbf{h}(t)$ by the transform of $\mathbf{u}(t)$ as in (7.228).

Similarly, since $\mathbf{C}\boldsymbol{\Phi}(s)$ is the transfer function matrix from $\mathbf{x}(0)$ to the zero-input $\mathbf{Y}(s)$,

$$\text{Zero-input } Y(s) = [\mathbf{C}\boldsymbol{\Phi}(s)]\mathbf{x}(0). \tag{7.229}$$

Mason's rule may be used to compute the transfer function from $x_j(0)$ to $Y_i(s)$, which is the (i, j)th element of the matrix $\mathbf{C}\boldsymbol{\Phi}(s)$ for all i and j.

Example 7-9. Consider the network in Figure 7.1 described by (7.5). Let v_L be the output which is given by (7.7). The network parameters are $R_1 = 1, R_2 = 1, L = 1, C = 1$, so that

$$\mathbf{A} = \begin{bmatrix} -1 & -1 \\ 1 & -1 \end{bmatrix} \tag{7.230}$$

$$\mathbf{B} = \begin{bmatrix} 1 \\ 0 \end{bmatrix} \tag{7.231}$$

$$\mathbf{C} = [-1 \quad -1] \tag{7.232}$$

$$\mathbf{D} = [1]. \tag{7.233}$$

The $\mathbf{A}$ matrix is identical to the one used in Example 7-6 [see (7.184)], and the $\boldsymbol{\Phi}(s)$ matrix is given by (7.186). The transform of the zero-input response is

$$\mathbf{Y}(s) = [-1 \quad -1]\begin{bmatrix} \dfrac{s+1}{(s+1)^2+1} & \dfrac{-1}{(s+1)^2+1} \\[2ex] \dfrac{1}{(s+1)^2+1} & \dfrac{s+1}{(s+1)^2+1} \end{bmatrix}\begin{bmatrix} x_1(0) \\ x_2(0) \end{bmatrix} \tag{7.234}$$

or

$$Y(s) = \begin{bmatrix} \dfrac{-s-2}{(s+1)^2+1} & \dfrac{-s}{(s+1)^2+1} \end{bmatrix} \begin{bmatrix} x_1(0) \\ x_2(0) \end{bmatrix}. \tag{7.235}$$

The inverse transform of (7.235) yields the zero-input response:

$$\begin{aligned} y(t) = &-x_1(0)[\epsilon^{-t}\cos t + \epsilon^{-t}\sin t]u(t) \\ &-x_2(0)[\epsilon^{-t}\cos t - \epsilon^{-t}\sin t]u(t). \end{aligned} \tag{7.236}$$

For the zero-state response, suppose the input is a step function, that is, zero for $t < 0$ and constant for $t \geqslant 0$. The input is the voltage v in Figure 7.1. Let $v = V$ (constant) for $t \geqslant 0$. Then $U(s) = V/s$. The zero-state response transform is

$$\mathbf{Y}(s) = \mathbf{C}\boldsymbol{\Phi}(s)\mathbf{B}\mathbf{U}(s) + \mathbf{D}\mathbf{U}(s). \tag{7.237}$$

The product $\mathbf{C}\boldsymbol{\Phi}$ is already available from (7.235), so

$$\begin{aligned} Y(s) &= \begin{bmatrix} \dfrac{-s-2}{(s+1)^2+1} & \dfrac{-s}{(s+1)^2+1} \end{bmatrix} \begin{bmatrix} 1 \\ 0 \end{bmatrix} \frac{V}{s} + \frac{V}{s} \\ &= \left(\frac{-s-2}{(s+1)^2+1} + 1 \right) \frac{V}{s} = \frac{V(s+1)}{(s+1)^2+1}. \end{aligned} \tag{7.238}$$

Hence the zero-state response is

$$\mathbf{y}(t) = [V\epsilon^{-t}\cos t]u(t) \tag{7.239}$$

where $u(t)$ is the unit step function. The complete response is the sum of the zero-input response in (7.236) and the zero-state response in (7.239). This complete response is

$$\begin{aligned} y(t) = [&-x_1(0)\epsilon^{-t}(\cos t + \sin t) - x_2(0)\epsilon^{-t}(\cos t - \sin t) \\ &+ V\epsilon^{-t}\cos t]u(t) \end{aligned} \tag{7.240}$$

in terms of $x_1(0)$, $x_2(0)$, and V. ▲

Example 7-10. Consider the single-axis reaction-wheel attitude control system discussed in Section 3.4. The state equations are given by (3.100), (3.101), and (3.103), which are reproduced below for convenience:

$$\dot{\theta} = \omega \tag{7.241}$$

$$\dot{\omega} = \frac{1}{J_V}\left(D + \frac{K^2 i_f^2}{R_a}\right)\Omega + \frac{1}{J_V}\tau_d - \frac{K i_f}{J_V R_a} V_a \tag{7.242}$$

and

$$\dot{\Omega} = -\left(\frac{1}{J_V} + \frac{1}{J_W}\right)\left(\frac{K^2 i_f^2}{R_a} + D\right)\Omega + \frac{K i_f}{R_a}\left(\frac{1}{J_V} + \frac{1}{J_W}\right)V_a - \frac{1}{J_V}\tau_d \tag{7.243}$$

where $\theta(t)$ is the attitude angle, $\omega(t)$ is the attitude rate, $\Omega(t)$ is the speed of the reaction wheel relative to the spacecraft, V_a is the armature voltage for the motor driving the reaction wheel, and τ_d is a disturbance torque. The system is third order because three initial conditions are needed to solve for all the signals completely. The variables θ, ω, and Ω are suitable state variables. The outputs are θ and ω, and the inputs are V_a and τ_d. The input τ_d is not at our disposal, but V_a is. In this example, V_a is assumed to be an independent time function which we could choose. In later chapters, V_a may be chosen as a function of the state variables.

Figure 7.6 shows an all-integrator block diagram that represents Equations 7.241, 7.242, and 7.243, with θ and ω as outputs, and V_a and τ_d as inputs. The initial conditions θ_0, ω_0, and Ω_0 are also indicated. The constants K_1, K_2, K_3, and K_4 are

$$K_1 = \frac{K i_f}{R_a} \tag{7.244}$$

$$K_2 = \frac{1}{J_W} + \frac{1}{J_V} \tag{7.245}$$

$$K_3 = D + \frac{K^2 i_f^2}{R_a} \tag{7.246}$$

$$K_4 = \frac{1}{J_V}. \tag{7.247}$$

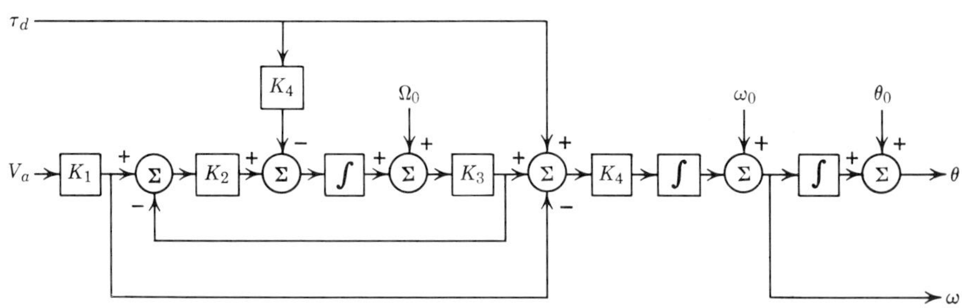

Figure 7.6 All-integrator block-diagram representation for reaction-wheel attitude-control system with θ and ω as outputs and V_a and τ_d as inputs.

Figure 7.7 shows the transfer-function block diagram where the signals Ω_0/s, ω_0/s, and θ_0/s are realized by considering Ω_0, ω_0, and θ_0 as inputs to the integrator blocks with transfer function $1/s$. Denote the state vector $[\theta\ \omega\ \Omega]^T$ by $\mathbf{x}$,

$$\mathbf{x}(t) = \begin{bmatrix} x_1(t) \\ x_2(t) \\ x_3(t) \end{bmatrix} = \begin{bmatrix} \theta(t) \\ \omega(t) \\ \Omega(t) \end{bmatrix}, \tag{7.248}$$

the input by $\mathbf{u}(t)$,

$$\mathbf{u}(t) = \begin{bmatrix} u_1(t) \\ u_2(t) \end{bmatrix} = \begin{bmatrix} V_a(t) \\ \tau_d(t) \end{bmatrix}, \tag{7.249}$$

and the output by $\mathbf{y}(t)$,

$$\mathbf{y}(t) = \begin{bmatrix} y_1(t) \\ y_2(t) \end{bmatrix} = \begin{bmatrix} \theta(t) \\ \omega(t) \end{bmatrix}. \tag{7.250}$$

The Laplace transform $\mathbf{\Phi}(s)$ of the transition matrix $\boldsymbol{\phi}(t)$ may be determined by applying Mason's rule for determining the nine transfer functions from θ_0, ω_0, and Ω_0 to the transforms of the state variables $\hat{\theta}(s)$, $\hat{\omega}(s)$, and $\hat{\Omega}(s)$. By inspection of Figure 7.7, $\Phi_{11}(s) = 1/s$, $\Phi_{21}(s) = 0$, $\Phi_{31}(s) = 0$, $\Phi_{12}(s) = 1/s^2$, $\Phi_{22}(s) = 1/s$, $\Phi_{32}(s) = 0$,

$$\Phi_{13}(s) = \frac{K_3K_4/s^3}{1 + K_2K_3/s}$$

$$\Phi_{23}(s) = \frac{K_3K_4/s^2}{1 + K_2K_3/s}$$

$$\Phi_{33}(s) = \frac{1/s}{1 + K_2K_3/s}.$$

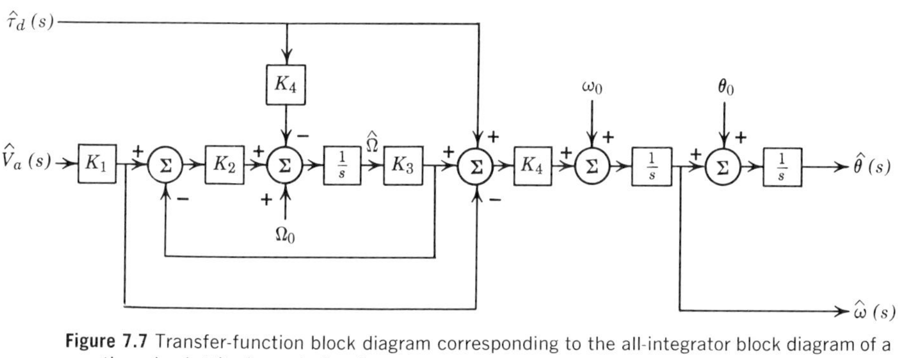

Figure 7.7 Transfer-function block diagram corresponding to the all-integrator block diagram of a reaction-wheel attitude-control system.

Hence

$$\Phi(s) = \begin{bmatrix} \frac{1}{s} & \frac{1}{s^2} & \frac{K_3K_4}{s^2(s+K_2K_3)} \\ 0 & \frac{1}{s} & \frac{K_3K_4}{s(s+K_2K_3)} \\ 0 & 0 & \frac{1}{(s+K_2K_3)} \end{bmatrix}. \tag{7.251}$$

The matrix **C** is

$$\mathbf{C} = \begin{bmatrix} 1 & 0 & 0 \\ 0 & 1 & 0 \end{bmatrix}, \tag{7.252}$$

B is

$$\mathbf{B} = \begin{bmatrix} 0 & 0 \\ -K_1K_4 & K_4 \\ K_1K_2 & -K_4 \end{bmatrix}, \tag{7.253}$$

and **D** is

$$\mathbf{D} = \begin{bmatrix} 0 & 0 \\ 0 & 0 \end{bmatrix}. \tag{7.254}$$

The transfer-function matrix $\mathbf{C}\Phi(s)\mathbf{B}+\mathbf{D}$ may be formed from (7.251), (7.252), (7.253), and (7.254). ▲

★ 7.9 Relationship of State Description to Transfer Functions

From a given state description such as (7.215), (7.216), and (7.218) involving the matrices **A**, **B**, **C**, and **D**, the output and state Laplace transforms $\mathbf{Y}(s)$ and $\mathbf{X}(s)$ are related to the initial state $\mathbf{x}_0$ and the input Laplace transform $\mathbf{U}(s)$ by (7.226) and (7.227). The matrices $\Phi(s)$, $\Phi(s)\mathbf{B}$, $\mathbf{C}\Phi(s)$, and $\mathbf{C}\Phi(s)\mathbf{B}+\mathbf{D}$ are transfer-function matrices. These transfer-function matrices are uniquely determined by the matrices **A**, **B**, **C**, and **D**. In analyzing a system whose all-integrator simulation block diagram is given, the integrator outputs may be chosen as state variables. For this choice, the **A**, **B**, **C**, and **D** matrices are obtained from the diagram. Thus the various transfer-function matrices may be computed. From the state description, it is straightforward to obtain the unique set of transfer-function matrices.

If the transfer-function matrices $\Phi(s)$, $\Phi(s)\mathbf{B}$, $\mathbf{C}\Phi(s)$, and $\mathbf{C}\Phi(s)\mathbf{B}+\mathbf{D}$ are all given, we shall show that the matrices **A**, **B**, **C**, and **D** can be computed, and thus the corresponding unique state equations can be obtained.

Suppose that the transition matrix Laplace transform $\mathbf{\Phi}(s)$ is given. Since

$$\mathbf{\Phi}(s) = (s\mathbf{I} - \mathbf{A})^{-1}, \tag{7.255}$$

then

$$(s\mathbf{I} - \mathbf{A}) = \mathbf{\Phi}^{-1}(s), \tag{7.256}$$

and

$$\mathbf{A} = -\lim_{s \to 0} \mathbf{\Phi}^{-1}(s). \tag{7.257}$$

Thus **A** can be obtained from the given $\mathbf{\Phi}(s)$. Equation (7.257) requires one matrix inversion. A computationally simpler formula is now derived. First, notice that since $\dot{\boldsymbol{\phi}}(t) = \mathbf{A}\boldsymbol{\phi}(t)$ as given in (7.179), and since $\boldsymbol{\phi}(0) = \mathbf{I}$,

$$\lim_{t \to 0} [\dot{\boldsymbol{\phi}}(t)] = \mathbf{A}. \tag{7.258}$$

From the initial value theorem,

$$\lim_{t \to 0} [\dot{\boldsymbol{\phi}}(t)] = \lim_{s \to \infty} s\mathscr{L}[\dot{\boldsymbol{\phi}}(t)]. \tag{7.259}$$

But

$$\mathscr{L}[\dot{\boldsymbol{\phi}}(t)] = s\mathbf{\Phi}(s) - \boldsymbol{\phi}(0) = s\mathbf{\Phi}(s) - \mathbf{I}. \tag{7.260}$$

Hence, from (7.258), (7.259), and (7.260)

$$\mathbf{A} = \lim_{s \to \infty} [s^2\mathbf{\Phi}(s) - s\mathbf{I}]. \tag{7.261}$$

The computation in (7.261) is much simpler than that in (7.257) if $\mathbf{\Phi}(s)$ is given.

Example 7-11. For the $\mathbf{\Phi}(s)$ given in (7.210) of Example 7-8,

$$\begin{aligned}
\mathbf{A} &= \lim_{s \to \infty} \begin{bmatrix} \dfrac{(s+6)s^2}{s^2+6s+5} - s & \dfrac{s^2}{s^2+6s+5} \\[2ex] \dfrac{-5s^2}{s^2+6s+5} & \dfrac{s^3}{s^2+6s+5} - s \end{bmatrix} \\
&= \lim_{s \to \infty} \begin{bmatrix} \dfrac{s^3+6s^2-s^3-6s^2-5s}{s^2+6s+5} & \dfrac{s^2}{s^2+6s+5} \\[2ex] \dfrac{-5s^2}{s^2+6s+5} & \dfrac{s^3-s^3-6s^2-5s}{s^2+6s+5} \end{bmatrix} \\
&= \begin{bmatrix} 0 & 1 \\ -5 & -6 \end{bmatrix}.
\end{aligned} \tag{7.262}$$

This is verified from the given **A** matrix in Example 7-8. ▲

Example 7-12. For $\mathbf{\Phi}(s)$ given in (7.186) of Example 7-6,

$$\mathbf{A} = \lim_{s\to\infty}\begin{bmatrix} \dfrac{(s+1)s^2}{(s+1)^2+1} - s & \dfrac{-s^2}{(s+1)^2+1} \\ \dfrac{s^2}{(s+1)^2+1} & \dfrac{(s+1)s^2}{(s+1)^2+1} - s \end{bmatrix}$$

$$= \begin{bmatrix} \dfrac{s^3+s^2-s^3-2s^2-2s}{(s+1)^2+1} & \dfrac{-s^2}{(s+1)^2+1} \\ \dfrac{s^2}{(s+1)^2+1} & \dfrac{s^3+s^2-s^3-2s^2-2s}{(s+1)^2+1} \end{bmatrix}$$

$$= \begin{bmatrix} -1 & -1 \\ 1 & 1 \end{bmatrix}, \tag{7.263}$$

which checks with the $\mathbf{A}$ matrix in (7.184). ▲

From the $\mathbf{A}$ matrix, the all-integrator simulation diagram relating $\mathbf{x}_0$ to $\mathbf{x}$ can be drawn directly.

From a given

$$\mathbf{H}_1(s) = \mathbf{\Phi}(s) \tag{7.264}$$

and a given

$$\mathbf{H}_2(s) = \mathbf{\Phi}(s)\mathbf{B}, \tag{7.265}$$

$\mathbf{A}$ can be determined from $\mathbf{H}_1(s)$ as described by (7.261) and $\mathbf{B}$ can be determined from

$$\mathbf{B} = \mathbf{H}_1^{-1}(s)\mathbf{H}_2(s). \tag{7.266}$$

If

$$\mathbf{H}_3(s) = \mathbf{C\Phi}(s) \tag{7.267}$$

is given also, then

$$\mathbf{C} = \mathbf{H}_3(s)\mathbf{H}_1^{-1}(s). \tag{7.268}$$

Finally, suppose

$$\mathbf{H}_4(s) = \mathbf{C\Phi}(s)\mathbf{B} + \mathbf{D}$$

is given. Since

$$\mathbf{\Phi}(s) = [s\mathbf{I} - \mathbf{A}]^{-1} = \left[(s\mathbf{I})\left(\mathbf{I} - \frac{1}{s}\mathbf{A}\right)\right]^{-1}$$

$$= \left(\mathbf{I} - \frac{1}{s}\mathbf{A}\right)^{-1}(s\mathbf{I})^{-1} = \frac{1}{s}\left(\mathbf{I} - \frac{1}{s}\mathbf{A}\right)^{-1}, \tag{7.269}$$

then

$$\lim_{s\to\infty} \mathbf{\Phi}(s) = \lim \frac{1}{s}\left(\mathbf{I} - \frac{1}{s}\mathbf{A}\right)^{-1} = \mathbf{0}. \tag{7.270}$$

Thus

$$\lim_{s\to\infty} \mathbf{H}_4(s) = \lim_{s\to\infty} [\mathbf{C\Phi}(s)\mathbf{B} + \mathbf{D}] = \mathbf{D}$$

or

$$\mathbf{D} = \lim_{s\to\infty} [\mathbf{H}_4(s)]. \tag{7.271}$$

From (7.261), (7.266), (7.268), and (7.271) the matrices **A**, **B**, **C**, and **D** can be determined if the four transfer-function matrices $\mathbf{H}_1(s)$, $\mathbf{H}_2(s)$, $\mathbf{H}_3(s)$, and $\mathbf{H}_4(s)$ are given. $\mathbf{H}_1(s)$ is the transfer-function matrix from initial state to state, $\mathbf{H}_2(s)$ is the transfer-function matrix from input to state, $\mathbf{H}_3(s)$ is the transfer-function matrix from initial state to output, and $\mathbf{H}_4(s)$ is the transfer-function matrix from input to output.

If only $\mathbf{H}_4(s)$ is given, it is not possible to obtain a unique set of matrices **A**, **B**, **C**, and **D**. However, from $\mathbf{H}_4(s)$, it is possible to determine the minimum-order n (dimension of **A**), which is also the number of integrators in a simulation diagram, with which $\mathbf{H}_4(s)$ can be realized. Even with this minimum order, there would be infinitely many different sets of **A**, **B**, **C**, and **D** that would yield the transfer function matrix $\mathbf{H}_4(s)$. There are several reasons for desiring a relization of $\mathbf{H}_4(s)$ using a minimum number of integrators. The first is concerned with economy and simplicity. A second reason deals with controllability and observability, which are discussed in Chapter 11. If $\mathbf{H}_4(s)$ is a scalar, that is, if the system has one input and one output only, then the input-output differential equation is directly obtainable, as discussed in Chapter 6. But from the differential equation, the method described in Chapter 4 may be used for drawing an all-integrator block diagram. For an nth-order differential equation, n independent initial conditions can be specified. Hence the state vector must be n-dimensional, and the resulting diagram using n integrators realizes $H_4(s)$ with a minimum-order n.

We shall state without proof a method due to Gilbert[3] that realizes a given transfer-function matrix with a minimum-order n for the **A** matrix. It is assumed that all the poles of $\mathbf{H}_4(s)$ are simple. The procedure involves determining the rank of a matrix, which we now define. A matrix **Q** is said to have rank r if **Q** has a submatrix $r \times r$ which is nonsingular and if no submatrix with dimension larger than $r \times r$ is nonsingular. Thus, if **R** has rank r, the largest dimension of any nonsingular submatrix of **R** is $r \times r$.

[3]E. G. Gilbert, "Controllability and Observability in Multivariable Control Systems," *J.S.I.A.M. Control*, Series A, **2**(1), 128–151 (1963).

Let the $q \times p$ matrix transfer function $\mathbf{H}(s)$ be expanded in partial fractions, that is, each element of $\mathbf{G}(s)$ is expanded in partial fractions Then, if each element of $\mathbf{H}(s)$ has a finite number of finite simple poles at $s = \lambda_1, \lambda_2, \ldots, \lambda_m$, $\mathbf{H}(s)$ can be written as

$$\mathbf{H}(s) = \sum_{i=1}^{m} \frac{\mathbf{K}_i}{s-\lambda_i} + \mathbf{D} \tag{7.272}$$

where $\mathbf{K}_i$ is the matrix of coefficients for the pole $s = \lambda_i$, and $\mathbf{D}$ is the limit of $\mathbf{H}(s)$ for $s \to \infty$. Let us illustrate this matrix partial fraction expansion.

Example 7-13. Let us expand the following $\mathbf{H}(s)$

$$\mathbf{H}(s) = \begin{bmatrix} \dfrac{\frac{1}{2}s + \frac{1}{36}}{s^2 + \frac{1}{6}s} & \dfrac{1}{6s+1} \\[2ex] \dfrac{24s+8}{36s^2+24s+3} & \dfrac{12s+1}{36s^2+6s} \\[2ex] \dfrac{36s+14}{36s^2+24s+3} & \dfrac{18s+7}{36s^2+24s+3} \end{bmatrix} \tag{7.273}$$

for a system with two inputs and three outputs. Since the limit of each element of $\mathbf{H}(s)$ for $s \to \infty$ is zero, $\mathbf{D}$ is the zero matrix

$$\mathbf{D} = \begin{bmatrix} 0 & 0 \\ 0 & 0 \\ 0 & 0 \end{bmatrix}. \tag{7.274}$$

Expanding each element of $\mathbf{H}(s)$ in partial fractions, we obtain

$$H(s) = \begin{bmatrix} \dfrac{\frac{1}{6}}{s} + \dfrac{\frac{1}{3}}{s+\frac{1}{6}} & \dfrac{\frac{1}{6}}{s+\frac{1}{6}} \\[2ex] \dfrac{\frac{1}{3}}{s+\frac{1}{6}} + \dfrac{\frac{1}{3}}{s+\frac{1}{2}} & \dfrac{\frac{1}{6}}{s} + \dfrac{\frac{1}{6}}{s+\frac{1}{6}} \\[2ex] \dfrac{\frac{2}{3}}{s+\frac{1}{6}} + \dfrac{\frac{1}{3}}{s+\frac{1}{2}} & \dfrac{\frac{1}{3}}{s+\frac{1}{6}} + \dfrac{\frac{1}{6}}{s+\frac{1}{2}} \end{bmatrix}. \tag{7.275}$$

Denote the three simple poles by $s_1 = 0$, $s_2 = -\frac{1}{6}$, and $s_3 = -\frac{1}{2}$. The corresponding coefficient matrices are

$$\mathbf{K}_1 = \begin{bmatrix} \frac{1}{6} & 0 \\ 0 & \frac{1}{6} \\ 0 & 0 \end{bmatrix}, \tag{7.276}$$

$$\mathbf{K}_2 = \begin{bmatrix} \frac{1}{3} & \frac{1}{6} \\ \frac{1}{3} & \frac{1}{6} \\ \frac{2}{3} & \frac{1}{3} \end{bmatrix}, \tag{7.277}$$

and

$$\mathbf{K}_3 = \begin{bmatrix} 0 & 0 \\ \frac{1}{3} & 0 \\ \frac{1}{3} & \frac{1}{6} \end{bmatrix}. \tag{7.278}$$

By inspection, $\mathbf{K}_1$ has rank 2, $\mathbf{K}_2$ has rank 1, and $\mathbf{K}_3$ has rank 2. ▲

The next step in the realization procedure is to express each matrix $\mathbf{K}_i$ as a product of two matrices $\mathbf{C}_i\mathbf{B}_i$. There are infinitely many ways of writing a matrix as a product of two matrices. Let us illustrate this by the following example.

Example 7-14. For the matrix $\mathbf{K}_1$ in (7.276), we may write it as the product

$$\mathbf{K}_1 = \begin{bmatrix} \frac{1}{6} & 0 \\ 0 & \frac{1}{6} \\ 0 & 0 \end{bmatrix} = \begin{bmatrix} 1 & 0 & 0 \\ 0 & 1 & 0 \\ -6 & -6 & 1 \end{bmatrix} \begin{bmatrix} \frac{1}{6} & 0 \\ 0 & \frac{1}{6} \\ 1 & 1 \end{bmatrix} \tag{7.279}$$

which is readily verified. The following product is also readily verified

$$\mathbf{K}_1 = \begin{bmatrix} \frac{1}{6} & 0 \\ 0 & \frac{1}{6} \\ 0 & 0 \end{bmatrix} = \begin{bmatrix} \frac{1}{6} & 0 \\ 0 & 1 \\ 0 & 0 \end{bmatrix} \begin{bmatrix} 1 & 0 \\ 0 & \frac{1}{6} \end{bmatrix}. \tag{7.280}$$

It should be noted that in forming $\mathbf{C}_i\mathbf{B}_i$, the number of columns of $\mathbf{C}_i$ must equal the number of rows of $\mathbf{B}_i$. Let the number of columns of $\mathbf{C}_i$ be denoted by r_i. Then any product $\mathbf{C}_i\mathbf{B}_i$ can also be written as $\mathbf{C}_i I_{r_i} \mathbf{B}_i$ where I_{r_i} is the identity matrix of order r_i. For example, the matrix product of (7.279) may be written as

$$\begin{bmatrix} 1 & 0 & 0 \\ 0 & 1 & 0 \\ -6 & -6 & 1 \end{bmatrix} \begin{bmatrix} \frac{1}{6} & 0 \\ 0 & \frac{1}{6} \\ 1 & 1 \end{bmatrix} = \begin{bmatrix} 1 & 0 & 0 \\ 0 & 1 & 0 \\ -6 & -6 & 1 \end{bmatrix} \begin{bmatrix} 1 & 0 & 0 \\ 0 & 1 & 0 \\ 0 & 0 & 1 \end{bmatrix} \begin{bmatrix} \frac{1}{6} & 0 \\ 0 & \frac{1}{6} \\ 1 & 1 \end{bmatrix} \tag{7.281}$$

and the matrix product in (7.280) may be written as

$$\begin{bmatrix} \frac{1}{6} & 0 \\ 0 & 1 \\ 0 & 0 \end{bmatrix} \begin{bmatrix} 1 & 0 \\ 0 & \frac{1}{6} \end{bmatrix} = \begin{bmatrix} \frac{1}{6} & 0 \\ 0 & 1 \\ 0 & 0 \end{bmatrix} \begin{bmatrix} 1 & 0 \\ 0 & 1 \end{bmatrix} \begin{bmatrix} 1 & 0 \\ 0 & \frac{1}{6} \end{bmatrix} \blacktriangle. \tag{7.282}$$

Then $\mathbf{K}_i/(s-\lambda_i)$ is written as

$$\frac{\mathbf{K}_i}{s-\lambda_i}=\frac{\mathbf{C}_i\mathbf{B}_i}{s-\lambda_i}=\frac{\mathbf{C}_i\mathbf{I}_{r_i}\mathbf{B}_i}{s-\lambda_i}=\mathbf{C}_i\left[\frac{1}{s-\lambda_i}\mathbf{I}_{r_i}\right]\mathbf{B}_i. \tag{7.283}$$

Example 7-15. For $\mathbf{K}_1$ expanded, (7.281), and $\lambda_1=0$ of Example 7-13, we have

$$\frac{\mathbf{K}_1}{s}=\begin{bmatrix}1&0&0\\0&1&0\\-6&-6&1\end{bmatrix}\left\{\frac{1}{s}\begin{bmatrix}1&0&0\\0&1&0\\0&0&1\end{bmatrix}\right\}\begin{bmatrix}\frac{1}{6}&0\\0&\frac{1}{6}\\1&1\end{bmatrix}$$

$$=\begin{bmatrix}1&0&0\\0&1&0\\-6&-6&1\end{bmatrix}\begin{bmatrix}1/s&0&0\\0&1/s&0\\0&0&1/s\end{bmatrix}\begin{bmatrix}\frac{1}{6}&0\\0&\frac{1}{6}\\1&1\end{bmatrix}. \tag{7.284}$$

The matrix $\mathbf{K}_1/s$ may be realized as a transfer-function matrix for a system consisting of a cascade connection of three systems whose transfer-function matrices are $\mathbf{T}_1(s)$, $\mathbf{T}_2(s)$, and $\mathbf{T}_3(s)$, as shown in Figure 7.8. The overall transfer-function matrix of the system is $\mathbf{T}_3(s)\,\mathbf{T}_2(s)\,\mathbf{T}_1(s)$. The ordering of the matrices is important. In general, since matrix multiplication is not commutative, it is not possible to interchange the positions of the matrices and obtain the same product. In (7.284), we identify

$$\mathbf{T}_3(s)=\begin{bmatrix}1&0&0\\0&1&0\\-6&-6&1\end{bmatrix} \tag{7.285}$$

$$\mathbf{T}_2(s)=\begin{bmatrix}1/s&0&0\\0&1/s&0\\0&0&1/s\end{bmatrix} \tag{7.286}$$

$$\mathbf{T}_1(s)=\begin{bmatrix}\frac{1}{6}&0\\0&\frac{1}{6}\\1&1\end{bmatrix}. \tag{7.287}$$

Figure 7.9 shows a realization of $\mathbf{T}_1(s)$, $\mathbf{T}_2(s)$, $\mathbf{T}_3(s)$, and $\mathbf{T}_3(s)\mathbf{T}_2(s)\mathbf{T}_1(s)$. The block diagram of Figure 7.9 realizes $\mathbf{K}_1/s$, using three integrators.

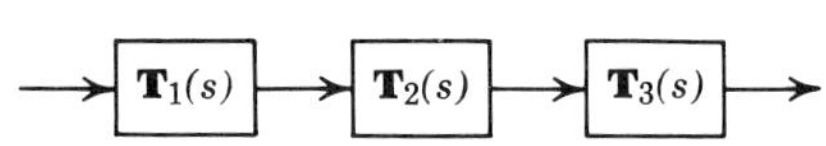

Figure 7.8 Transfer-function matrix of the cascade is $\mathbf{T}_3(s)\,\mathbf{T}_2(s)\,\mathbf{T}_1(s)$.

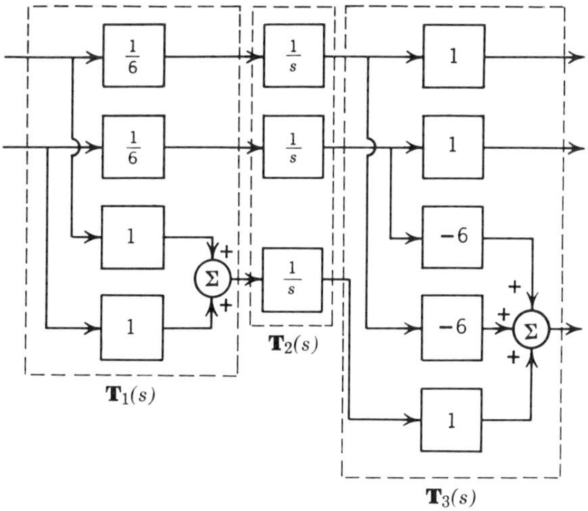

Figure 7.9 Realization of $\mathbf{T}_3(s)\mathbf{T}_2(s)\mathbf{T}_1(s)$, which is also $\mathbf{K}_1/s$ in (7.284).

Let us try another realization of $\mathbf{K}_1/s$, using the expansion in (7.282). Thus

$$
\begin{aligned}
\frac{\mathbf{K}_1}{s} &= \begin{bmatrix} \frac{1}{6} & 0 \\ 0 & 1 \\ 0 & 0 \end{bmatrix} \left\{ \frac{1}{s} \begin{bmatrix} 1 & 0 \\ 0 & 1 \end{bmatrix} \right\} \begin{bmatrix} 1 & 0 \\ 0 & \frac{1}{6} \end{bmatrix} \\
&= \begin{bmatrix} \frac{1}{6} & 0 \\ 0 & 1 \\ 0 & 0 \end{bmatrix} \begin{bmatrix} 1/s & 0 \\ 0 & 1/s \end{bmatrix} \begin{bmatrix} 0 & 0 \\ 0 & \frac{1}{6} \end{bmatrix}. \qquad (7.288)
\end{aligned}
$$

This is realized by the cascade structure shown in Figure 7.10 using two integrators only. ▲

Each term of the matrix partial fraction expansion of (7.272) may be expressed as in (7.283), and each such term is realized as a transfer-function matrix involving r_i integrators. Then the systems are connected in parallel, that is, the system inputs are made common, and

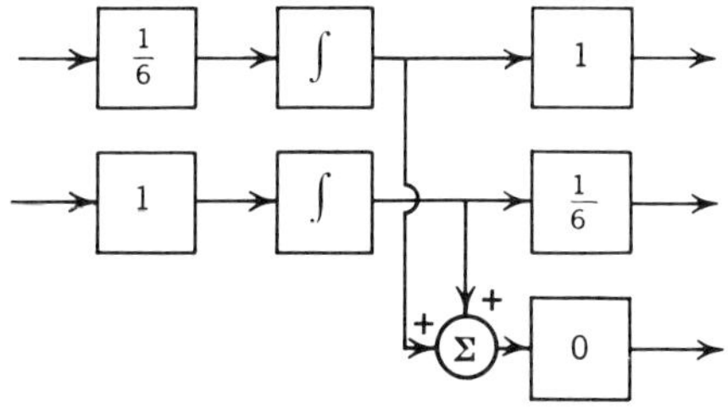

Figure 7.10 Alternative realization for $\mathbf{K}_1/s$ using two integrators only.

the outputs are added. Such a realization involves $r_1 + r_2 + \cdots r_m$ integrators. In Gilbert's theorem, it is proved that the minimum number of integrators required for the realization of $\mathbf{K}_i/(s-\lambda_i)$ is equal to the rank of $\mathbf{K}_i$, and the minimum number of integrators for realizing the total $\mathbf{H}(s)$ is equal to the sum of the ranks of $\mathbf{K}_i$ for all i. The product $\mathbf{C}_i\mathbf{I}_{r_i}\mathbf{B}_i$ can always be written so that the order of I_{r_i} equals the rank of $\mathbf{K}_i$. The product $\mathbf{C}_i[(1/s-\lambda_i)\mathbf{I}_{r_i}]\mathbf{B}_i$ can be realized as a cascade structure where the middle structure involves r_i integrators and the end structures are matrices of constant scalors. Observe that

$$\frac{1}{s-\lambda_i}\mathbf{I}_{r_i} = \begin{bmatrix} \frac{1}{s-\lambda_i} & 0 & 0 \cdots 0 \\ 0 & \frac{1}{s-\lambda_i} & 0 \cdots 0 \\ \vdots & & \\ 0 & 0 \quad \cdots & \frac{1}{s-\lambda_i} \end{bmatrix} \tag{7.289}$$

is a square diagonal matrix of order r_i, which can be realized as shown in Figure 7.11.

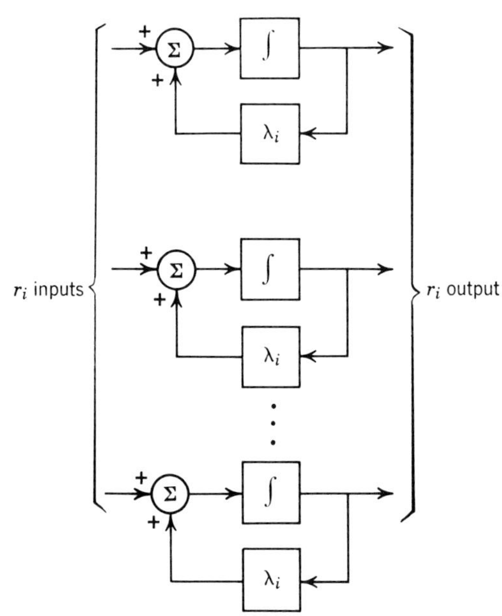

Figure 7.11 Realization of (7.289) using r_i identical feedback structures.

The procedure for determining $\mathbf{C}_i$ and $\mathbf{B}_i$ is as follows. For each $\mathbf{K}_i$, determine the rank of $\mathbf{K}_i$. This is performed by checking various submatrices of $\mathbf{K}_i$ to see if they are nonsingular. The largest nonsingular submatrix contained in $\mathbf{K}_i$ has a dimension equal to the rank of $\mathbf{K}_i$. There may be several such submatrices but choose any one. Pick the columns of $\mathbf{K}_i$ where the elements of the chosen nonsingular submatrix appear and form $\mathbf{C}_i$ as a matrix of these columns. If the rank of $\mathbf{K}_i$ is r_i, then $\mathbf{C}_i$ has r_i columns. Repeat the construction for all $\mathbf{K}_i$. Now form

$$\mathbf{C}_i\mathbf{B}_i = \mathbf{K}_i \tag{7.290}$$

where $\mathbf{B}_i$ is an unknown matrix to be determined. To solve for $\mathbf{B}_i$, premultiply (7.290) by $\mathbf{C}_i^*$ the complex conjugate transpose of $\mathbf{C}_i$,

$$\mathbf{C}_i^*\mathbf{C}_i\mathbf{B}_i = \mathbf{C}_i^*\mathbf{K}_i. \tag{7.291}$$

Note that C_i^* is $r_i \times q$ and $\mathbf{C}_i^*\mathbf{C}_i$ is $r_i \times r_i$. From matrix theory, if $\mathbf{C}_i$ has rank r_i then $\mathbf{C}_i^*\mathbf{C}_i$ has rank r_i. Since $\mathbf{C}_i^*\mathbf{C}_i$ is $r \times r$, it must be nonsingular. Hence

$$\mathbf{B}_i = [\mathbf{C}_i^*\mathbf{C}_i]^{-1}\mathbf{C}_i^*\mathbf{K}_i. \tag{7.292}$$

With this construction, $\mathbf{I}_{r_i}$ has a dimension equal to the rank of $\mathbf{K}_i$.

Example 7-16. For $\mathbf{K}_1$ in Example 7-13, since it is of rank 2 and has only two columns, we choose all the columns of $\mathbf{K}_1$ for forming $\mathbf{C}_1$:

$$\mathbf{C}_1 = \begin{bmatrix} \frac{1}{6} & 0 \\ 0 & \frac{1}{6} \\ 0 & 0 \end{bmatrix}. \tag{7.293}$$

Since $\mathbf{C}_1 = \mathbf{K}_1$, obviously

$$\mathbf{C}_1\mathbf{B}_1 = \mathbf{K}_1 \tag{7.294}$$

yields

$$\mathbf{B}_1 = \begin{bmatrix} 1 & 0 \\ 0 & 1 \end{bmatrix}. \tag{7.295}$$

However, to illustrate the use of (7.292), let us form

$$\mathbf{B}_1 = \left[\begin{bmatrix} \frac{1}{6} & 0 & 0 \\ 0 & \frac{1}{6} & 0 \end{bmatrix}\begin{bmatrix} \frac{1}{6} & 0 \\ 0 & \frac{1}{6} \\ 0 & 0 \end{bmatrix}\right]^{-1}\begin{bmatrix} \frac{1}{6} & 0 & 0 \\ 0 & \frac{1}{6} & 0 \end{bmatrix}\begin{bmatrix} \frac{1}{6} & 0 \\ 0 & \frac{1}{6} \\ 0 & 0 \end{bmatrix}$$

$$= \begin{bmatrix} \frac{1}{36} & 0 \\ 0 & \frac{1}{36} \end{bmatrix}^{-1}\begin{bmatrix} \frac{1}{36} & 0 \\ 0 & \frac{1}{36} \end{bmatrix} = \begin{bmatrix} 1 & 0 \\ 0 & 1 \end{bmatrix}. \tag{7.296}$$

Next, since $\mathbf{K}_2$ has rank 1, we pick any column of $\mathbf{K}_2$ to form $\mathbf{C}_2$. Choose the second column

$$\mathbf{C}_2 = \begin{bmatrix} \frac{1}{6} \\ \frac{1}{6} \\ \frac{1}{3} \end{bmatrix}. \tag{7.297}$$

Using (7.292),

$$\begin{aligned} \mathbf{B}_2 &= \left[\begin{bmatrix} \frac{1}{6} & \frac{1}{6} & \frac{1}{3} \end{bmatrix} \begin{bmatrix} \frac{1}{6} \\ \frac{1}{6} \\ \frac{1}{3} \end{bmatrix} \right]^{-1} \begin{bmatrix} \frac{1}{6} & \frac{1}{6} & \frac{1}{3} \end{bmatrix} \begin{bmatrix} \frac{1}{3} & \frac{1}{6} \\ \frac{1}{3} & \frac{1}{6} \\ \frac{2}{3} & \frac{1}{3} \end{bmatrix} \\ &= \left(\tfrac{1}{36}+\tfrac{1}{36}+\tfrac{1}{9}\right)^{-1}\left[\tfrac{1}{18}+\tfrac{1}{18}+\tfrac{2}{9} \quad \tfrac{1}{36}+\tfrac{1}{36}+\tfrac{1}{9}\right] \\ &= 6\left[\tfrac{1}{3} \quad \tfrac{1}{6}\right] = [2 \quad 1]. \end{aligned} \tag{7.298}$$

Since $\mathbf{K}_3$ has rank 2 and only two columns, use all the columns to form $\mathbf{C}_3$.

$$\mathbf{C}_3 = \begin{bmatrix} 0 & 0 \\ \frac{1}{3} & 0 \\ \frac{1}{3} & \frac{1}{6} \end{bmatrix} \tag{7.299}$$

Then

$$\begin{aligned} \mathbf{B}_3 &= \left[\begin{bmatrix} 0 & \frac{1}{3} & \frac{1}{3} \\ 0 & 0 & \frac{1}{6} \end{bmatrix} \begin{bmatrix} 0 & 0 \\ \frac{1}{3} & 0 \\ \frac{1}{3} & \frac{1}{6} \end{bmatrix} \right]^{-1} \begin{bmatrix} 0 & \frac{1}{3} & \frac{1}{3} \\ 0 & 0 & \frac{1}{6} \end{bmatrix} \begin{bmatrix} 0 & 0 \\ \frac{1}{3} & 0 \\ \frac{1}{3} & \frac{1}{6} \end{bmatrix} \\ &= \begin{bmatrix} \frac{2}{9} & \frac{1}{18} \\ \frac{1}{18} & \frac{1}{36} \end{bmatrix} \begin{bmatrix} \frac{2}{9} & \frac{1}{18} \\ \frac{1}{18} & \frac{1}{36} \end{bmatrix}^{-1} = \begin{bmatrix} 1 & 0 \\ 0 & 1 \end{bmatrix} \quad \blacktriangle. \end{aligned} \tag{7.300}$$

Using the factorization of $\mathbf{K}_i$ as described above, each $\mathbf{K}_i/(s-\lambda_i)$ may be realized using r_i integrators. Then $\mathbf{H}(s)$ of (7.272) is realized as a parallel connection of the m sets of systems and of the constant system represented by $\mathbf{D}$ if it is nonzero. The outputs of the systems whose transfer-function matrices are $\mathbf{I}_{r_i}/(s-\lambda_i)$ are outputs of integrators. If these integrator outputs are chosen as state variables then $\mathbf{A}$, $\mathbf{B}$, $\mathbf{C}$, and $\mathbf{D}$ of the state description of the realization are

$$\mathbf{A} = \begin{bmatrix} \lambda_1 \mathbf{I}_{r_1} & 0 & \cdots & \\ 0 & \lambda_2 \mathbf{I}_{r_2} & \cdots & \\ \vdots & & & \\ 0 & \cdots & & \lambda_m \mathbf{I}_{r_m} \end{bmatrix}, \tag{7.301}$$

$$\mathbf{C} = [\mathbf{C}_1 \mid \mathbf{C}_2 \mid \cdots \mid \mathbf{C}_m], \tag{7.302}$$

$$\mathbf{B} = \begin{bmatrix} \mathbf{B}_1 \\ \mathbf{B}_2 \\ \cdot \\ \cdot \\ \cdot \\ \mathbf{B}_m \end{bmatrix} \tag{7.303}$$

and $\mathbf{D}$ equals the $\mathbf{D}$ obtained in (7.272). The total number of integrators (and hence the dimension of $\mathbf{A}$) is $r_1+r_2+\cdots+r_m$. $\mathbf{C}$ has q rows and $(r_1+r_2+\cdots+r_m)$ columns. $\mathbf{B}$ has $(r_1+r_2+\cdots+r_m)$ rows and p columns.

Example 7-17. From Example 7-16, $\mathbf{H}(s)$ of Example 7-13 may be written as

$$\mathbf{H}(s) = \frac{\mathbf{K}_1}{s} + \frac{\mathbf{K}_2}{s-\frac{1}{6}} + \frac{\mathbf{K}_3}{s-\frac{1}{2}} \tag{7.304}$$

where

$$\frac{\mathbf{K}_1}{s} = \begin{bmatrix} \frac{1}{6} & 0 \\ 0 & \frac{1}{6} \\ 0 & 0 \end{bmatrix} \begin{bmatrix} 1/s & 0 \\ 0 & 1/s \end{bmatrix} \begin{bmatrix} 1 & 0 \\ 0 & 1 \end{bmatrix} \tag{7.305}$$

$$\frac{\mathbf{K}_2}{s-\frac{1}{6}} = \begin{bmatrix} \frac{1}{6} \\ \frac{1}{6} \\ \frac{1}{3} \end{bmatrix} \begin{bmatrix} \dfrac{1}{s-\frac{1}{6}} \end{bmatrix} [2 \quad 1] \tag{7.306}$$

$$\frac{\mathbf{K}_3}{s-\frac{1}{2}} = \begin{bmatrix} 0 & 0 \\ \frac{1}{3} & 0 \\ \frac{1}{3} & \frac{1}{6} \end{bmatrix} \begin{bmatrix} \dfrac{1}{s-\frac{1}{2}} & 0 \\ 0 & \dfrac{1}{s-\frac{1}{2}} \end{bmatrix} \begin{bmatrix} 1 & 0 \\ 0 & 1 \end{bmatrix}. \tag{7.307}$$

The realizations for $\mathbf{K}_1/s$, $\mathbf{K}_2/(s-\frac{1}{6})$, and $\mathbf{K}_3/(s-\frac{1}{2})$ may be connected in parallel, and the composite system realizes $\mathbf{H}(s)$. The all-integrator block diagram is shown in Figure 7.12. The matrices $\mathbf{A}$, $\mathbf{B}$, $\mathbf{C}$, and $\mathbf{D}$ are obtained according to (7.301), (7.302), (7.303), and (7.274). Thus

$$\mathbf{A} = \begin{bmatrix} 0 & 0 & 0 & 0 & 0 \\ 0 & 0 & 0 & 0 & 0 \\ 0 & 0 & -\frac{1}{6} & 0 & 0 \\ 0 & 0 & 0 & -\frac{1}{2} & 0 \\ 0 & 0 & 0 & 0 & -\frac{1}{2} \end{bmatrix}, \tag{7.308}$$

$$\mathbf{C} = \begin{bmatrix} \frac{1}{6} & 0 & \frac{1}{6} & 0 & 0 \\ 0 & \frac{1}{6} & \frac{1}{6} & \frac{1}{3} & 0 \\ 0 & 0 & \frac{1}{3} & \frac{1}{3} & \frac{1}{6} \end{bmatrix}, \tag{7.309}$$

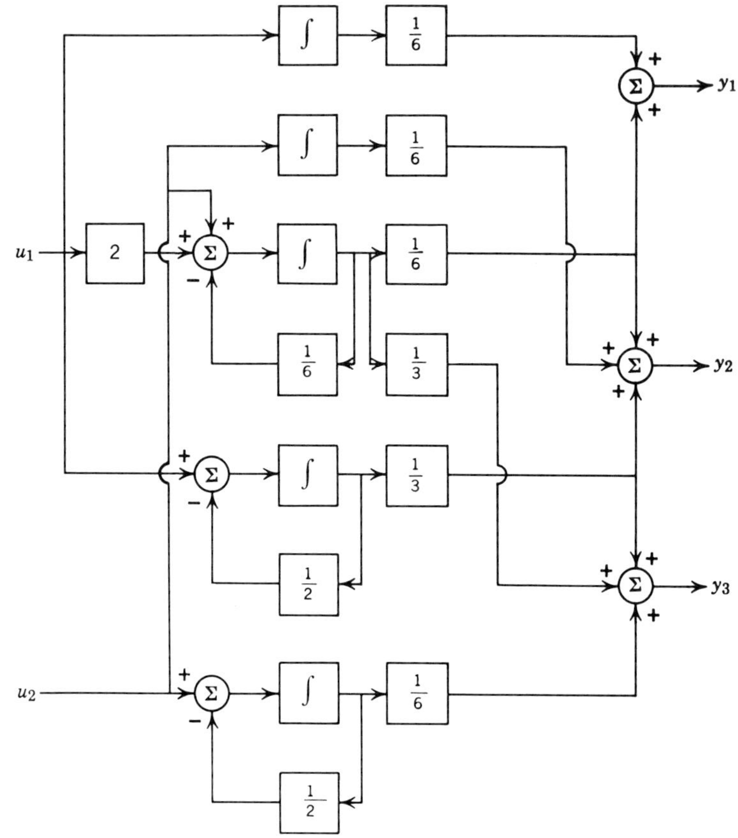

Figure 7.12 All-integrator block-diagram realization of **H**(s) in (7.273) using the minimum number of integrators.

$$\mathbf{B} = \begin{bmatrix} 1 & 0 \\ 0 & 1 \\ 2 & 1 \\ 1 & 0 \\ 0 & 1 \end{bmatrix}, \tag{7.310}$$

and

$$\mathbf{D} = \begin{bmatrix} 0 & 0 \\ 0 & 0 \\ 0 & 0 \end{bmatrix}. \tag{7.311}$$

These matrices are verified by inspection of Figure 7.12. The realization involves five integrators, which is the minimum number required to realize $\mathbf{H}(s)$ in (7.273). ▲

Let us summarize Gilbert's procedure for obtaining a minimum-order state realization for a transfer-function matrix $\mathbf{H}(s)$. It is assumed that $\mathbf{H}(s)$ is matrix of rational functions which have simple finite poles and that the limit of each element of $\mathbf{H}(s)$ for $s \to \infty$ is a constant (including zero). Thus the elements of $\mathbf{H}(s)$ have no poles at infinity.

Step 1. Expand $\mathbf{H}(s)$ in partial fractions as in (7.272) where $\mathbf{D}$ is the limit of $\mathbf{H}(s)$ for $s \to \infty$.

Step 2. Check the rank of each $\mathbf{K}_i$. Denote the rank of $\mathbf{K}_i$ by r_i. In checking the rank of $\mathbf{K}_i$, a nonsingular $r_i \times r_i$ submatrix is encountered. Form $\mathbf{C}_i$ whose columns are the columns of $\mathbf{K}_i$, which contain the elements of the $r_i \times r_i$ nonsingular submatrix. Note that $\mathbf{C}_i$ has r_i columns. Repeat for each $\mathbf{K}_i$.

Step 3. Compute $\mathbf{B}_i$ from

$$\mathbf{B}_i = [\mathbf{C}_i^* \mathbf{C}_i]^{-1} \mathbf{C}_i^* \mathbf{K}_i$$

where $\mathbf{C}_i^*$ is the complex conjugate transpose of $\mathbf{C}_i$.

Step 4. Form

$$\mathbf{A} = \begin{bmatrix} \lambda_1 \mathbf{I}_1 & 0 & & \\ 0 & \lambda_2 \mathbf{I}_2 & 0 & \\ 0 & 0 & & \lambda_m \mathbf{I}_m \end{bmatrix}$$

where λ_i is the ith pole of $\mathbf{H}(s)$, and I_{r_i} is an identity matrix of order r_i.

Step 5. Form

$$\mathbf{C} = [\mathbf{C}_1 \;\vdots\; \mathbf{C}_2 \;\vdots\; \cdots \;\vdots\; \mathbf{C}_m]$$

Step 6. Form

$$\mathbf{B} = \begin{bmatrix} \mathbf{B}_1 \\ \mathbf{B}_2 \\ \vdots \\ \mathbf{B}_m \end{bmatrix}$$

Step 7. With $\mathbf{A}$, $\mathbf{B}$, $\mathbf{C}$, and $\mathbf{D}$ obtained as above, $\mathbf{H}(s)$ is realized as

$$\mathbf{H}(s) = \mathbf{C}[s\mathbf{I} - \mathbf{A}]^{-1}\mathbf{B} + \mathbf{D}$$

using an all-integrator block diagram whose state equations are

$$\dot{\mathbf{x}} = \mathbf{A}\mathbf{x} + \mathbf{B}\mathbf{u}$$

$$\mathbf{y} = \mathbf{C}\mathbf{x} + \mathbf{D}\mathbf{u}$$

where $\mathbf{x}$ is the state vector, $\mathbf{y}$ is the output vector, and $\mathbf{u}$ is the input vector. The realization is shown as an all-integrator matrix block diagram in Figure 7.13.

If some of the poles are complex then the complex conjugate of these poles are poles also. In general, $\mathbf{K}_i$ corresponding to a complex λ_i is complex. Complex conjugate terms may be combined to produce a block diagram using real scalors only.

From a minimum-order state realization of $\mathbf{H}(s)$, infinitely many other minimum-order realizations can be obtained by simply changing the choice of state variables. Suppose that $\mathbf{H}(s)$ is realized by a system described by the state equations

$$\dot{\mathbf{x}} = \mathbf{A}\mathbf{x} + \mathbf{B}\mathbf{u} \tag{7.312}$$

$$\mathbf{y} = \mathbf{C}\mathbf{x} + \mathbf{D}\mathbf{u} \tag{7.313}$$

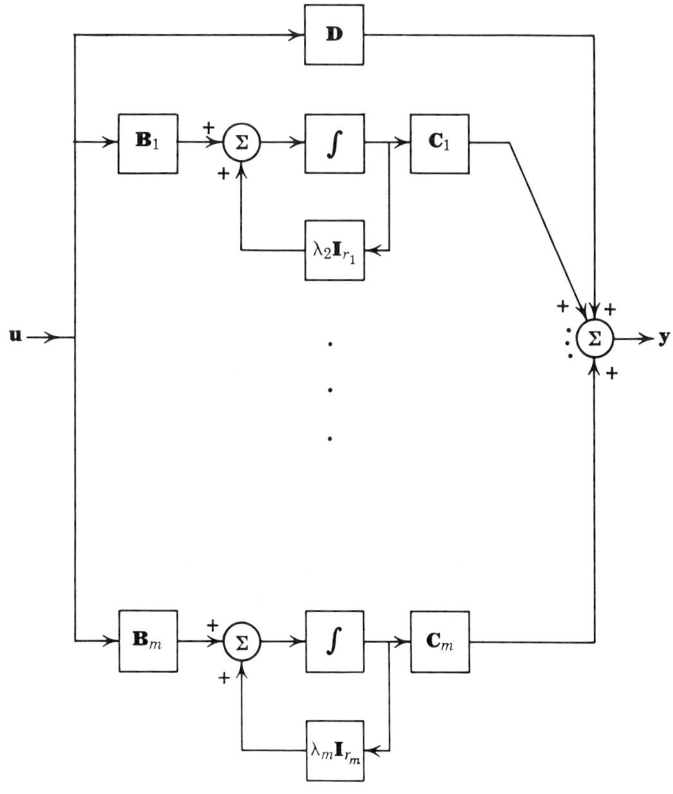

Figure 7.13 All-integrator matrix block diagram for (7.272).

where $\mathbf{x}$ is the n-dimensional state order, $\mathbf{y}$ is the output, and $\mathbf{u}$ is the input. Let $\mathbf{T}$ be *any* nonsingular $n \times n$ matrix and define an n-vector $\mathbf{z}$ by

$$\mathbf{z} = \mathbf{T}\mathbf{x} \tag{7.314}$$

Clearly, $\mathbf{z}$ qualifies as a state vector, since knowledge of $\mathbf{z}$ uniquely determines $\mathbf{x}$ and knowledge of $\mathbf{x}$ uniquely determines $\mathbf{z}$. From (7.314),

$$\mathbf{x} = \mathbf{T}^{-1}\mathbf{z}. \tag{7.315}$$

Substituting the expression for $\mathbf{x}$ from (7.315) in (7.312),

$$\mathbf{T}^{-1}\dot{\mathbf{z}} = \mathbf{A}\mathbf{T}^{-1}\mathbf{z} + \mathbf{B}\mathbf{u}$$

or

$$\dot{\mathbf{z}} = \mathbf{T}\mathbf{A}\mathbf{T}^{-1}\mathbf{z} + \mathbf{T}\mathbf{B}\mathbf{u}. \tag{7.316}$$

Similarly, (7.313) becomes

$$\mathbf{y} = \mathbf{C}\mathbf{T}^{-1}\mathbf{z} + \mathbf{D}\mathbf{u}. \tag{7.317}$$

Equations (7.316) and (7.317) is a different state description using the state vector $\mathbf{z}$ for the *same* system. However, an all-integrator block diagram different from that based on (7.312) and (7.313) can be used to represent the system. The block diagram is based on

$$\dot{\mathbf{z}} = \mathbf{A}'\mathbf{z} + \mathbf{B}'\mathbf{u} \tag{7.318}$$

$$\mathbf{y} = \mathbf{C}'\mathbf{z} + \mathbf{D}'\mathbf{u} \tag{7.319}$$

where

$$\mathbf{A}' = \mathbf{T}\mathbf{A}\mathbf{T}^{-1} \tag{7.320}$$

$$\mathbf{B}' = \mathbf{T}\mathbf{B} \tag{7.321}$$

$$\mathbf{C}' = \mathbf{C}\mathbf{T}^{-1} \tag{7.322}$$

and

$$\mathbf{D}' = \mathbf{D} \tag{7.323}$$

Since there are infinitely many choices for an $n \times n$ nonsingular matrix T, infinitely many different all-integrator block diagrams can be used to represent the system. Let us verify that an all-integrator block diagram based on (7.318) and (7.319) yields the same transfer-function matrix as one based on (7.312) and (7.313). Thus for (7.318) and (7.319), with the relations in (7.320), (7.321), (7.322), and (7.323), we have

$$\begin{aligned}\mathbf{H}(s) &= \mathbf{C}'(s\mathbf{I}-\mathbf{A}')^{-1}\mathbf{B}'+\mathbf{D}' \\ &= \mathbf{CT}^{-1}(s\mathbf{I}-\mathbf{TAT}^{-1})^{-1}\mathbf{TB}+\mathbf{D}.\end{aligned} \tag{7.324}$$

Since $\mathbf{TT}^{-1}=\mathbf{I}$, (7.324) may be rewritten as

$$\begin{aligned}\mathbf{H}(s) &= \mathbf{CT}^{-1}(s\mathbf{TT}^{-1}-\mathbf{TAT}^{-1})^{-1}\mathbf{TB}+\mathbf{D} \\ &= \mathbf{CT}^{-1}[(s\mathbf{T}-\mathbf{TA})\mathbf{T}^{-1}]^{-1}\mathbf{TB}+\mathbf{D} \\ &= \mathbf{CT}^{-1}[\mathbf{T}^{-1}]^{-1}[s\mathbf{T}-\mathbf{TA}]^{-1}\mathbf{TB}+\mathbf{D} \\ &= \mathbf{CT}^{-1}\mathbf{T}[\mathbf{T}(s\mathbf{I}-\mathbf{A})]^{-1}\mathbf{TB}+\mathbf{D} \\ &= \mathbf{C}[s\mathbf{I}-\mathbf{A}]^{-1}[\mathbf{T}]^{-1}\mathbf{TB}+\mathbf{D} \\ &= \mathbf{C}[s\mathbf{I}-\mathbf{A}]^{-1}\mathbf{B}+\mathbf{D}.\end{aligned} \tag{7.325}$$

But the transfer-function matrix corresponding to (7.312) and (7.313) is

$$\mathbf{H}(s) = \mathbf{C}[s\mathbf{I}-\mathbf{A}]^{-1}\mathbf{B}+\mathbf{D}. \tag{7.326}$$

Hence the two transfer matrices are equal. This also shows that the transfer-function matrix is independent of the choice of state.

★ 7.10 Time-Varying Linear Systems

One of the most important reasons for studying methods for the analysis of linear time-varying systems is that there are practical systems for which a linear time-varying model is appropriate and adequate. In addition, for some nonlinear systems, whether time-invariant or not, the approximate change in behavior due to small perturbations in initial conditions or system parameters can be obtained from a linearized set of differential equations which are generally time-varying. As mentioned earlier, such sensitivity analyses are very important in engineering because of the uncertainties in the system models as well as uncertainties in measurement. In addition, such sensitivity analyses are useful for design via parameter adjustment. Details of this type of linearization will be discussed in Chapter 9.

For a linear time-varying system whose behavior is described by ordinary differential equations, the state representation is

$$\dot{x}(t) = \mathbf{A}(t)\mathbf{x}(t) + \mathbf{B}(t)\mathbf{u}(t) \tag{7.327}$$

$$\mathbf{y}(t) = \mathbf{C}(t)\mathbf{x}(t) + \mathbf{D}(t)\mathbf{u}(t) \tag{7.328}$$

where $\mathbf{x}(t)$ is an n-dimensional state vector, $\mathbf{y}(t)$ is a q-dimensional output vector, and $\mathbf{u}(t)$ is the p-dimensional input vector. The matrices $\mathbf{A}(t)$, $\mathbf{B}(t)$, $\mathbf{C}(t)$, and $\mathbf{D}(t)$ are functions of time. In the special case when all the matrices are constant matrices the linear system is time-invariant. If at least one of the matrices is time-varying the system is called time-varying. Before we proceed, let us exhibit an example of a system modeled by (7.327) and (7.328).

Example 7-18. The Euler dynamic equations for the angular motion of a spacecraft were derived in Chapter 5. For the case when the control torques are supplied by reaction wheels driven by electric motors, the equations are given by (5.105) reproduced below:

$$J_{V_1}\dot{\omega}_1 = (J_{V_2} - J_{V_3})\omega_2\omega_3 + J_W(\omega_3\Omega_2 - \omega_2\Omega_3 - \dot{\omega}_1 - \dot{\Omega}_1) + \tau'_{d_1} \tag{7.329}$$

$$J_{V_2}\dot{\omega}_2 = (J_{V_3} - J_{V_1})\omega_3\omega_1 + J_W(\omega_1\Omega_3 - \omega_3\Omega_1 - \dot{\omega}_2 - \dot{\Omega}_2) + \tau'_{d_2} \tag{7.330}$$

$$J_{V_3}\dot{\omega}_3 = (J_{V_1} - J_{V_2})\omega_1\omega_2 + J_W(\omega_2\Omega_1 - \omega_1\Omega_2 - \dot{\omega}_3 - \dot{\Omega}_3) + \tau'_{d_3} \tag{7.331}$$

where ω_1, ω_2, and ω_3 are the spacecraft angular velocities, and Ω_1, Ω_2, and Ω_3 are reaction wheel speeds relative to the spacecraft. There are other equations relating wheel speeds to armature voltage, and spacecraft angular velocities to Euler angles for the spacecraft. The above equations are nonlinear.

In this example, we consider the following problem. Suppose that it is desired to achieve certain specified angular velocities $\omega_1(t)$, $\omega_2(t)$, and $\omega_3(t)$, in the absence of disturbance torques τ'_{d_1}, τ'_{d_2}, and τ'_{d_3}. The problem is to determine the reaction wheel speeds $\Omega_1(t)$, $\Omega_2(t)$, and $\Omega_3(t)$ in order to attain the desired spacecraft angular velocities $\omega_1(t)$, $\omega_2(t)$, and $\omega_3(t)$.

If the known time functions $\omega_1(t)$, $\omega_2(t)$, and $\omega_3(t)$ are substituted in (7.329), (7.330), and (7.331), the result is a set of linear time-varying differential equations in $\Omega_1(t)$, $\Omega_2(t)$, and $\Omega_2(t)$. Thus

$$\dot{\Omega}_1(t) = \omega_3(t)\Omega_2(t) - \omega_2(t)\Omega_3(t) + \frac{J_{V_2} - J_{V_3}}{J_W}\omega_2(t)\omega_3(t) - \left(1 + \frac{J_{V_1}}{J_W}\right)\dot{\omega}_1(t) \tag{7.332}$$

$$\dot{\Omega}_2(t) = -\omega_3(t)\Omega_1(t) + \omega_1(t)\Omega_3(t) + \frac{J_{V_3} - J_{V_1}}{J_W}\omega_3(t)\omega_1(t) - \left(1 + \frac{J_{V_2}}{J_W}\right)\dot{\omega}_2(t) \tag{7.333}$$

and

$$\dot{\Omega}_3(t) = \omega_2(t)\Omega_1(t) - \omega_1(t)\Omega_2(t) + \frac{J_{V_1} - J_{V_2}}{J_W}\omega_1(t)\omega_2(t) - \left(1 + \frac{J_{V_3}}{J_W}\right)\dot{\omega}_3(t) \tag{7.334}$$

Equations (7.332), (7.333), and (7.334) are first-order linear differential equations with time-varying coefficients, as in (7.327). The $\mathbf{A}(t)$ matrix is

$$\mathbf{A}(t) = \begin{bmatrix} 0 & \omega_3(t) & -\omega_2(t) \\ -\omega_3(t) & 0 & \omega_1(t) \\ \omega_2(t) & -\omega_1(t) & 0 \end{bmatrix}, \tag{7.335}$$

for a state vector

$$\mathbf{x}(t) = \begin{bmatrix} \Omega_1(t) \\ \Omega_2(t) \\ \Omega_3(t) \end{bmatrix}. \tag{7.336}$$

The equivalent input is

$$\mathbf{u}(t) = \begin{bmatrix} \dfrac{J_{V_2} - J_{V_3}}{J_W}\omega_2(t)\omega_3(t) - \left(1 + \dfrac{J_{V_1}}{J_W}\right)\dot{\omega}_1(t) \\ \dfrac{J_{V_3} - J_{V_1}}{J_W}\omega_3(t)\omega_1(t) - \left(1 + \dfrac{J_{V_2}}{J_W}\right)\dot{\omega}_2(t) \\ \dfrac{J_{V_1} - J_{V_2}}{J_W}\omega_1(t)\omega_2(t) - \left(1 + \dfrac{J_{V_3}}{J_W}\right)\dot{\omega}_3(t) \end{bmatrix} \tag{7.337}$$

and

$$\mathbf{B} = \begin{bmatrix} 1 & 0 & 0 \\ 0 & 1 & 0 \\ 0 & 0 & 1 \end{bmatrix}. \tag{7.338}$$

Notice that $\mathbf{A}(t)$ is equal to $-\mathbf{P}$ in (5.27). Thus the required wheel speeds are solutions of (7.332), (7.333), and (7.334) for any initial conditions. ▲

As in the time-invariant case, the zero-input response may be expressed in terms of the initial state and a transition matrix. We define the *state-transition matrix* corresponding to the vector state equation in (7.327) as the matrix that satisfies the matrix differential equation

$$\dot{\mathbf{X}}(t, t_0) = \mathbf{A}(t)\mathbf{X}(t, t_0) \tag{7.339}$$

and the initial condition

$$\mathbf{X}(t_0, t_0) = \mathbf{I} \tag{7.340}$$

where $\mathbf{X}(t,\ t_0)$ is $n \times n$ if $\mathbf{A}(t)$ is $n \times n$, $\mathbf{I}$ is the $n \times n$ identity matrix. It is assumed that the elements of $\mathbf{A}(t)$ are continuous in t and that t_0 is an arbitrary initial instant. The existence and uniqueness of a solution for (7.339) and (7.340) can be proved.[4] However, no closed form analytical solution can be obtained in general. A numerical method for determining the solution approximately is to solve the sequence of matrix differential equations

$$\dot{\mathbf{X}}_{n+1}(t, t_0) = \mathbf{A}(t)\mathbf{X}_n(t, t_0) \tag{7.341}$$

which satisfies

$$\mathbf{X}_{n+1}(t_0, t_0) = \mathbf{I} \tag{7.342}$$

for $n = 0, 1, 2, \ldots$ starting from an arbitrary $\mathbf{X}_0(t,\ t_0)$ which satisfies $\mathbf{X}_0(t_0, t_0) = \mathbf{I}$. Integrating (7.341), we have

$$\mathbf{X}_{n+1}(t, t_0) - \mathbf{X}_{n+1}(t_0, t_0) = \int_{t_0}^{t} \mathbf{A}(\tau)\mathbf{X}_n(\tau, t_0)\, d\tau. \tag{7.343}$$

Substituting (7.342) in (7.343), we have

$$\mathbf{X}_{n+1}(t, t_0) = \mathbf{I} + \int_{t_0}^{t} \mathbf{A}(\tau)\mathbf{X}_n(\tau, t_0)\, d\tau. \tag{7.344}$$

Clearly $\mathbf{X}_{n+1}(t, t_0)$ in (7.344) satisfies (7.341) and (7.342). The sequence defined by (7.344) converges uniformly to the solution of (7.339) and (7.340). This convergence will not be proved here. The solution of (7.339) and (7.340) is also nonsingular for all finite t, which we assume here without proof.[5] Denote the solution of (7.339) and (7.340) by $\boldsymbol{\phi}(t, t_0)$. Then

$$\boldsymbol{\phi}(t, t_0) \approx \mathbf{X}_n(t, t_0)$$

[4] See R. Bellman, "Introduction to Matrix Analysis," McGraw-Hill, New York, 1960, pp. 163–165.

[5] For proofs, see, for example, R. Bellman, "Introduction to Matrix Analysis," McGraw-Hill, New York, 1960, pp. 163–168, and L. S. Pontriagin, "Ordinary Differential Equations," Addison-Wesley, Reading, Mass., 1962, pp. 127–144.

for n large. A simple choice for $\mathbf{X}_0(t, t_0)$ in (7.343) is simply $\mathbf{I}$. Then

$$\mathbf{X}_1(t, t_0) = \mathbf{I} + \int_{t_0}^{t} \mathbf{A}(\tau)\, d\tau \tag{7.345}$$

$$\mathbf{X}_2(t, t_0) = \mathbf{I} + \int_{t_0}^{t} \mathbf{A}(\tau)\Big[\mathbf{I} + \int_{t_0}^{\tau} \mathbf{A}(\zeta)\, d\zeta\Big]\, d\tau \tag{7.346}$$

Other matrices in the sequence are obtained using the recursion formula in (7.344).

Based on the assumption that $\boldsymbol{\phi}(t, t_0)$ has an inverse for all finite t, where $\boldsymbol{\phi}(t, t_0)$ is the unique solution of (7.339) and (7.340), we shall prove several other properties of the transition matrix. First we note that

$$\mathbf{X}(t, t_0) = \boldsymbol{\phi}(t, t_0)\mathbf{C} \tag{7.347}$$

where $\mathbf{C}$ is any constant $n \times n$ matrix satisfies (7.339) and

$$\mathbf{X}(t_0, t_0) = \mathbf{C}. \tag{7.348}$$

To prove this, we differentiate (7.347):

$$\frac{d}{dt}\mathbf{X}(t, t_0) = \frac{d}{dt}\boldsymbol{\phi}(t, t_0)\mathbf{C} = \mathbf{A}(t)\boldsymbol{\phi}(t, t_0)\mathbf{C} \tag{7.349}$$

because $\dot{\boldsymbol{\phi}}(t, t_0) = \mathbf{A}(t)\boldsymbol{\phi}(t, t_0)$. Next, form

$$\mathbf{A}(t)\mathbf{X}(t, t_0) = \mathbf{A}(t)\boldsymbol{\phi}(t, t_0)\mathbf{C}. \tag{7.350}$$

Comparing (7.349) and (7.350), it is seen that $\mathbf{X}(t, t_0)$ in (7.347) satisfies (7.339). Next, set $t = t_0$ in (7.347) and obtain

$$\mathbf{X}(t_0, t_0) = \boldsymbol{\phi}(t_0, t_0)\mathbf{C} = \mathbf{C} \tag{7.351}$$

Thus, $\mathbf{X}(t, t_0)$ in (7.347) satisfies (7.348). The initial time in (7.347) and (7.348) is arbitrary. For example,

$$\mathbf{X}(t, t_1) = \boldsymbol{\phi}(t, t_1)\mathbf{C} \tag{7.352}$$

satisfies (7.339) and

$$\mathbf{X}(t_1, t_1) = \mathbf{C}. \tag{7.353}$$

Since **C** is any arbitrary constant, suppose

$$\mathbf{C} = \boldsymbol{\phi}(t_1, t_0). \tag{7.354}$$

Then

$$\mathbf{X}(t, t_1) = \boldsymbol{\phi}(t, t_1)\mathbf{C} = \boldsymbol{\phi}(t, t_1)\boldsymbol{\phi}(t_1, t_0) \tag{7.355}$$

satisfies (7.339) and

$$\mathbf{X}(t_1, t_1) = \boldsymbol{\phi}(t_1, t_0). \tag{7.356}$$

Now consider

$$\mathbf{Y}(t, t_0) = \boldsymbol{\phi}(t, t_0). \tag{7.357}$$

Clearly, $\mathbf{Y}(t, t_0)$ in (7.357) satisfies (7.339). Furthermore, for $t = t_1$,

$$\mathbf{Y}(t_1, t_0) = \boldsymbol{\phi}(t_1, t_0). \tag{7.358}$$

Therefore, $\mathbf{X}(t, t_1)$ and $\mathbf{Y}(t, t_0)$ in (7.355) and (7.357) satisfy the same differential equation (7.339) and the same initial condition at time $t = t_1$, as given by (7.356) and (7.358). From the uniqueness of the solution of a differential equation satisfying a specific initial condition, we conclude that the solutions are equal. Hence, from (7.355) and (7.357),

$$\boldsymbol{\phi}(t, t_0) = \boldsymbol{\phi}(t, t_1)\boldsymbol{\phi}(t_1, t_0) \tag{7.359}$$

where $\boldsymbol{\phi}(t, t_j)$ is the transition matrix satisfying (7.339) and

$$\mathbf{X}(t_j, t_j) = \mathbf{I} \tag{7.360}$$

for any t_j. Equation (7.359) expresses the semigroup property of the transition matrix. From (7.359), if we choose $t = t_0$,

$$\boldsymbol{\phi}(t_0, t_0) = \mathbf{I} = \boldsymbol{\phi}(t_0, t_1)\boldsymbol{\phi}(t_1, t_0). \tag{7.361}$$

Since the inverse exists (this was asserted earlier without proof), (7.361) yields

$$\boldsymbol{\phi}^{-1}(t_1, t_0) = \boldsymbol{\phi}(t_0, t_1). \tag{7.362}$$

The above properties hold for an arbitrary time-varying $\mathbf{A}(t)$. They hold also for the particular case when $\mathbf{A}(t)$ is a constant matrix. Recall that the transition matrix for a time-invariant **A** is $\epsilon^{(t-t_0)}$,

$$\boldsymbol{\phi}(t, t_0) = \epsilon^{\mathbf{A}(t-t_0)}. \tag{7.363}$$

The function of two variables t and t_0 becomes a function of the difference $(t-t_0)$ only. That is, for a time-invariant $\mathbf{A}$, $\phi(t, t_0)$ reduces to a function $\hat{\boldsymbol{\phi}}(t-t_0)$. In the previous section we have used the symbol $\boldsymbol{\phi}(t-t_0)$ to denote the transition matrix for a time-invariant $\mathbf{A}$. We shall continue to use the latter notation when no confusion arises. From (7.359), it is seen that

$$\epsilon^{\mathbf{A}(t-t_0)} = \epsilon^{\mathbf{A}(t-t_1)}\epsilon^{\mathbf{A}(t_1-t_0)}. \tag{7.364}$$

In particular, t_1 could be zero and (7.364) becomes

$$\epsilon^{\mathbf{A}(t-t_0)} = \epsilon^{\mathbf{A}t}\epsilon^{-\mathbf{A}t_0}. \tag{7.365}$$

Similarly, from (7.362) we obtain

$$[\epsilon^{\mathbf{A}(t_1-t_0)}]^{-1} = \epsilon^{\mathbf{A}(t_0-t_1)}. \tag{7.366}$$

For $t_0 = 0$, (7.366) becomes

$$[\epsilon^{\mathbf{A}t_1}]^{-1} = \epsilon^{-\mathbf{A}t_1}. \tag{7.367}$$

Let us now return to the vector state equation (7.327). For $\mathbf{u} = \mathbf{0}$, it is easy to verify that

$$\mathbf{x}(t) = \boldsymbol{\phi}(t, t_0)\mathbf{x}_0 \tag{7.368}$$

satisfies

$$\dot{\mathbf{x}} = \mathbf{A}(t)\mathbf{x}(t) \tag{7.369}$$

and

$$\mathbf{x}(t_0) = \mathbf{x}_0, \tag{7.370}$$

by direct substitution of $x(t)$ of (7.368) in (7.369) and (7.370). Similarly, the solution of (7.327) that satisfies (7.370) is

$$\mathbf{x}(t) = \boldsymbol{\phi}(t, t_0)\mathbf{x}_0 + \int_{t_0}^{t} \boldsymbol{\phi}(t, \tau)\mathbf{B}(\tau)\mathbf{u}(\tau)\, d\tau. \tag{7.371}$$

Again, this may be verified by subtituting $\mathbf{x}(t)$ of (7.371) in (7.327) and setting $t = t_0$ to verify (7.370).

If the output of the system is related to $\mathbf{x}(t)$ and $\mathbf{u}(t)$ by (7.328), then the complete output is

$$\begin{aligned}\mathbf{y}(t) &= \mathbf{C}(t)\boldsymbol{\phi}(t, t_0)\mathbf{x}(t_0) + \int_{t_0}^{t} \mathbf{C}(t)\boldsymbol{\phi}(t,\tau)\mathbf{B}(\tau)\mathbf{u}(\tau)d\tau + \mathbf{D}(t)\mathbf{u}(t) \\ &= \mathbf{C}(t)\boldsymbol{\phi}(t, t_0)\mathbf{x}_0 + \int_{t_0}^{t} [\mathbf{C}(t)\boldsymbol{\phi}(t,\tau)\mathbf{B}(\tau) + \mathbf{D}(\tau)\,\delta(t-\tau)]\mathbf{u}(\tau)\, d\tau.\end{aligned} \tag{7.372}$$

As in the time-invariant case, the output or response when there is no input is called the *zero-input response*. From (7.372),

$$\text{zero-input response} = \mathbf{C}(t)\boldsymbol{\phi}(t, t_0)\mathbf{x}_0 \tag{7.373}$$

The response when the initial state is zero is called the *zero-state response*. From (7.372),

$$\text{zero-state response} = \int_{t_0}^{t} [\mathbf{C}(t)\boldsymbol{\phi}(t,\tau)\mathbf{B}(\tau) + \mathbf{D}(\tau)\,\delta(t-\tau)]\mathbf{u}(\tau)\,d\tau. \tag{7.374}$$

The sum of the zero-input response and the zero-state response gives the total response.

★7.11 Linearity and Time-Invariance

In previous chapters, a system described by linear ordinary differential equations has been called a linear system. In particular, if the equations are in state form, we defined the system to be linear if the equations are in the form of (7.8) and (7.9) where **A**, **B**, **C**, and **D** are either constant matrices or matrices of time functions. If the coefficients or coefficient matrices are constants, the system is said to be a *time-invariant system.*

In this section, more general definitions of linear systems and time-invariant systems will be given, which will enlarge the class of systems considered to include those that cannot be described by ordinary differential equations. For example, some dynamic systems are described by partial differential equations. Other systems are described by difference equations. We shall first state these more general definitions and then show that systems described by ordinary differential equations, which were defined earlier in the less general sense as linear or time-invariant, are also linear or time-invariant in the sense of the more general definitions.

The following will be defined: decomposition, homogeneity, additivity, zero-input linear system, zero-state linear system, linear system, zero-state time-invariant system, zero-input time-invariant system, and time-invariant system. These definitions will be stated in terms of the response or output at time t due to an input function $\mathbf{u}(t)$, an initial state $\mathbf{x}_0$, and initial time t_0. Since responses for different inputs, different initial states, and different initial times must be distinguished in the definitions, it is convenient to adopt the following notation for the output. Let $\mathbf{y}(t; \mathbf{u}(t), \mathbf{x}_0, t_0)$ denote the system output at time t, when the input function is $\mathbf{u}(t)$, the initial state is $\mathbf{x}_0$, and the initial time is t_0, for any $t \geqslant t_0$. For $\mathbf{u}(t) = 0$, the output $\mathbf{y}(t; \mathbf{0}, \mathbf{x}_0, t_0)$ is defined as the zero-input response. For $\mathbf{x}_0 = 0$, the output $\mathbf{y}(t; \mathbf{u}(t), \mathbf{0}, t_0)$ is defined as the *zero-state response.*

Definition 7-1. Decomposition. *Decomposition* is defined as the following property of a system:

$$\mathbf{y}(t; \mathbf{u}(t), \mathbf{x}_0, t_0) = \mathbf{y}(t; \mathbf{0}, \mathbf{x}_0, t_0) + \mathbf{y}(t; \mathbf{u}(t), \mathbf{0}, t_0) \qquad (7.375)$$

for any initial state $\mathbf{x}_0$, any initial instant t_0, and any input-function $\mathbf{u}(t)$. A system that satisfies (7.375) is said to possess the decomposition property. ▲

For example, for the systems discussed in the previous section, this property is satisfied. That is, the total response is always equal to the zero-input response plus the zero-state response, for any initial state and any input function. For example, if a system is described by

$$\frac{dy}{dt} = u^2 \qquad (7.376)$$

where y is output, u is input, and y is also the state, then, by simple integration, we have

$$y(t) = y(t_0) + \int_{t_0}^{t} u^2(\tau)\, d\tau. \qquad (7.377)$$

By our previous definition of a linear system, this is not a linear system, but it possesses the decomposition property.

Definition 7-2. Homogeneity. *Homogeneity with respect to input* is defined as the following property of a system:

$$\mathbf{y}(t; k\mathbf{u}(t), \mathbf{0}, t_0) = k\mathbf{y}(t; \mathbf{u}(t), \mathbf{0}, t_0) \qquad (7.378)$$

for any input-function $\mathbf{u}(t)$, any number k, and any initial time t_0. A system for which (7.378) is satisfied is said to possess the property of homogeneity with respect to input. ▲

For example, let a system be described by

$$\frac{d^2y}{dt^2} + y^2 = u^2 \qquad (7.379)$$

where y is the output and u is the input. For zero initial state, if y_1 denotes the output when the input is u_1, that is, y_1 satisfies (7.379) when $u = u_1$, it is easily verified that $y = ky_1$ satisfies (7.379) when the input is $u = ku_1$, so that the output is ky_1. From our definition of a linear system in earlier chapters, this system is not linear.

Homogeneity with respect to initial state is defined as the following property of a system:

$$\mathbf{y}(t; \mathbf{0}, k\mathbf{x}_0, t_0) = k\mathbf{y}(t; \mathbf{0}, \mathbf{x}_0, t_0) \tag{7.380}$$

for any initial state $\mathbf{x}_0$ and any initial time t_0. A system for which (7.380) is satisfied is said to possess the property of homogeneity with respect to initial state.

Definition 7-3. Additivity. *Additivity with respect to input* is defined as the following property of a system:

$$\mathbf{y}(t; \mathbf{u}_1(t) + \mathbf{u}_2(t), \mathbf{0}, t_0) = \mathbf{y}(t; \mathbf{u}_1(t), \mathbf{0}, t_0) + \mathbf{y}(t; \mathbf{u}_2(t), \mathbf{0}, t_0) \tag{7.381}$$

for all $\mathbf{u}_1(t)$ and $\mathbf{u}_2(t)$, and all t_0. A system for which (7.381) is satisfied is said to possess the property of additivity with respect to input.

Additivity with respect to initial state is defined as the following property of a system:

$$\mathbf{y}(t; \mathbf{0}, \mathbf{x}_{01} + \mathbf{x}_{02}, t_0) = \mathbf{y}(t; \mathbf{0}, \mathbf{x}_{01}, t_0) + \mathbf{y}(t; \mathbf{0}, \mathbf{x}_{02}, t_0) \tag{7.382}$$

for all $\mathbf{x}_{01}$, all $\mathbf{x}_{02}$, and all t_0. A system that satisfies (7.382) is said to possess the property of additivity with respect to initial state. ▲

Definition 7-4. Zero-Input Linear System. A system that possesses the homogeneity and additivity properties with respect to initial state is called a *zero-input linear system.* ▲

For example, a system described by (7.376) is a zero-input linear system because the zero-input response satisfies the homogeneity and additivity properties with respect to initial state.

Definition 7-5. Zero-State Linear. A system that possesses the homogeneity and additivity properties with respect to input is called a *zero-state linear system.* ▲

Definition 7-6. Linear System. A system that is zero-input linear and zero-state linear, and that possesses the decomposition property is called a *linear system.* ▲

Let us show that the system described by the state differential equations

$$\dot{\mathbf{x}}(t) = \mathbf{A}(t)\mathbf{x}(t) + \mathbf{B}(t)\mathbf{u}(t) \qquad (7.383)$$

$$\mathbf{y}(t) = \mathbf{C}(t)\mathbf{x}(t) + \mathbf{D}(t)\mathbf{u}(t), \qquad (7.384)$$

where $\mathbf{x}(t)$ is the state vector, $\mathbf{u}(t)$ is the input vector, $\mathbf{y}(t)$ is the output vector, and $\mathbf{A}(t)$, $\mathbf{B}(t)$, $\mathbf{C}(t)$, and $\mathbf{D}(t)$ are matrices of constants or functions of time, is a linear system. Recall that the solution of (7.383) and (7.384) is

$$\mathbf{y}(t) = \mathbf{C}(t)\boldsymbol{\phi}(t, t_0)\mathbf{x}_0 + \int_{t_0}^{t} [\mathbf{C}(t)\boldsymbol{\phi}(t, \tau)\mathbf{B}(\tau) + \mathbf{D}(\tau)\, \delta(t-\tau)]\mathbf{u}(\tau)\, d\tau \qquad (7.385)$$

Clearly, the system possesses the decomposition property.

The zero-state response, using our expanded notation for $\mathbf{y}$, is

$$\mathbf{y}(t; \mathbf{u}(t), \mathbf{0}, t_0) = \int_{t_0}^{t} [\mathbf{C}(t)\boldsymbol{\phi}(t, \tau)\mathbf{B}(\tau) + \mathbf{D}(\tau)\, \delta(t-\tau)]\mathbf{u}(\tau)\, d\tau. \qquad (7.386)$$

From (7.386), it is seen that (7.381) is satisfied. Furthermore, it is seen that (7.378) is satisfied. Hence the system is zero-state linear.

The zero-input response is

$$\mathbf{y}(t; \mathbf{0}, \mathbf{x}_0, t_0) = \mathbf{C}(t)\boldsymbol{\phi}(t, t_0)\mathbf{x}_0. \qquad (7.387)$$

From (7.387), it is easily seen that (7.382) is satisfied and (7.380) is satisfied. Hence the system is also zero-input linear. Since it is zero-input linear and zero-state linear, and it possesses the decomposition property, it is a linear system in the sense of Definition 7-6.

Definition 7-7. Zero-Input Time-Invariant System. A system that satisfies

$$\mathbf{y}(t; \mathbf{0}, \mathbf{x}_0, t_0 + T) = \mathbf{y}(t-T; \mathbf{0}, \mathbf{x}_0, t_0) \qquad (7.388)$$

for all $\mathbf{x}_0$, a given t_0, for all positive T, and for all $t \geq t_0 + T$, is said to be *zero-input time-invariant after* t_0. If (7.388) holds for all t_0, the system is called a *zero-input time-invariant system.* ▲

Recall from our notation that $\mathbf{y}(t; \mathbf{0}, \mathbf{x}_0, t_0 + T)$ denotes the response at time t for all $t \geq t_0 + T$, due to an initial state $\mathbf{x}_0$ specified at time $t_0 + T$. Moreover, $\mathbf{y}(t-T; \mathbf{0}, \mathbf{x}_0, t_0)$ denotes the response of the same system at time $t-T$, for all $t-T \geq t_0$, due to an initial state $\mathbf{x}_0$ specified at time t_0. But if $\mathbf{y}(t; \mathbf{0}, \mathbf{x}_0, t_0)$ is known, then $\mathbf{y}(t-T; \mathbf{0}, \mathbf{x}_0, t_0)$ is simply obtained from $\mathbf{y}(t; \mathbf{0}, \mathbf{x}_0, t_0)$ by replacing t by $(t-T)$ wherever t appears. If $\mathbf{y}(t; \mathbf{0}, \mathbf{x}_0, t_0)$ is given as a plot versus t for $t \geq t_0$, then the plot of $\mathbf{y}(t-T, \mathbf{0}, \mathbf{x}_0, t_0)$ versus t is obtained by shifting the plot of $\mathbf{y}(t; \mathbf{0}, \mathbf{x}_0, t_0)$ to the right by an amount T, where the plot is now valid for $t \geqslant t_0+T$. Hence the plot of $\mathbf{y}(t; \mathbf{0}, \mathbf{x}_0, t_0+T)$ for a zero-input time-invariant system is obtained from that of $\mathbf{y}(t; \mathbf{0}, \mathbf{x}_0, t_0)$ by shifting the latter to the right by an amount T (see Figure 7.14).

Definition 7-8. Zero State Time-invariant System. A system that satisfies

$$\mathbf{y}[t; \mathbf{u}(t-T), \mathbf{0}, t_0+T] = \mathbf{y}[t-T; \mathbf{u}(t-T), \mathbf{0}, t_0] \tag{7.389}$$

for all positive T, for all functions $\mathbf{u}(t)$, a specified t_0 and for all $t \geqslant t_0+T$ is said to be *zero-state time-invariant after* t_0. If (7.389) holds for all t_0, the system is called a *zero-state time-invariant system.* ▲

The notation $\mathbf{y}(t; \mathbf{u}(t-T), \mathbf{0}, t_0+T)$ means the output at time t for an input $\mathbf{u}(t-T)$ which is delayed by T compared to $\mathbf{u}(t)$, with zero initial state at the initial time t_0+T, for all $t \geqslant t_0+T$. The notation $\mathbf{y}(t-T; \mathbf{u}(t-T), \mathbf{0}, t_0)$ means the output at time $t-T$ due to an input $\mathbf{u}(t-T)$, with zero initial state at the initial time t_0, for all $t-T \geqslant t_0$. But $\mathbf{y}(t-T); \mathbf{u}(t-T), \mathbf{0}, t_0)$ is simply obtained from $\mathbf{y}(t; \mathbf{u}(t), \mathbf{0}, t_0)$ by replacing t by $t-T$ wherever t appears. The effect of this substitution is a shift to the right by an amount T. Hence the plot of $\mathbf{y}(t; \mathbf{u}(t-T), \mathbf{0}, t_0+T)$ is obtainable from the plot of $\mathbf{y}(t; \mathbf{u}(t), \mathbf{0}, t_0)$ by shifting the latter plot to the right by an amount T if the system is zero-state time-invariant. The shifted plot is valid for $t \geqslant t_0+T$.

Definition 7-9. Time-Invariant System. A system that satisfies

$$\mathbf{y}(t; \mathbf{u}(t-T), \mathbf{x}_0, t_0+T) = \mathbf{y}(t-T; u(t-T), \mathbf{x}_0, t_0) \tag{7.390}$$

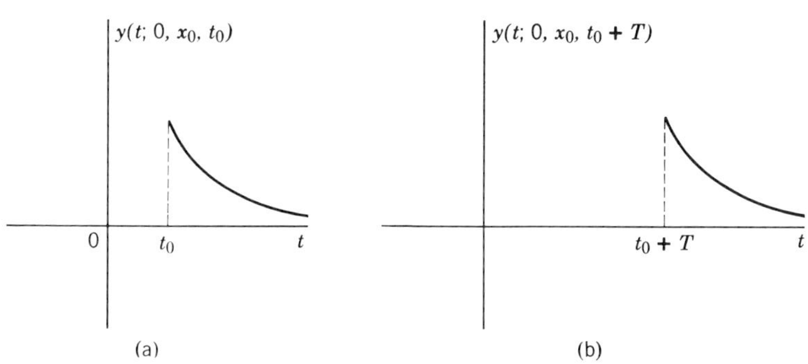

Figure 7.14 Plots illustrating zero-input time-invariance.

for all $\mathbf{u}(t-T)$, all $\mathbf{x}_0$, all t_0, all positive T, and for all $t \geq t_0+T$ is called a *time-invariant system.* ▲

This is a generalization of zero-state time-invariance where $\mathbf{x}_0$ is arbitrary instead of zero. Let us show that the system described by

$$\dot{\mathbf{x}}(t) = \mathbf{A}\mathbf{x}(t) + \mathbf{B}\mathbf{u}(t) \tag{7.391}$$

$$\mathbf{y}(t) = \mathbf{C}\mathbf{x}(t) + \mathbf{D}\mathbf{u}(t) \tag{7.392}$$

for all t, is a time-invariant system. From earlier discussions, recall that the complete response for an initial state $\mathbf{x}_0$ at initial time t_0 and input function $\mathbf{u}_1(t)$ is

$$\mathbf{y}(t; \mathbf{u}_1(t), x_0, t_0) = \mathbf{C}\epsilon^{(t-t_0)}\mathbf{x}_0 + \int_{t_0}^{t} \mathbf{C}\epsilon^{(t-\tau)}\mathbf{B}\mathbf{u}_1(\tau)\, d\tau + \mathbf{D}\mathbf{u}_1(t), \tag{7.393}$$

for $t \geq t_0$. For an input function

$$\mathbf{u}_2(t) = \mathbf{u}_1(t-T) \tag{7.394}$$

and initial state $\mathbf{x}_0$ at initial time t_0+T, the response is

$$\mathbf{y}(t; \mathbf{u}_2(t), x_0, t_0+T) = \mathbf{C}\epsilon^{(t-T-t_0)}\mathbf{x}_0 + \int_{t_0+T}^{t} \mathbf{C}\epsilon^{\mathbf{A}(t-\tau)}\mathbf{B}\mathbf{u}_2(\tau)\, d\tau + \mathbf{D}\mathbf{u}_2(t)$$

$$= \mathbf{C}\epsilon^{(t-T-t_0)}\mathbf{x}_0 + \int_{t_0+T}^{t} \mathbf{C}\epsilon^{\mathbf{A}(t-\tau)}\mathbf{B}\mathbf{u}_1(\tau-T)\, d\tau + \mathbf{D}\mathbf{u}_1(t-T) \tag{7.395}$$

for $t \geq t_0+T$. Let us make a change of variables in the integral. Let

$$\tau - T = \zeta, \tag{7.396}$$

then

$$d\tau = d\zeta. \tag{7.397}$$

For $\tau = t_0+T$, (7.396) gives $\zeta = t_0$. For $\tau = t$, (7.396) gives $\zeta = t-T$. Hence, (7.395) becomes

$$\mathbf{y}(t; \mathbf{u}_1(t-T), \mathbf{x}_0, t_0+T) = \mathbf{C}\epsilon^{\mathbf{A}(t-T-t_0)}\mathbf{x}_0$$
$$+ \int_{t_0}^{t-T} \mathbf{C}\epsilon^{\mathbf{A}(t-T-\zeta)}\mathbf{B}\mathbf{u}_1(\zeta)\, d\zeta + \mathbf{D}\mathbf{u}_1(t-T) \tag{7.398}$$

Comparing (7.393) and (7.398), it is noted that if t in (7.393) is replaced by $(t-T)$, the result is identical to (7.398). Hence

$$\mathbf{y}(t; \mathbf{u}_1(t-T), \mathbf{x}_0, t_0+T) = \mathbf{y}(t-T; \mathbf{u}_1(t-T), \mathbf{x}_0, t_0), \tag{7.399}$$

for all $t \geqslant t_0 + T$. Since $u_1(t-T)$, x_0, and t_0 are arbitrary, and since (7.399) holds for all positive T, then the system described by (7.391) and (7.392) is time-invariant.

The general definitions enumerated in this section are intended to provide a wider scope for the types of systems that can be treated. Since the systems described by ordinary differential equations defined as linear or time-invariant earlier are also linear or time-invariant systems according to the wider definitions here, we have, in fact, derived additional properties of state equations which have coefficient matrices that are either constants or functions of time.

Problems

7-1. For the network in Figure 7.1, derive a condition among the parameters so that the eigenvalues are both real.

7-2. Suppose that $[i_L \; v_C]^T$ is chosen as the state vector for the network in Figure 7.1, where the state equation is given in (7.5). Let $R_1 = R_2 = 10$ ohms, $L = 0.01$ henry, and $C = 0.01$ microfarad. Compute the state-transition matrix.

7-3. For the network in Figure P7-3, choose $[v_1 \; v_2]^T$ as the state vector.

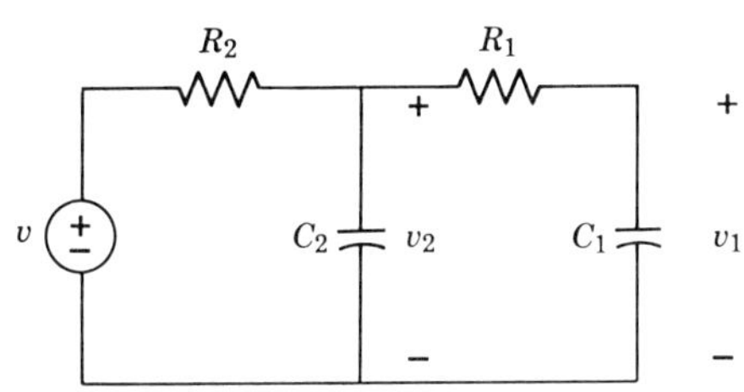

Figure P7-3

(a) Write the state equation.

(b) Suppose that $R_1 = 1000$ ohms, $R_2 = 100$ ohms, $C_1 = C_2 = 0.01$ microfarad. Determine the eigenvalues and the eigenvectors of the $\mathbf{A}$ matrix of the state equation. Sketch the eigenvectors on a phase plane.

(c) Express the state vector as a sum of the eigenvectors, when the initial conditions are $v_1(0) = 1$ volt and $v_2(0) = 1$ volt, and when the input $v(t)$ equals zero.

7-4. Suppose that for the accelerometer in Example 7-5 and with the choice of state variables in (7.148) and (7.149), it is known that $\omega_n = 1$ and $\zeta = 1/\sqrt{2}$. Determine $\mathbf{\Phi}(s)$.

7-5. A second-order system with zero input is described by

$$\dot{x}_1 = x_2$$
$$\dot{x}_2 = -2x_1 - 3x_2.$$

(a) Determine the eigenvalues and eigenvectors for the **A** matrix.
(b) Sketched a few trajectories on the phase plane.
(c) Determine the transition matrix for the system.

7-6. Find an all-integrator block diagram whose transfer function is

$$H(s) = \frac{(s+2)}{(s+1)(s+3)}$$

and determine a state representation for your system, that is, determine the matrices **A**, **B**, **C**, and **D** for your system.

7-7. Repeat Problem 7-6 for

$$H(s) = \frac{(s+4)(s+2)}{(s+1)(s+3)}.$$

7-8. The Laplace transform $\mathbf{\Phi}(s)$ of the transition matrix for a system is

$$\mathbf{\Phi}(s) = \begin{bmatrix} \frac{1}{s+1} & 0 \\ 0 & \frac{1}{s+2} \end{bmatrix}$$

Determine the **A** matrix for the system.

7-9. Repeat Problem 7-8 for

$$\mathbf{\Phi}(s) = \begin{bmatrix} \frac{s+3}{s^2+3s+2} & \frac{-2}{s^2+3s+2} \\ \frac{1}{s^2+3s+2} & \frac{s}{s^2+3s+2} \end{bmatrix}.$$

7-10. Repeat Problem 7-6 for

$$\mathbf{H}(s) = \begin{bmatrix} \frac{1}{s+2} & \frac{-1}{s+1} \\ \frac{2}{s+1} & \frac{3}{s+1} \end{bmatrix}.$$

7-11. For the system in Figure P7-11, choose an appropriate set of state variables and write a vector state equation. Determine $\mathbf{\Phi}(s)$ for your choice of state variables, using Mason's formula.

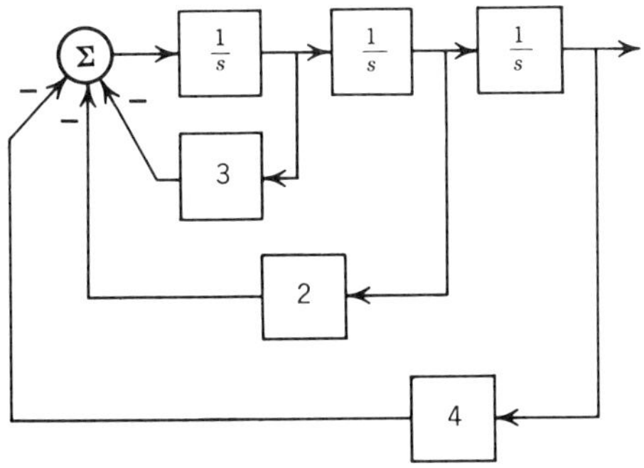

Figure P7-11

7-12. Determine the impulse response for the system

$$\begin{bmatrix} \dot{x}_1 \\ \dot{x}_2 \end{bmatrix} = \begin{bmatrix} 0 & 1 \\ -5 & -3 \end{bmatrix} \begin{bmatrix} x_1 \\ x_2 \end{bmatrix} + \begin{bmatrix} 1 \\ 2 \end{bmatrix} u$$

$$\begin{bmatrix} y_1 \\ y_2 \end{bmatrix} = \begin{bmatrix} 1 & -4 \\ 2 & -1 \end{bmatrix} \begin{bmatrix} x_1 \\ x_2 \end{bmatrix}$$

7-13. Sketch the phase portrait for the following systems:

(a) $\dot{x}_1 = x_2$
$\dot{x}_2 = -x_1$

(b) $\dot{x}_1 = x_2$
$\dot{x}_2 = x_1 - 5x_2$

(c) $\dot{x}_1 = x_2$
$\dot{x}_2 = x_1 - x_2$

(d) $\dot{x}_1 = x_2$
$\dot{x}_2 = x_1 + 5x_2$

(e) $\dot{x}_1 = x_2$
$\dot{x}_2 = x_1 + x_2$

7-14. The state representation for a system is given by

$$\begin{aligned} \dot{x}_1 &= x_2 + u_1 \\ \dot{x}_2 &= -5x_1 - 6x_2 + u_2 \\ y_1 &= x_1 \\ y_2 &= x_2 + u_1. \end{aligned}$$

The initial state is $x_1(0) = 3$, $x_2(0) = 2$. The inputs are $u_1(t) = 1$ for $t \geqslant 0$, $u_2(t) = 1$ for $t \geqslant 0$. Determine the complete response $y_1(t)$ and $y_2(t)$ for $t \geqslant 0$.

7-15. (a) Determine the transition matrix for the system in Problem 7-14.
(b) Show that $\boldsymbol{\phi}(t)$ in (a) is nonsingular for all t.
(c) Verify that $\boldsymbol{\phi}(t)\boldsymbol{\phi}(-t) = \mathbf{I}$ for the transition matrix in (a).
(d) Verify that $\dot{\boldsymbol{\phi}}(t)|_{t_0} = \mathbf{A}$ for the transition matrix in (a).

7-16. Repeat Problem 7-15 for the system in Problem 7-13(a).

7-17. Repeat Problem 7-15 for the system in Problem 7-13(b).

7-18. Repeat Problem 7-15 for the system in Problem 7-13(c).

7-19. Sketch the phase portrait for the following systems:

(a) $\dot{x}_1 = x$
$\dot{x}_2 = -4x_2$

(b) $\dot{x}_1 = x_2$
$\dot{x}_2 = 4x_2$

(c) $\dot{x}_1 = x_2$
$\dot{x}_2 = -x_1 - 2x_2.$

7-20. (a) Determine the transfer function $Y(s)/R(s)$ for the structure in Figure P7-20.

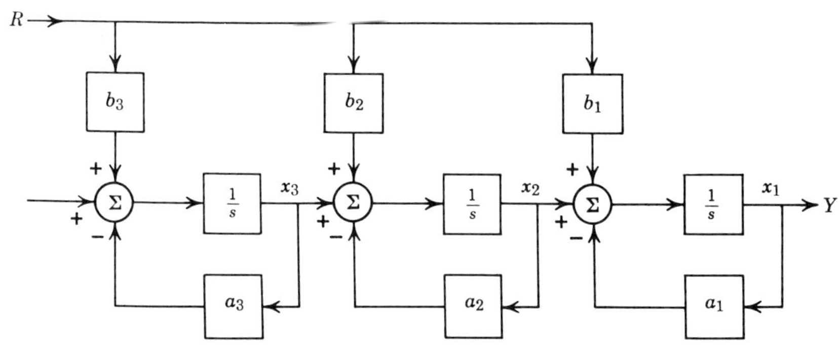

Figure P7-20

(b) It is desired to realize

$$H(s) = \frac{s^2 + 6s + 5}{(s+2)(s+3)(s+4)}$$

in the form as shown in Figure P7-20. Determine the parameters a_1, a_2, a_3, b_1, b_2, and b_3.
(c) For the system in (b), and using the state variables x_1, x_2, and x_3 as indicated in Figure P7-20, write the state equations for your system, where $r(t)$ is the input and $y(t)$ is the output.
(d) Verify that $\mathbf{C}\boldsymbol{\Phi}(s)\mathbf{B} + \mathbf{D}$ is equal to the given transfer function.

7-21. (a) Determine the transfer-function matrix of the reaction-wheel attitude system in Example 7-10.

(b) Determine the zero-state response due to a unit impulse disturbance torque $\tau_d(t) = \delta(t)$ and $v_a(t) = 0$.

(c) Determine the zero-state response due to a unit step voltage

$$v_a(t) = \begin{cases} 1 & \text{for} \quad t \geq 0 \\ 0 & \text{for} \quad t < 0 \end{cases}$$

and $\tau_d(t) = 0$.

7-22. The transfer function of a system is given as

$$H(s) = \frac{10(s+6)}{(s+2)(s+5)}$$

Determine a state representation such that the **A** matrix is diagonal.

7-23. Repeat Problem 7-22 for

$$H(s) = \frac{5(s+4)}{(s+1)(s+2)(s+3)}$$

7-24. Find the eigenvalues and eigenvectors of

$$\mathbf{A} = \begin{bmatrix} 2 & -5 \\ 3 & -7 \end{bmatrix}.$$

7-25. (a) Expand the transfer function

$$H(s) = \frac{(s+1)(s+3)}{(s+10)(s+2)(s+5)}$$

in partial fractions and draw an all-integrator block diagram to realize $H(s)$ as a sum of terms in the expansion.

(b) Determine the state equations and compute $\mathbf{C\Phi}(s)\mathbf{B}+\mathbf{D}$. Compare your answer to $H(s)$.

7-26. For the reaction-wheel attitude system in Example 7-10 and Figure 7.7, determine the zero-input response for the initial state

$$[\theta_0 \omega_0 \Omega_0]^T.$$

7-27. Determine the transfer-function matrix $\mathbf{H}(s)$ for the system in Problem.7-12.

7-28. Find a state representation for the system in Figure P7-28.

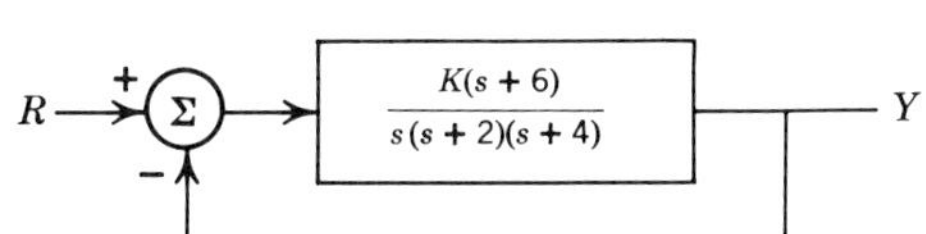

Figure P7-28

7-29. Find an all-integrator block diagram for the system in Figure P7-29 that yields the same transfer function. K, τ, K_1, and σ are constant parameters. Determine **A**, **B**, **C**, and **D** for your state representation of the system. Assume that $r(t)$ is the input and $y(t)$ is the output.

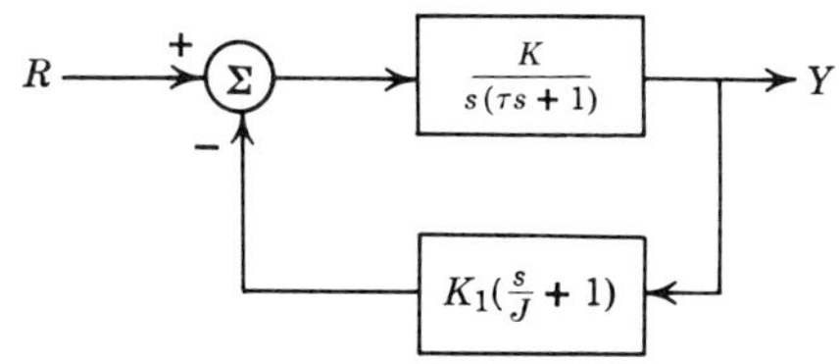

Figure P7-29

7-30. One problem that arises in simulating a dynamic system whose input signal is $r(t)$ is the generation of the signal $r(t)$. For many engineering problems, it is adequate to approximate $r(t)$ by a sum of exponentials and exponentially damped sinusoids. Such a signal can be generated as a zero-input response of a linear time-invariant system. That is, the signal $r(t)$ can be generated as the output of a linear time-invariant system with an appropriate set of initial conditions and zero input.

Find an all-integrator block diagram and specify the initial conditions on the integrators so that

$$r(t) = 5\epsilon^{-t} - 5\epsilon^{-3t}$$

is generated as a signal.

7-31. Repeat Problem 7-30 for $r(t) = \sin 10t, t \geq 0$.

7-32. Repeat Problem 7-30 for $r(t) = \cos 10t, t \geq 0$.

7-33. Repeat Problem 7-30 for $r(t) = \sin(10t + \pi/4), t \geq 0$.

7-34. Repeat Problem 7-30 for $r(t) = [1 - \epsilon^{-\alpha t} \cos \omega t], t \geq 0$, where α and ω are constant parameters.

7-35. Repeat Problem 7-30 for $r(t) = 6t\epsilon^{-t}, t \geq 0$.

CHAPTER 8

Stability of Linear Time-Invariant Systems

In many cases, a practical system must have the property that its zero-input response should decay to zero as time increases. Moreover, its zero-state response should be bounded provided the input itself is bounded. These requirements are related to the notion of *stability*.

There is no single definition of stability. This is because of the richness of possible system behavior—almost anything can happen. Several of the various definitions are explored in Chapter 9. We concentrate in this chapter on linear time-invariant (LTI) systems. For the purposes of this chapter we adopt the following definition.

Definition 8-1. A linear time-invariant system having a $n \times n$ state-transition matrix $\boldsymbol{\phi}(t)$ with elements of $\phi_{ij}(t)$, $i, j = 1, \ldots, n$, is said to be an *asymptotically stable system* if

$$\lim_{t \to \infty} \phi_{ij}(t) = 0$$

for $i, j = 1, \ldots, n$.

Definition 8.1 clearly implies that the zero-input response decays to zero as time increases.

We now develop several tests for indirectly determining when a system satisfies Definition 8.1.

8.1 Characteristic Equation

It is convenient for the case of LTI systems to test stability indirectly in the frequency domain rather than directly in the time domain. The connection between the two domains is established from

$$\mathscr{L}\{\boldsymbol{\phi}(t)\} = (s\mathbf{I} - \mathbf{A})^{-1} = \boldsymbol{\Phi}(s) \tag{8.1}$$

as was derived in Chapter 7. Thus

$$\mathbf{\Phi}(s)=\frac{\text{adj }(s\mathbf{I}-\mathbf{A})}{\det\,(s\mathbf{I}-\mathbf{A})}. \tag{8.2}$$

The state-transition matrix can be found by taking the inverse Laplace transform of (8.1). In matrix form, the partial fraction expansion of (8.1) is

$$\mathbf{\Phi}(s)=\left[\sum_{i=1}^{n'}\sum_{r=1}^{m_i}\mathbf{B}_{ri}\frac{1}{(s-s_i)^r}\right] \tag{8.3}$$

where the s_i ($i=1,2,\ldots,n'\leqslant n$, n the dimension of $\mathbf{x}$) are the n' different roots of

$$\det\,(s\mathbf{I}-\mathbf{A})=0, \tag{8.4}$$

and m_i and their respective multiplicities. The matrix $\mathbf{B}_{ri}$ is the matrix of partial fraction coefficients. The inverse transform of (8.3) is

$$\boldsymbol{\phi}(t)=\sum_{i=1}^{n'}\epsilon^{s_it}\sum_{r=1}^{m_i}\mathbf{B}_{ri}\frac{t^{r-1}}{(r-1)!}. \tag{8.5}$$

Example 8-1. Consider a system where

$$\mathbf{X}(s)=\mathbf{\Phi}(s)\mathbf{x}_0,$$

with

$$\mathbf{\Phi}(s)=\frac{1}{(s+1+j1)(s+1-j1)}\begin{bmatrix}s+1 & 1\\ -1 & s+1\end{bmatrix}. \tag{8.6}$$

Partial fraction expansion yields

$$\mathbf{\Phi}(s)=\frac{1}{s+1+j1}\begin{bmatrix}\frac{1}{2} & \frac{1}{2}j\\ -\frac{1}{2}j & \frac{1}{2}\end{bmatrix}+\frac{1}{s+1-j1}\begin{bmatrix}\frac{1}{2} & -\frac{1}{2}j\\ \frac{1}{2}j & \frac{1}{2}\end{bmatrix}. \tag{8.7}$$

Thus, for this case, there are no repeated roots, so

$$\begin{aligned}s_1&=-1-j1\\ s_2&=-1+j1\\ m_1&=m_2=1\\ n'&=n=2.\end{aligned} \tag{8.8}$$

From (8.4) and (8.8), the partial fraction coefficient matrices are

$$\mathbf{B}_{11} = \begin{bmatrix} \frac{1}{2} & \frac{1}{2}j \\ -\frac{1}{2}j & \frac{1}{2} \end{bmatrix}$$

$$\mathbf{B}_{12} = \begin{bmatrix} \frac{1}{2} & -\frac{1}{2}j \\ \frac{1}{2}j & \frac{1}{2} \end{bmatrix}. \tag{8.9}$$

Finally, from (8.2),

$$\boldsymbol{\phi}(t) = \epsilon^{(-1-j1)t}\mathbf{B}_{11} + \epsilon^{(-1+j1)t}\mathbf{B}_{12}. \blacktriangle \tag{8.10}$$

We notice from (8.5) that terms of the form $\epsilon^{s_i t}$ determine the general character of $\boldsymbol{\phi}(t)$. The roots of (8.4) thus have very special importance. The polynomial det $(s\mathbf{I}-\mathbf{A})$ is called the *characteristic polynomial* of the system. Equation (8.4) is called the *characteristic equation* of the system. The roots s_i of the characteristic equation are known in matrix theory as the *eigenvalues* of the matrix **A**. They are also called the *characteristic roots* of the system, or the system *eigenvalues*.

Stability of LTI systems is closely connected to the values of the characteristic roots. The situation is slightly complicated by the fact that there may be cancellations between the numerator and denominator polynomials in (8.2) or, equivalently,

$$\mathbf{B}_{qi} = \mathbf{B}_{q+1,i} = \cdots = \mathbf{B}_{m_i i} = \mathbf{0}, \tag{8.11}$$

$q \geqslant 1$, for one or more of the characteristic roots s_i. This means that the system state-transition matrix, and hence $\mathbf{X}(s)$, will appear to be of lower order than the characteristic equation. If all possible cancellations in (8.2) are performed, the resulting denominator polynomial is called the *minimal polynomial*. It is possible to show, using other results in matrix theory, that every eigenvalue of **A** appears as a zero of the minimal polynomial. That is, every simple eigenvalue will appear as a simple zero of the minimal polynomial and every multiple eigenvalue of **A** will appear as a zero (perhaps with reduced multiplicity) of the minimal polynomial.

The relationship between the system characteristic roots and stability is established by the following basic theorem.

Theorem 8.1. The system $\dot{\mathbf{x}} = A\mathbf{x}$ is asymptotically stable if and only if *all* the system characteristic roots have negative real parts.

The proof of this theorem follows directly from (8.5) and is left as an exercise (see Problems).

The characteristic equation can be formed by actually computing the determinant of $(s\mathbf{I}-\mathbf{A})$, as the basic definition of (8.4) suggests. Another method that is often useful is to apply Mason's rule to an all-integrator transfer-function block diagram of the system, with initial conditions shown as inputs. (An all-integrator diagram can be obtained easily from a simulation diagram, as was discussed in Chapter 6.) Then, because

$$\mathbf{X}(s) = \mathbf{\Phi}(s)\mathbf{x}_0, \tag{8.12}$$

we may interpret $\mathbf{\Phi}(s)$ as the transfer-function matrix from $\mathbf{x}_0$ to $\mathbf{X}(s)$. This transfer-function matrix may be found using Mason's rule, as was discussed in Chapter 6. All entries in $\mathbf{\Phi}(s)$ contain the same $\Delta(s)$ from Mason's rule:

$$\Phi_{ij}(s) = \frac{\Sigma P_k{}^{ij}\Delta_k{}^{ij}}{\Delta} \tag{8.13}$$

Since, in general, the numerator and denominator of the Mason rule expression are not polynomials, but rather ratios of polynomials, we may write

$$\Phi_{ij}(s) = \frac{\dfrac{N_1(s)}{D_1(s)}}{\dfrac{N_2(s)}{D_2(s)}} = \frac{N_1(s)D_2(s)}{N_2(s)D_1(s)}, \tag{8.14}$$

where N_1, D_1, N_2, D_2 are polynomials defined by

$$\sum P_k{}^{ij}\Delta_k{}^{ij} = \frac{N_1(s)}{D_1(s)} \tag{8.15}$$

and

$$\Delta = \frac{N_2(s)}{D_2(s)}. \tag{8.16}$$

Since blocks having transfer functions of $1/s$ are the only dynamic elements, D_2 and D_1 will have the form

$$\begin{aligned} D_2 &= s^n \\ D_1 &= s^m, \end{aligned} \tag{8.17}$$

where $m \leqslant n$ (or else an *all-integrator* diagram could not be drawn). Thus D_1 will disappear because of cancellations, and

$$\Phi_{ij}(s) = \frac{s^{n-m}N_1(s)}{N_2(s)}. \tag{8.18}$$

There may be further cancellations between N_1 and N_2, but these correspond to cancellations between numerator and denominator of (8.2). We may now state that

$$\det(s\mathbf{I}-\mathbf{A}) = N_2(s).$$

Thus the zeros of $s^n\Delta(s)$ are the characteristic roots of the system and must lie in the left half of the s-plane (LHP) for asymptotic stability.

We have now interpreted the stability question, expressed initially as a time-domain definition, in terms of the location of the characteristic roots in the s-plane. The location of these roots can be found by actually solving (by hand or computer) for the roots of the characteristic equation. This direct solution is tedious or even impossible by hand for higher-order systems, and even a computer solution becomes unwieldy if not all system parameters are given numerical values (which is the case in system *design*, for example). The following three sections of this chapter are devoted to a discussion of several stability techniques that do not require the direct calculation of the system characteristic roots.

8.2 Routh Stability Test

We now describe the *Routh test* for determining whether an equation such as the system characteristic equation has any right-half-plane (RHP) roots. The test will be stated here without proof. If $\mathbf{A}$ is an $n\times n$ matrix, the characteristic polynomial is of the nth degree, and the characteristic equation may be written in the form

$$s^n + a_{n-1}s^{n-1} + \cdots + a_0 = 0. \tag{8.19}$$

The first step is to arrange the coefficients of (8.19) in two rows where the first row contains the first, third, fifth, ... coefficients and the second row contains the second, fourth, sixth, ... coefficients.

$$\begin{array}{llllll} \textit{Row 1} & 1 & a_{n-2} & a_{n-4} & a_{n-6} & \cdots \\ \textit{Row 2} & a_{n-1} & a_{n-3} & a_{n-5} & a_{n-7} & \cdots \end{array}$$

Now, form a third row

$$b_{n-1} \quad b_{n-3} \quad b_{n-5} \quad b_{n-7} \quad \cdots$$

where

$$b_{n-1} = -\frac{1}{a_{n-1}} \begin{vmatrix} 1 & a_{n-2} \\ a_{n-1} & a_{n-3} \end{vmatrix} \tag{8.20}$$

$$b_{n-3} = -\frac{1}{a_{n-1}} \begin{vmatrix} 1 & a_{n-4} \\ a_{n-1} & a_{n-5} \end{vmatrix} \tag{8.21}$$

$$b_{n-5} = -\frac{1}{a_{n-1}} \begin{vmatrix} 1 & a_{n-6} \\ a_{n-1} & a_{n-7} \end{vmatrix} \tag{8.22}$$

Next, form a fourth row

$$c_{n-1} \quad c_{n-3} \quad c_{n-5} \quad c_{n-7} \quad \cdots$$

where

$$c_{n-1} = -\frac{1}{b_{n-1}} \begin{vmatrix} a_{n-1} & a_{n-3} \\ b_{n-1} & b_{n-3} \end{vmatrix}$$

$$c_{n-3} = -\frac{1}{b_{n-1}} \begin{vmatrix} a_{n-1} & a_{n-5} \\ b_{n-1} & b_{n-5} \end{vmatrix} \tag{8.23}$$

Continue this procedure of forming a row from the preceding two rows until only zero elements are obtained. (Premature appearance of zeros will be discussed below.)

The Routh test is the following: all the roots of (8.17) have negative real parts if and only if the elements in the first column of the array

$$\begin{matrix} 1 & a_{n-2} & a_{n-4} & a_{n-6} & \cdots \\ a_{n-1} & a_{n-3} & a_{n-5} & a_{n-7} & \cdots \\ b_{n-1} & b_{n-3} & b_{n-5} & b_{n-7} & \cdots \\ c_{n-1} & c_{n-3} & c_{n-5} & c_{n-7} & \cdots \end{matrix} \tag{8.24}$$

have the same algebraic sign. The number of roots in the right half plane is equal to the number of alternations of sign in the sequence 1, $a_{n-1}, b_{n-1}, c_{n-1}, \ldots$.

Example 8-2. Let us determine whether the equation

$$s^4 + 3s^3 + 8s^2 + s + 9 = 0 \tag{8.25}$$

has roots in the right half plane or not.

First we form the two rows

$$\begin{matrix} 1 & 8 & 9 \\ 3 & 1 & \end{matrix}$$

Then we form the third row

$$b_3 \quad b_1$$

where

$$b_3 = -\frac{1}{3}\begin{vmatrix} 1 & 8 \\ 3 & 1 \end{vmatrix} = \frac{23}{3} \tag{8.26}$$

and

$$b_1 = -\frac{1}{3}\begin{vmatrix} 1 & 9 \\ 3 & 0 \end{vmatrix} = 9. \tag{8.27}$$

Next we form the fourth row

$$c_3 \quad c_1$$

where

$$c_3 = -\frac{1}{23/3}\begin{vmatrix} 3 & 1 \\ \frac{23}{3} & 9 \end{vmatrix} = -\frac{58}{23} \tag{8.28}$$

and

$$c_1 = -\frac{1}{23/3}\begin{vmatrix} 3 & 0 \\ \frac{23}{3} & 0 \end{vmatrix} = 0. \tag{8.29}$$

Finally, the last row is

$$d_3 = \frac{1}{58/23}\begin{vmatrix} \frac{23}{3} & 9 \\ -\frac{58}{23} & 0 \end{vmatrix} = 9. \tag{8.30}$$

Thus the Routh-Hurwitz array is

$$\begin{array}{ccc} 1 & 8 & 9 \\ 3 & 1 & \\ \frac{23}{3} & 9 & \\ -\frac{58}{23} & & \\ 9 & & \end{array} \tag{8.31}$$

We observe two sign changes in the first column, so there must be exactly two RHP roots of (8.25). ▲

Example 8-3. Consider the transfer function block diagram for a feedback system shown in Figure 8.1. From an examination of the diagram, we obtain the following equations.

$$\dot{x}_1 = -2x_1 + x_2 \tag{8.32}$$

$$\dot{x}_2 = 6x_1 - x_2 + u \tag{8.33}$$

$$y = -3x_1 + x_2, \tag{8.34}$$

or

$$\mathbf{A} = \begin{bmatrix} -2 & 1 \\ 6 & -1 \end{bmatrix}. \tag{8.35}$$

The characteristic equation is

$$\det\,[s\mathbf{I} - \mathbf{A}] = \begin{vmatrix} s+2 & -1 \\ -6 & s+1 \end{vmatrix} = (s+2)(s+1) - 6 = s^2 + 3s - 4 \tag{8.36}$$

The array is

$$\begin{matrix} 1 & -4 \\ 3 & \\ -4 & \end{matrix} \tag{8.37}$$

Thus there is one RHP root, as is evident from factoring (8.36). The characteristic equation can also be formed using Mason's rule:

$$\Delta(s) = 1 + \frac{1}{s} + \frac{2}{s} - \frac{6}{s^2} + \frac{2}{s^2}$$

$$= \frac{s^2 + 3s - 4}{s^2}. \tag{8.38}$$

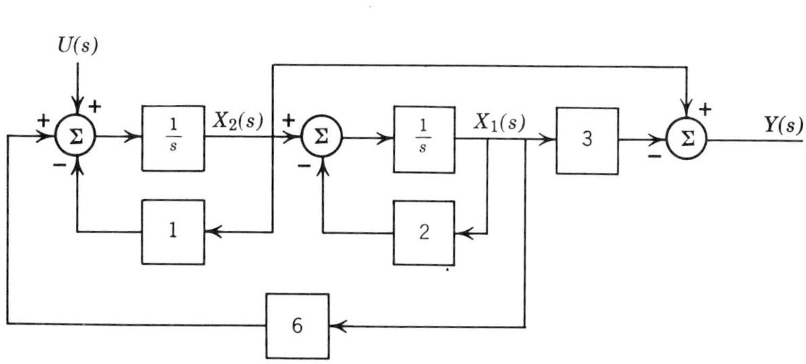

Figure 8.1 Block diagram for Example 8-3.

From the discussion in Section 8.1, the characteristic equation is the numerator of $\Delta(s)$, which is (8.36), as before. Notice that the inputs and outputs have nothing to do with stability, as we have defined system stability in terms of the zero-input response of the state variables. ▲

Two special cases may occur in forming the Routh array.

Case I. The first term in a row is zero, but not all the other terms in that row are zero. The array may be computed by either of the following procedures:

(a) Replace the zero by a small quantity ϵ. (The sign of ϵ does not affect stability.) Complete the array as before, then let $\epsilon \to 0$.

(b) Replace s by z^{-1} in the characteristic equation. Then apply the Routh test to the new characteristic equation.

Example 8-4. Consider the characteristic equation

$$s^4 + s^3 + 2s^2 + 2s + 3 = 0. \tag{8.39}$$

The array is

$$\begin{array}{ccc} 1 & 2 & 3 \\ 1 & 2 & \\ \epsilon & 3 & \\ 2-\dfrac{3}{\epsilon} & & \\ 3 & & \end{array} \tag{8.40}$$

For small ϵ, $-3/\epsilon$ dominates the fourth entry in the first column. Thus we have two sign changes (independent of the sign of ϵ) and therefore, two RHP roots.

Alternatively,

$$\frac{1}{z^4} + \frac{1}{z^3} + \frac{2}{z^2} + \frac{2}{z} + 3 = 0, \tag{8.41}$$

or

$$3z^4 + 2z^3 + 2z^2 + z + 1 = 0 \tag{8.42}$$

Then

$$\begin{array}{ccc} 3 & 2 & 1 \\ 2 & 1 & \\ \frac{1}{2} & 1 & \\ -3 & & \\ 1 & & \end{array} \tag{8.43}$$

There are two sign changes in the first columns and, hence, two RHP roots. ▲

The second special case occurs when an entire row is zero. A zero row indicates roots symmetric about the origin, including $j\omega$-axis roots, real roots of equal magnitude but opposite sign, roots with quadrantal symmetry, etc. The procedure is illustrated by the following example.

Example 8-5. Consider the characteristic equation

$$s^4+3s^3+4s^2+6s+4=0. \tag{8.44}$$

Beginning the array,

$$\begin{matrix} 1 & 4 & 4 \\ 3 & 6 & \\ 2 & 4 & \\ 0 & & \end{matrix} \tag{8.45}$$

We have encountered a row of zeros. We form an *auxiliary equation* using as coefficients the entries 2 and 4 in the row just preceding the row of zeros. In general, the auxiliary equation contains only *even* powers of s and is of order $(n-j+2)$, where n is the order of the characteristic equation, and j is the row of zeros. Here

$$n-j+2=4-4+2=2. \tag{8.46}$$

So the auxiliary equation is

$$2s^2+4=0. \tag{8.47}$$

The roots of this equation, $s=\pm j\sqrt{2}$, are roots of the characteristic equation. The array is continued by inserting the coefficients of the derivative of the auxiliary equation in place of the row of zeros. Thus

$$\begin{matrix} 1 & 4 & 4 \\ 3 & 6 & \\ 2 & 4 & \\ 4 & 4 & \\ 2 & & \\ 4 & & \end{matrix} \tag{8.48}$$

The first column is inspected as before for sign changes indicating RHP poles. Thus, (8.44) has two $j\omega$-axis roots and two LHP roots. The system is not asymptotically stable. ▲

We now state the general rule for Case II.

Case II. Entire row of zeros. Suppose the jth row is a row of zeros. Form the auxiliary equation using the entries of the $(j-1)$st row as coefficients. The auxiliary equation contains only even powers of s

and is of order $(n-j+2)$. The roots of the auxiliary equation are system eigenvalues. Differentiate the auxiliary equation and use the coefficients of the differentiated equation in the array in place of the jth row. Complete the Routh test as before. This procedure is repeated if a row of zeros is encountered again in forming the array.

We complete this section with reaction-wheel control system example.

Example 8.6. The all-integrator transfer-function block diagram for a single-axis reaction-wheel attitude-control system was developed in Chapter 6, and is repeated here in modified form as Figure 8.2. Figure 8.2 also shows *feedback* from $\theta(t)$ and $\dot{\theta}(t) = \omega(t)$. That is, the armature voltage $v_a(t)$ is formed from a linear combination of quantities proportional to the attitude and attitude rate. We suppose the parameters K_1, K_2, K_3, and K_4 are known, and we wish to study the system stability as a function of the "feedback gains" k_1 and k_2.

This system has five loops, with loop-transfer functions

$$-\frac{K_2K_3}{s},\ \frac{k_2K_4}{s},\ -\frac{k_2K_1K_2K_3K_4}{s^2},\ -\frac{k_1K_1K_2K_3K_4}{s^2},\ \frac{k_1K_4}{s^2}. \tag{8.49}$$

From Mason's rule,

$$\Delta(s) = 1+\frac{K_2K_3}{s}-\frac{k_2K_4}{s}+\frac{k_2K_1K_2K_3K_4}{s^2}-\frac{k_1K_4}{s^2}+\frac{k_1K_1K_2K_3K_4}{s^3}$$
$$+\left(-\frac{K_2K_3}{s}\right)\left(\frac{k_2K_4}{s}\right)+\left(-\frac{K_2K_3}{s}\right)\left(\frac{k_1K_4}{s^2}\right). \tag{8.50}$$

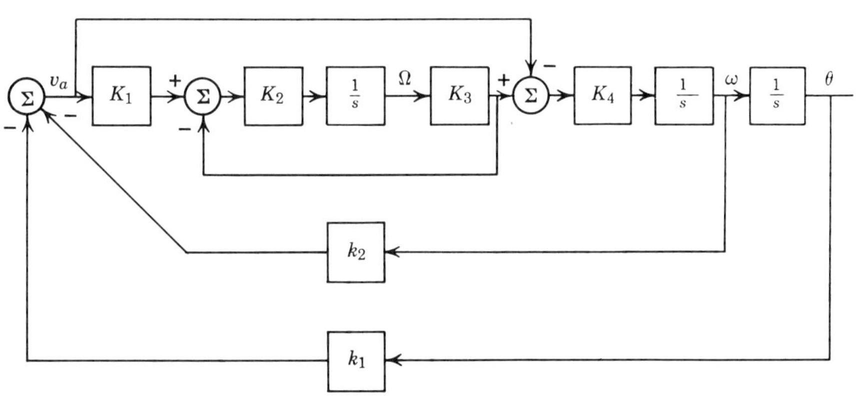

Figure 8.2 Feedback control of reaction-wheel attitude control system.

Thus the characteristic equation is

$$s^3 + [K_2K_3 - k_2K_4]s^2 + [k_2K_1K_2K_3K_4 - k_1K_4 - k_2K_2K_3K_4]s + [k_1K_1K_2K_3K_4 - k_1K_2K_3K_4] = 0. \tag{8.51}$$

This equation has the form

$$s^3 + As^2 + Bs + C = 0. \tag{8.52}$$

The Routh array is

$$\begin{array}{cc} 1 & B \\ A & C \\ \dfrac{AB-C}{A} & \\ C & \end{array} \tag{8.53}$$

Thus, for asymptotic stability, we require (K_1, K_2, K_3, K_4 all positive)

$$A = K_2K_3 - k_2K_4 > 0 \tag{8.54}$$

or

$$k_2 < \frac{K_2K_3}{K_4}; \tag{8.55}$$

$$C = k_1[K_1K_2K_3K_4 - K_2K_3K_4] > 0 \tag{8.56}$$

or

$$k_1[K_1 - 1] > 0, \tag{8.57}$$

which implies

$$k_1 > 0 \text{ if } K_1 > 1 \text{ } or \text{ } k_1 < 0 \text{ if } K_1 < 1. \tag{8.58}$$

Finally, we require

$$AB - C > 0, \tag{8.59}$$

or

$$B = \{k_2[K_1K_2K_3K_4 - K_2K_3K_4] - k_1K_4\} > \frac{C}{A}. \blacktriangle \tag{8.60}$$

8.3 Root Locus Method

The Routh test indicates whether the system is stable or not, but does not give any additional information about the response. Several graphical methods have been developed that not only answer the stability question but yield additional information about the response as well. We discuss two of these methods in this section and the next.

The root locus method was developed by W. R. Evans (American) in 1949. The method indicates the locus in the s-plane of the roots of the characteristic equation (system eigenvalues) as a linear function of a parameter K. The essential ideas of the procedure are indicated by the following example.

Example 8-7. Consider a system represented by the transfer-function block diagram of Figure 8.3. The characteristic equation is

$$s^2 + 100s + 500K = 0. \tag{8.61}$$

This can be rewritten as

$$1 + \frac{500K}{s(s+100)} = 0. \tag{8.62}$$

This equation is of the form

$$1 + KG(s) = 0. \tag{8.63}$$

Some special points on the locus can be found immediately. When $K = 0$, the roots of the characteristic equation are the roots of

$$s(s+100) = 0, \tag{8.64}$$

that is, they are the finite *poles* of $G(s)$. These are indicated by X in Figure 8.4. Furthermore, as $K \to \infty$, the roots of (8.63) are given by the zeros of $G(s)$ which, in this case, are at $s = \infty$.

The locus can be completed by noting that, from (8.63),

$$KG(s) = -1. \tag{8.65}$$

Consider a "test point" $s = s_0$. The test point lies on the locus (that is, it is a system eigenvalue) if and only if

$$KG(s_0) = -1. \tag{8.66}$$

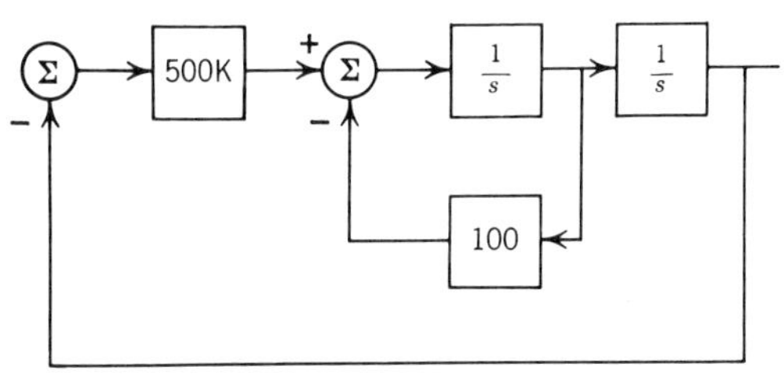

Figure 8.3 System for Example 8-7.

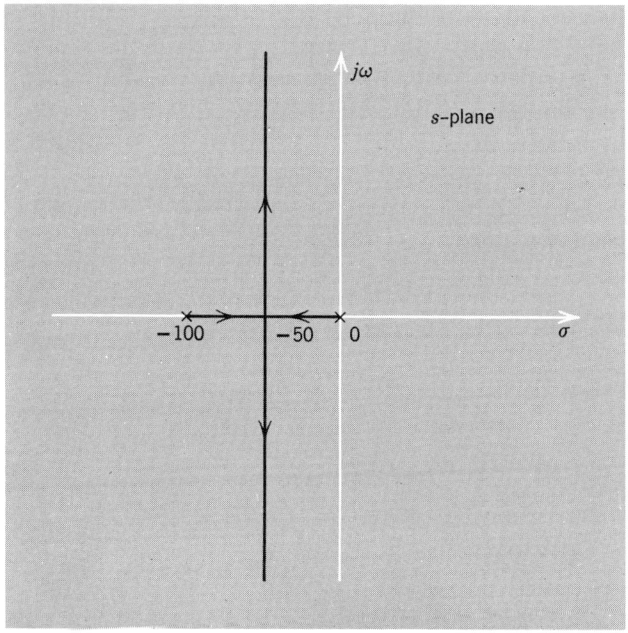

Figure 8.4 Root locus for Example 8-7.

This means that all points $s = s_0$, for which the total phase angle of $KG(s_0)$ is 180°, are points on the locus. The magnitude of K corresponding to that point is that value of K making the magnitude of the left-hand side of (8.66) equal to unity. Points on the locus can be found graphically by angle measurement with a protractor. The corresponding magnitude of K can be found from length measurement, using dividers, for example. Figure 8.5 illustrates these calculations for this example. The completed locus is shown in Figure 8.4. Notice that, as K increases, the two roots move along the real axis, meeting at $s = -50$. The roots then become complex and move along the vertical line $s = -50$. (Since the characteristic equation has real coefficients, complex roots occur in conjugate pairs. Thus the locus is symmetrical about the real axis.) ▲

Thus, drawing the root locus is a trial-and-error procedure.[1] Fortunately, several rules have been developed for quickly sketching the approximate locus. We discuss the most useful of these rules now. We assume $K \geq 0$. The rules are easily modified for $K < 0$.

[1]A circular protractor and a logarithmic magnitude scale comprise the *Spirule*, a commercially available aid for root locus construction.

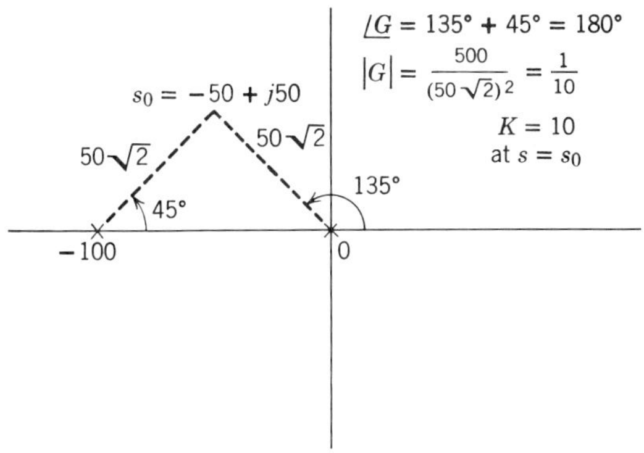

Figure 8.5 Calculation of magnitude and angle at a test point.

Rule 1. Continuous curves, which comprise the root locus for (8.63), start ($K=0$) at each pole of $G(s)$ and terminate on zeros of $G(s)$. (Poles and zeros may be at infinity.) ▲

PROOF. Let

$$1+KG(s)=1+K\frac{N(s)}{D(s)}=0,$$

where $N(s)$ and $D(s)$ are, respectively, the numerator and denominator polynomials of $G(s)$. So

$$D(s)+KN(s)=0. \tag{8.67}$$

For $K=0$, $D(s)=0$ unless $N(s)\to\infty$. The case $N(s)\to\infty$ is included in the *asymptotic behavior* of the locus, to be discussed in later rules. Thus the roots start at the poles of $G(s)$. For K large, we write

$$\frac{D(s)}{K}+N(s)=0. \tag{8.68}$$

As $K\to\infty$, $N(s)\to 0$ unless $D(s)\to\infty$. The case $D(s)\to\infty$ is part of the asymptotic behavior of the locus to be discussed in later rules. ▲

Rule 2. The locus includes points on the real axis to the left of an *odd* number of real-axis poles and zeros of $G(s)$. ▲

PROOF. The validity of this rule follows from consideration of the angle contributions of poles and zeros of $G(s)$. A typical case of a real-axis test point $s=s_0$ is illustrated in Figure 8.6. It is clear that complex

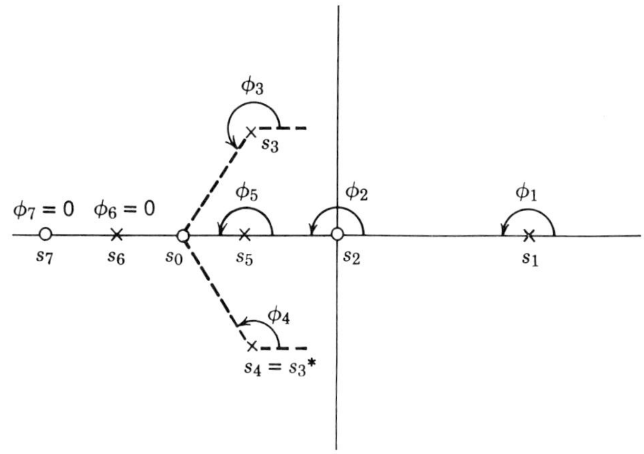

Figure 8.6 $\angle G(s_0) = \phi_2 + \phi_7 - \phi_1 - \phi_3 - \phi_4 - \phi_5 - \phi_6 = 180° + 0° - 180° - \phi_3 - \phi_4 - 180° - 0° = 180°$.

poles and zeros contribute multiples of 360° to the angle of $G(s_0)$, since complex poles and zeros occur in conjugate pairs. Zeros and poles to the right of a test point contribute ±180°. If an odd number of poles and zeros lie to the right, the angle of $G(s_0)$ will be an *odd* multiple of 180°, which proves the rule. ▲

Rule 3. Let $G(s) = N(s)/D(s)$, where $N(s)$ is an mth-order polynomial and $D(s)$ is an nth-order polynomial. For large K, m of the loci tend toward the finite zeros of $G(s)$ (see Rule 1) and $(n-m) = r$ loci tend toward infinity along *asymptotes*. The angles of the asymptotes are given by

$$\frac{(2k+1)180°}{n-m}. \quad ▲ \tag{8.69}$$

PROOF. For very large K, either $N(s) \to 0$ (see Rule 1) or $s \to \infty$ and

$$G(s) = \frac{N(s)}{D(s)} \sim \frac{1}{s^{n-m}} = \frac{1}{s^r}. \tag{8.70}$$

Let $\angle s = \theta$. Then $\angle s^r = r\theta$. Thus $r\theta = (2k+1)180°$ for $K \to \infty$ and $s \to \infty$. This proves the rule. ▲

Rule 4. The center of the asymptotes is given by

$$\alpha = \frac{(\text{sum of finite poles}) - (\text{sum of finite zeros})}{n-m}. \quad ▲ \tag{8.71}$$

PROOF. Figure 8.7 represents the approximation of a root locus by straight-line asymptotes originating at $s = \alpha$. This approximate locus is the *exact* locus for

$$G_a(s) = \frac{1}{(s-\alpha)^r}. \tag{8.72}$$

In our case,

$$G(s) = \frac{N(s)}{D(s)} = \frac{s^m + a_{m-1}s^{m-1} + \cdots + a_1 s + a_0}{s^n + b_{n-1}s^{n-1} + \cdots + b_0} \tag{8.73}$$

By division, we may write (8.73) as

$$G(s) = [s^r - (a_{m-1} - b_{n-1})s^{r-1} + \cdots]^{-1}, \tag{8.74}$$

and (8.72) may be written as

$$G(s) = [s^r - r\alpha s^{r-1} + \cdots]^{-1}. \tag{8.75}$$

Thus for large s, we require

$$r\alpha = (a_{m-1} - b_{n-1}) \tag{8.76}$$

or

$$\alpha = \frac{-b_{n-1} + a_{m-1}}{r}. \tag{8.77}$$

But, from the theory of equations,

$$b_{n-1} = -(\text{sum of finite poles of } G)$$
$$a_{m-1} = -(\text{sum of finite zeros of } G).$$

Thus the rule is proved. ▲

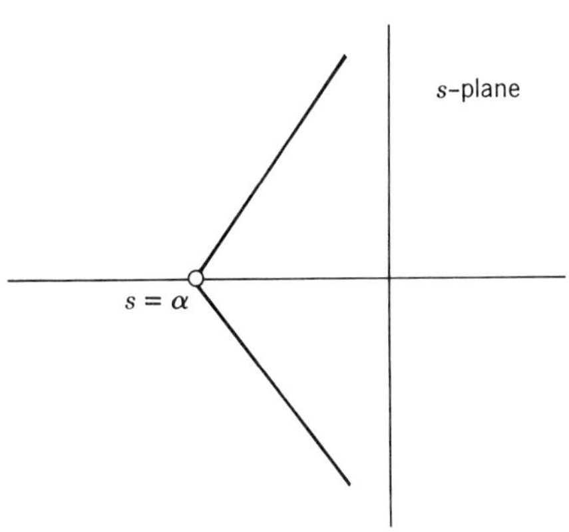

Figure 8.7 Asymptotic locus, $r = 3$.

A few definitions are needed in order to state the next rule. Points for which

$$\frac{dG(s)}{ds} = 0 \tag{8.78}$$

are called *saddle points*. Points where multiple roots of the characteristic equation occur are *crossings* of the root locus.

Rule 5. All locus crossings occur at saddle points, and all saddle points lying on the locus are locus crossings. ▲

PROOF. We first show that all locus crossings (that is, points at which multiple roots occur) are saddle points. Let k roots occur at $s = s_1$. Then

$$1 + KG(s) = (s - s_1)^k F(s) = 0. \tag{8.79}$$

Thus

$$\frac{d}{ds}[1 + KG(s)] = (s - s_1)^k \frac{dF(s)}{ds} + k(s - s_1)^{k-1} F(s). \tag{8.80}$$

If $s = s_1$, and $k > 1$, then

$$\left.\frac{dG(s)}{ds}\right|_{s=s_1} = 0, \tag{8.81}$$

and thus $s = s_1$ is a saddle point.

We next consider a saddle point $s = s_1$ lying on the locus. That is,

$$\left.\frac{dG(s)}{ds}\right|_{s=s_1} = 0 \tag{8.82}$$

and

$$[1 + KG(s)]_{s=s_1} = 0. \tag{8.83}$$

Since $G(s)$ is rational, (8.83) implies

$$1 + KG(s) = (s - s_1)^k F(s) \tag{8.84}$$

for some positive integer k. Then

$$\frac{d}{ds}[1 + KG(s)] = k(s - s_1)^{k-1} F(s) + (s - s_1)^k \frac{dF(s)}{ds}. \tag{8.85}$$

Suppose $k=1$. Then (8.85) yields

$$\frac{d}{ds}[1+KG(s)]|_{s=s_1}=K\frac{dG(s)}{ds}\bigg|_{s=s_1}=F(s_1)\neq 0. \tag{8.86}$$

Thus $s=s_1$ cannot be a saddle point. But this contradicts the hypothesis. Therefore $k\neq 1$. Suppose $k>1$. Then, from (8.85),

$$K\frac{dG(s)}{ds}\bigg|_{s=s_1}=0. \tag{8.87}$$

Thus k must be an integer greater than one, that is, $s=s_1$ must be a *multiple* root of the characteristic equation. ▲

Let us apply these rules to the system of Example 8.7.

Example 8.8.

$$1+KG(s)=1+K\frac{500}{s(s+100)}. \tag{8.88}$$

Rule 1. Locus starts at $s=0$, $s=-100$. Locus terminates at ∞ (all branches).

Rule 2. Locus includes the segment of the real axis between $s=0$ and $s=-100$.

Rule 3. $n-m=2-0=2$. Therefore,

$$\theta=\frac{(2k+1)180^\circ}{2}=(2k+1)90^\circ=90^\circ,-90^\circ. \tag{8.89}$$

Rule 4.

$$\alpha=\frac{(0-100)-0}{2}=-50. \tag{8.90}$$

See Figure 8.4 for illustration of these rules. ▲

The techniques of rapid locus sketching are mastered through practice. We present one further example here (see Figure 8.8).

Example 8-9. Consider

$$1+\frac{K}{s(s+1)(s+5)}=0. \tag{8.91}$$

Application of the rules is illustrated by Figure 8.9.

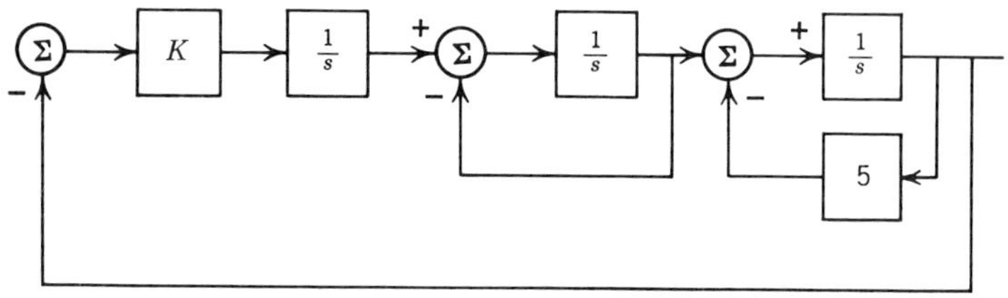

Figure 8.8 System for Example 8-9.

Rule 1. Locus starts at $s = 0, -1, -5$.

Rule 2. The real axis between 0 and -1 and between -5 and ∞ is part of the locus.

Rule 3.

$$\theta = \frac{(2k+1)180°}{3} = 60°, 180°, -60°.$$

Rule 4.

$$\alpha = \frac{0-1-5}{3} = \frac{-6}{3} = -2.$$

Rule 5.

$$\frac{d}{ds}\left[\frac{1}{s^3+6s^2+5s}\right] = \frac{3s^2+12s+5}{-(s^3+6s^2+5s)^2} = 0.$$

Thus, possible intersections occur for

$$3s^2+12s+5=0$$
$$(s+3.5)(s+0.5)=0.$$

Of these, $s=-0.5$ is on the locus. Thus the locus can be sketched as shown in Figure 8.9.

Crossings of the $j\omega$-axis can be determined from the Routh test. Applying the test to (8.91), multiplied out,

$$s^3+6s^2+5s+K=0, \tag{8.92}$$

we obtain the array

$$\begin{array}{cc} 1 & 5 \\ 6 & K \\ \dfrac{K-30}{6} & \\ K & \end{array} \tag{8.93}$$

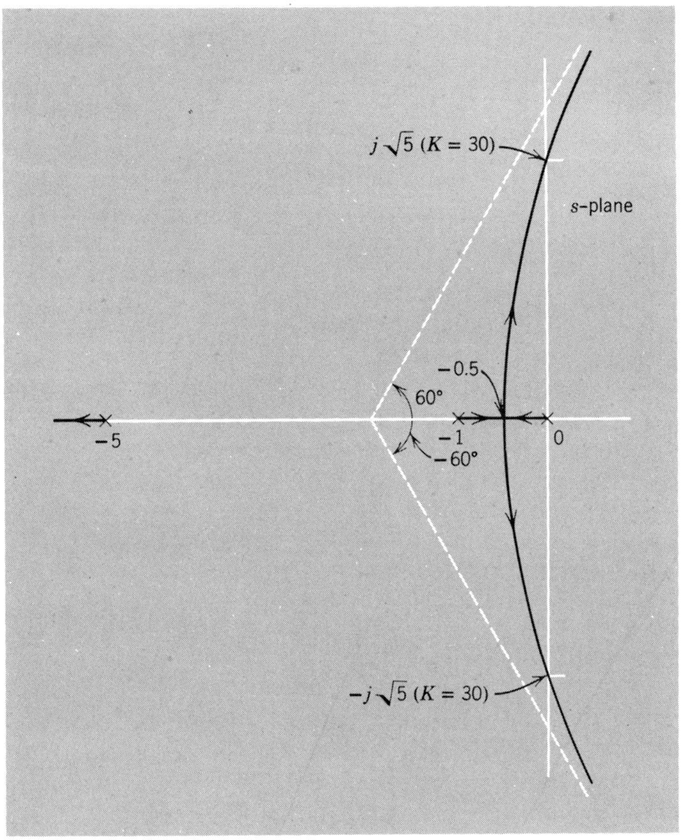

Figure 8.9 Root locus for Example 8-9.

Thus the system is stable for

$$0 < K < 30. \tag{8.94}$$

For $K = 30$, there is a zero row in the array, the third row. Forming the auxiliary equation,

$$6s^2 + 30 = 0, \tag{8.95}$$

or

$$s = \pm j\sqrt{5}. \tag{8.96}$$

Thus we have j-axis roots at $\pm j\sqrt{5}$ for $K = 30$. For $K > 30$, the system is unstable. ▲

8.4 Nyquist Stability Criterion

In this section, we discuss a second widely used stability test—the Nyquist criterion.

At the heart of the test is a *mapping* from the s-plane to the $F(s)$ plane. For each value s_i of the complex number s, there is a corresponding value of the complex number $F(s_i)$. Tracing the contour C_1 in the s-plane produces a corresponding contour C_2 in the $F(s)$ plane, as indicated in Figure 8.10. We say $F(s)$ constitutes a *mapping* from the s-plane to the $F(s)$-plane.

An interesting feature of this mapping is whether the origin in the $F(s)$-plane is encircled or not. Suppose

$$F(s) = \frac{K(s-z_1)}{(s-p_1)(s-p_2)}. \tag{8.97}$$

Consider a contour C_1 in the s-plane encircling the zero z_1, but not the poles p_1 or p_2 (see Figure 8.11).

Suppose we start at $s = s_a$ and traverse C_1 in a clockwise (CW) direction, returning to s_a. The phasor $(s-z_1)$ makes one CW revolution. The phasors $(s-p_1)$ and $(s-p_2)$ make no *net* revolutions. From (8.97), the phase angle of F is the angle of $(s-z_1)$ minus the angles of $(s-p_1)$ and $(s-p_2)$. Thus if $(s-z_1)$ undergoes a clockwise rotation of 360°, the phasor $F(s)$ will likewise make one clockwise rotation in the $F(s)$ plane, and C_2 will encircle the origin (see Figure 8.12). In a similar manner, it can be shown that if a single pole is enclosed by an s-plane contour, then $F(s)$ will undergo one counterclockwise (CCW) rotation and C_2 encircles the origin. Moreover, if neither a pole nor a zero is enclosed, no rotation will occur in the $F(s)$ plane, and thus the origin will not be encircled.

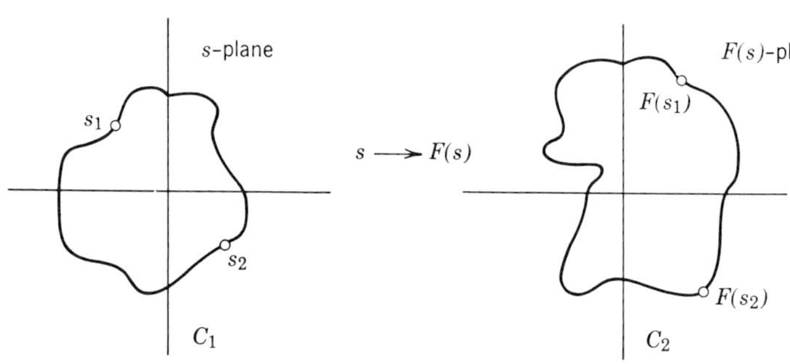

Figure 8.10 Mapping of contour in s-plane to contour in $G(s)$ plane.

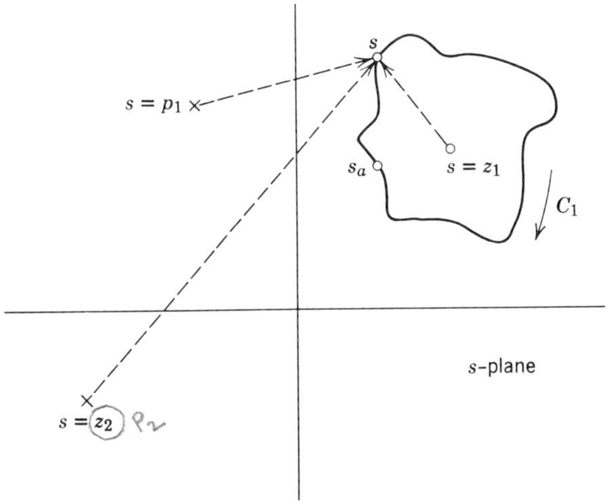

Figure 8.11 Contour encircling a zero.

These observations are summarized in the *Principle of the Argument*:

A CW encirclement of a region in the s-plane results in $(m-n)$ CW encirclements of the origin in the $F(s)$-plane, where m is the number of zeros in the s-plane region, and n is the number of poles in the s-plane region.

$$\text{CW encirclements of origin in } F\text{-plane} = m - n. \tag{8.98}$$

In this statement negative CW encirclements, corresponding to $n > m$, means CCW encirclement.

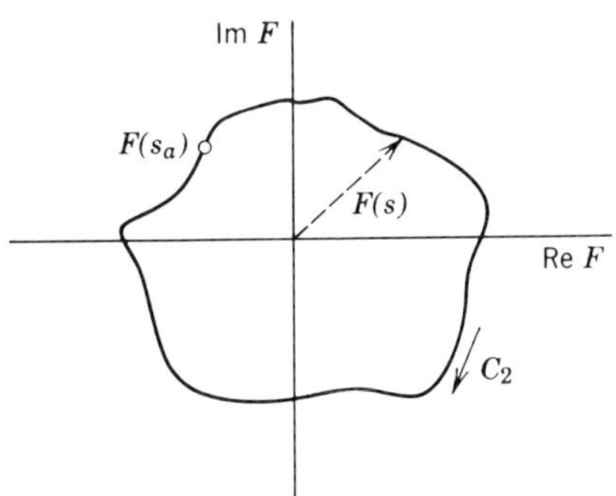

Figure 8.12 Encirclement of the origin in the $F(s)$ plane.

The Principle of the Argument can be used to investigate system stability by encircling the *entire right-half* s-plane with the contour C_1 and by using as a mapping the rational function

$$F(s) = 1 + KG(s). \tag{8.99}$$

Then, by (8.98), the number of CW encirclements of the origin in the $1+KG$ plane equals the number of RHP system eigenvalues minus the number of RHP poles of F. The poles of F are the poles of G. Thus, if we know a priori the number of RHP poles of G, the stability of a system may be found by the Principle of the Argument.

We note from (8.99) that the $KG(s)$ and the $1+KG(s)$ planes differ only by a translation. In particular, the origin in the $1+KG(s)$ plane corresponds to the point $-1+j0$ in the $KG(s)$. Thus, stability can be investigated by noting encirclements of the point $-1+j0$ in the $KG(s)$, rather than encirclements of the origin in the $1+KG$ plane. This modification is often useful in system analysis and design.

The process of encircling the entire RHP by a contour deserves careful discussion, as the region encircled is a semi-infinite region. We shall assume for our purposes here that $G(s) \to 0$ as $s \to \infty$. This allows us to "encircle" a finite portion of the RHP by a semicircular contour, as shown in Figure 8.13, and then let $R \to \infty$ to enclose the entire RHP. Some technical difficulties arise if $G(s)$ has poles on the $j\omega$-axis. In this case, small detours in C_1 skirting these poles must be considered. It is customary, although not necessary, to detour into the RHP as shown in Figure 8.14.
An enlargement of the semicircular "detour" is shown in Figure 8.15. Near $s = s_a$, $KG(s)$ may be approximated by

$$KG(s) \sim \frac{K}{(s-s_a)^r}, \tag{8.100}$$

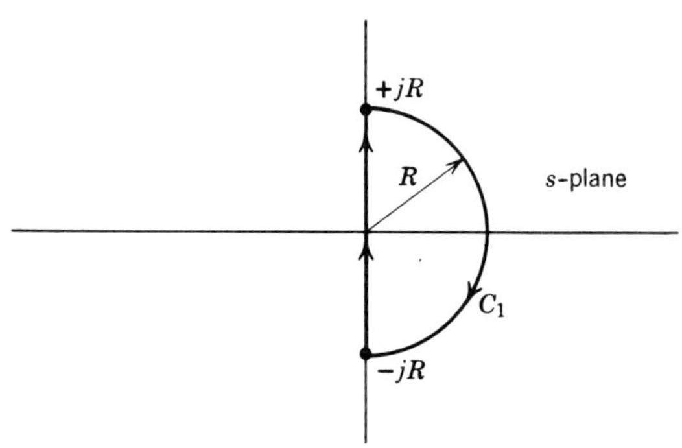

Figure 8.13 Contour for encircling RHP.

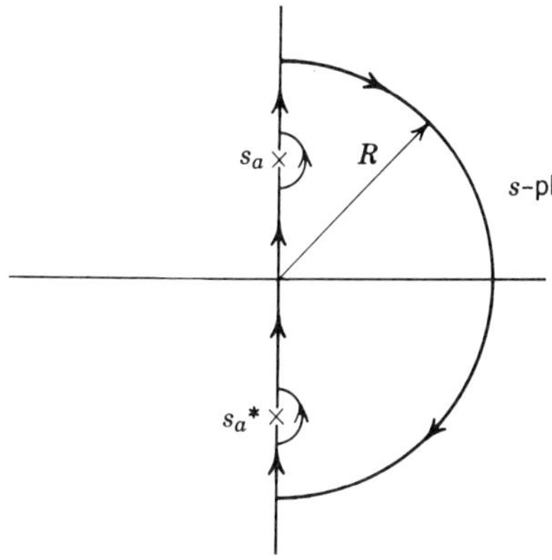

Figure 8.14 Contour for $j\omega$-axis poles.

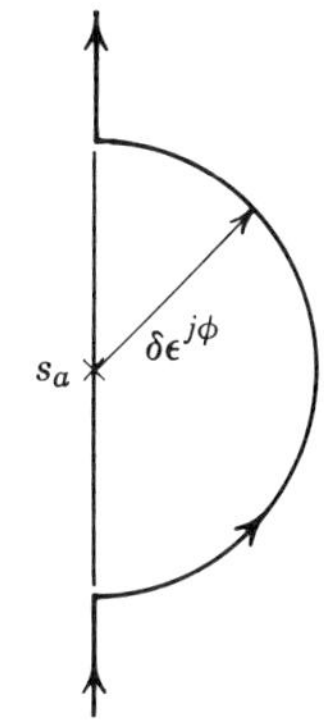

Figure 8.15 Enlargement of contour in vicinity of a $j\omega$-axis pole.

where r is the multiplicity of the pole. Since, from Figure 8.15,

$$s = s_a + \delta\epsilon^{j\phi}, \tag{8.101}$$

$$KG(s) \sim \frac{1}{\delta^r \epsilon^{jr\phi}} = \frac{1}{\delta^r}\epsilon^{-jr\phi}. \tag{8.102}$$

We now shrink the detour, with $\delta \to 0$ while ϕ goes from $-90°$ to $90°$. Thus, $G(s)$ becomes very large in magnitude and goes from $90r$ degrees to $-90r$ degrees in phase, that is, a $j\omega$-axis pole of $G(s)$ of multiplicity r results in r CW semicircles of infinite radius in the G plane. This situation is illustrated by the following example.

Example 8-10. Suppose

$$G(s) = \frac{1}{s(s+1)}. \tag{8.103}$$

Note that $G(s) \to 0$ for $s \to \infty$. There is a pole on the $j\omega$ axis ($s = 0$). Thus $G(j\omega) \to \infty$ for $\omega \to 0$. At $G(j\omega) = \infty$ there is one semicircle, as the phase angle of $G(j\omega)$ goes CW from $90°$ to $-90°$. The plot is shown in Figure 8.16. Since there are no encirclements of $-1+j0$, and no poles of $G(s)$ in RHP, a system with the characteristic equation

$$s(s+1) + K = 0 \tag{8.104}$$

is asymptotically stable. ▲

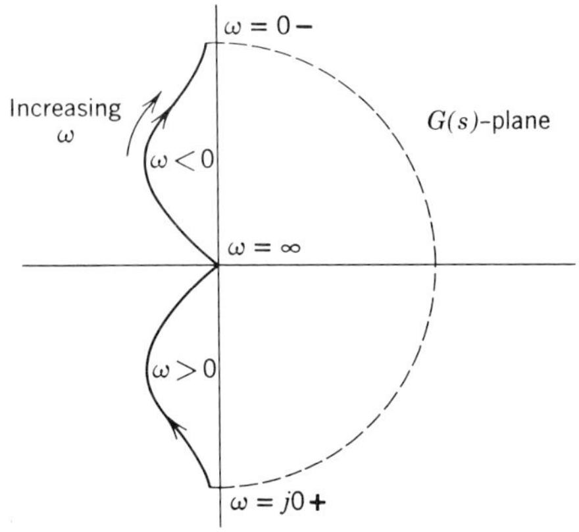

Figure 8.16 Nyquist plot for Example 8-10.

Sometimes it may appear to be difficult to determine whether a diagram encircles $-1+j0$. Consider Figure 8.17. Encirclements of $-1+j0$ can be determined easily by the following homespun approach. Imagine the $G(j\omega)$ plot formed by a string. Hold one of your fingers at $-1+j0$. If the string can be pulled away, the point $-1+j0$ is not encircled. If the string is wound around your finger, the number of turns is the number of encirclements. In Figure 8.17, there are no encirclements.

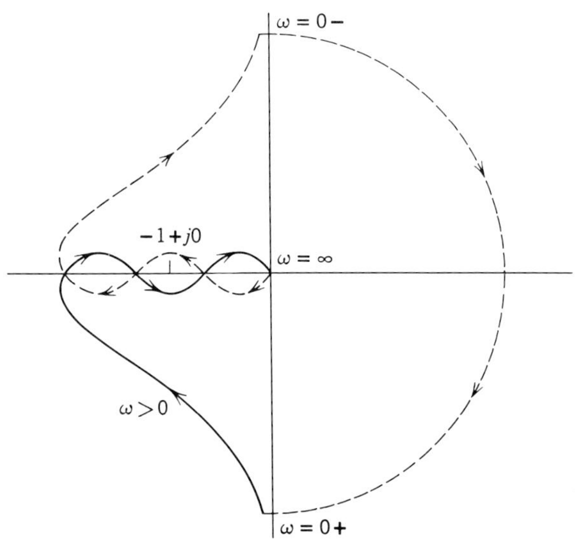

Figure 8.17 $G(j\omega)$ plot.

In summary, we state the *Nyquist stability criterion*: Consider a linear, lumped, time-invariant system whose characteristic roots are zeros of

$$1+KG(s)=0$$

with P poles of $G(s)$ in the RHP. The system is asymptotically stable if and only if the polar plot of $G(j\omega)$ encircles in a *counterclockwise* direction the point $-1+j0$ *exactly* P times.

8.5 Input-Output Stability

Thus far we have considered system stability only in terms of the zero-input response. Another important notion concerns stability of system outputs. These two stability concepts are different, in general. A system may have an output that does not grow with time and yet may be an unstable system in the sense of Definition 8.1. This situation is illustrated by the following example.

Example 8-11. Consider once again the system for Figure 8.1. The characteristic equation is

$$s^2+3s-4=(s-1)(s+4)=0. \tag{8.105}$$

The system is unstable. However, if we examine the transfer function from U to Y, an application of Mason's rule yields

$$\frac{Y}{U}=\frac{-\dfrac{3}{s^2}+\dfrac{1}{s}\left(1+\dfrac{2}{s}\right)}{1+\dfrac{1}{s}+\dfrac{2}{s}-\dfrac{6}{s^2}+\left(\dfrac{1}{s}\right)\left(\dfrac{2}{s}\right)}=\frac{s-1}{s^2+3s-4}=\frac{(s-1)}{(s-1)(s+4)}=\frac{1}{s+4}. \tag{8.106}$$

There is only one pole in the transfer function and the pole is a characteristic root. The other characteristic root, which happens to be in the right-half plane, is not a pole of the transfer function. Thus, if $u(t)$ is not itself growing with time, $y(t)$ will not grow with time. The system has a decreasing output response for a nongrowing input, although the zero-input response is increasing. In this section, we shall formally define an input-output stability notion, and explore some connections between input-output stability and zero input stability. ▲

To put this notion precisely, we need the concept of *norm*. Norm is a generalization of the notion of distance.

Definition 8-2. Norm. A norm of $\mathbf{x}$, denoted by $\|\mathbf{x}\|$, must satisfy all four of the following postulates:

(1) $\|\mathbf{x}\| \geq 0$ for all $\mathbf{x}$.
(2) $\|\mathbf{x}\| = 0$ if and only if $\mathbf{x} = \mathbf{0}$.
(3) $\|a\mathbf{x}\| = |a| \, \|\mathbf{x}\|$ for all scalars a and all $\mathbf{x}$.
(4) $\|\mathbf{x}+\mathbf{y}\| \leq \|\mathbf{x}\| + \|\mathbf{y}\|$ the triangle inequality. ▲

Example 8-12. Euclidean Norm. Consider a state vector having two components,

$$\mathbf{x} = \begin{bmatrix} x_1 \\ x_2 \end{bmatrix}.$$

The Euclidean norm for this case is defined by

$$\|\mathbf{x}\| = \sqrt{x_1^2 + x_2^2},$$

where the positive square root is taken. Notice that this is very similar to the distance from the origin to a point (x_1, x_2) in a two-dimensional Euclidean coordinate system. It is easy to verify that this proposed norm satisfies the conditions of Definition 8.2 (see Problems). ▲

Loosely speaking, a bounded time function is one that never becomes infinite. We now give a formal definition of boundedness.

Definition 8-3. A function $\mathbf{f}(t)$ is said to be bounded if there exists a constant M such that

$$\|\mathbf{f}(t)\| < M < \infty \tag{8.107}$$

for all t. ▲

Notice that an exponentially decaying time function $K\epsilon^{-\alpha t}$, $\alpha > 0$, is bounded by its initial value K. A time tunction consisting of a sum of decaying exponentials is also bounded.

Definition 8-4. A system with input $\mathbf{u}(t)$ and output $\mathbf{y}(t)$ is said to be *zero-state bounded-input bounded-output (BIBO) stable* if, for $\mathbf{x}_0 = \mathbf{0}$ and for all bounded inputs $\mathbf{u}(t)$, bounded on $t_0 < t < \infty$, the output $\mathbf{y}(t)$ is bounded on $t_0 \leq t < \infty$. ▲

This definition can be applied to nonlinear as well as linear systems.

If we specialize our interests to linear time-invariant systems, we may express BIBO stability in terms of transfer functions. The transform $\mathbf{Y}(s)$ of the zero-state output response $\mathbf{y}(t)$ of a linear time-invariant system is related to the transform $\mathbf{U}(s)$ of the input $\mathbf{u}(t)$ by the transfer function matrix $\mathbf{H}(s)$, as was discussed in Chapter 6.

$$\mathbf{Y}(s) = \mathbf{H}(s)\mathbf{U}(s). \tag{8.108}$$

If we denote the elements of $\mathbf{H}$ by $H_{ij}(s)$, we may state the following.

Theorem 8.2. A linear time-invariant system is zero-state BIBO stable if and only if all poles of every transfer function $H_{ij}(s)$ lie (strictly) in the left-half plane, and if the number of zeros of $H_{ij}(s)$ does not exceed the number of poles for all $H_{ij}(s)$. ▲

The proof of this theorem is left to the reader (see Problems).

Zero-state BIBO stability does not imply stability of the zero-input response, as is shown by Example 8-12. However, the stability of the zero-input response does guarantee BIBO stability. This follows because, from Chapter 7, we may express the transfer function of the system

$$\dot{\mathbf{x}}(t) = \mathbf{A}\mathbf{x}(t) + \mathbf{B}\mathbf{u}(t) \tag{8.109}$$

$$\mathbf{y}(t) = \mathbf{C}\mathbf{x}(t) + \mathbf{D}\mathbf{u}(t) \tag{8.110}$$

by

$$\mathbf{H}(s) = \mathbf{C}(s\mathbf{I} - \mathbf{A})^{-1}\mathbf{B} + \mathbf{D}. \tag{8.111}$$

Thus a system having all characteristic roots in the LHP, determined by the Routh test, the root locus, or the Nyquist criterion, will also be zero-state BIBO stable for *any* choice of input and output terminals.

8.6 Parameter Variations and Stability

It is important that practical systems maintain stability for a range of parameter variations. The root locus, for example, shows the variations in characteristic root location as a function of a single parameter. A first-order estimate of the change in root location for a parameter change can be obtained from a Taylor's series expansion of the characteristic root as a function of a parameter α:

$$s(\alpha) - s(\alpha_0) = \Delta s \approx \frac{\partial s}{\partial \alpha}(\alpha - \alpha_0) = \frac{\partial s}{\partial \alpha}\Delta\alpha \tag{8.112}$$

where α_0 is the nominal value of the parameter α. The quantity $\partial s/\partial\alpha$ may be called the *root sensitivity* and is equal to the slope of the root locus at $\alpha = \alpha_0$.

Consider the equation

$$1+\alpha G(s) = 0. \tag{8.113}$$

We denote by $s = s_0$ a characteristic root of (8.113) corresponding to $\alpha = \alpha_0$. The root sensitivity of s_0 at α_0 may be found by differentiating (8.113) and by evaluating the terms at $s = s_0$, $\alpha = \alpha_0$.

$$\alpha_0 \left.\frac{dG}{ds}\right|_{s=s_0} \left.\frac{\partial s}{\partial\alpha}\right|_{\alpha_0} + G(s_0) = 0. \tag{8.114}$$

But, since from (8.113) $G(s_0) = -1/\alpha_0$, (8.114) yields

$$\left.\frac{ds}{d\alpha}\right|_{\alpha_0} = \frac{1}{\alpha_0{}^2} \frac{1}{\left.\dfrac{dG}{ds}\right|_{s=s_0}}. \tag{8.115}$$

Equation 8.115 shows that the root sensitivity is infinite at saddle points on the root locus.

Problems

8-1. The transfer function block diagrams of Figure P8-1 represent several systems. Find the characteristic equation of each by Mason's rule.

8-2. Find the characteristic equation of each system in Figure P8-1 by selecting suitable state variables and forming $\det(s\mathbf{I}-\mathbf{A})$.

8-3. Suppose a system is described by $\dot{\mathbf{x}} = \mathbf{A}\mathbf{x}$. Consider a new choice of state variables, $\mathbf{z}$, related to $\mathbf{x}$ by $\mathbf{x} = \mathbf{T}\mathbf{z}$, where $\mathbf{T}$ is a constant non-singular matrix. Show that the system described in terms of $\mathbf{z}$ has the same eigenvalues (characteristic roots) as the $\mathbf{x}$ system.

8-4. Prove Theorem 8.1.

8-5. Using the Routh stability test, determine the values of K for which the systems of Figure P8-1 are asymptotically stable.

8-6. Suppose, for the attitude control system of Example 8-6, that

$$K_1 = 2.0 \qquad K_3 = 1.0$$
$$K_2 = 1.0 \qquad K_4 = 1.0$$

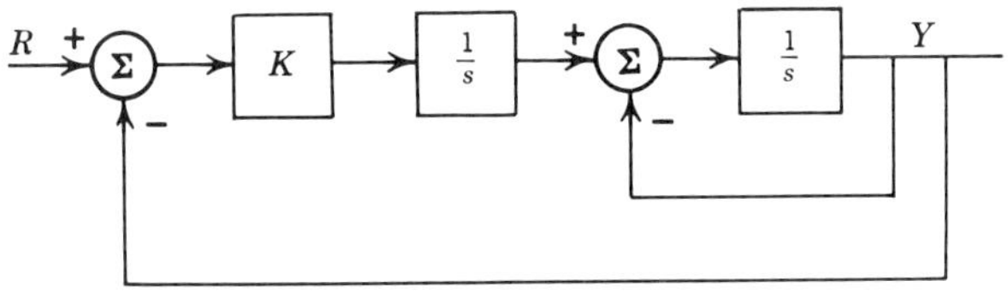

(a)

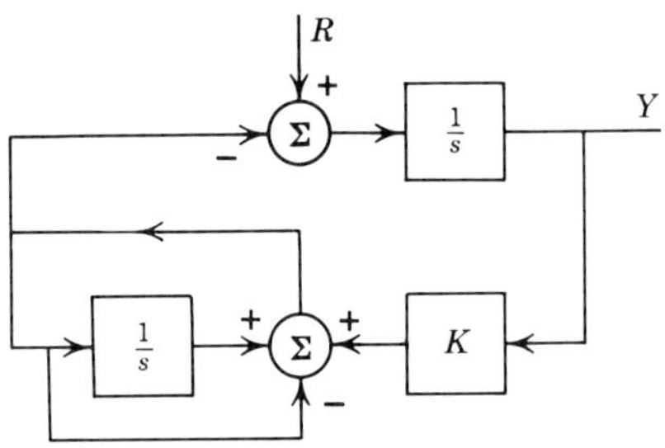

(b)

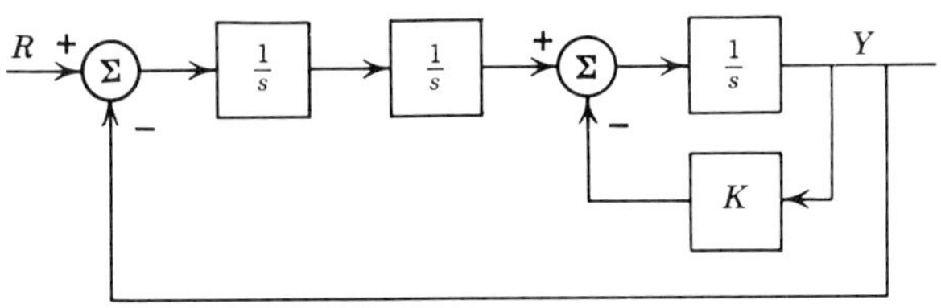

(c)

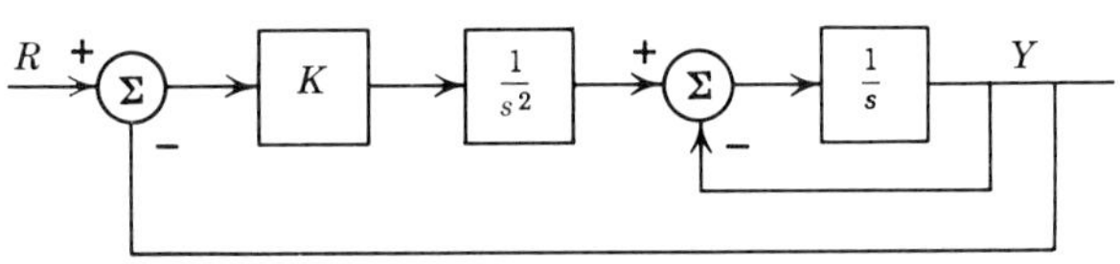

(d)

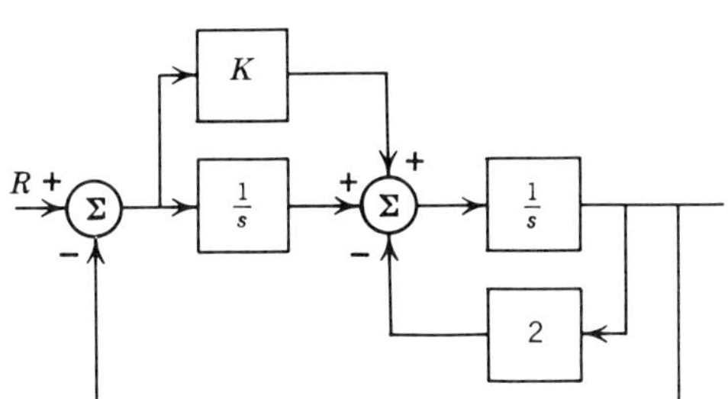

(e)

Figure P8-1

Show the stability region in the (k_1, k_2) plane, that is, the locus of values of k_1 and k_2 yielding an asymptotically stable attitude control system.

8-7. Sketch the root locus as a function of K, $K > 0$, for each system of Figure P8-1.

8-8. A simplified single-axis attitude control system is represented by the block diagram of Figure P8-8. Sketch the root locus as a function of K, assuming a and J to be fixed.

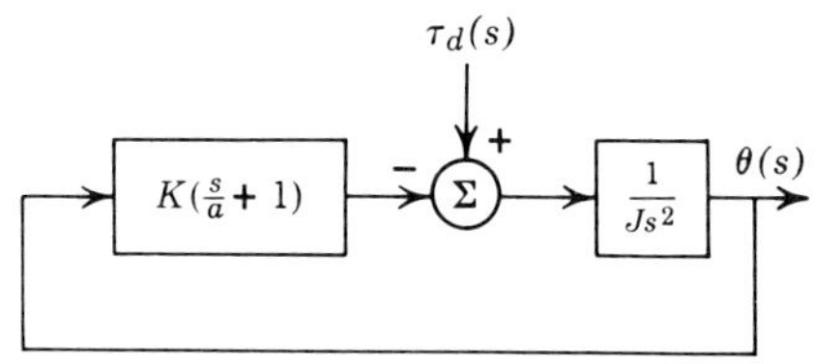

Figure P8-8

8-9. Discuss the general nature of the response $\theta(t)$ to impulse disturbances $\tau_d(t) = \delta(t)$ for various values of K, making use of the root locus drawn in Problem 8-8.

8-10. A somewhat more accurate representation of a single-axis attitude-control system than that of Figure P8-8 is given in Figure P8-10.

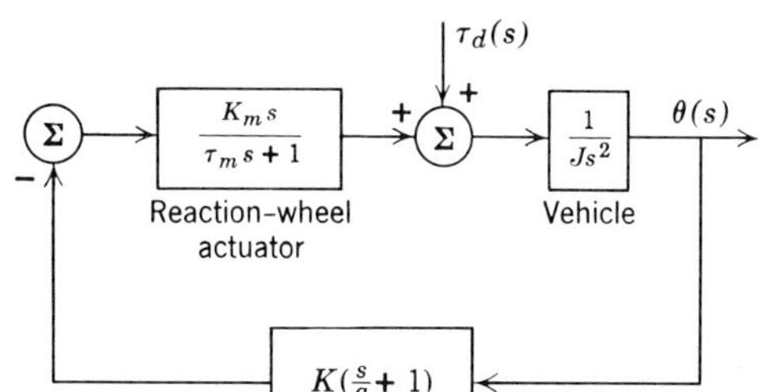

Figure P8-10

Sketch root loci for the parameter $K' = KK_m/Ja\tau_m$ as the variable parameter, with a and τ_m fixed, for the following three cases:

(1) $a > \dfrac{1}{\tau_m}$

(2) $a < \dfrac{1}{\tau_m}$

(3) $a \to \infty$ (no rate feedback).

8-11. Discuss the general nature of the responses to impulse disturbance inputs for the three cases of Problem 8-10.

8-12. Extend the root locus sketching rules to include $K < 0$.

8-13. In *MKS* units,

$M_1 = 1.5$
$M_2 = 0.6$
$K_1 = 1200$
$K_2 = 840.$

Sketch the root locus as a function of D for the system of Figure P8-13.

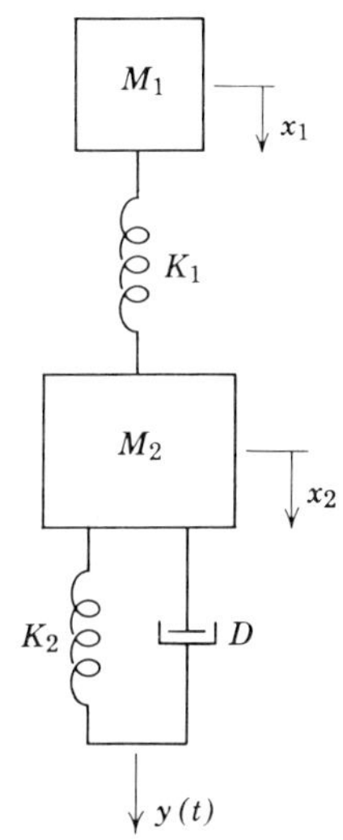

Figure P8-13

8-14. Use your root locus of Problem 8-13 to select the value of D to maximize the damping ratio of the least damped complex conjugate characteristic roots.

8-15. Sketch a root locus (K variable) for the system of Figure P8-15. For what values of K is this system asymptotically stable?

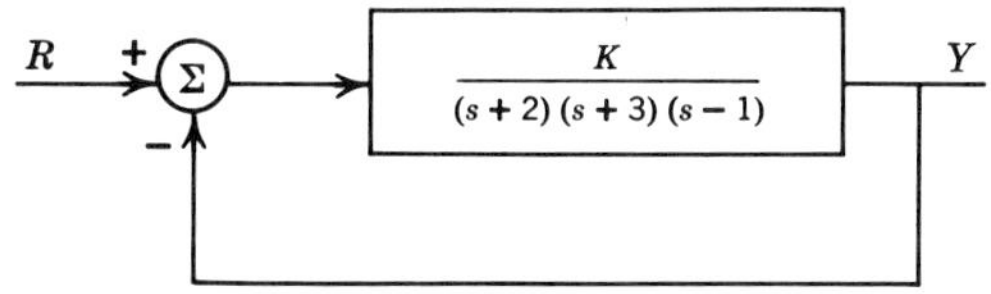

Figure P8-15

8-16. Consider the system with "state feedback" shown in Figure P8-16. Assume K_2 is fixed, K_1 variable. In what regions of the s-plane may the system characteristic roots be found? Show a root locus.

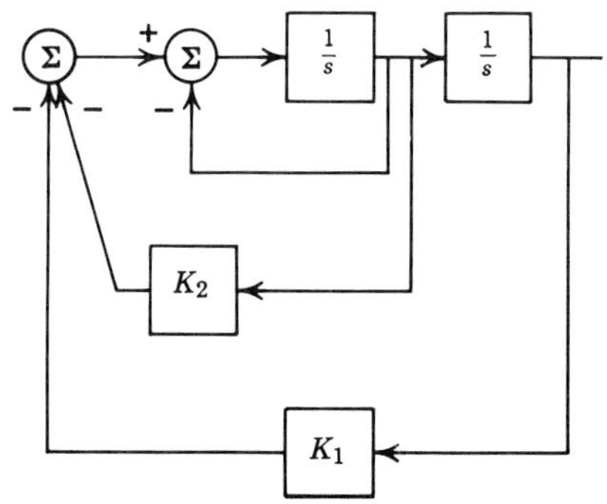

Figure P8-16

8-17. Repeat Problem 8-16 with K_1 fixed, K_2 variable.

8-18. Repeat Problem 8-16 with *both* K_1 and K_2 variable. Show a family of root loci ("root contours").

8-19. Consider the Nyquist plots of Figure P8-19. The direction of increasing ω is indicated by the arrows, with $G(j\omega)$ for $\omega > 0$ shown as a solid line, and $G(j\omega)$ for $\omega < 0$ as a broken line. #P indicates the number of RHP poles of $G(s)$. Which systems are asymptotically stable?

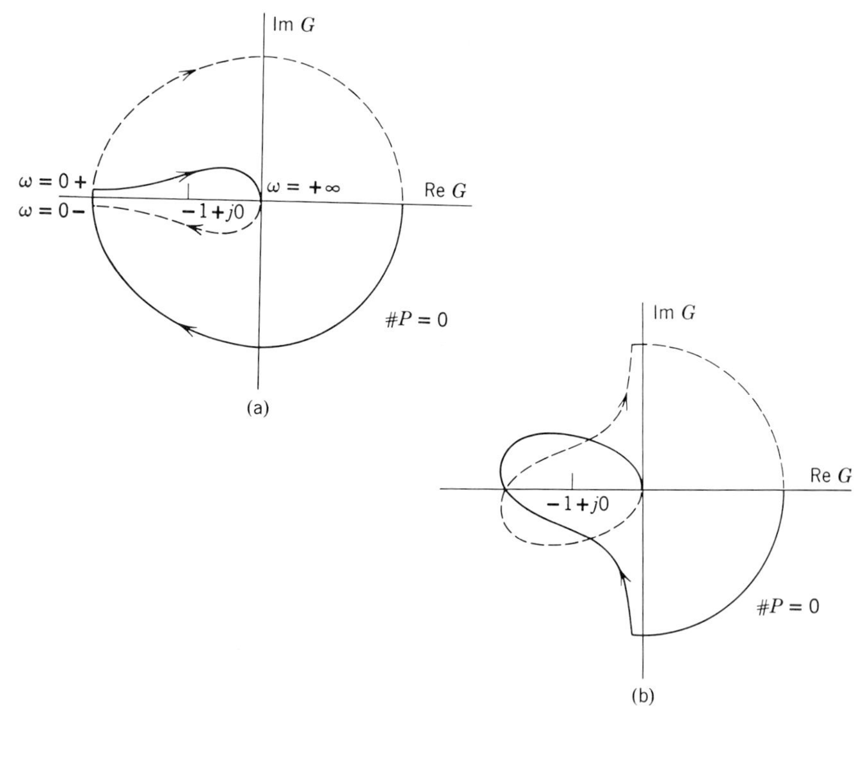

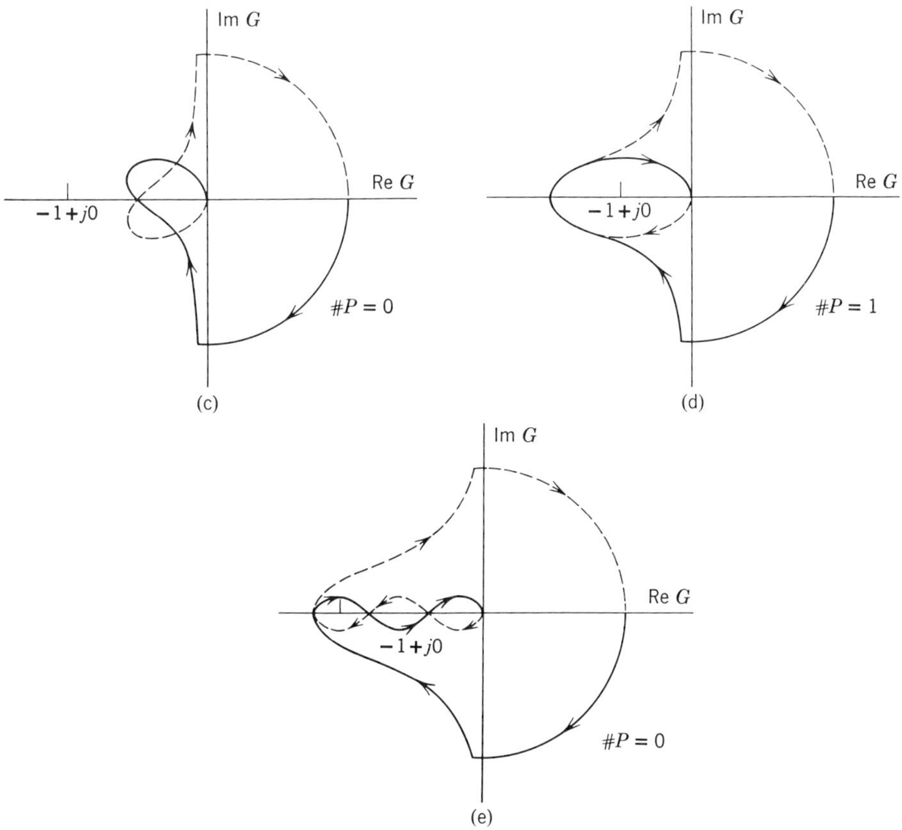

Figure P8-19

8-20. Investigate the stability of the systems of Figure P8-1 using the Nyquist criterion. (Suggestion: Sketch Bode gain and phase diagrams first using asymptotic approximations with corrections, then transfer this information to a polar plot.)

8-21. Consider the feedback system of Figure P8-21. The system transfer function, for $s = j\omega$, is

$$T(j\omega) = \frac{Y(j\omega)}{R(j\omega)} = \frac{G(j\omega)}{1 + G(j\omega)}.$$

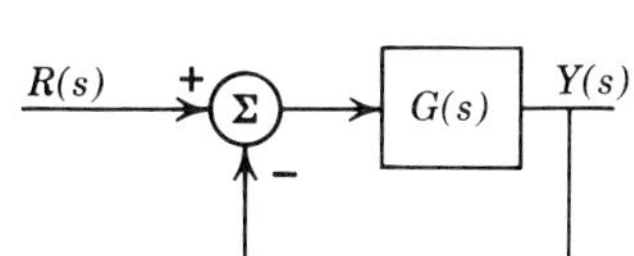

Figure P8-21

In polar form,

$$T(j\omega) = M\epsilon^{jN},$$

where $M = |T(j\omega)|$ and $N = \angle T(j\omega)$. Show that in the $G(j\omega)$-plane the contours of constant M are circles. (Suggestion: Write $G(j\omega) = x + jy$. Solve for M in terms of x and y.)

8-22. Show that in the $G(j\omega)$-plane the contours of constant N are circles.

8-23. Complete Example 8-13 by verifying that $\sqrt{x_1^2 + x_2^2}$ satisfies the definition of a norm, Definition 8.2.

8-24. In the paragraph following (8.107), it is asserted that "a time function consisting of a sum of decaying exponentials is . . . bounded." Prove this assertion.

8-25. Prove Theorem 8.2.

8-26. Show that

$$\|\mathbf{x}\| = \max_i |x_i|,$$

where x_i is the ith component of $\mathbf{x}$, satisfies the definition of a norm.

8-27. Consider the system of Figure P8-27. Is the system asymptotically stable? Is the system BIBO stable?

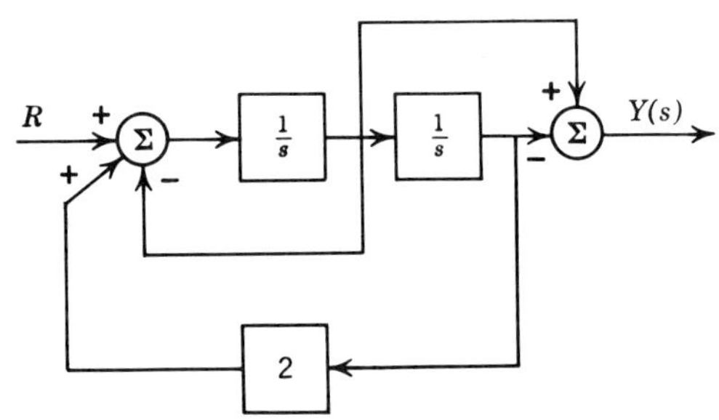

Figure P8-27

CHAPTER 9

Response of Nonlinear Systems

We have observed that system models are nonlinear, in general. In Chapters 6, 7, and 8 we have discussed analysis techniques that can be applied when the system model can be linearized. In this chapter, we shall discuss the analysis of nonlinear systems, that is, systems containing at least one component that cannot be modeled realistically by a linear model.

We can discuss only a few of the many interesting and important nonlinear problems in this chapter, of course. We shall concentrate on methods useful for our design example—spacecraft attitude control. Fortunately, these methods can be used for many other problems also.

9.1 Piecewise-Linear Systems

The phase-plane method applied in Chapter 7 to linear systems is easily applied to the class of nonlinear systems for which the nonlinearities can be represented by connected straight-line segments. Such nonlinearities are called *piecewise-linear*. The idea of the procedure is to decompose the phase plane into regions corresponding to the various linear pieces of the nonlinearities. The methods of Chapter 7 then can be applied to the various regions in which the system is linear.

These ideas are illustrated by the following attitude-control example.

Example 9-1. In this example, we consider a single-axis gas-jet attitude-control problem. It was shown in Chapter 2 that

$$J\ddot{\theta} = \tau_c, \tag{9.1}$$

where τ_c is the applied control torque. (Disturbance torques will be neglected in this example.) Suppose the jets have an operating characteristic as shown in Figure 9.1. The input signal to the gas-jet unit is denoted by u. Notice that the jets are modeled by a piecewise-linear nonlinearity.

Suppose $x_1 = \theta$, $x_2 = \dot{\theta}$ are chosen as state variables. Then

$$\dot{x}_1 = x_2 \tag{9.2}$$

$$\dot{x}_2 = \frac{1}{J}\tau_c. \tag{9.3}$$

Consider the phase plane trajectories for this system for the two possible control torques, $\tau_c = +T$ and $\tau_c = -T$. The slope (isocline) equation, including both cases, is

$$\frac{dx_2}{dx_1} = \frac{\pm T}{Jx_2}.$$

This integrates readily to yield the family of parabolas

$$\frac{x_2^2}{2} - \frac{x_2^2(0)}{2} = \pm\frac{T}{J}[x_1 - x_1(0)]. \tag{9.4}$$

The trajectories are shown in the phase-plane plot of Figure 9.2. Notice that, if the jets are left in either position indefinitely, the spacecraft continues to rotate, that is, there is no equilibrium point. The problem, then, is to choose m in which the jets are switched so that desirable attitude control is achieved. Suppose, for example, that we

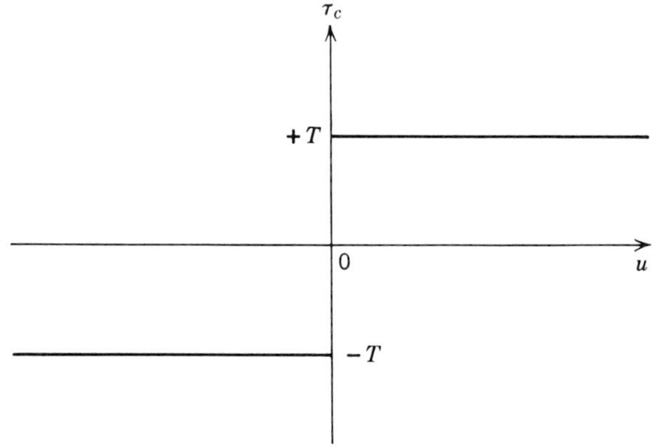

Figure 9.1 Gas-jet operating characteristic.

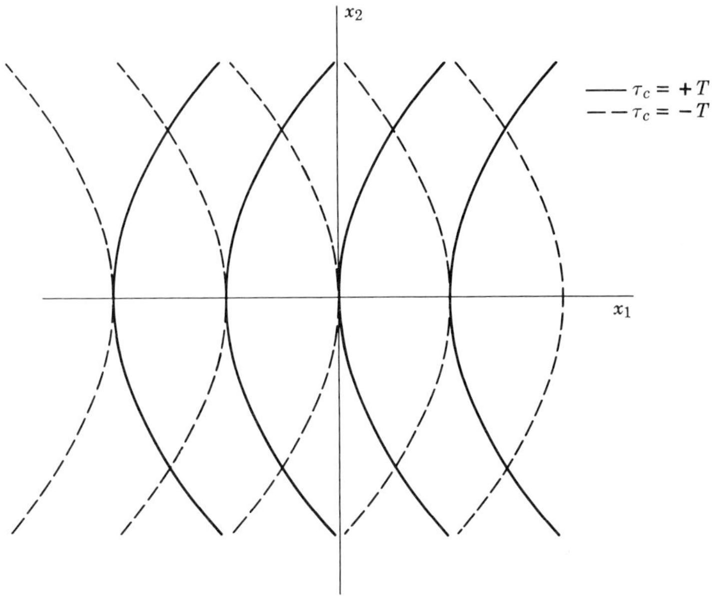

Figure 9.2 Trajectories for Example 9-1.

wish the response to an initial condition to be reduced to zero in the shortest possible time. We notice that one parabola from each family of trajectories passes through the origin. Thus we may reach the origin by moving along a trajectory that intersects the parabolas through the origin, then moving along these parabolas to the origin. How do we choose our trajectories to accomplish this in minimum time? Although, in general, it is difficult to see the effects of the time parameter on the phase plane, in this special case, integration of the $\dot{x}_2$ state equation leads to

$$t = \pm \frac{J}{T}[x_2 - x_2(0)]. \tag{9.5}$$

Thus, time is proportional to the rotational velocity of the spacecraft (which makes physical sense!). Hence, we pick a trajectory which results in the shortest path between the initial velocity and the desired final velocity, $x_2 = 0$, measured along the x_2 axis. So our intuitive argument indicates that the trajectory of Figure 9.3*a* represents shorter response time than that of Figure 9.3*b* or Figure 9.3*c*. Indeed, using modern control system optimization theory (which we shall not discuss here), we can *prove* that our intuition is correct. We should

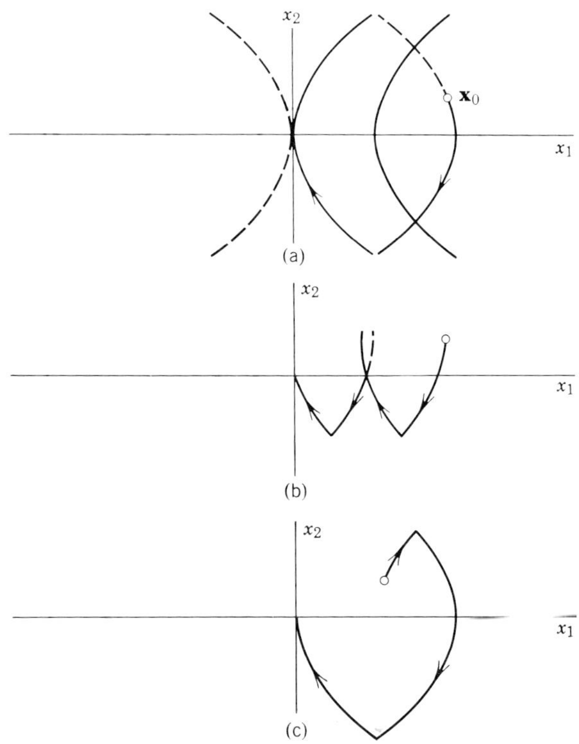

Figure 9.3 Various phase-plane trajectories for Example 9-1.

proceed as shown in Figure 9.3*a*, switching when the trajectories cross the "switching line" given by

$$2x_1 = x_2|x_2|. \tag{9.6}$$

To implement this minimum-time attitude control, we need to observe x_1 and x_2, and check (using logic circuits, for example) to see which switching region we are in. A pictorial representation of the control minimum-time gas-jet control system is shown in Figure 9.4. ▲

In the preceding example, the gas jets are on all the time, supplying either $+T$ or $-T$ torques to the spacecraft. Clearly, such operation involves large fuel consumption. In other missions, it is preferable to conserve fuel by using a control strategy that permits the jets to be off most of the time. Speed of response is decreased, in general, by such a strategy. Also, such a strategy often leads to a steady-state oscillation, called a *limit cycle*, rather than to be the origin in the state space. These notions are illustrated by the following example.

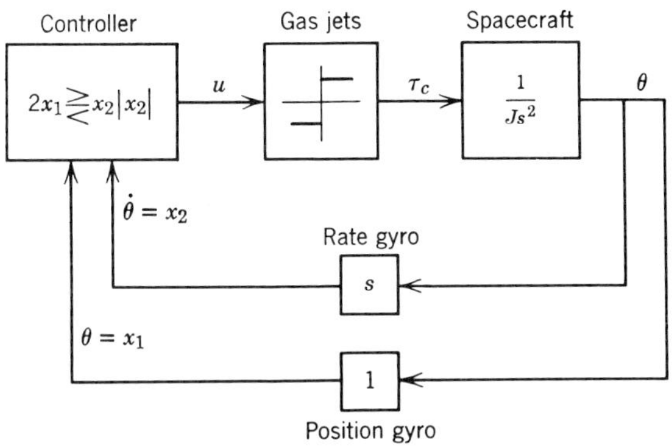

Figure 9.4 Minimum-time gas-jet attitude control system.

Example 9-2. Suppose that the gas jets have a piecewise-linear operating characteristic as shown in Figure 9.5. Note that the torque can be zero for nonzero jet input u. The region of zero torque is called the *dead zone*. Notice also that the switching points for the torque are not the same for torque-on as for torque-off. This phenomenon is called *hysteresis*.

Gas jets described in Figure 9.5 are used in the feedback system of Figure 9.6. Attitude and attitude rate are fed back to form u. No external inputs are present.

The first step in sketching the phase plane is to establish the boundaries in the $x_1 = \theta$, $x_2 = \dot{\theta}$ plane. These boundaries are referred to as switching lines. We observe from Figure 9.6 that

$$u = -\theta - K\theta. \tag{9.7}$$

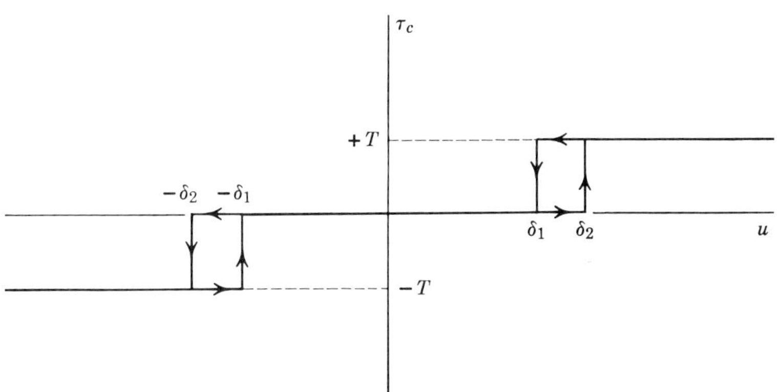

Figure 9.5 Gas-jet operating characteristic with dead zone and hysteresis.

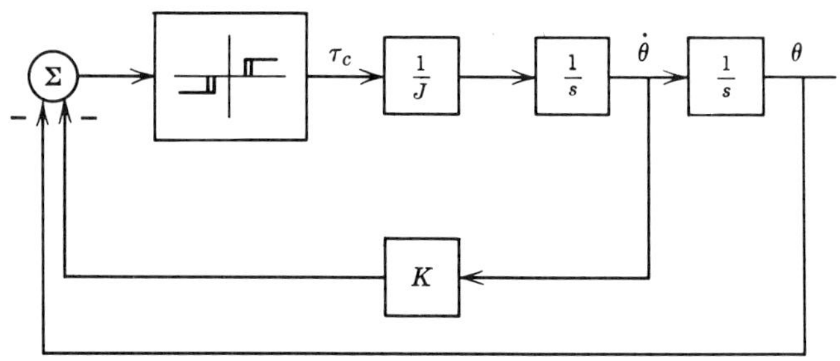

Figure 9.6 Unity feedback gas-jet attitude control system.

From Figure 9.5 and (9.7), it is clear that the switchings are defined by

$$\text{Torque on: } -\theta - K\dot\theta = \pm\,\delta_2$$
$$\text{Torque off: } -\theta - K\dot\theta = \pm\,\delta_1. \qquad (9.8)$$

These are straight lines in the phase plane with slope $-1/K$. See Figure 9.7.

Thus the $(\theta, \dot\theta)$ plane is divided into three regions corresponding to $\tau_c = +T$, 0, and $-T$. In each region the system is linear, and the methods of Chapter 7 may be used to sketch trajectories. Consider first the region corresponding to $\tau_c = 0$.

$$\dot x_1 = x_2 \qquad (9.9)$$
$$\dot x_2 = 0. \qquad (9.10)$$

There is no single equilibrium point; rather, $x_2 = 0$ is an equilibrium line. Since $\dot x_2 = 0$, $x_2(t) = x_{20} = \dot\theta_0$. Therefore, the trajectories are parallel to the x_1 axis (isoclines have zero slope), with x_2 intercept equal to the value of $x_2(t)$ at the point where the trajectory enters the region.

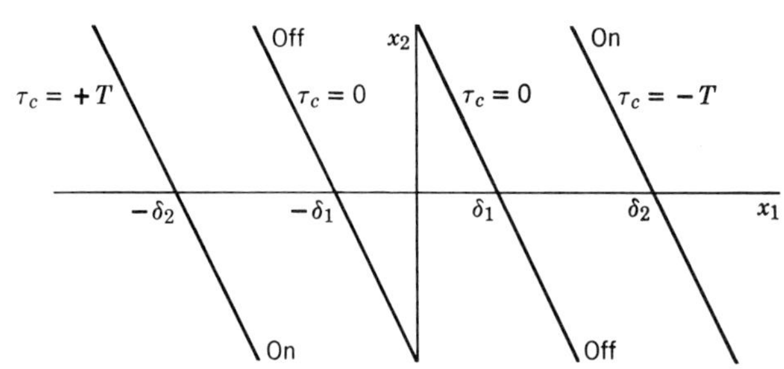

Figure 9.7 Switching lines for Example 9-2.

In the region corresponding to $\tau_c = +T$, an analysis similar to that of Example 9-1 reveals that the trajectories are parabolic:

$$\dot{x}_1 = x_2, \tag{9.11}$$

$$\dot{x}_2 = \frac{T}{J}. \tag{9.12}$$

So

$$\frac{dx_2}{dx_1} = \frac{T}{Jx_2} \tag{9.13}$$

or

$$\frac{x_2^2}{2} - \frac{x_{20}^2}{2} = \frac{T}{J}(x_1 - x_{10}). \tag{9.14}$$

Similar parabolic trajectories apply when $\tau_c = -T$. Figure 9.8 shows some typical trajectories for various initial conditions. ▲

Examination of Figure 9.8 shows the existence of a closed-path trajectory. This is an example of a limit cycle. Formally, a limit cycle is an *isolated periodic solution* of the system equations. The amplitude and the period of the limit cycle can be found (approximately)

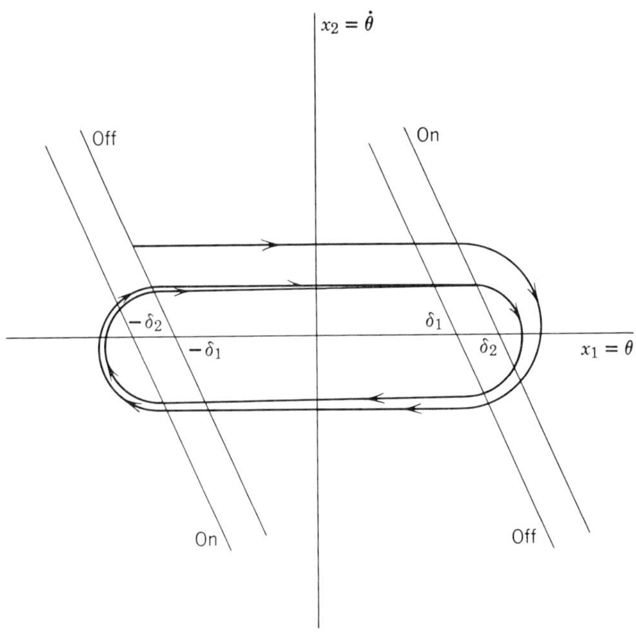

Figure 9.8 Trajectories for Example 9-2.

from a carefully drawn phase portrait or calculated analytically from the geometry of such a portrait, together with the trajectory equations. We indicate the latter procedure in the following example:

Example 9-3. The limit cycle of Example 9-2 is redrawn in Figure 9.9. We wish to compute the amplitude and period for this limit cycle as a function of the actuator parameters, K, δ_1, δ_2, and T, and the spacecraft moment of inertia J.

We first find the coordinates of the switching points A, B, C, D, on the limit cycle. Observe the symmetry of these points, as shown in Figure 9.9. Since the points lie on switching lines, and the equations of the switching lines are known, the quantities θ_1, $\dot{\theta}_1$ can be found readily. From Figure 9.9,

$$\theta_1 + K\dot{\theta}_1 = \delta_2 \tag{9.15}$$

$$\theta_1 - K\dot{\theta}_1 = \delta_1. \tag{9.16}$$

Thus

$$\theta_1 = \frac{\delta_2 + \delta_1}{2} \tag{9.17}$$

$$\dot{\theta}_1 = \frac{\delta_2 - \delta_1}{2K}. \tag{9.18}$$

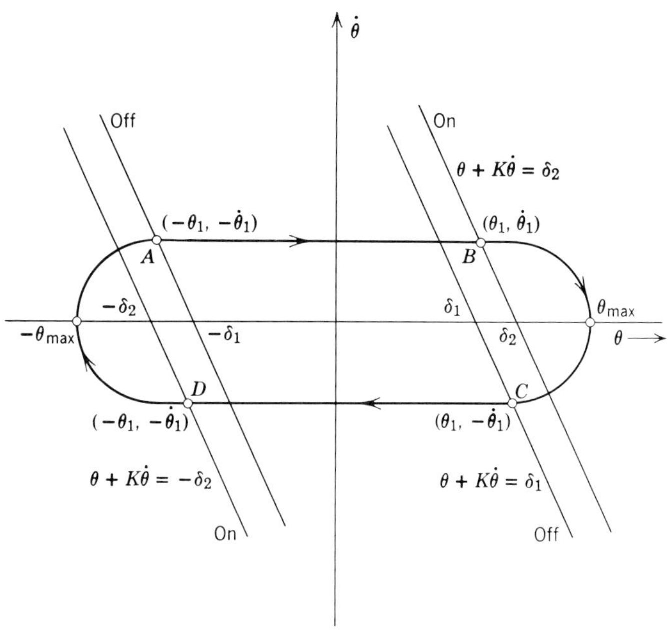

Figure 9.9 Geometry of limit cycle for Example 9.3.

The period of the limit cycle can be found from the times responses in the various regions. For example, the trajectory from A to B is governed by (9.9) and (9.10). Integrating these equations with initial conditions of point A,

$$\theta_0 = -\theta_1 \tag{9.19}$$

$$\dot{\theta}_0 = \dot{\theta}_1, \tag{9.20}$$

we obtain

$$x_1(t) = \dot{\theta}_1 t - \theta_1. \tag{9.21}$$

If $t = 0$ corresponds to the trajectory starting at A, and $t = t_{AB}$ corresponds to the trajectory arriving at B, then, from (9.21),

$$x_1(t_{AB}) = \theta_1 = \dot{\theta}_1 t_{AB} - \theta_1, \tag{9.22}$$

or

$$t_{AB} = \frac{2\theta_1}{\dot{\theta}_1}. \tag{9.23}$$

Substituting (9.17) and (9.18) in (9.23),

$$t_{AB} = 2K\left(\frac{\delta_2 + \delta_1}{\delta_2 - \delta_1}\right). \tag{9.24}$$

The trajectory from B to C is governed by (9.11) and (9.12), but with $\tau_c = T$. Integrating these with the coordinates of B as initial conditions, we obtain

$$x_1(t) = -\frac{T}{J}\frac{t^2}{2} + \dot{\theta}_1 t + \theta_1. \tag{9.25}$$

Denoting the time of arrival of the trajectory at C as t_{BC}, we have

$$x_1(t_{BC}) = \theta_1 = -\frac{T}{J}\frac{t_{BC}^2}{2} + \dot{\theta}_1 t + \theta_1, \tag{9.26}$$

or

$$t_{BC} = \frac{2J\dot{\theta}_1}{T}. \tag{9.27}$$

Substituting (9.17) and (9.18), (9.27) becomes

$$t_{BC} = \frac{J}{KT}(\delta_2 - \delta_1). \tag{9.28}$$

Thus

$$\begin{aligned}\text{period} &= 2(t_{AB}+t_{BC})\\ &= 4K\left(\frac{\delta_2+\delta_1}{\delta_2-\delta_1}\right)+\frac{2J}{KT}(\delta_2-\delta_1). \end{aligned} \tag{9.29}$$

We now compute the limit cycle amplitude, $\theta_{\max}$. From (9.12), with $-T$ instead of T, and the coordinates of B as initial conditions,

$$x_2(t) = \frac{-T}{J}t+\dot{\theta}_1. \tag{9.30}$$

When $x_2(t) = 0$, we are at $x_1 = \theta_{\max}$. Denoting this time by t, (9.30) yields

$$t_1 = \frac{J\dot{\theta}_1}{T}. \tag{9.31}$$

Substituting (9.31) in (9.25) yields

$$x_1(t_1) = \theta_{\max} = -\frac{T}{2J}\frac{J^2\theta_1^2}{T^2}+\dot{\theta}_1\frac{J\dot{\theta}_1}{T}+\theta_1. \tag{9.32}$$

Thus

$$\theta_{\max} = \frac{J\dot{\theta}_1^2}{2T}+\theta_1. \tag{9.33}$$

Substituting (9.17) and (9.18) in (9.33),

$$\theta_{\max} = \frac{J}{8TK^2}(\delta_2-\delta_1)^2+\left(\frac{\delta_2+\delta_1}{2}\right). \ \blacktriangle \tag{9.34}$$

As was mentioned above, a design for which $\tau_c = 0$ for an appreciable portion of a period results in low fuel consumption. We define *duty cycle* as

$$\text{duty cycle} = \frac{t_{\text{on}}}{t_{\text{on}}+t_{\text{off}}}, \tag{9.35}$$

where t_{on} is the number of seconds per period the jets are on (delivering either $+T$ or $-T$ torque), and $t_{\text{on}}+t_{\text{off}}$ is the period of the limit cycle.

Example 9-4. We wish to find the duty cycle for the system of Example 9-2 and Example 9-3. It is convenient to express this duty cycle in terms of quantities usually specified in design. There are three such parameters, the maximum allowable error θ_{max}, the angular acceleration due to control T/J, and the minimum on-time t_{min} for the jets. (There is a minimum time-on for the jets because the jets cannot be *reliably* switched on, and then off, in an arbitrarily short time.) Using the minimum on-time,

$$t_{on} = 2t_{min}, \tag{9.36}$$

since the jets are on twice per limit cycle period. Then (9.35) becomes

$$\text{duty cycle} = \frac{2t_{min}}{2t_{min} + t_{off}}. \tag{9.37}$$

But, form (9.23), (9.27), and (9.33),

$$t_{off} = 2t_{AB} = \frac{4\theta_1}{\dot{\theta}_1} \tag{9.38}$$

$$t_{on} = 2t_{min} = 2t_{BC} = \frac{4J\theta_1}{T} \tag{9.39}$$

$$\theta_{max} = \frac{J\dot{\theta}_1^{\,2}}{2T} + \theta_1. \tag{9.40}$$

Solving,

$$t_{off} = \frac{8J\theta_{max}}{T\,t_{min}} - t_{min}. \tag{9.41}$$

Thus, (9.37) can be written

$$\text{duty cycle} = \frac{2}{\dfrac{8J\theta_{max}}{T\,t_{min}^{\,2}} + 1}. \tag{9.42}$$

Suppose, for example, that

$$\theta_{max} = 1 \text{ degree}$$

$$\frac{T}{J} = 10 \text{ deg/sec}^2$$

$$t_{min} = 10 \text{ milliseconds} = 10^{-2} \text{ sec}. \tag{9.43}$$

Then

$$\text{duty cycle} = 2.5 \times 10^{-4}. \tag{9.44}$$

This duty cycle is quite good—the jets are on only 0.025 percent of the time! Each period, the vehicle coasts for 40 seconds, thrusts for

10 milliseconds, coasts another 40 seconds, and thrusts 10 milliseconds. And this low-fuel operation is acheived with a maximum attitude error (in the steady-state) of only 1 degree. ▲

Example 9-5. Some additional flexibility is achieved by introducing adjustable gains for both position and rate feedback, as shown in Figure 9.10, instead of for only rate feedback, as was done in Figure 9.6. For this case, calculations similar to those of Examples 9-2, 9-3, and 9-4 establish that

$$t_{\text{on}} = \frac{2J}{T}\left(\frac{\delta_2 - \delta_1}{K_2}\right) \tag{9.45}$$

$$t_{\text{off}} = \frac{4K_2}{K_1}\left(\frac{\delta_2 + \delta_1}{\delta_2 - \delta_1}\right) = \frac{8J(\delta_1 + \delta_2)}{K_1 T t_{\text{on}}} \tag{9.46}$$

Furthermore,

$$\theta_m = \frac{J}{2T}\left(\frac{\delta_2 - \delta_1}{2K_2}\right)^2 + \frac{\delta_1 + \delta_2}{2K_1} \tag{9.47}$$

or

$$\theta_m = \frac{1}{32}\frac{T}{J}t_{\text{on}}^2 + \frac{1}{16}\frac{T}{J}t_{\text{on}}t_{\text{off}}. \tag{9.48}$$

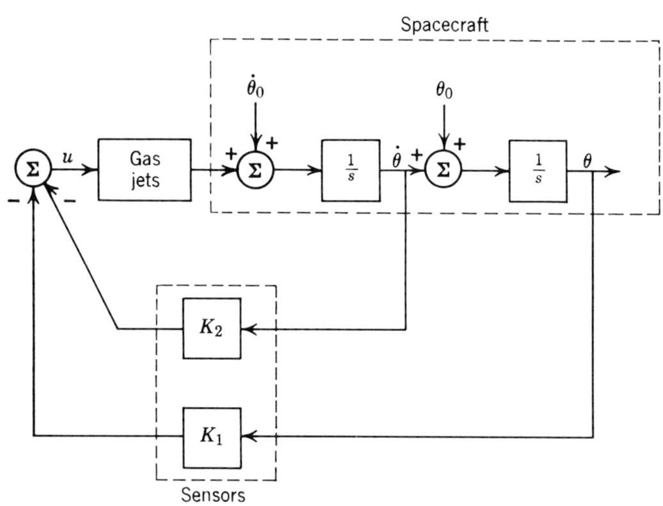

Figure 9.10 Single-axis gas-jet control system with adjustable position and rate feedback.

Thus specifying θ_m, T/J, and t_{on} fixes t_{off}, by (9.48). The duty cycle (9.35) is also fixed by this specification:

$$D = \frac{2}{1 + \dfrac{\theta_m}{\dfrac{1}{32}\dfrac{T}{J}t_{on}^2}}. \tag{9.49}$$

Thus far, we have considered steady-state operation only. The parameters K_1, K_2, δ_1, and δ_2 may be chosen for a specified minimum t_{on} and θ_m, and some freedom still remains for designing satisfactory transient response. For any chosen design, one may multiply K_1 and K_2 by a constant and also multiply δ_1 and δ_2 by the same constant without disturbing the control strategy. This means that one parameter can be chosen arbitrarily. The transient response may be varied by changing the slope K_2/K_1 of the switching lines. Thus δ_1, δ_2, and K_2/K_1 may be regarded as the design parameters to be chosen from a specified θ_m, minimum t_{on}, and satisfactory transient response. Figure 9.11

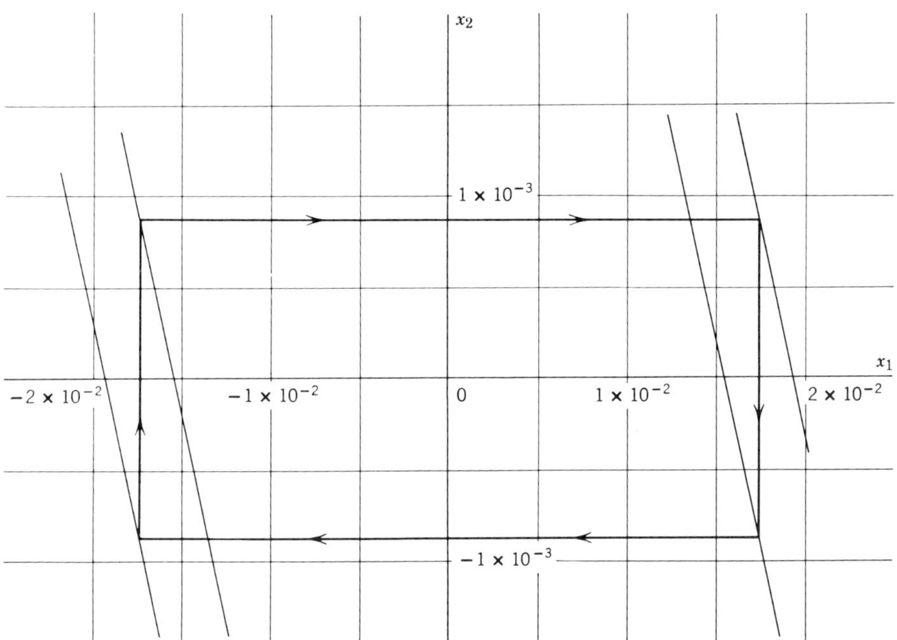

(a)

Figure 9.11(*a*) Trajectory for $K_1 = 1.688$, $K_2 = 3.75$, $\delta_1 = 2.62 \times 10^{-2}$, $\delta_2 = 3.27 \times 10^{-2}$.

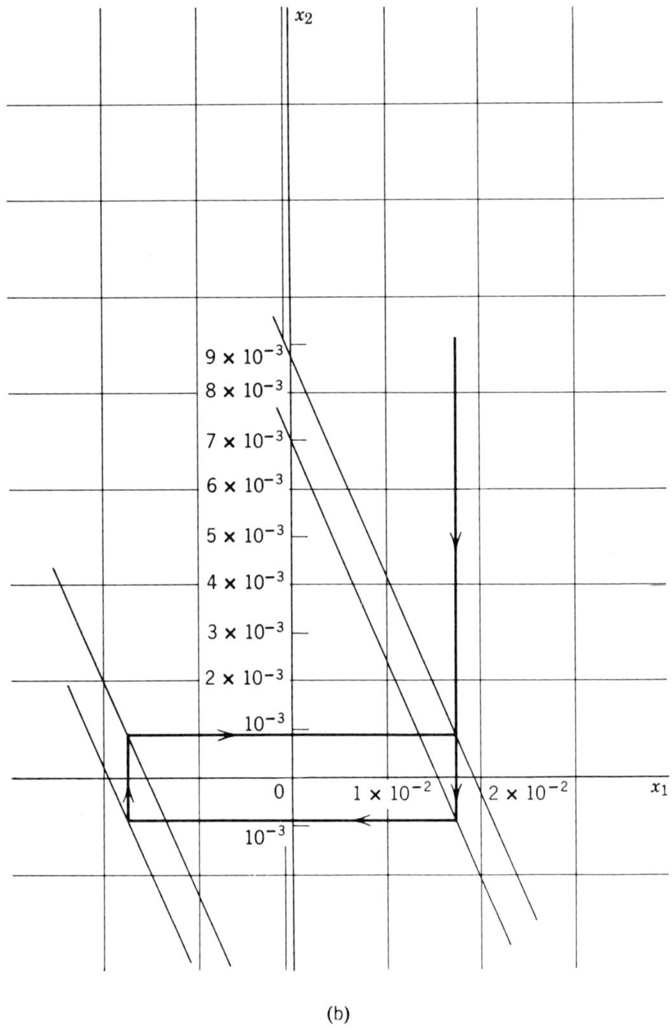

Figure 9.11(*b*)

shows trajectories for $K_1 = 1.688$, $K_2 = 3.75$, $\delta_1 = 2.62 \times 10^{-2}$, and $\delta_2 = 3.27 \times 10^{-2}$, corresponding to $\theta_m = 1°$ and $D = 2.5 \times 10^{-4}$, for three different initial conditions, obtained by simulation.[1] Figure 9.12 shows a trajectory for a different K_2/K_1 ratio. Since the system is nonlinear, the response shapes will be different if the initial conditions are scaled by a constant. ▲

[1]This simulation was performed by Mr. Cheng-I Chen.

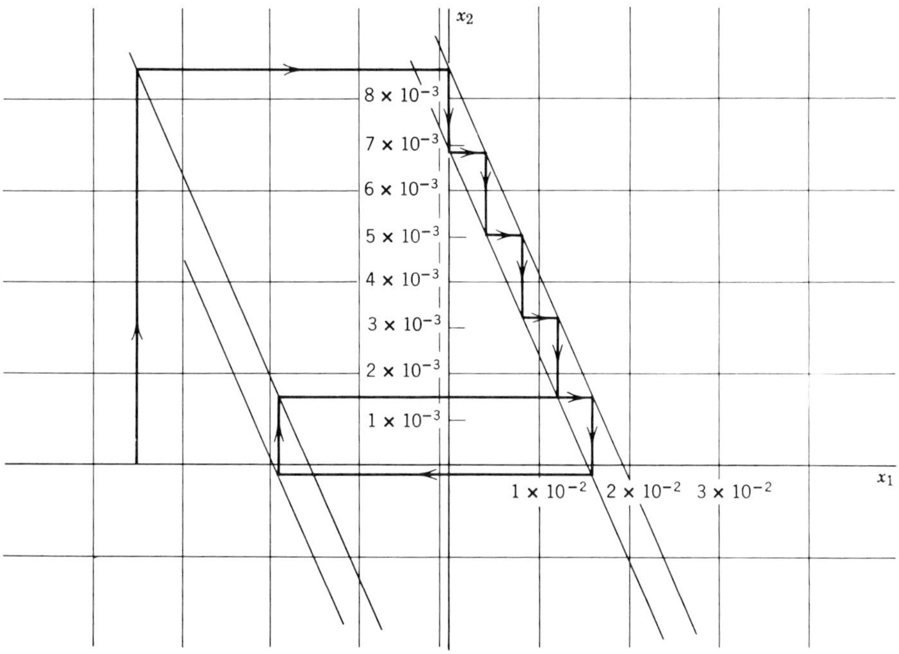

(c)

Figure 9.11(*c*)

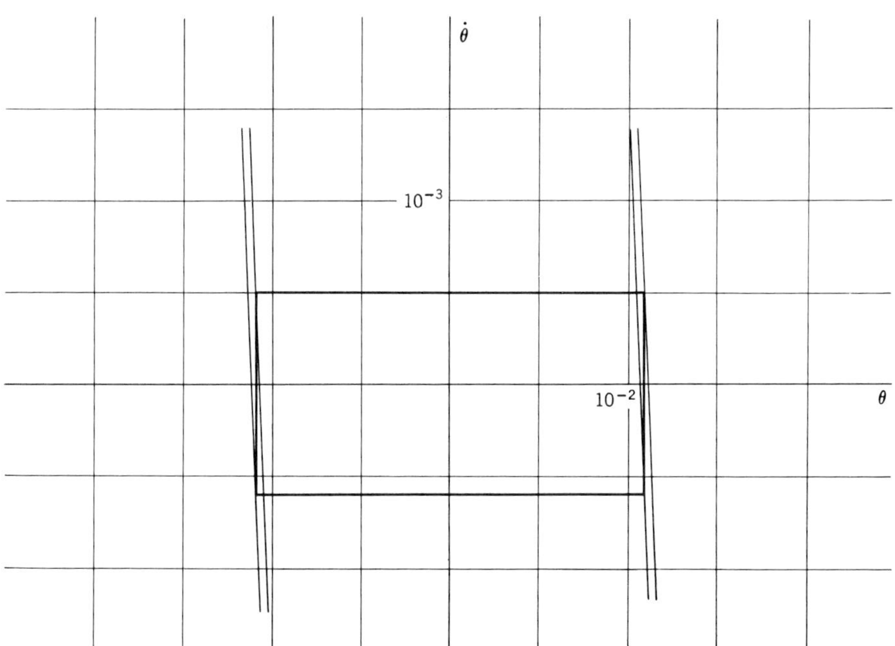

Figure 9.12 Trajectory for $K_1 = 2.45$, $K_2 = 1$, $\delta_1 = 0.026$, $\delta_2 = 0.0273$.

9.2 The Describing Function

The phase-plane method suffers from several disadvantages. Most severe, perhaps, is its limitation to second-order systems.

A method that overcomes the limitation to second-order systems is the method of the *describing function*. This method is approximate in that it attempts to assign a kind of transfer function to the system of nonlinearity. The approach in its simplest form (which is all we shall discuss here) makes use of a Fourier series expansion of the signal at the output of the nonlinearity, and neglects effects of all harmonics except the fundamental. The main application of the method is in investigating limit cycles.

The method supposes that the system under investigation contains one nonlinearity and can be put into the form shown in Figure 9.13. We *assume* that the input to the nonlinearity is a sinusoid:

$$e(t) = E \sin \omega t. \tag{9.50}$$

The output of the nonlinearity can be expressed in a Fourier series:

$$f(e) = F_1 \sin (\omega t + \phi_1) + F_2 \sin (2\omega t + \phi_2) + \cdots \tag{9.51}$$

(We assume for convenience that the nonlinearity is *odd*, so $F_0 = 0$). The F_i and ϕ_i can be found from

$$F_i = \sqrt{A_i^2 + B_i^2}, \tag{9.52}$$

$$\phi_i = \tan^{-1}\frac{B_i}{A_i}, \tag{9.53}$$

where

$$A_i = \frac{\omega}{\pi} \int_{-\pi/\omega}^{\pi/\omega} f(E \sin \omega t) \sin i\omega t \, dt \tag{9.54}$$

$$B_i = \frac{\omega}{\pi} \int_{-\pi/\omega}^{\pi/\omega} f(E \sin \omega t) \cos i\omega t \, dt. \tag{9.55}$$

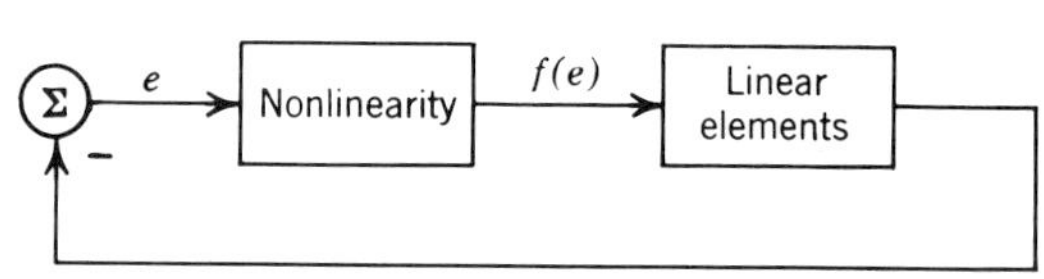

Figure 9.13 Single-loop system with a single nonlinearity.

The describing function is defined as

$$N(j\omega, E) = \frac{F_1 \epsilon^{j\phi_1}}{E}. \tag{9.56}$$

Notice that, for *linear* systems, the describing function is just the ordinary transfer function for $s = j\omega$.

Example 9-6. Calculation of the describing function for the gas-jet actuator of Figure 9.1. The assumed sinusoidal input and resulting "square-wave" output for this device are shown in Figure 9.14.

Thus

$$B_1 = \frac{\omega}{\pi}\left[\int_{-\pi/\omega}^{0} (-T) \cos \omega t \, dt + \int_{0}^{\pi/\omega} T \cos \omega t \, dt\right] = 0 \tag{9.57}$$

$$A_1 = \frac{\omega}{\pi}\left[\int_{-\pi/\omega}^{0} (-T) \sin \omega t \, dt + \int_{0}^{\pi/\omega} T \sin \omega t \, dt\right] \tag{9.58}$$

$$= \frac{4T}{\pi} \tag{9.59}$$

Hence, $F_1 = A_1$, $\phi_1 = 0$, and

$$N(j\omega, E) = \frac{4T}{\pi E}. \ \blacktriangle \tag{9.60}$$

Example 9-7. This example exhibits the calculation of the describing function for the nonlinearity of Figure 9.5. The output of the nonlinearity for a sinusoidal input is shown in Figure 9.15.

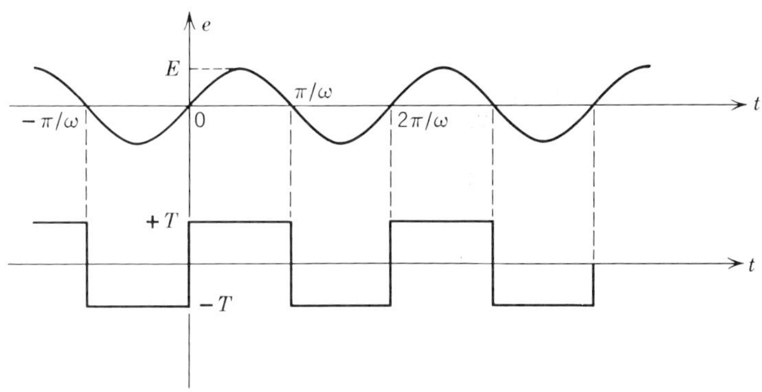

Figure 9.14 Gas-jet input and output.

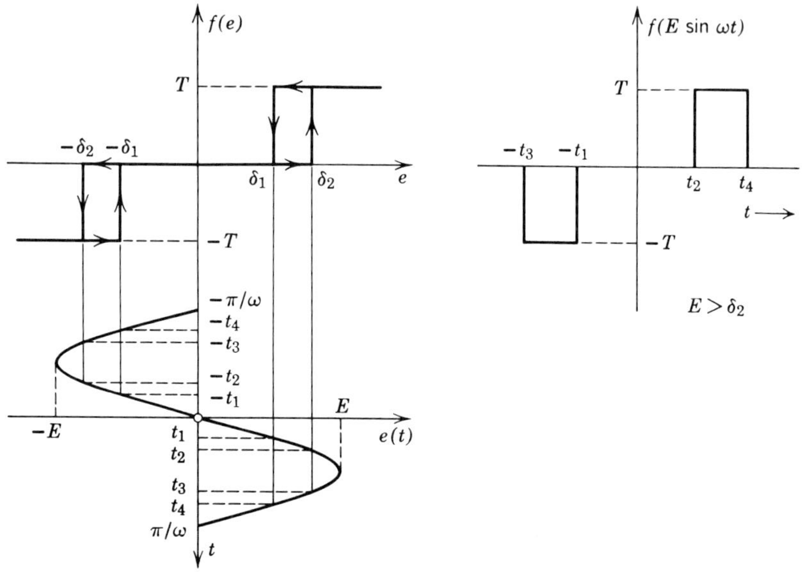

Figure 9.15 Construction of gas-jet nonlinearity output.

$$A_1 = \frac{\omega}{\pi}\left[\int_{-t_3}^{-t_1}(-T)\sin\omega t\, dt + \int_{t_2}^{t_4} T\sin\omega t\, dt\right]$$

$$A_1 = \frac{2T}{\pi}(\cos\omega t_1 + \cos\omega t_2) \tag{9.61}$$

where

$$\cos\omega t_1 = \cos\left(\sin^{-1}\frac{\delta_1}{E}\right) = \sqrt{1-\left(\frac{\delta_1}{E}\right)^2} \tag{9.62}$$

$$\cos\omega t_2 = \cos\left(\sin^{-1}\frac{\delta_2}{E}\right) = \sqrt{1-\left(\frac{\delta_2}{E}\right)^2} \tag{9.63}$$

$$B_1 = \frac{\omega}{\pi}\left[\int_{-t_3}^{-t_1}(-T)\cos\omega t\, dt + \int_{t_2}^{t_4} T\cos\omega t\, dt\right] \tag{9.64}$$

$$B_1 = \frac{2T}{\pi}(\sin\omega t_1 - \sin\omega t_2) \tag{9.65}$$

where

$$\sin\omega t_1 = \frac{\delta_1}{E} \tag{9.66}$$

$$\sin\omega t_2 = \frac{\delta_2}{E}. \blacktriangle \tag{9.67}$$

The main usefulness of the describing function is in the study of limit cycles. Reliable estimates of the amplitude and frequency of limit-cycle oscillations can be obtained in many cases of engineering importance. The technique is based on a *pseudocharacteristic equation* for the system of Figure 9.13:

$$1+N(j\omega, E)G(j\omega) = 0, \tag{9.68}$$

where $G(j\omega)$ is the ordinary transfer function of the linear portion of the system. The values of E and ω (if any) satisfying (9.68) approximate the amplitudes and frequencies of oscillations (limit cycles). This method is approximate, of course, since the effects of higher harmonics have been neglected. The procedure is often (surprisingly!) accurate, however.

Solutions of (9.68) can be found graphically. We recast (9.68) as

$$G(j\omega) = -\frac{1}{N(j\omega, E)}. \tag{9.69}$$

Plots of $G(j\omega)$ and $1/N(j\omega, E)$ can be constructed in the complex plane. Intersections indicate solutions. See Figure 9.16. Several applications of this approach are given in the problems.

9.3 Lyapunov's Direct Method

As was remarked in Chapter 8, there are many kinds of stability. Stability analysis of linear, time-invariant systems was discussed in

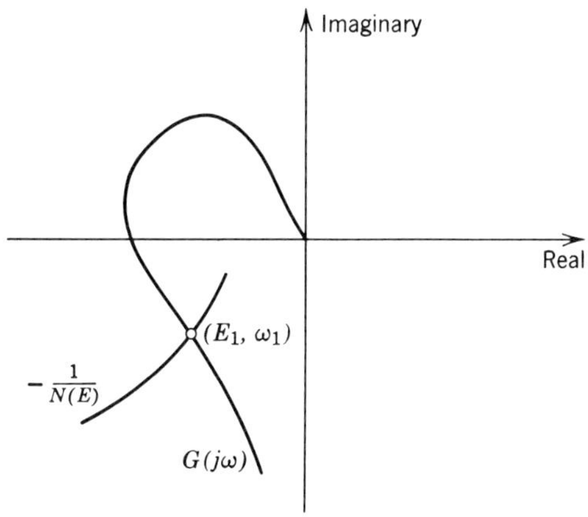

Figure 9.16 Describing function analysis showing a possible limit cycle with amplitude E_1, frequency ω_1.

Chapter 8. In this section we discuss some kinds of stability for nonlinear systems described by

$$\dot{\mathbf{x}} = \mathbf{f}(\mathbf{x}). \tag{9.70}$$

We shall suppose that f is *not* an explicit function of t, although many of the results to be discussed here can be extended to that case. Furthermore, we shall consider only *unforced* systems, $\mathbf{u} \equiv \mathbf{0}$. Stability of forced nonlinear systems is an exceptionally complex topic and is beyond the scope of this book. Finally, we consider only the stability of *equilibrium points*. As has been mentioned earlier, the system (9.65) is said to have an equilibrium point at $\mathbf{x} = \mathbf{x}_e$ if

$$\mathbf{f}(\mathbf{x}_e) = \mathbf{0}. \tag{9.71}$$

Note that an equilibrium point (the system may have several) can always be shifted to the origin in state space by a translation in coordinates. Because of the many kinds of stability, even for equilibrium points, it will be necessary to define concepts rather carefully, and to establish major results in the form of theorems. We shall need an extended notation for the solution to (9.70) similar to that employed in Chapter 7. We shall denote the solution to (9.70) with initial condition $\mathbf{x}_0$ at time $t = 0$ (notice that (9.70) is time-invariant) by $\boldsymbol{\psi}(t; \mathbf{x}_0)$. We are now ready for some definitions.

Definition 9-1. An equilibrium point $\mathbf{x}_e$ of the system (9.70) is said to be *stable* in the sense of Lyapunov (S) if, for all $\epsilon > 0$ there exists a $\delta > 0$, so that if

$$\|\mathbf{x}_0 - \mathbf{x}_e\| < \delta \tag{9.72}$$

then

$$\|\boldsymbol{\psi}(t; \mathbf{x}_0) - \mathbf{x}_e\| < \epsilon \tag{9.73}$$

for all $t \geqslant 0$. ▲

Recalling that norms are an extension of the idea of distance, Definition 9-1 conveys the idea that an equilibrium point is stable if and only if the trajectory $\boldsymbol{\psi}$ remains within a specified "distance" from $\mathbf{x}_e$ *provided* the starting point $\mathbf{x}_0$ is "close enough" to $\mathbf{x}_e$. This idea is expressed geometrically in Figure 9.17, where a typical trajectory near a stable equilibrium point is shown for a choice of ϵ. The Euclidean norm $\|\mathbf{x}\| = \sqrt{x_1^2 + x_2^2}$ is used.

Frequently, we wish the trajectory actually to approach the equili-

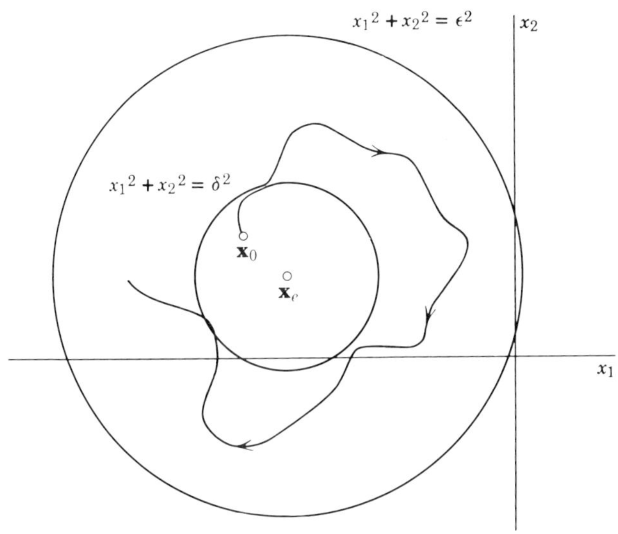

Figure 9.17 A typical trajectory near a stable equilibrium point.

brium point as $t \to \infty$, rather than just to remain in some specified neighborhood. Thus we have

Definition 9-2. An equilibrium point $\mathbf{x}_e$ of the system (9.70) is said to be *asymptotically stable* (AS) if

(a) it is stable
and if, for $\|\mathbf{x}_0 - \mathbf{x}_e\| < \delta$, (9.74)
(b) $\lim_{t\to\infty} \boldsymbol{\psi}(t; \mathbf{x}_0) = \mathbf{x}_e$. ▲

Note that Part (b) alone is not enough: Figure 9.18 shows a time response of a one-dimensional system satisfying the limit requirement of Part (b), but does not satisfy Part (a) and, thus, is not AS. It is also not *stable*!

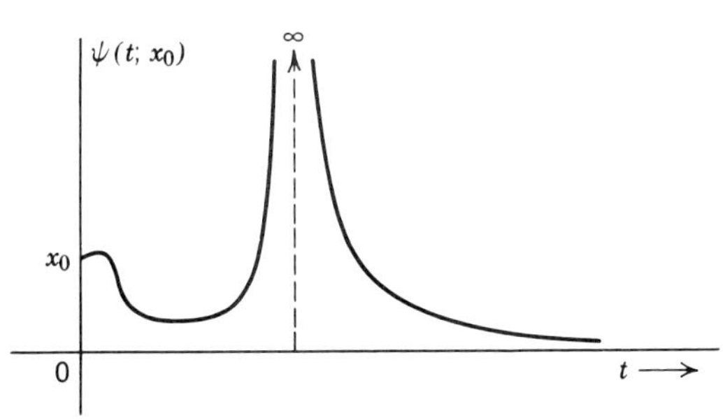

Figure 9.18 Unstable response for which $\lim_{t\to\infty} \psi(t; x_0) \to x_e = 0$.

Finally, we have

Definition 9-3. An equilibrium point $\mathbf{x}_e$ of the system (9.70) is said to be *asymptotically stable in the large* (ASL) if it is AS with Part (b) satisfied for *any* initial condition $\mathbf{x}_0$. ▲

These stability definitions are expressed in terms of the time-domain solution of (9.70). Since, in most cases, (9.70) cannot be solved analytically, we need an indirect method of investigating stability. Such a method was proposed by the Russian mathematician A. M. Lyapunov (1857–1918) at about the turn of the century. Although this method provides only sufficient but not necessary conditions for stability, it represents the best general approach to stability currently available. The method is based on an examination of a scalar function of the state vector, and will be denoted here by $V(\mathbf{x})$. Before stating Lyapunov's main stability theorem, we consider two motivations for the role of $V(\mathbf{x})$ in stability examinations.

Example 9-8. $V(\mathbf{x})$ **as Stored Energy.** In the system below (see Figure 9.19), state variables x_1 and x_2, the total stored energy is

$$V(\mathbf{x}) = \tfrac{1}{2}x_1^2 + \tfrac{1}{2}x_2^2. \tag{9.75}$$

Intuitively, we would expect the equilibrium point to be stable (or even AS or ASL) if the stored energy were always decreasing as a function of time, that is, if

$$\dot{V} = \frac{dV}{dt} < 0 \tag{9.76}$$

for all $t > 0$. In this example,

$$\dot{V} = x_1\dot{x}_1 + x_2\dot{x}_2. \tag{9.77}$$

The state variable derivatives can be evaluated from the state equations:

$$\dot{x}_1 = -x_1 - x_2 \tag{9.78}$$

$$\dot{x}_2 = x_1 - x_2. \tag{9.79}$$

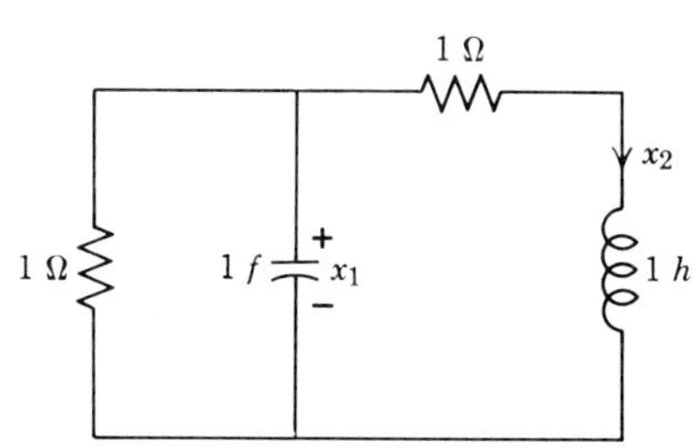

Figure 9.19 Circuit for Example 9-8.

Then

$$\dot{V} = -x_1^2 - x_2^2, \tag{9.80}$$

which is negative everywhere except at the equilibrium point

$$\mathbf{x}_e = \begin{bmatrix} 0 \\ 0 \end{bmatrix}. \tag{9.81}$$

Thus the stored energy (a nonnegative quantity) is always decreasing and, thus, we would expect stability. ▲

Example 9-9. $V(\mathbf{x})$ as "Distance." Consider the same network as in Example 9-8. In the state plane (x_1, x_2), the contours of constant distance from the equilibrium point (origin) are given by

$$V(\mathbf{x}) = x_1^2 + x_2^2 = M^2. \tag{9.82}$$

Intuitively, we would expect the equilibrium point to be stable (or even AS or ASL) if the trajectory is such that the distance to the origin is deceasing, that is, if

$$\dot{V} < 0.$$

From Example 9-8,

$$\dot{V} = -2x_1^2 - 2x_2^2 < 0, \tag{9.83}$$

and thus the trajectory gets nearer the origin as time increases. ▲

The purpose of the preceding two examples is to suggest an interpretation of $V(\mathbf{x})$ in terms of energy or distance. Such a viewpoint is often helpful in applications. However, $V(\mathbf{x})$ need not correspond to stored energy or to distance (or even to a norm). The stored energy might *increase* for some time interval $t_1 < t < t_2$, for example, and then finally decay to zero.

To state the stability theorem precisely, we need the notion of a positive definite function. A continuous scalar $V(\mathbf{x})$ is said to be positive definite if

$$\text{(a)}\ \ V(\mathbf{x}) = 0 \text{ for } \mathbf{x} = \mathbf{0}. \tag{9.84}$$

$$\text{(b)}\ \ V(\mathbf{x}) > 0 \text{ for } \mathbf{x} \neq \mathbf{0}. \tag{9.85}$$

For example, the function

$$V(\mathbf{x}) = \tfrac{1}{2}x_1^2 + \tfrac{1}{2}x_2^2 \tag{9.86}$$

is positive definite, while

$$V(\mathbf{x}) = \tfrac{1}{2}x_1^2 + \tfrac{1}{2}(x_1 - x_2)^2 \tag{9.87}$$

is not. [(9.87) is said to be positive semidefinite.] We shall frequently choose *quadratic forms* as V-functions. A quadratic form is the scalar function given by

$$\mathbf{x}^T\mathbf{P}\mathbf{x} \tag{9.88}$$

where $\mathbf{P}$ is a constant matrix. For example, in the two-dimensional case,

$$\begin{aligned}\mathbf{x}^T\mathbf{P}\mathbf{x} &= [x_1 \quad x_2]\begin{bmatrix} p_{11} & p_{12} \\ p_{21} & p_{22}\end{bmatrix}\begin{bmatrix} x_1 \\ x_2\end{bmatrix} \\ &= p_{11}x_1^2 + (p_{12}+p_{21})x_1x_2 + p_{22}x_2^2.\end{aligned} \tag{9.89}$$

Notice that (9.75) and (9.80), (9.82), (9.83), and (9.87) are quadratic forms.

A useful test exists for determining whether or not a quadratic form is positive definite. We state, without proof, Sylvester's criterion for positive definiteness of a quadratic form:

Theorem 9.1. A quadratic form $\mathbf{x}^T\mathbf{P}\mathbf{x}$, where P is a symmetric $n \times n$ matrix, is positive definite if and only if all *principal minors* of $\mathbf{P}$ are positive, that is,

$$p_{11} > 0, \begin{vmatrix} p_{11} & p_{12} \\ p_{21} & p_{22}\end{vmatrix} > 0,$$

$$\begin{vmatrix} p_{11} & p_{12} & p_{13} \\ p_{21} & p_{22} & p_{23} \\ p_{31} & p_{32} & p_{33}\end{vmatrix} > 0, \ldots, \det \mathbf{P} > 0. \blacktriangle \tag{9.90}$$

Example 9-10. From (9.75),

$$V(\mathbf{x}) = \tfrac{1}{2}x_1^2 + \tfrac{1}{2}x_2^2 = [x_1 \quad x_2]\begin{bmatrix} \frac{1}{2} & 0 \\ 0 & \frac{1}{2}\end{bmatrix}\begin{bmatrix} x_1 \\ x_2\end{bmatrix}. \tag{9.91}$$

Thus, since $\frac{1}{2} > 0$, $(\frac{1}{2})(\frac{1}{2}) > 0$, $V(\mathbf{x})$ is positive definite. ▲

We now consider the main result of this section, a stability theorem due to Lyapunov. As has been mentioned previously, we restrict our attention to the time-invariant system

$$\dot{\mathbf{x}} = \mathbf{f}(\mathbf{x}), \tag{9.92}$$

although similar results can be established for $\dot{\mathbf{x}} = \mathbf{f}(\mathbf{x}, t)$. In stating the theorem, we shall consider the equilibrium point whose stability is being investigated to be at the origin, that is,

$$\mathbf{f}(\mathbf{x}_e) = \mathbf{f}(\mathbf{0}) = \mathbf{0}. \tag{9.93}$$

This can always be arranged by a translation of coordinates and thus produces no loss of generality.

Theorem 9.2. Consider a system described by

$$\dot{\mathbf{x}} = \mathbf{f}(\mathbf{x})$$

with

$$\mathbf{f}(\mathbf{0}) = \mathbf{0}.$$

Suppose there exists a continuous scalar function $V(\mathbf{x})$ with continuous first partial derivatives having the following properties:

(1) $V(\mathbf{x})$ is positive definite for all $\mathbf{x}$.
(2) $-\dot{V}(\mathbf{x})$ is positive definite for all $\mathbf{x}$.
(3) $V(\mathbf{x}) \to \infty$ as $\|\mathbf{x}\| \to \infty$.

Then $\mathbf{x}_e = \mathbf{0}$ is asymptotically stable in the large. ▲

We consider two examples.

Example 9-11. In Example 9-8,

$$\dot{x}_1 = -x_1 - x_2$$
$$\dot{x}_2 = x_1 - x_2.$$

Clearly, $\mathbf{0}$ is an equilibrium point. Try $V(\mathbf{x}) = \frac{1}{2}x_1^2 + \frac{1}{2}x_2^2$. As was shown in Example 9-10, the quadratic form is positive definite. Moreover,

$$\dot{V} = x_1\dot{x}_1 + x_2\dot{x}_2 = -x_1^2 - x_2^2.$$

Thus

$$-\dot{V} = [x_1 \quad x_2]\begin{bmatrix}1 & 0\\ 0 & 1\end{bmatrix}\begin{bmatrix}x_1\\ x_2\end{bmatrix},$$

which is also positive definite. Thus, by Theorem 9.2, for this system, $\mathbf{x}_e = \mathbf{0}$ is asymptotically stable in the large. ▲

Example 9-12. In Chapter 5 it was shown by (5.36) that, using principal axes as body axes, spacecraft motion is described by

$$\begin{aligned} J_1\dot{\omega}_1 &= (J_2 - J_3)\omega_2\omega_3 + \tau_1' \\ J_2\dot{\omega}_2 &= (J_3 - J_1)\omega_3\omega_1 + \tau_2' \\ J_3\dot{\omega}_3 &= (J_1 - J_2)\omega_1\omega_2 + \tau_3'. \end{aligned} \tag{9.94}$$

Suppose that, by means of feedback, the torque is arranged to be

$$\tau' = \begin{bmatrix} K_1\omega_1 \\ K_2\omega_2 \\ K_3\omega_3 \end{bmatrix}. \tag{9.95}$$

Clearly, $\omega_1 = \omega_2 = \omega_3 = 0$ is then an equilibrium point of (9.94). To investigate the stability of this equilibrium point, choose

$$V(\mathbf{x}) = V(\omega_1, \omega_2, \omega_3) = \tfrac{1}{2}J_1\omega_1^{\,2} + \tfrac{1}{2}J_2\omega_2^{\,2} + \tfrac{1}{2}J_3\omega_3^{\,2} \tag{9.96}$$

Plainly, (9.96) is positive definite (the moments of inertia are positive, of course). Then

$$\begin{aligned} -\dot{V} &= -J_1\omega_1\dot{\omega}_1 - J_2\omega_2\dot{\omega}_2 - J_3\omega_3\dot{\omega}_3 \\ &= -K_1\omega_1^{\,2} - K_2\omega_2^{\,2} - K_3\omega_3^{\,2}. \end{aligned} \tag{9.97}$$

Thus $-\dot{V}$ of (9.97) will be positive definite if

$$K_1 < 0,$$

and

$$K_2 < 0,$$

and

$$K_3 < 0. \tag{9.98}$$

If K_1, K_2, and K_3 are chosen as in (9.98), the origin will be asymptotically stable in the large, regardless of the values of the spacecraft moments of inertia. ▲

We now consider the geometric aspects of proof of Theorem 9.2. We shall not give a rigorous proof of this theorem. We must establish asymptotic stability in the large. Reviewing Definitions 9.1, 9.2, and 9.3, we see that ASL requires first *stability*, and then $\lim \boldsymbol{\psi}(t; \mathbf{x}_0) = \mathbf{0}$. From Definition 9.1, the equilibrium point $\mathbf{x}_e = \mathbf{0}$ is stable if for all $\epsilon > 0$ there exists *a* $\delta > 0$ such that if $\|\mathbf{x}_0\| < \delta$, then $\|\boldsymbol{\psi}(t; \mathbf{x}_0\| < \epsilon$ for all $t \geqslant 0$. Notice that since, by hypothesis, $\dot{V}(\mathbf{x}) < 0$ (Condition (2) of Theorem 9.2),

$$V(\boldsymbol{\psi}(t; \mathbf{x}_0)) < V(\mathbf{x}_0)$$

for any solution $\boldsymbol{\psi} \neq \mathbf{x}_0$. Moreover, from Condition (1), $V(\mathbf{x})$ is zero only at $\mathbf{x} = \mathbf{0}$. Thus the geometry of the V-space is as shown by the "cup" in Figure 9.20. (The geometry for $n = 2$ is pictured. The situation is similar for $n > 2$). Suppose that any value of ϵ is chosen.

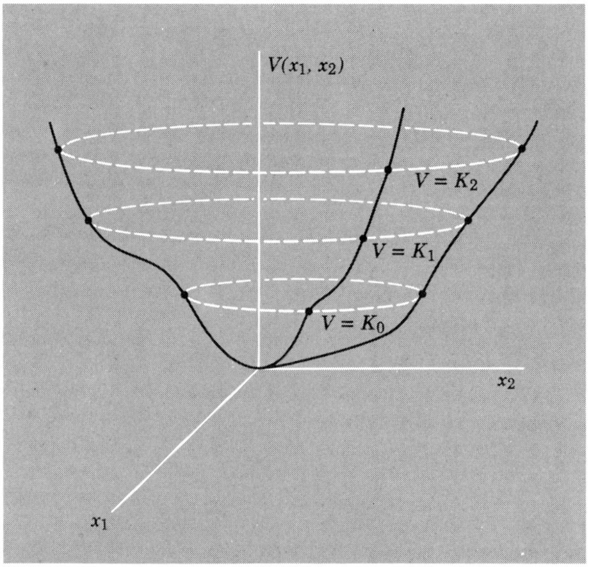

Figure 9.20 The "cup" geometry of $V(\mathbf{x})$.

Figure 9.17 indicates how a contour of constant V (shown as an ellipse in this case) is selected to just contain the region $\|\mathbf{x}\| = \epsilon$ (using the Euclidean norm). Since $\dot{V}$ is negative, any trajectory originating with the contour $V = K_0$ will remain within this contour and, hence, within $\|\mathbf{x}\| = \epsilon$. An appropriate choice of δ is indicated in Figure 9.21.

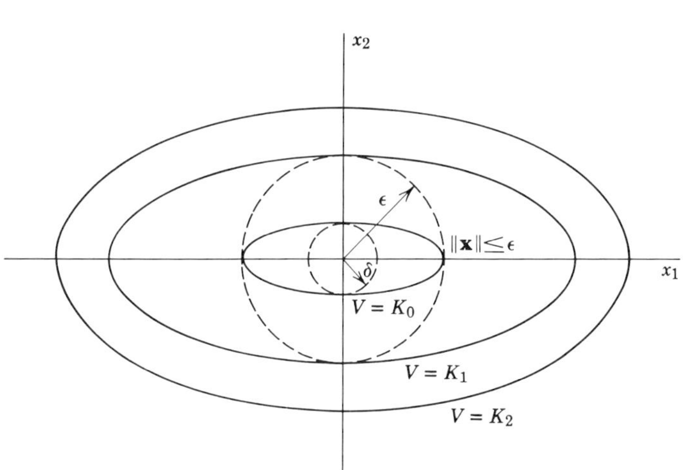

Figure 9.21 Projection of the cup of Figure 9.20 on the (x_1, x_2) plane.

A scalar function of $\mathbf{x}$ satisfying the three conditions of Theorem 9.2 is called a *Lyapunov function*. Unfortunately, no general method exists for the construction of Lyapunov functions, except for the case of linear time-invariant systems, as is discussed in the next section. Hence the examination of stability of nonlinear systems is a trial-and-error procedure. Some useful guidelines have been developed for synthesizing possible trial Lyapunov functions, however. See D. R. Ingwerson, "A Modified Lyapunov Method for Nonlinear Stability Analysis," *IRE Trans. on Automatic Control*, AC-6, pp. 199–210, May 1961; and D. G. Schultz and J. E. Gibson, "The Variable Gradient Method for Generating Liapunov Functions," *Trans. AIEE*, Vol. 81, Part II, pp. 203–210, 1962. Quadratic forms have been found to be most useful in practice. Fortunately, it has been established (by Massera) that at least one Lyapunov function exists for every system having an asymptotically stable equilibrium point.

Notice also that Theorem 9.2 merely provides a sufficient condition. Failure to find a suitable Lyapunov function does not imply instability.

9.4 Stability of Linear Systems via Lyapunov's Direct Method

A complete theory of stability using Lyapunov functions does exist for linear time-invariant systems. In particular, *necessary and sufficient* conditions can be stated, and suitable Lyapunov functions can *always* be generated. We shall consider these results in this section, both as a specific application of the general results of the previous section and as a point of connection of Lyapunov stability with the types of stability discussed in Chapter 8.

We first connect the stability of equilibrium points with the notion of the "asymptotically stable system" of Chapter 8. To establish this, we need that the notion of the *norm of a matrix*.

Definition 9.4. The norm of the matrix $\mathbf{M}$, denoted by $\|\mathbf{M}\|$, is defined by

$$\|\mathbf{M}\| = \sup_{\mathbf{x}\neq\mathbf{0}} \frac{\|\mathbf{M}\mathbf{x}\|}{\|\mathbf{x}\|}, \tag{9.99}$$

where "sup" denotes the *least upper bound* (*supremum*) of the ratio of vector norms. ▲

Thus, by Definition 9.4,

$$\|\mathbf{M}\mathbf{x}\| \leq \|\mathbf{M}\|\,\|\mathbf{x}\|. \tag{9.100}$$

This is all we need to know about matrix norms to establish

Theorem 9.3. The *equilibrium point* $\mathbf{x}_e = \mathbf{0}$ of the linear time-invariant system

$$\dot{\mathbf{x}} = \mathbf{A}\mathbf{x} \tag{9.101}$$

is ASL if and only if the *system* (9.101) is an asymptotically stable system in the sense of Definition 8.1, that is, if and only if

$$\lim_{t\to\infty} \phi_{ij}(t) = 0. \blacktriangle \tag{9.102}$$

PROOF. Suppose that (9.102) holds. Since

$$\mathbf{x}(t) = \boldsymbol{\phi}(t)\mathbf{x}_0, \tag{9.103}$$

then, from (9.102) and (9.100),

$$\lim_{t\to\infty} \mathbf{x}(t) = [\lim_{t\to\infty} \boldsymbol{\phi}(t)]\mathbf{x}_0 = \mathbf{0}. \tag{9.104}$$

Thus an "asymptotically stable system" implies ASL of $\mathbf{x}_e = \mathbf{0}$.
Now suppose $\mathbf{x}_e = \mathbf{0}$ is ASL. This means

$$\lim_{t\to\infty} \mathbf{x}(t) = \mathbf{0}, \text{ or} \tag{9.105}$$

$$\lim_{t\to\infty} [\boldsymbol{\phi}(t)\mathbf{x}_0] = \mathbf{0} \text{ for all } \mathbf{x}_0.$$

Thus

$$\lim_{t\to\infty} \phi_{ij}(t) = 0. \blacktriangle \tag{9.106}$$

Therefore, ASL is equivalent to a linear time-invariant system being asymptotically stable in the sense that one notion implies the other. Moreover, the tests for an asymptotically stable system given in Chapter 8 are also tests for ASL of $\mathbf{x}_e = \mathbf{0}$.
The main result of this section is the following.

Theorem 9.4. The equilibrium point $\mathbf{x}_e = \mathbf{0}$ of (9.101) is asymptotically stable in the large if and only if, given any positive definite symmetric matrix $\mathbf{Q}$, there exists a positive definite symmetric matrix $\mathbf{P}$ such that

$$\mathbf{A}^T\mathbf{P} + \mathbf{P}\mathbf{A} = -\mathbf{Q}. \tag{9.107}$$

Furthermore, $\mathbf{x}^T\mathbf{P}\mathbf{x}$ is a Lyapunov function for the system of (9.101). $\blacktriangle$

PROOF. First, suppose there exists a positive definite solution $\mathbf{P}$ to (9.107) for a positive definite $\mathbf{Q}$. Then

$$V(\mathbf{x}) = \mathbf{x}^T\mathbf{P}\mathbf{x}$$

satisfies conditions (1) and (3) of Theorem 9.2. To examine condition (2), we form

$$\begin{aligned} -\dot{V}(\mathbf{x}) &= -\dot{\mathbf{x}}^T\mathbf{P}\mathbf{x} - \dot{\mathbf{x}}^T\mathbf{P}\mathbf{x} \\ &= -\mathbf{x}^T\mathbf{A}^T\mathbf{P}\mathbf{x} - \mathbf{x}^T\mathbf{P}\mathbf{A}\mathbf{x} \\ &= \mathbf{x}^T\mathbf{Q}\mathbf{x}. \end{aligned} \tag{9.108}$$

Thus, $-\dot{V}$ is positive definite, and all conditions of Theorem 9.2 are satisfied. Hence a solution $\mathbf{P}$ to (9.100) implies ASL.

Next, we assume the equilibrium point is ASL and show that a positive definite solution to (9.107) exists. Consider the *matrix* equation

$$\dot{\mathbf{X}} = \mathbf{A}^T\mathbf{X} + \mathbf{X}\mathbf{A}, \quad \mathbf{X}(0) = \mathbf{Q}, \tag{9.109}$$

where $\mathbf{X}$ is an $n \times n$ matrix. It is easily verified that

$$\mathbf{X} = \boldsymbol{\epsilon}^{\mathbf{A}^Tt}\mathbf{Q}\boldsymbol{\epsilon}^{\mathbf{A}t} \tag{9.110}$$

is the solution to (9.109), assuming uniqueness. Notice that the initial condition is clearly satisfied and, moreover, that the differential equation is satisfied (recall that $(d/dt)\boldsymbol{\epsilon}^{\mathbf{A}t} = \mathbf{A}\boldsymbol{\epsilon}^{\mathbf{A}t}$).

Integrating both sides of (9.109),

$$\mathbf{X}(\infty) - \mathbf{X}(0) = \mathbf{A}^T\left[\int_0^\infty \mathbf{X}\,dt\right] + \left[\int_0^\infty \mathbf{X}\,dt\right]\mathbf{A}. \tag{9.111}$$

Because of stability and Theorem 9.3, $\mathbf{X}(\infty) = \mathbf{0}$. Furthermore, we adopt the notation

$$\mathbf{P} = \int_0^\infty \mathbf{X}\,dt = \int_0^\infty \boldsymbol{\epsilon}^{\mathbf{A}^Tt}\mathbf{Q}\boldsymbol{\epsilon}^{\mathbf{A}t}\,dt. \tag{9.112}$$

Again because of stability, $\mathbf{P}$ exists. Observe also that $\mathbf{P}^T = \mathbf{P}$. To establish that $\mathbf{P}$ is positive definite, consider the quadratic form

$$V = \mathbf{x}_0{}^T\mathbf{P}\mathbf{x}_0 = \mathbf{x}_0{}^T\int_0^\infty \boldsymbol{\epsilon}^{\mathbf{A}^Tt}\mathbf{Q}\boldsymbol{\epsilon}^{\mathbf{A}t}\,dt\,\mathbf{x}_0 \tag{9.113}$$

for all $\mathbf{x}_0$. This becomes

$$V = \int_0^\infty \mathbf{x}^T(t)\,\mathbf{Q}\mathbf{x}(t)\,dt. \tag{9.114}$$

Since **Q** itself is positive definite, the integrand and hence the integral are positive definite. ▲

Example 9-13. Suppose a system is characterized by

$$A = \begin{bmatrix} 0 & 1 \\ -5 & -6 \end{bmatrix}. \tag{9.115}$$

Choosing $\mathbf{Q} = \mathbf{I}$, (9.107) becomes

$$\begin{bmatrix} 0 & -5 \\ 1 & -6 \end{bmatrix}\begin{bmatrix} p_{11} & p_{12} \\ p_{12} & p_{22} \end{bmatrix} + \begin{bmatrix} p_{11} & p_{12} \\ p_{12} & p_{22} \end{bmatrix}\begin{bmatrix} 0 & 1 \\ -5 & -6 \end{bmatrix} = \begin{bmatrix} -1 & 0 \\ 0 & -1 \end{bmatrix}, \tag{9.116}$$

where the symmetric character of **P** is used explicitly. Expanding (9.116) yields three simultaneous linear *algebraic* equations for the elements of **P**:

$$\begin{aligned} -10p_{12} &= -1 \\ p_{11} - 6p_{12} - 5p_{22} &= 0 \\ 2(p_{12} - 6p_{22}) &= -1. \end{aligned} \tag{9.117}$$

The solution is

$$\begin{aligned} p_{11} &= \tfrac{11}{10} \\ p_{12} &= \tfrac{1}{10} \\ p_{22} &= \tfrac{1}{10} \end{aligned} \tag{9.118}$$

Checking

$$\mathbf{P} = \begin{bmatrix} \frac{11}{10} & \frac{1}{10} \\ \frac{1}{10} & \frac{1}{10} \end{bmatrix}$$

for positive definiteness using Theorem 9.1, we have

$$\tfrac{11}{10} > 0, \qquad (\tfrac{11}{10})(\tfrac{1}{10}) - (\tfrac{1}{10})(\tfrac{1}{10}) > 0, \tag{9.119}$$

and thus **P** is positive definite. Therefore, by Theorem 9.4, the system of (9.117) is asymptotically stable in the large or, equivalently, $\mathbf{x}_e = \mathbf{0}$ is ASL. ▲

Note that (9.107) represents $n(n+1)/2$ simultaneous linear algebraic equations for the elements of **P**.

9.5 Sensitivity of Nonlinear Systems

We have remarked that there are no general methods of analysis for nonlinear systems. Indeed, the techniques discussed in this chapter clearly represent limited, although useful, approaches to the general problem of nonlinear system analysis. In this section, we consider computer simulation as an analysis tool. Special consideration is given to parameter variation problems.

Nonlinear systems can be simulated on computers, of course. In Chapter 4, some methods useful in nonlinear system simulation were discussed. These techniques are particularly useful if the response of a totally specified system to a specific input is desired. If some system parameters are not fixed numerically, however, the procedure becomes more complicated. One approach is to make several computer runs, one for each "typical" combination of parameter values. Although such an approach is often very tedious and "unscientific," even with computer assistance, it is frequently used, primarily because there may be essentially no alternatives.

An alternative that may be easier in some cases involves the use of sensitivity functions. As has been pointed out in previous chapters, sensitivity functions provide first-order estimates of the effect on responses of parameter changes. To be specific, consider a nonlinear system described by

$$\dot{\mathbf{x}} = \mathbf{f}(\mathbf{x}, \mathbf{u}, \boldsymbol{\alpha}, t), \tag{9.120}$$

where $\mathbf{x}$ is the state, $\mathbf{u}$ the input, and $\boldsymbol{\alpha}$ a set of p time-invariant parameters. Denote the solution of (9.115) by

$$\boldsymbol{\psi}(t; \mathbf{x}_0, t_0, \boldsymbol{\alpha}), \tag{9.121}$$

where the dependence of the solution on the initial state $\mathbf{x}_0$, the initial time t_0, and the parameter vector $\boldsymbol{\alpha}$, as well as on time t, has been indicated explicitly. We are interested in the sensitivity functions,

$$\left.\frac{\partial \boldsymbol{\psi}}{\partial \alpha_i}\right|_{\alpha_i^0} = \boldsymbol{\eta}^i(t), \tag{9.122}$$

$i = 1, \ldots, p$. Notice that the functions are evaluated at the normal parameter values. Differential equations for these functions can be found by differentiation of (9.120) with respect to α_i.

$$\begin{aligned} \frac{\partial}{\partial \alpha_i}\dot{\mathbf{x}} &= \frac{\partial}{\partial \alpha_i}\mathbf{f}(\mathbf{x}, \mathbf{u}, \boldsymbol{\alpha}, t) \\ &= \left[\frac{\partial \mathbf{f}}{\partial \mathbf{x}}\right]\frac{\partial}{\partial \alpha_i}\mathbf{x} + \left[\frac{\partial \mathbf{f}}{\partial \mathbf{a}}\right]\frac{\partial \alpha}{\partial \alpha_i}, \end{aligned} \tag{9.123}$$

where the Jacobian matrices are

$$\left[\frac{\partial \mathbf{f}}{\partial \mathbf{x}}\right]_{ij} = \frac{\partial f_i}{\partial x_j}$$

$$\left[\frac{\partial \mathbf{f}}{\partial \boldsymbol{\alpha}}\right]_{ij} = \frac{\partial f_i}{\partial \alpha_j}, \tag{9.124}$$

and it has been assumed that the system input $\mathbf{u}$ does not depend on $\boldsymbol{\alpha}$. We assume further that the time derivative and α_i derivative may be interchanged. Recalling that all quantities are evaluated for nominal parameters and the nominal solution, we may write (9.123) as

$$\dot{\boldsymbol{\eta}}^i = \mathbf{A}(t)\boldsymbol{\eta} + \mathbf{B}(t), \tag{9.125}$$

where

$$\mathbf{A}(t) = \left[\frac{\partial \mathbf{f}}{\partial \mathbf{x}}\right]_{\text{nominal}} \tag{9.126}$$

and

$$\mathbf{B}(t) = \left[\frac{\partial \mathbf{f}}{\partial \boldsymbol{\alpha}} \frac{\partial \boldsymbol{\alpha}}{\partial \alpha_i}\right]_{\text{nominal}} \tag{9.127}$$

Thus the sensitivity functions are solutions to a linear, time-varying ordinary differential equation. These equations can be used to generate system responses for small deviations from a nominal response.

Problems

9-1. The input to the piecewise-linear system of Figure P9-1 is a unit step. Sketch some typical trajectories in the $(c, \dot{c})$ plane.

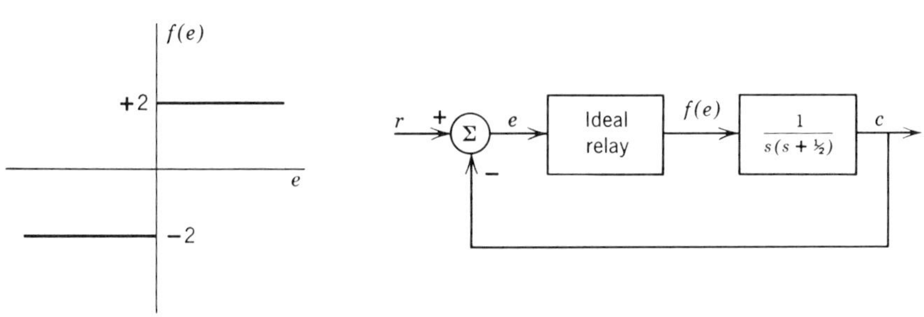

Figure P9-1

9-2. For the piecewise-linear system of Figure P9-2, sketch some trajectories in the $(c, \dot{c})$ plane.

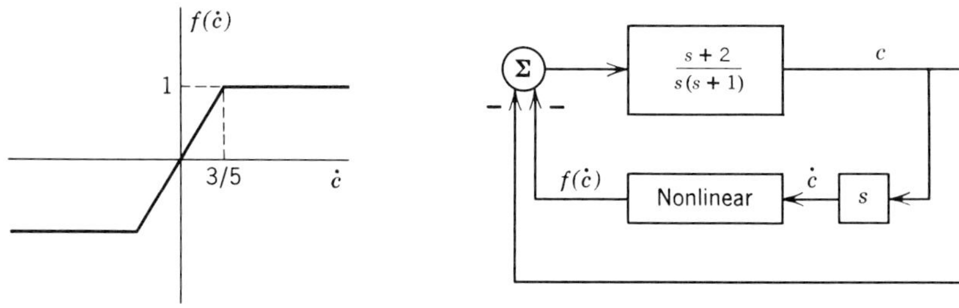

Figure P9-2

9-3. (Computer Problem) Develop an analog computer simulation of the controller of Figure 9.4, Example 9-1.

9-4. (Computer Problem) Develop a digital computer algorithm for simulation of the controller of Figure 9.4, Example 9-1.

9-5. Figure 9.8 shows a possible trajectory for the attitude-control system of Example 9-2. Sketch some other possible trajectories for various initial conditions and various values of the parameters K and T/J. In particular, discuss the case of large positive initial conditions combined with large T and large K.

9-6. In practice, there will be some time delay associated with the switching of the gas jets with dead zone and hysteresis described in Example 9-2. This delay may be modeled by introducing a pure time delay in Figure 9.6, as shown in Figure P9-6. Sketch the resulting switching lines replacing those of Figure 9.7.

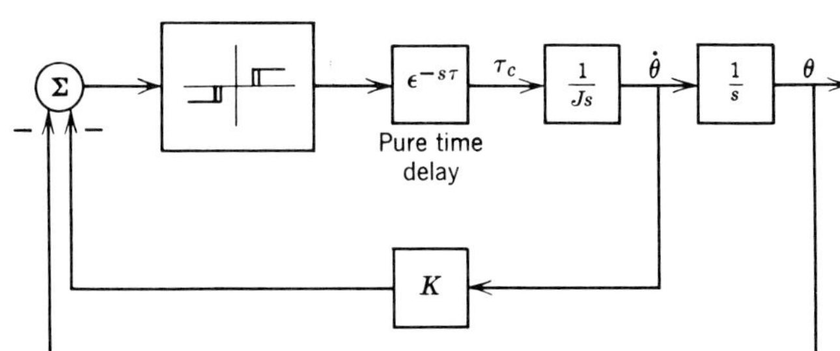

Figure P9-6

9-7. Sketch some trajectories for the system of Figure P9-6 for various values ot τ, T/J, K, and initial conditions. Interpret your sketch physically.

9-8. Plot $|N(j\omega, E)|$ vs E and $\phi(j\omega, E)$ vs E for the describing function of the "ideal relay" gas-jet actuator of Example 9.6, Equation (9.60).

9-9. Repeat Problem 9-8 for the describing function of the hysteresis-dead-zone characteristic of Example 9.7.

9-10. Use the describing function method to find the duty cycle of the attitude control system of Example 9.4. Use the numerical values given in that example for your calculation.

9-11. Calculate the describing function for the saturation nonlinearity of Figure P9-11. Sketch the magnitude and phase vs E.

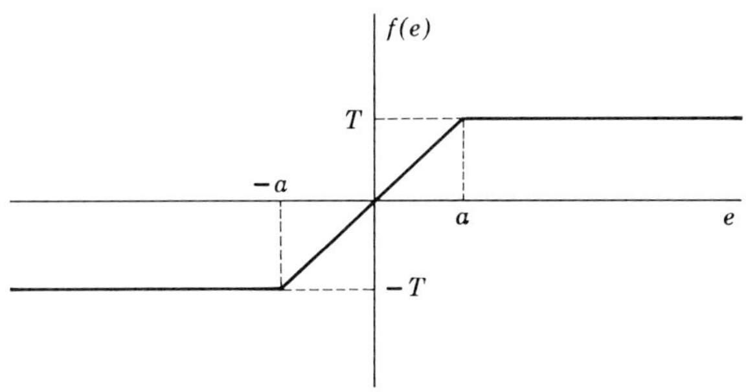

Figure P9-11

9-12. Investigate the existence of limit cycles in the system of Figure P9-12 using the describing function method.

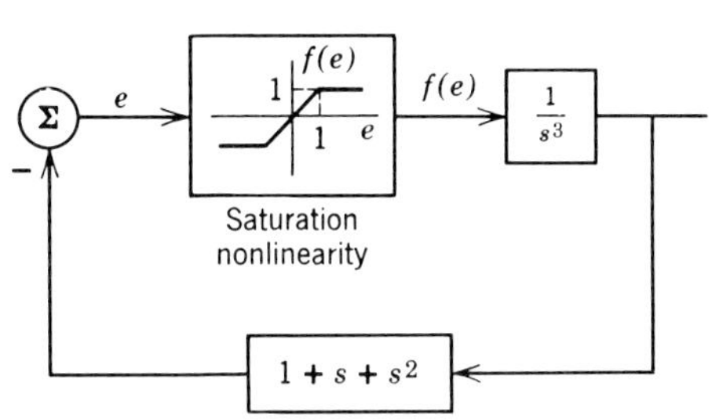

Figure P9-12

9-13. Investigate the stability of the point $\mathbf{x}_e = \begin{bmatrix} 0 \\ 0 \end{bmatrix}$ for the system

$$\dot{x}_1 = x_2 - ax_1(x_1^2 + x_2^2)$$
$$\dot{x}_2 = -x_1 - ax_2(x_1^2 + x_2^2)$$

using $V(\mathbf{x}) = x_1^2 + x_2^2$ as a Lyapunov function.

9-14. Euler's dynamical equations describing rigid body motion with control torque inputs are:

$$\dot{\omega}_1 = \left[\frac{J_2 - J_3}{J_2}\right]\omega_2\omega_3 + u_1$$

$$\dot{\omega}_2 = \left[\frac{J_3 - J_1}{J_2}\right]\omega_3\omega_1 + u_2$$

$$\dot{\omega}_3 = \left[\frac{J_1 - J_2}{J_3}\right]\omega_1\omega_2 + u_3$$

$$u_1 = K_1\omega_1 + K_{13}\omega_1^3$$

$$u_2 = K_2\omega_2 + K_{23}\omega_2^3$$

$$u_3 = K_3\omega_3 + K_{33}\omega_3^3$$

Discuss the stability of the equilibrium point $\omega = \mathbf{0}$ using an appropriate Lyapunov function.

9-15. Select some state variables and use Theorem 9.4 to study the stability of the equilibrium point of the state representation of

$$\dddot{y} + a_1\ddot{y} + a_2\dot{y} + a_3y = 0.$$

Are the stability conditions equivalent to those obtained by the Routh test?

9-16. Verify that (9.110) is the solution to (9.109), assuming uniqueness.

9-17. Draw an all-integrator block diagram showing the generation of sensitivity functions for nonlinear systems by implementing the equations of Section 9.5.

9-18. A nonlinear system is described by

$$\dot{x}_1 = x_2$$

$$\dot{x}_2 = -(1 - \alpha x_1^2)x_2 - x_1 + u(t).$$

Show a simulation diagram for obtaining $\partial x_1/\partial\alpha$.

9-19. (Computer problem) Using the simulation of Problem 9-18, obtain $x_1(t)$, $x_2(t)$, and $\partial x_1(t)/\partial\alpha$ for $u(t)$ = unit step, $x_{10} = 1$, $x_{20} = 0$.

CHAPTER 10

Analysis of Discrete-Time Systems

The digital computer has had a major role in systems engineering. For example, in Chapter 2 the role of the computer in simulation was discussed. Digital computers perform calculations in *discrete time*. For instance, the computer output typically is a sequence of numbers appearing at discrete-time instants t_k. Systems whose inputs and outputs are functions of a discrete-time variable t_k are called *discrete-time systems*. Discrete-time systems (such as computers) often occur as subsystems of larger systems containing continuous-time subsystems also. These combination systems are called *hybrid* or *sampled-data* systems.

The topics of modeling, simulation, and analysis discussed in the previous chapters in terms of continuous-time systems can be extended to discrete-time systems. The purpose of this chapter is to develop many of these extensions.

10.1 System Modeling Using Difference Equations

In the previous chapters, we have modeled continuous-time systems by differential equations. Discrete-time systems, in contrast, are modeled naturally by *difference equations*. We consider two examples of how difference equations arise in modeling.

Example 10-1. Consider a *first-order* continuous-time system described by the nonlinear differential equation

$$\frac{dx}{dt} = \dot{x} = f[x(t), u(t)]. \tag{10.1}$$

Suppose that we wish to "solve" (10.1) using a digital computer. The derivative $\dot{x}$ must be replaced by some discrete-time approximation, since the computer can manipulate only discrete-time data. A wealth of approximations is available. A very simple method, called Euler's procedure, is

$$\dot{x}(t_k) \approx \frac{x(t_{k+1}) - x(t_k)}{t_{k+1} - t_k}, \tag{10.2}$$

where t_{k+1} and t_k are time instants. The limit of (10.2) as $\Delta t = t_{k+1} - t_k$ approaches zero is exactly $\dot{x}$. However, for any nondifferential Δt, the expression is approximate. Using (10.2), (10.1) can be written in the form

$$\begin{aligned} x(t_{k+1}) &= x(t_k) + (t_{k+1} - t_k) f[x(t_k), u(t_k)] \\ &= g(x(t_k), u(t_k), t_{k+1}, t_k). \end{aligned} \tag{10.3}$$

Equation 10.3 is a *difference equation*. The variable x at the discrete-time instant t_{k+1} is expressed in terms of x and u at the previous time instant t_k, as well as the instants t_{k+1} and t_k. ▲

Although Euler's procedure does converge under suitable circumstances [that is, the solution of (10.3) approaches the solution of (10.1)], the rate of convergence is frequently too slow. This is because $\Delta t = (t_{k+1} - t_k)$ must be small so that the straight-line approximation of (10.3) is sufficiently close to the actual solution of (10.1). Thus, many iterations will be required. Moreover, if Δt is chosen too small, round-off errors (the computer makes computations using only a finite number of significant figures) will become significant after a large number of iterations.

For these reasons, more complicated approximations for the derivative are used, permitting a given approximation error to be achieved with fewer iterations than are needed for Euler's procedure. Several widely used procedures are called *Runge-Kutta procedures*. (Several such procedures were developed by C. Runge and W. Kutta in 1895–1901.) For example, the *second-order Runge-Kutta* difference equation formula is

$$x(t_{k+1}) = x(t_k) + a_1 p + a_2 q$$

where

$$\begin{aligned} p &= \Delta t f[x(t_k), u(t_k)] \\ q &= \Delta t f[x(t_k) + \alpha p, u(t_k + \alpha \Delta t)]. \end{aligned}$$

The constants satisfy

$$a_1 + a_2 = 1$$
$$a_2\alpha = \tfrac{1}{2}.$$

With these constants, the approximation (truncation) error is of the order $(\Delta t)^3$. This procedure is called *second-order* because two values of x are used. As with Euler's procedure, the second-order Runge-Kutta procedure is described by a first-order difference equation. ▲

Another problem leading to difference equations is the modeling of a continuous-time system having a piecewise-constant input. Such problems arise in a sampled-data control system.

Example 10-2. Consider a mass-damper mechanical system having the network diagram of Figure 10.1. A (continuous-time) state representation for this system is

$$\dot{x}(t) = -\frac{D}{M}x(t) + \frac{1}{M}u(t). \quad (10.4)$$

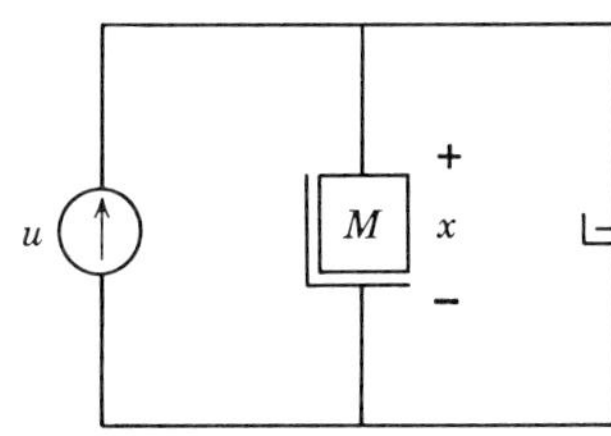

Figure 10.1 Mechanical network diagram for Example 10.2.

Suppose that the external force input to this system is *piecewise-constant*,

$$u(t) = U_k,\ t_k < t \leq t_{k+1}.$$

A sketch of a piecewise-constant signal appears in Figure 10.2. In the interval $t_k < t \leq t_{k+1}$, the solution of (10.4), obtained by the usual analysis methods for continuous-time systems, is

$$x(t) = \frac{U_k}{D} + \left[x(t_k) - \frac{U_k}{D}\right]\epsilon^{-(t-t_k)(D/M)}, \quad (10.5)$$

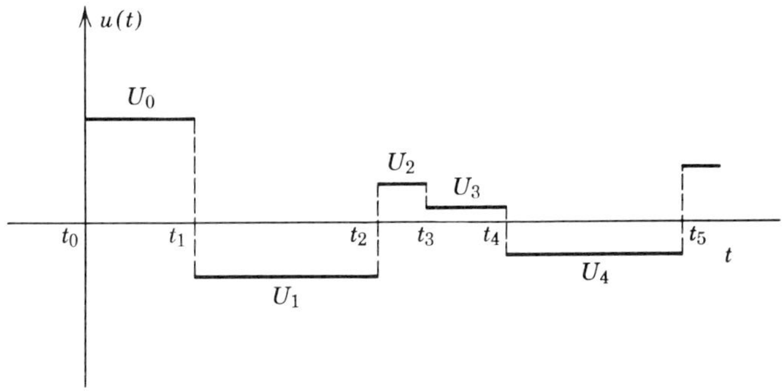

Figure 10.2 A piecewise-constant signal.

where $x(t_k)$ is the value of x at the beginning of the interval, that is, at $t = t_k$. At $t = t_{k+1}$, (10.5) yields

$$x(t_{k+1}) = \epsilon^{-(t_{k+1}-t_k)(D/M)}x(t_k) + [1-\epsilon^{-(t_{k+1}-t_k)(D/M)}]\frac{U_k}{D}. \qquad (10.6)$$

Equation 10.6 is a difference equation, relating the system state at time t_{k+1} to the state and input at the previous instant, t_k. ▲

These two examples are special cases of the representation of discrete-time systems by difference equations. In general, we shall represent discrete-time systems in state form by simultaneous first-order difference equations,

$$\mathbf{x}(t_{k+1}) = \mathbf{g}(\mathbf{x}(t_k), \mathbf{u}(t_k), t_k), \qquad k = 0, 1, 2, \ldots \qquad (10.7)$$

If $\mathbf{g}$ is linear,

$$\mathbf{x}(t_{k+1}) = \mathbf{F}(t_k)\mathbf{x}(t_k) + \mathbf{G}(t_k)\mathbf{u}(t_k), \qquad (10.8)$$

then we have a linear time-varying discrete-time system. If t_k does not appear explicitly in $\mathbf{F}$ and $\mathbf{G}$, *and* if the discrete-time instants are equally spaced,

$$t_k = kT, \qquad (10.9)$$

where T is a constant, then (10.8) becomes

$$\mathbf{x}[(k+1)T] = \mathbf{F}\mathbf{x}(kT) + \mathbf{G}\mathbf{u}(kT). \qquad (10.10)$$

Equation 10.10 describes a linear, time-invariant discrete-time system. The restriction to equally spaced discrete-time instants is necessary for time-invariance because time shifts must be meaningfully defined (review the discussion of time-invariance in Chapter 7).

As with continuous systems, the system output may not be the state. Thus, output equations must be added to the state equations. For example, an appropriate output equation for a system described by (10.10) is

$$\mathbf{y}(kT) = \mathbf{C}\mathbf{x}(kT) + \mathbf{D}\mathbf{u}(kT). \qquad (10.11)$$

10.2 Digital Simulation

The basic building blocks for simulation of discrete-time systems are discussed in this section. The development follows closely the simulation of continuous-time systems discussed in Chapter 4. The first

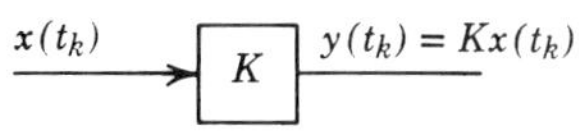

Figure 10.3 Multiplication by a constant.

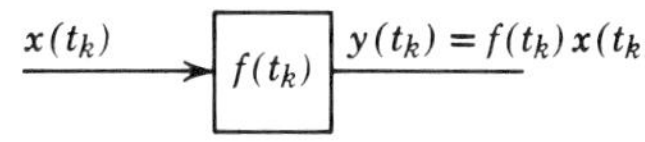

Figure 10.4 Multiplication by a time-varying coefficient.

element we consider is multiplication by a constant. The simulation block diagram for this operation, a scalor, is shown in Figure 10.3. Multiplication of a function by a time-varying coefficient $f(t_k)$ is represented by the simulation block diagram shown in Figure 10.4. Addition of signals is represented by a summer in a manner analogous to the continuous-time summer. Figure 10.5 shows the symbol

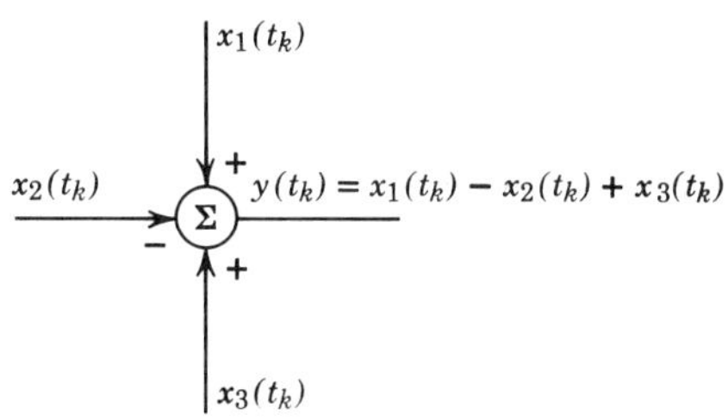

Figure 10.5 Addition of signals by a summer.

for a summer. Any number of inputs may be indicated for a summer. Instead of an integration that is useful for the simulation of differential equations, we use a *delayor* as a key element in simulating systems of difference equations. The defining equation for the delayor is

$$w(t_k) = z(t_{k-1})$$

with (10.12)

$$w(t_0) = w_0,$$

where w is the output, z is the input, $t_k - t_{k-1}$ is the delay time, and k is an integer. Notice that the delayor output is defined only at discrete instants of time and that the time delay $t_k - t_{k-1} = T_k$ may be different for different values of k. A delayor block diagram is shown in Figure 10.6. The boundary condition w_0 plays the same role as the initial condition in an integrator. Nonlinear equations are represented by simulation block diagrams similar to continuous-time simulation. For instance, for multiplication and division, we use the diagrams in Figure 10.7.

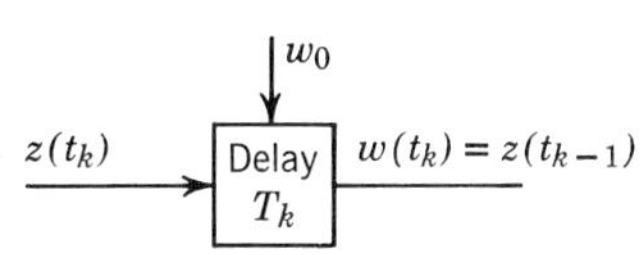

Figure 10.6 Simulation block diagram for delayor.

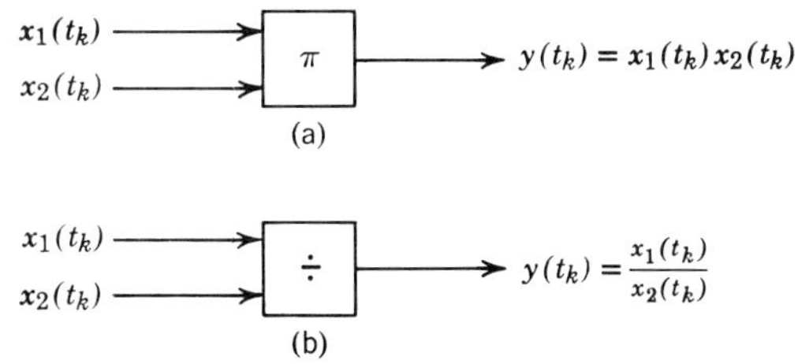

Figure 10.7 Block diagrams for some nonlinear operations. (*7*) Multiplication. (*b*) Division.

Example 10-3. Let us examine a simple feedback system in which the output is subtracted from the input to produce an error signal. The error signal is delayed to produce the output. The block diagram is shown in Figure 10.8. Uniform spacing of discrete-time instants is assumed.

From the block diagram, it is clear that

$$e(kT) = r(kT) - Kc(kT) \tag{10.13}$$

and

$$c(kT) - e(kT-T). \tag{10.14}$$

Combining (10.13) and (10.14), we obtain

$$c(kT) = r(kT-T) - Kc(kT-T). \tag{10.15}$$

If k is replaced by another variable $n+1$ where n is also an integer, we obtain

$$c(nT+T) = -Kc(nT) + r(nT). \tag{10.16}$$

If K is a constant, (10.16) is a linear constant coefficient first-order difference equation. One boundary condition is required to determine the output, given the input over the range of n for which the input is specified. ▲

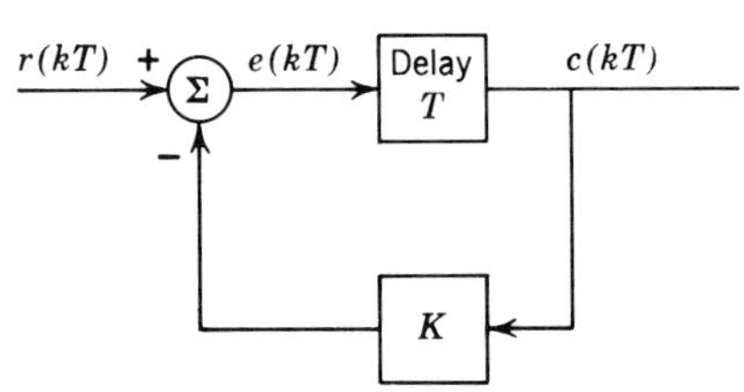

Figure 10.8 A simple feedback system.

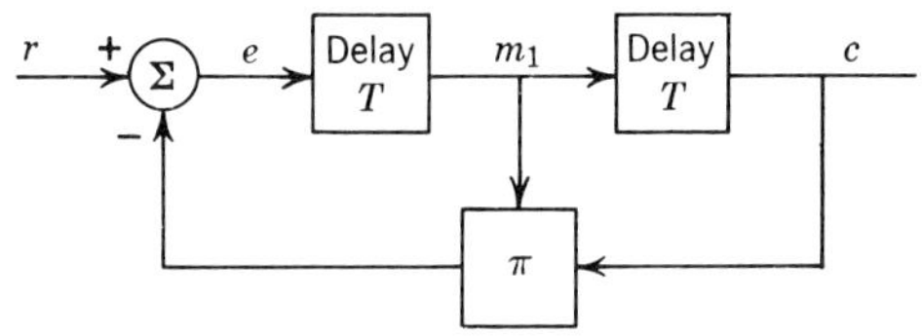

Figure 10.9 Nonlinear system describable by a difference equation.

Example 10-4. Let us determine the set of difference equations for the block diagram in Figure 10.9. From the simulation block diagram, the following equations are obtained:

$$c[(k+1)T] = m_1(kT) \tag{10.17}$$

and

$$m_1[(k+1)T] = e(kT) = -m_1(kT)c(kT) + r(kT). \tag{10.18}$$

Two boundary conditions are required, such as $m_1(0)$ and $c(0)$. The outputs of delayors may be chosen as state variables just as outputs of integrators may be chosen as state variables for the time-continuous case. Equations 10.17 and 10.18 provide the algorithm for generating the sequences $m_1(kT)$ and $c(kT)$ for integral values of k, starting from known initial values of $c(0)$ and $m_1(0)$. The arithmetical operations specified by the algorithm can be carried out readily by a digital computer. ▲

Next we discuss some methods for constructing the simulation block diagram for a system modeled by a set of difference equations, using delayors as the only dynamic elements. When the set of equations consists of first-order difference equations, the diagram construction is readily made from an inspection of the equations. The situation is similar to that for obtaining an all-integrator block diagram to represent a set of first-order differential equations as described in Chapter 2. We demonstrate the simplicity of the procedure by way of an example.

Example 10-5. Let a system be modeled by the difference equations

$$x_1(n+1) = -5x_1(n) + x_2(n) + 2r(n) \tag{10.19}$$

$$x_2(n+1) = -9x_1(n) + x_3(n) + r(n) \tag{10.20}$$

$$x_3(n+1) = 4r(n) \tag{10.21}$$

$$c(n) = x_1(n) \tag{10.22}$$

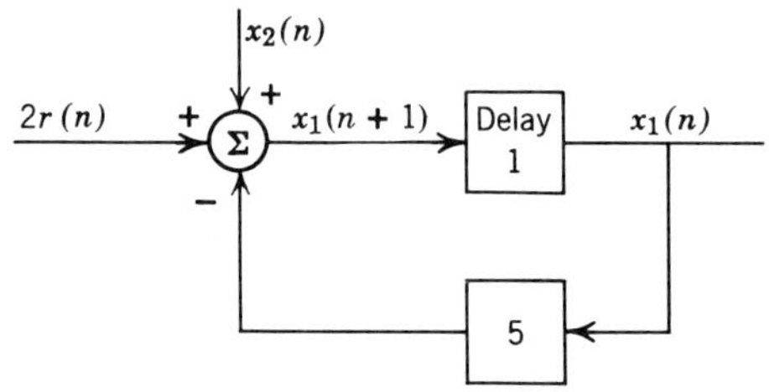

Figure 10.10 Block diagram for (10.19).

where $r(n)$ is the system input sequence, $c(n)$ is the system output sequence, and $x_1(n)$, $x_2(n)$, and $x_3(n)$ are state variables. Notice that uniform discrete-time spacing is assumed and the time unit is normalized to unity. Equations 10-19 to 10-21 are the state equations of the system. Given the present state and the present input (at time n), the state equations provide the algorithm for computing the values of the state variables at the next instant (at time $n+1$). Equation 10.22 expresses the output in terms of the state variable $x_1(n)$. Noting that since inputs of delayors at the present instant are the outputs of the delayors at the next instant, then if $x(n)$ is the output of a delayor, $x(n+1)$ is the input of the delayor. Let the outputs of three delayors be $x_1(n)$, $x_2(n)$, and $x_3(n)$, respectively. Their corresponding inputs are $x_1(n+1)$, $x_2(n+1)$, and $x_2(n+1)$. Equations 10.19 to 10.21 provide the rules for forming the input signals to the three delayors in terms of the outputs of the delayors and the input $r(n)$. Equation 10.19 is represented by the simulation block diagram in Figure 10.10. The signal $x_1(n+1)$ is generated by adding $-5x_1(n)$, $x_2(n)$, and $2r(n)$ as shown. Similarly, (10.20) has a simulation block-diagram representation as shown in Figure 10.11. Finally, (10.21) is represented by

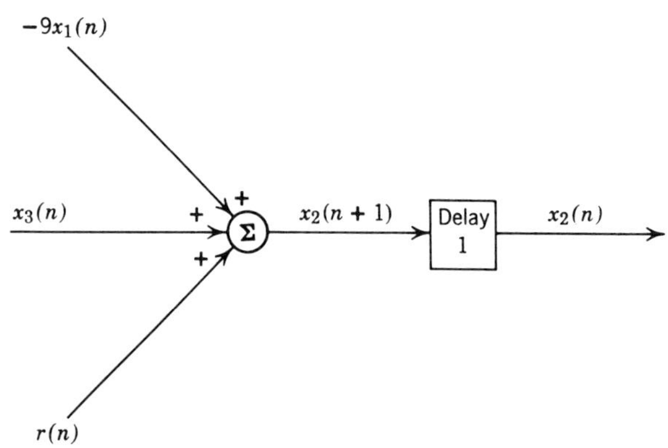

Figure 10.11 Block diagram of (10.20).

the simulation diagram in Figure 10.12. Combining these diagrams, we have Figure 10.13. Initial conditions for $x_1(n)$, $x_2(n)$, and $x_3(n)$ are needed in the simulation. ▲

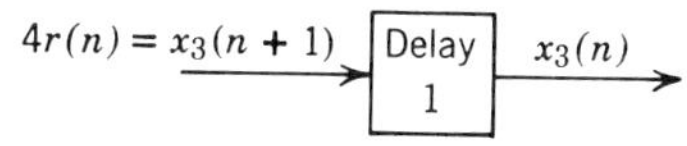

Figure 10.12 Block diagram of (10.21).

From Example 10-5, we see that from a set of first-order difference equations, an all-delayor simulation diagram may be obtained by inspection. Although the example describe a linear and time-invariant system, the same procedure applies to nonlinear and time-varying systems.

Next, let us consider the digital simulation of higher-order difference equations, using only delayors as the dynamic elements. Again, the problem is analogous to that of finding an all-integrator block diagram to represent a high-order differential equation. There are several ways of converting a single nth-order difference equation into n first-order difference equations. The first-order difference equations are simulated using the method described above. Let us adapt the method described in Chapter 4 to difference equations.

Example 10-6. It is desired to represent a system described by the difference equation

$$y(n+3)+6y(n+2)+11y(n+1)+6y(n)$$

$$=r(n+2)+9r(n+1)+20r(n) \qquad (10.23)$$

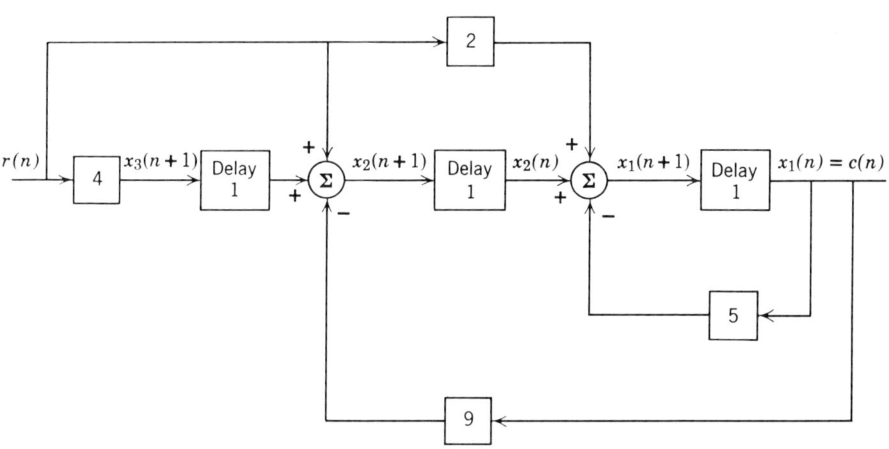

Figure 10.13 All-delayor representation for Example 10-5.

by an all-delayor block diagram. Consider, first, an associated system with input $r(n)$ and output $c(n)$:

$$c(n+3)+6c(n+2)+11c(n+1)+6c(n)=r(n). \qquad (10.24)$$

This system may be simulated easily, since $c(n)$, $c(n+1)$, and $c(n+2)$ may be identified as delayor outputs and, from (10.24),

$$c(n+3)=-6c(n+2)-11c(n+1)-6c(n)+r(n). \qquad (10.25)$$

Figure 10.14 shows the resulting simulation diagram. We now use linearity and time-invariance to obtain $y(n)$, the output of the system of (10.23). Since $c(n)$ is the response to $r(n)$, $20c(n)$ is the response to $20r(n)$. Moreover, $c(n+1)$ is, by time-invariance, the response to $r(n+1)$. Thus $9c(n+1)$ is the response to $9r(n+1)$. Finally, $c(n+2)$ is the response to $r(n+2)$. Thus $y(n)$ of (10.23), which is the response to $r(n+2)+9r(n+1)+20r(n)$, is, by superposition, the sum of the responses to $r(n+2)$, $9r(n+1)$, and $20r(n)$. That is,

$$y(n)=c(n+2)+9c(n+1)+20c(n). \qquad (10.26)$$

Hence a simulation diagram for (10.23) is as shown in Figure 10.15. ▲

10.3 Response of Discrete-Time Systems via Transfer Functions

In Chapter 6 we observed that the Laplace transform and corresponding transfer functions were very useful in the analysis of linear time-invariant continuous-time systems. Analogously, there exists a useful transform technique for linear time-invariant discrete-time

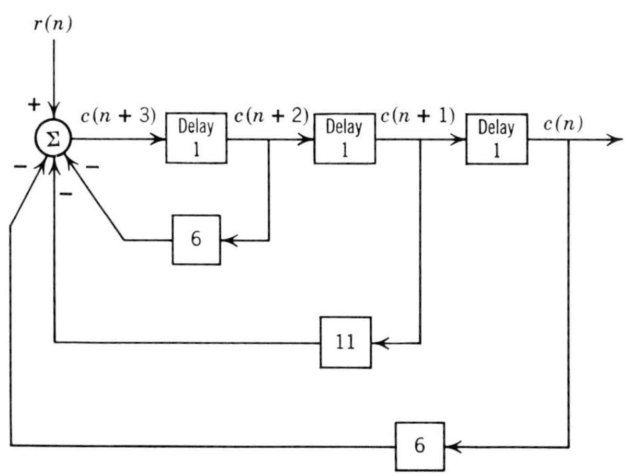

Figure 10.14 Simulation diagram for (10.24).

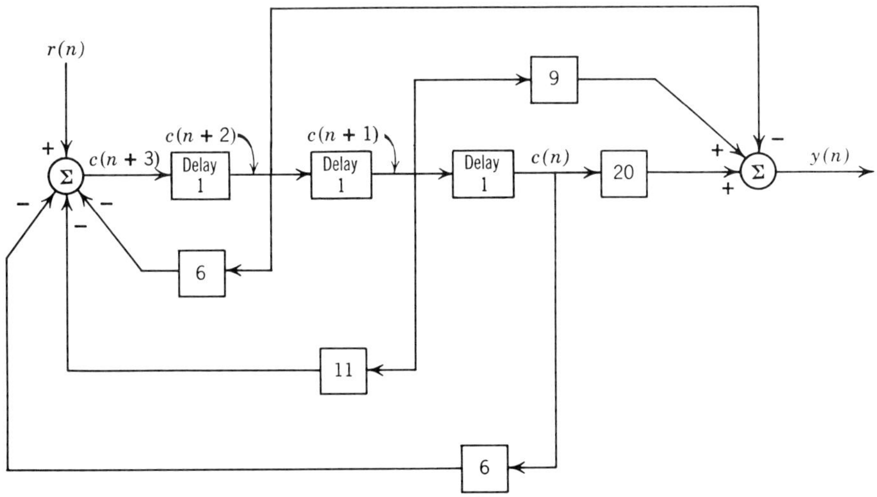

Figure 10.15 Simulation diagram for Example 10-6.

systems, the z transform. In this section we develop a basic facility in using this transform for discrete-time system analysis.

The z transform of a discrete-time signal $f(nT)$ is defined as

$$F(z) = \sum_{n=0}^{\infty} f(nT)z^{-n}. \tag{10.27}$$

Thus the transform is a power series in the complex variable z^{-1}, having as coefficients the signal values at discrete-time instants.

Example 10-7. Suppose that, for $T = 1$,

$$f(n) = a^n, \qquad n = 0, 1, 2, \ldots \tag{10.28}$$

Then

$$F(z) = \sum_{n=0}^{\infty} a^n z^{-n}. \tag{10.29}$$

Noting that this is a geometric series, we may sum the infinite series to obtain

$$F(z) = \frac{1}{1 - az^{-1}} = \frac{z}{z - a} \quad \text{for} \quad |z| > a. \blacktriangle \tag{10.30}$$

A short table of z transforms is given as Table 10.1. More extensive tables may be found in E. I. Jury, *Theory and Application of the z-Transformation Method*, Wiley, 1964, or J. R. Ragazzini and G. F. Franklin, *Sampled-Data Control Systems*, McGraw-Hill, 1958.

Table 10.1

$f(n), n \geqslant 0$	$F(z)$
a^n	$\dfrac{1}{1-az^{-1}}$
nT	$\dfrac{Tz^{-1}}{(1-z^{-1})^2}$
$(nT)^2$	$\dfrac{T^2z^{-1}(1+z^{-1})}{(1-z^{-1})^3}$
$\cos \omega nT$	$\dfrac{1-z^{-1}\cos \omega T}{1-2z^{-1}\cos \omega T+z^{-2}}$
$\sin \omega nT$	$\dfrac{z^{-1}\sin \omega T}{1-2z^{-1}\cos \omega T+z^{-2}}$

Several useful properties of the z transform are summarized in Table 10.2. In particular, we consider the transform of $x[(n+1)T]$.

Table 10.2

$f(n), n \geqslant 0$	$F(z)$
$f(n+k)$	$z^kF(z)-z^k\sum_{m=0}^{k-1} f(m)z^{-m}$
$af(n)+bg(n)$ (linearity)	$aF(z)+bG(z)$
$y(n)=\sum_{k=0}^{n} h(n-k)r(k)$ (convolution)	$Y(z)=H(z)R(z)$
$a^nf(n)$	$F(a^{-1}z)$

Example 10-8. Denoting the z transform of $x[(n+1)T]$ by $\mathscr{Z}\{x[(n+1)T]\}$, we have

$$\mathscr{Z}\{x[(n+1)T]\}=\sum_{n=0}^{\infty} x[(n+1)T]z^{-n}. \tag{10.31}$$

Letting $m=n+1$ in (10.31),

$$\begin{aligned}\mathscr{Z}\{x[(n+1)T]\} &= \sum_{m=1}^{\infty} x(mT)z^{-m+1}\\ &= z\Big[\sum_{m=0}^{\infty} x(mT)z^{-m}-x(0)\Big]\\ &= zX(z)-zx(0). \blacktriangle \end{aligned} \tag{10.32}$$

This result is a special case of the first entry in Table 10.2.

Application of the z transform to the solution of difference equations is illustrated by the following example.

Example 10-9. Suppose that a system is described by the second-order difference equation

$$x(n+2)+a_1x(n+1)+a_0(n)=r(n). \tag{10.33}$$

Using the linearity property [see Table 10.2],

$$\mathscr{Z}\{x(n+2)\}+a_1\mathscr{Z}\{x(n+1)\}+a_0\mathscr{Z}\{x(n)\}=\mathscr{Z}\{r(n)\}. \tag{10.34}$$

Then, using the first entry in Table 10.2, and assuming $x(0)=x(1)=0$,

$$z^2X(z)+a_1zX(z)+a_0X(z)=R(z), \tag{10.35}$$

or

$$X(z)=\frac{1}{z^2+a_1z+a_0}R(z)\cdot \blacktriangle \tag{10.36}$$

The next step in Example 10-9 is to invert $X(z)$, that is, to find $x(nT)$. There are several methods for z transform inversion. For example, an inversion integral,

$$f(nT)=\frac{1}{2\pi j}\oint_\Gamma z^{n-1}F(z)\,dz, \tag{10.37}$$

where Γ is a contour enclosing all poles of $F(z)$, can be used. As with Laplace transforms, it is usually more convenient to use *partial fraction expansion* and tables, rather than the inversion integral. The procedure is very similar to a Laplace transform partial fraction expansion, Appendix B. The basic term in the expansion is $K_i/(1-a_iz^{-1}) = K_iz/(z-a_i)$, corresponding to a pole of $F(z)$ at $z=a$. The following example illustrates the procedure.

Example 10-10.

$$F(z)=\frac{-z(z-4)}{z^2-3z+2}\cdot \tag{10.38}$$

It is convenient to first divide both sides of (10.38) by z, so that basic partial fraction terms have the form $K_i/(z-a_i)$, rather than $K_iz/(z-a)$. The K_i can then be found in exactly the same way as the coefficients in a Laplace transform partial fraction expansion. For this case

$$\frac{F(z)}{z}=\frac{-z+4}{(z-2)(z-1)}=\frac{K_1}{z-2}+\frac{K_2}{z-1}\cdot \tag{10.39}$$

But

$$K_1 = \left.\frac{(z-2)F(z)}{z}\right|_{z=2} = 2 \tag{10.40}$$

$$K_2 = \left.\frac{(z-1)F(z)}{z}\right|_{z=1} = -3. \tag{10.41}$$

Thus

$$F(z) = \frac{2z}{z-2} - \frac{3z}{z-1}. \tag{10.42}$$

Finally, from Table 10.1:

$$f(n) = 2(2)^n - 3(1)^n$$
$$f(n) = (2)^{n+1} - 3, \quad n = 0, 1, 2, \ldots \blacktriangle \tag{10.43}$$

Z transform transfer functions can be defined analogously to Laplace transform transfer functions*, that is, as the ratio of the output signal transform to the input signal transform. For instance, in the case of the problem of Example 10-9, the input-output transfer function is

$$H(z) = \frac{X(z)}{R(z)} = \frac{1}{z^2 + a_1 z + a_0}. \tag{10.44}$$

This representation can obviously be generalized to multiple-input multiple-output systems using matrices, as was done for the continuous-time case. Moreover, transfer-function block diagrams can be drawn, and system transfer functions can be obtained using Mason's rule. Two examples illustrate the procedure.

Example 10-11. Consider the system of Example 10-5, represented by the simulation diagram of Figure 10.13. This diagram is easily converted to a transfer-function block diagram by noting that the transforms of the output $x(n)$ and the input $r(n) = x(n+1)$ of a delayor are related by

$$R(z) = zX(z) - zx(0), \tag{10.45}$$

or

$$z^{-1}R(z) + x(0) = X(z). \tag{10.46}$$

An appropriate transfer-function block diagram is Figure 10.16. Using the delayor representation of Figure 10.16, the transfer function block diagram version of Figure 10.13 is Figure 10.17.

Mason's rule is exactly the same as for a Laplace transform transfer-

*See Chapter 6.

function block diagram. From Figure 10.17, with $x_1(0) = x_2(0) = x_3(0) = 0$:

$$\frac{C(z)}{R(z)} = \frac{4z^{-3}+z^{-2}+2z^{-1}}{1+5z^{-1}+9z^{-2}}. \ \blacktriangle \qquad (10.47)$$

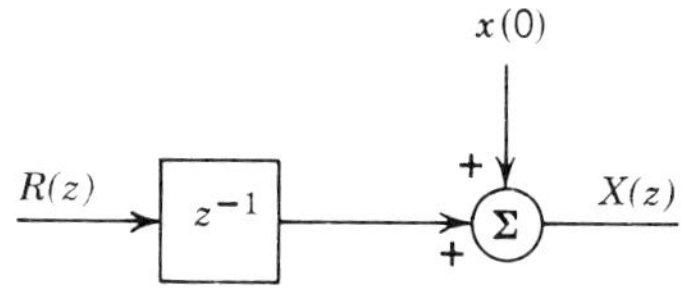

Figure 10.16 Transfer-function block-diagram representation of unit delayor.

Example 10-12. Consider a system described by (10.23), repeated here as (10.48)

$$y(n+3)+6y(n+2)+11y(n+1)+6y(n) = r(n+2)+9r(n+1)+20r(n). \qquad (10.48)$$

Assuming zero boundary conditions,

$$(z^3+6z^2+11z+6)Y(z) = (z^2+9z+20)R(z),$$

or

$$\frac{Y(z)}{R(z)} = \frac{z^2+9z+20}{z^3+6z^2+11z+6}. \qquad (10.49)$$

From these equations, the transfer-function block diagram of Figure 10.18 is drawn. (Zero initial conditions are assumed.) Then, from Mason's rule,

$$\frac{Y(z)}{R(z)} = \frac{+20z^{-3}+9z^{-2}+z^{-1}}{1+6z^{-1}+11z^{-2}+6z^{-3}} = \frac{z^2+9z+20}{z^3+6z^2+11z+6}, \qquad (10.50)$$

which is the same as (10.49). ▲

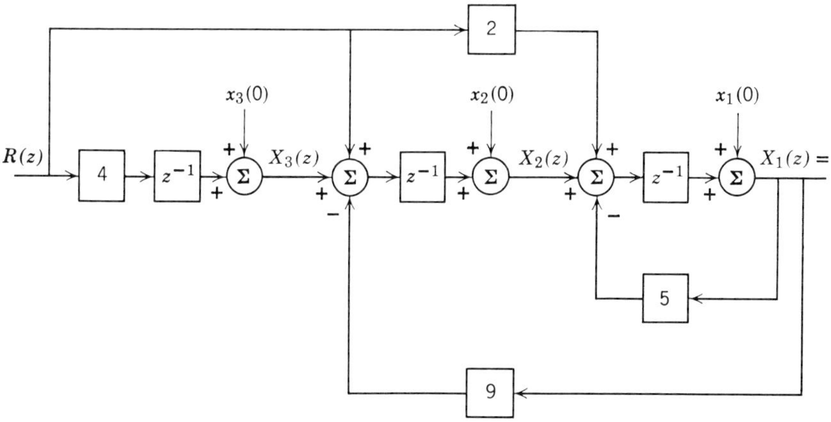

Figure 10.17 Transfer-function block diagram for Example 10-11.

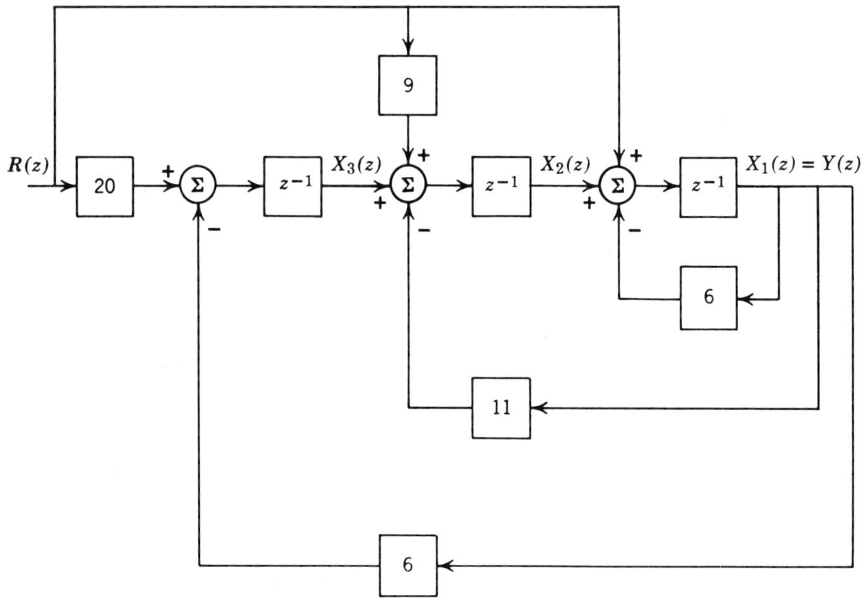

Figure 10.18 Block diagram for Example 10-12.

10.4 Response of Discrete-Time Systems via State Equations

As was observed in Section 10.1, discrete-time systems can be described in state-variable form by a set of simultaneous first-order difference equations,

$$\mathbf{x}(t_{k+1}) = \mathbf{g}(\mathbf{x}(t_k), \mathbf{u}(t_k), t_k). \tag{10.51}$$

This set of equations can always be solved by *recursion*. Suppose the initial condition is

$$\mathbf{x}(t_0) = \mathbf{x}_0. \tag{10.52}$$

Then the suceeding $\mathbf{x}(t_k)$ can be formed easily from (10.51):

$$\begin{aligned} \mathbf{x}(t_1) &= \mathbf{g}[\mathbf{x}(t_0), \mathbf{u}(t_0), t_0] \\ \mathbf{x}(t_2) &= \mathbf{g}[\mathbf{x}(t_1), \mathbf{u}(t_1), t_1] \\ &= \mathbf{g}\{\mathbf{g}[\mathbf{x}(t_0), \mathbf{u}(t_0), t_0], \mathbf{u}(t_1), t_1\} \\ &\qquad \text{etc.} \end{aligned} \tag{10.53}$$

More extensive results can be obtained if the function $\mathbf{g}$ is further restricted. Let us discuss in detail the linear time-invariant case,

$$\mathbf{x}[(k+1)T] = \mathbf{F}\mathbf{x}(kT) + \mathbf{G}\mathbf{u}(kT) \tag{10.54}$$

$$\mathbf{y}(kT) = \mathbf{C}\mathbf{x}(kT) + \mathbf{D}\mathbf{u}(kT). \tag{10.55}$$

We consider the zero-input case first, using the abbreviated notation

$$\mathbf{x}[(k+1)T] = \mathbf{x}_{k+1} \tag{10.56}$$

$$\mathbf{x}(kT) = \mathbf{x}_k, \tag{10.57}$$

so that, for $\mathbf{u}_k \equiv \mathbf{0}$, (10.54) becomes

$$\mathbf{x}_{k+1} = \mathbf{F}\mathbf{x}_k. \tag{10.58}$$

The initial condition is $\mathbf{x}(0) = \mathbf{x}_0$. By analogy to the continuous-time case, we define the (discrete-time) state-transition matrix as the matrix $\boldsymbol{\phi}_n$ satisfying the matrix-difference equation

$$\boldsymbol{\phi}_{n+1} = \mathbf{F}\boldsymbol{\phi}_n, \tag{10.59}$$

with

$$\boldsymbol{\phi}_0 = \mathbf{I}. \tag{10.60}$$

Note that

$$\mathbf{x}_n = \boldsymbol{\phi}_n\mathbf{x}_0 \tag{10.61}$$

is the solution to (10.58), since the initial condition is satisfied,

$$\mathbf{x}_0 = \boldsymbol{\phi}_0\mathbf{x}_0 = \mathbf{I}\mathbf{x}_0 = \mathbf{x}_0, \tag{10.62}$$

and (10.58) is satisfied,

$$\mathbf{x}_{k+1} = \boldsymbol{\phi}_{k+1}\mathbf{x}_0 = \mathbf{F}\boldsymbol{\phi}_k\mathbf{x}_0 = \mathbf{F}\mathbf{x}_k. \tag{10.63}$$

An explicit formula for $\boldsymbol{\phi}_n$ can be found from (10.58) by recursion:

$$\begin{aligned}
\mathbf{x}_1 &= \mathbf{F}\mathbf{x}_0 \\
\mathbf{x}_2 &= \mathbf{F}\mathbf{x}_1 = \mathbf{F}^2\mathbf{x}_0 \\
&\vdots \\
\mathbf{x}_n &= \mathbf{F}^n\mathbf{x}_0.
\end{aligned} \tag{10.64}$$

Comparison of (10.64) and (10.61) yields

$$\boldsymbol{\phi}_n = \mathbf{F}^n. \tag{10.65}$$

Another expression for $\boldsymbol{\phi}_n$ can be found by applying the z transform to (10.58):

$$z\mathbf{X}(z) - z\mathbf{x}_0 = \mathbf{F}\mathbf{X}(z), \qquad (10.66)$$

or

$$\mathbf{X}(z) = (\mathbf{I} - z^{-1}\mathbf{F})^{-1}\mathbf{x}_0. \qquad (10.67)$$

Comparison of (10.67) with the z transform of (10.61),

$$\mathbf{X}(z) = \mathbf{\Phi}(z)\mathbf{x}_0, \qquad (10.68)$$

where

$$\mathbf{\Phi}(z) \triangleq \mathscr{Z}\{\boldsymbol{\phi}_n\}, \qquad (10.69)$$

reveals that

$$\mathbf{\Phi}(z) = (\mathbf{I} - z^{-1}\mathbf{F})^{-1}. \qquad (10.70)$$

This result is the discrete-time system counterpart of $\mathbf{\Phi}(s) = (s\mathbf{I} - \mathbf{A})^{-1}$ for continuous-time systems.

A third method for finding the state-transition matrix for a discrete-time system is based on the use of Mason's rule and an all-unit-delayor transfer-function representation. From (10.68), the ith transformed state variable is given by

$$X_i(z) = \Phi_{ij}(z)x_j(0). \qquad (10.71)$$

In other words, $\Phi_{ij}(z)$ is the *transfer function* on an all-unit-delay diagram from $x_j(0)$ (which is applied at the output of the jth unit delayor in accordance with Figure 10.16) to $X_i(z)$. Mason's rule can be used to obtain this transfer function by inspection.

Example 10-13. Consider the system represented by Figure 10.17. By inspection of this diagram,

$$\Phi_{11}(z) = \frac{1}{1+5z^{-1}+9z^{-2}}$$

$$\Phi_{12}(z) = \frac{z^{-1}}{1+5z^{-1}+9z^{-2}}$$

$$\Phi_{13}(z) = \frac{z^{-2}}{1+5z^{-1}+9z^{-2}}.$$

Other entries in $\mathbf{\Phi}(z)$ may be found similarly. ▲

We now consider the total solution to (10.54). The time-domain solution may be obtained by recursion. Using abbreviated notation,

$$\mathbf{x}_1 = \mathbf{F}\mathbf{x}_0 + \mathbf{G}\mathbf{u}_0$$

$$
\begin{aligned}
\mathbf{x}_2 &= \mathbf{F}\mathbf{x}_1 + \mathbf{G}\mathbf{u}_1 \\
&= \mathbf{F}^2\mathbf{x}_0 + \mathbf{F}\mathbf{G}\mathbf{u}_0 + \mathbf{G}\mathbf{u}_1 \\
\mathbf{x}_3 &= \mathbf{F}\mathbf{x}_2 + \mathbf{G}\mathbf{u}_2 \\
&= \mathbf{F}^3\mathbf{x}_0 + \mathbf{F}^2\mathbf{G}\mathbf{u}_0 + \mathbf{F}\mathbf{G}\mathbf{u}_1 + \mathbf{G}\mathbf{u}_2 \\
&\vdots \\
\mathbf{x}_n &= \mathbf{F}^n\mathbf{x}_0 + \sum_{k=0}^{n-1} \mathbf{F}^{n-k-1}\mathbf{G}\mathbf{u}_k. \qquad (10.72)
\end{aligned}
$$

The summation is a *convolution sum*. Equation 10.72 may be expressed in terms of the state-transition matrix as

$$\mathbf{x}_n = \boldsymbol{\phi}_n\mathbf{x}_0 + \sum_{k=0}^{n-1} \boldsymbol{\phi}_{n-k-1}\mathbf{G}\mathbf{u}_k. \qquad (10.73)$$

The total solution may also be obtained, using z transforms. The z transform of (10.54) yields

$$\mathbf{X}(z) = \mathbf{\Phi}(z)\mathbf{x}_0 + z^{-1}\mathbf{\Phi}(z)\mathbf{G}\mathbf{U}(z). \qquad (10.74)$$

Equation 10.74 provides a connection between the state-variable description of this section and the transfer-function description of the previous section. From (10.55),

$$\mathbf{Y}(z) = \mathbf{C}\mathbf{X}(z) + \mathbf{D}\mathbf{U}(z). \qquad (10.75)$$

Substituting for $\mathbf{X}(z)$ from (10.74), assuming zero initial conditions,

$$\mathbf{Y}(z) = [z^{-1}\mathbf{C}\mathbf{\Phi}(z)\mathbf{G} + \mathbf{D}]\mathbf{U}(z). \qquad (10.76)$$

Thus the transfer-function matrix relating the input transform $\mathbf{U}(z)$ to the output transform $\mathbf{Y}(z)$ is

$$\mathbf{H}(z) = [z^{-1}\mathbf{C}\mathbf{\Phi}(z)\mathbf{G} + \mathbf{D}] \qquad (10.77)$$

Additional topics concerning analysis of discrete-time systems using state variables appear in the Problems.

10.5 Stability of Discrete-Time Systems

All definitions, theorems, and tests for stability discussed in Chapter 8 and 9 have counterparts for discrete-time systems. The development of these counterparts generally parallels the continuous-time case closely. The reader will be able to develop these himself readily, and we shall not pause here for that purpose, with one exception. The counterpart of the main Lyapunov stability theorem, Theorem 9.2, will be stated, and certain special results for stability of linear time-invariant discrete-time systems will be given. No proofs will be attempted, as the reader can construct these himself from continuous-time counterparts.

Definition 10-1. A state $\mathbf{x}_e$ of the zero-input system

$$\mathbf{x}_{k+1} = \mathbf{g}(\mathbf{x}_k, t_k) \tag{10.78}$$

is said to be an *equilibrium state* if

$$\mathbf{g}(\mathbf{x}_e, t_k) = \mathbf{x}_e \tag{10.79}$$

for all t_k. ▲

In the following, we shall assume $\mathbf{x}_e = \mathbf{0}$, which can always be arranged through coordinate transformation.

The definitions of stability, asymptotic stability, and asymptotic stability in the large of equilibrium points is the same as in Definitions 9.1, 9.2, and 9.3, with t replaced by the discrete-time variable t_k. The counterpart of Theorem 9.2 is:

Theorem 10.1. Consider a system described by

$$\mathbf{x}_{k+1} = \mathbf{g}(\mathbf{x}_k) \tag{10.80}$$

with

$$\mathbf{g}(\mathbf{0}) = \mathbf{0}.$$

Suppose there exists a scalar function $V(\mathbf{x}_k)$ having the following properties:

(1) $V(\mathbf{x}_k)$ is positive definite.
(2) $-\Delta V(\mathbf{x}) = -[V(\mathbf{x}_{k-1}) - V(\mathbf{x}_k)]$ is positive definite.
(3) $V(\mathbf{x}_k) \to \infty$ when $\|\mathbf{x}_k\| \to \infty$.

Then $\mathbf{x}_e = \mathbf{0}$ is asymptotically stable in the large. ▲

More detailed results can be obtained for linear time-invariant systems. Consider first the following simple example.

Example 10-14. Suppose that a system is described by the scalar first-order linear time-invariant difference equation

$$x_{k+1} = Fx_k, \tag{10.81}$$

where F is a scalar parameter. If the initial condition is denoted by x_0, then the solution to (10.81) is

$$x_n = F^n x_0. \tag{10.82}$$

Notice that the growth or decay of the response x_n is governed by F. In particular,

$$|x_n| = |F^n|\,|x_0|, \tag{10.83}$$

so $|x_n| \to 0$ for $n \to \infty$ if $|F| < 1$. Moreover, if $|F| > 1$, $|x_n|$ grows without bound for $n \to \infty$. Thus the system is asymptotically stable in the large if and only if $|F| < 1$. Note that the system (10.81) has the characteristic equation

$$z - F = 0. \tag{10.84}$$

Thus (10.82) is ASL if and only if the magnitude of the root of the characteristic equation is less than unity. ▲

Stability of the nth-order case is given by the following:

Theorem 10.2. The equilibrium point $\mathbf{x}_e = \mathbf{0}$ of the system

$$\mathbf{x}_{k+1} = \mathbf{F}\mathbf{x}_k$$

is ASL if and only if the eigenvalues of the system [that is, the eigenvalues of $\mathbf{F}$ or the roots of the characteristic equation $\det(\mathbf{I} - z^{-1}\mathbf{F}) = 0$] all are less than unity in magnitude, that is, they lie within the unit circle of the z-plane. ▲

Example 10-15. Consider the system of Figure 10.17. The characteristic equation is

$$z^2 + 5z + 9 = 0.$$

The eigenvalues are

$$z = \frac{-5 \pm j\sqrt{11}}{2},$$

and thus

$$|z| = 3 > 1.$$

The equilibrium point $\mathbf{x}_e = \mathbf{0}$ is unstable. ▲

The characteristic equation need not be factored to obtain the roots explicitly. Algebraic stability tests analogous to the Routh test exist for linear, time-invariant, discrete-time systems. One form, due to Jury,[1] will be presented here.

Jury's Stability Table. A necessary and sufficient condition for the polynomial

$$F(z) = a_0 + a_1 z + a_2 z^2 + \cdots + a_n z^n,\ a_n > 0, \tag{10.85}$$

to have all its roots inside the unit circle can be obtained from Table 10.3. The conditions for asymptotic stability are:

$$\begin{aligned} &F(1) > 0,\ (-1)^n F(-1) > 0 \\ &\left.\begin{aligned} &|a_0| < a_n \quad \text{or} \quad a_n > |a_0| \\ &|b_0| > |b_{n-1}| \\ &|c_0| > |c_{n-2}| \\ &|d_0| > |d_{n-3}| \\ &\quad\vdots \\ &|r_0| > |r_2| \end{aligned}\right\} (n-1) \text{ conditions} \end{aligned} \tag{10.86}$$

10.6 Sampled-Data Systems

As was observed in the introduction to this chapter, many systems are *sampled-data* systems, that is, they are interconnections of continuous-time and discrete-time subsystems. In this section, we consider a few basic notions useful in the analysis of such systems. The subject is a vast one, and only a few of the fundamentals are discussed here.

In the interconnection of discrete-time and continuous-time systems, special attention must be given to the *interface* or *data converters* linking the subsystems. A device that converts a continuous-time signal to a discrete-time signal will be called an analog-to-digital converter (A/D). A device that converts a discrete-time signal to a continuous-time signal is called a digital-to-analog converter (D/A). A sampled-data system employing data converters is pictured in Figure 10.19.

Many types of data converters are in use. We shall consider only the simplest model for the D/A converter. Denote the input to this device

[1]See E. I. Jury, *Theory and Application of the z-Transform Method*, Wiley, New York, 1964, pp. 97–101, for a discussion of this table and some modifications of it, as well as references to the original technical articles presenting the method.

Table 10.3

Row	z^0	z^1	z^2	$\cdots$	z^{n-k}	$\cdots$	z^{n-1}	z^n
1	a_0	a_1	a_2	$\cdots$	a_{n-k}	$\cdots$	a_{n-1}	a_n
2	a_n	a_{n-1}	a_{n-2}	$\cdots$	a_k	$\cdots$	a_1	a_0
3	b_0	b_1	b_2	$\cdots$	$\cdots$	$\cdots$	b_{n-1}	
4	b_{n-1}	b_{n-2}	b_{n-3}	$\cdots$	$\cdots$	$\cdots$	b_0	
5	c_0	c_1	c_2	$\cdots$	$\cdots$	c_{n-2}		
6	c_{n-2}	c_{n-3}	c_{n-1}	$\cdots$	$\cdots$	c_0		
$\vdots$	$\vdots$	$\vdots$	$\vdots$	$\vdots$	$\vdots$			
$2n-5$	s_0	s_1	s_2	s_3				
$2n-4$	s_3	s_2	s_1	s_0				
$2n-3$	r_0	r_1	r_2					

where

$$b_k = \begin{vmatrix} a_0 & a_{n-k} \\ a_n & a_k \end{vmatrix} \qquad d_k = \begin{vmatrix} c_0 & c_{n-2-k} \\ c_{n-2} & c_k \end{vmatrix}$$

$$c_k = \begin{vmatrix} b_0 & b_{n-1-k} \\ b_{n-1} & b_k \end{vmatrix} \qquad r_0 = \begin{vmatrix} s_0 & s_3 \\ s_3 & s_0 \end{vmatrix}$$

$$r_2 = \begin{vmatrix} s_0 & s_1 \\ s_3 & s_2 \end{vmatrix}$$

by $\mathbf{f}(t_k)$, and the output by $\mathbf{u}(t)$, as in Figure 10.19. Then the converter output is given by

$$\mathbf{u}(t) = \mathbf{f}(t_k), t_k < t \leqslant t_{k+1}. \tag{10.87}$$

Thus the input to the continuous-time subsystems is *piecewise constant*. We shall represent the A/D converter by a simple model also.

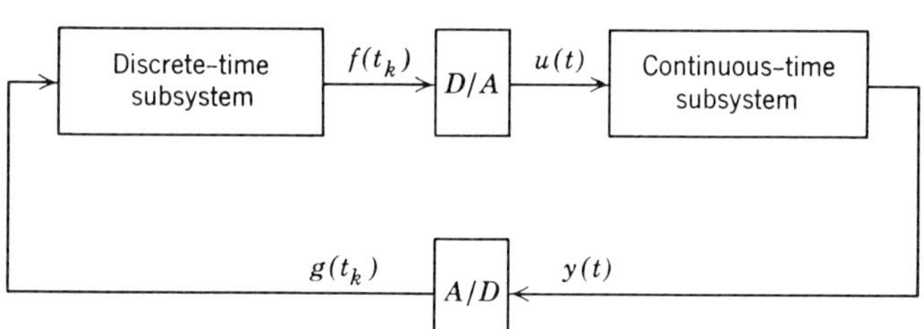

Figure 10.19 A sampled-data system.

Denote the converter output by $\mathbf{y}(t)$, as in Figure 10.19. Then the converter output is given by

$$\mathbf{g}(t_k) = \mathbf{y}(t_k) = \mathbf{y}(t)|_{t=t_k}. \tag{10.88}$$

The above converter models represent well actual converters except for the phenomenon of *quantization*. Because of the requirement of finite word length, no A/D converter can reproduce sample values of a continuous-time signal exactly. The difference between the actual and quantized sample values is called *quantization error*. Similarly, D/A converters do not produce a piecewise-constant output having values *exactly* equal to the discrete-time input signal at the instants t_k, again introducing a quantization error. These effects are nonlinear, as they are amplitude-dependent. Thus quantization introduces a *nonlinearity* accompanying digital-data processing. These nonlinearities may become significant in some applications. Limit cycles may occur, for example, due to amplitude quantization. Quantization errors may be made arbitrarily small at the cost of providing arbitrarily long digital word lengths, of course. In our discussion here, we shall neglect quantization effects. The reader may consult J. E. Bertram, "The Effect of Quantization in Sampled-Feedback Systems," *Trans. AIEE*, Vol. 77, Part 2, pp. 177–182, September 1958, for a discussion of *bounds* on quantization errors. The treatment of quantization errors as stochastic disturbance inputs has been proposed in B. Widrow, "Statistical Analysis of Amplitude Quantized Sampled-Data Systems," *AIEE Trans. on Appl. and Ind.*, No. 52, January 1961.

We assume that all continuous-time portions of the system are preceded by a D/A converter of the type described above. Thus all continuous-time subsystems are driven by piecewise-constant inputs. We next obtain expression for the time-domain response to such inputs.

Consider a linear, time-invariant continuous-time system described by

$$\dot{\mathbf{x}}(t) = \mathbf{A}\mathbf{x}(t) + \mathbf{B}\mathbf{u}(t). \tag{10.89}$$

The solution of (10.89) in the interval $t_k < t \leqslant t_{k+1}$ is given by

$$\mathbf{x}(t) = \boldsymbol{\phi}(t-t_k)\mathbf{x}(t_k) + \int_{t_k}^{t} \boldsymbol{\phi}(t-\tau)\mathbf{B}\mathbf{u}(\tau)\, d\tau, \tag{10.90}$$

where $t \leqslant t_{k+1}$ and $\boldsymbol{\phi}(t-t_k) = \boldsymbol{\epsilon}^{\mathbf{A}(t-t_k)}$. Since $\mathbf{u}(t)$ is constant in this time interval, (10.90) becomes

$$\mathbf{x}(t) = \boldsymbol{\phi}(t-t_k)\mathbf{x}(t_k) + \left[\int_{t_k}^{t} \boldsymbol{\phi}(t-\tau)\mathbf{B}\, d\tau\right]\mathbf{f}(t_k). \tag{10.91}$$

This equation describes the state of the continuous-time subsystem between the "sampling instants" t_k and t_{k+1}. In particular, if the behavior of the continuous-time system is required at sampling instants only, then set $t = t_{k+1}$ in (10.91):

$$\mathbf{x}(t_{k+1}) = \boldsymbol{\phi}(t_{k+1} - t_k)\mathbf{x}(t_k) + \left[\int_{t_k}^{t_{k+1}} \boldsymbol{\phi}(t_{k+1} - \tau)\mathbf{B}\, d\tau \right] \mathbf{f}(t_k). \tag{10.92}$$

In abbreviated notation,

$$\mathbf{x}_{k+1} = \mathbf{F}(T_k)\mathbf{x}_k + \mathbf{G}(T_k)\mathbf{f}(t_k), \tag{10.93}$$

where

$$\begin{aligned} \mathbf{F}(T_k) &= \boldsymbol{\phi}(T_k) \\ \mathbf{G}(T_k) &= \int_{t_k}^{t_{k+1}} \boldsymbol{\phi}(t_{k+1} - \tau) B\, d\tau \\ T_k &= t_{k+1} - t_k. \end{aligned} \tag{10.94}$$

Thus the response of the continuous-time system at sampling instants is described by a linear difference equation, (10.93). If the sampling is uniform, $T_k = T$, then the difference equation is linear and time-invariant:

$$\mathbf{x}_{k+1} = \mathbf{F}\mathbf{x}_k + \mathbf{G}\mathbf{f}_k. \tag{10.95}$$

We consider a simple example of the analysis of sampled-data systems.

Example 10-16. Figure 10.20 depicts a simplified sampled-data single-axis attitude control system. The spacecraft motion is modeled by a con-

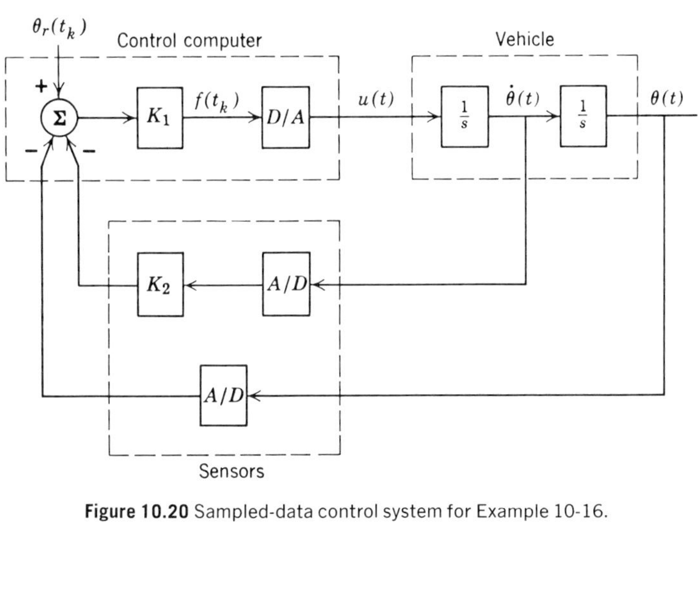

Figure 10.20 Sampled-data control system for Example 10-16.

tinuous-time system. The controller is a discrete-time system. (The spacecraft moment of inertia has been normalized to unity for convenience.)

We first model the spacecraft. Letting $x_1(t) = \theta(t)$, $x_2(t) = \dot{\theta}(t)$, we have

$$\begin{aligned}\dot{x}_1 &= x_2 \\ \dot{x}_2 &= u.\end{aligned} \tag{10.96}$$

Furthermore, it is clear from the diagram that

$$f(t_k) = K_1[\theta_r(t_k) - \theta(t_k)] - K_1K_2\dot{\theta}(t_k), \tag{10.97}$$

assuming the various converters are operating synchronously. Nonsynchronous operation can be described using the general approach discussed above. The equations become more complicated, of course. Using the state variables of (10.96), (10.97) becomes

$$f(t_k) = K_1[\theta_r(t_k) - x_1(t_k)] - K_1K_2x_2(t_k). \tag{10.98}$$

We wish to describe the system in the form of (10.91). A simple calculation reveals that

$$\boldsymbol{\phi}(t) = \begin{bmatrix} 1 & t \\ 0 & 1 \end{bmatrix}, \tag{10.99}$$

and

$$\int_{t_k}^{t} \phi(t-\tau)B\,d\tau = \begin{bmatrix} \frac{1}{2}(t-t_k)^2 \\ t-t_k \end{bmatrix}. \tag{10.100}$$

Thus (10.98) becomes, for $t_k < t_{k+1}$,

$$\begin{bmatrix} x_1(t) \\ x_2(t) \end{bmatrix} = \begin{bmatrix} 1 & (t-t_k) \\ 0 & 1 \end{bmatrix} \begin{bmatrix} x_1(t_k) \\ x_2(t_k) \end{bmatrix} + \begin{bmatrix} \frac{1}{2}(t-t_k)^2 \\ (t-t_k) \end{bmatrix} f(t_k), \tag{10.101}$$

where f_k is given by (10.98). Equations 10.101 and 10.98 can be combined to yield a matrix representation for the total system (including plant and feedback controller):

$$\begin{aligned}\begin{bmatrix} x_1(t) \\ x_2(t) \end{bmatrix} &= \begin{bmatrix} 1 - \dfrac{K_1}{2}(t-t_k)^2 & (t-t_k) - \dfrac{K_1K_2}{2}(t-t_k)^2 \\ -K_1(t-t_k) & 1 - K_1K_2(t-t_k) \end{bmatrix} \begin{bmatrix} x_1(t_k) \\ x_2(t_k) \end{bmatrix} \\ &\quad + \begin{bmatrix} \dfrac{K_1}{2}(t-t_k)^2 \\ K_1(t-t_k) \end{bmatrix} \theta_r.\end{aligned} \tag{10.102}$$

As a final step, suppose we have "conventional" sampling with $t_k = kT$, and we wish to obtain the state variables at the sampling instant. Then, substituting $t = t_{k+1} = (k+1)T$ in (10.102), we have

$$\begin{bmatrix} x_1[(k+1)T] \\ x_2[(k+1)T] \end{bmatrix} = \begin{bmatrix} 1-\dfrac{K_1T^2}{2} & T-\dfrac{K_1K_2T^2}{2} \\ -K_1T & 1-K_1K_2T \end{bmatrix} \begin{bmatrix} x_1(kT) \\ x_2(kT) \end{bmatrix} + \begin{bmatrix} \dfrac{K_1T^2}{2} \\ K_1T \end{bmatrix} \theta_r(kT) \tag{10.103}$$

as the difference equation corresponding to (10.95). ▲

Problems

10-1. *Newton's method* for finding a root of the equation $g(x) = 0$ is the recursion formula

$$x(t_{k+1}) = x(t_k) - \frac{g[x(t_k)]}{g'[x(t_k)]},$$

where $g' = dg/dx$. Under suitable conditions, $\lim_{k\to\infty} x(t_k)$ is a root of $g(x) = 0$. Suppose we seek the square root of b, so $g(x) = x^2 - b$. Find the difference equation for using Newton's method to solve $x^2 - b = 0$. What *order* is the equation? Is it linear? Time-invariant?

10-2. (Computer Problem) Write a computer program (in FORTRAN or your favorite language) to integrate a first-order differential equation using Euler's procedure, Equation 10.2. What modifications of the program would be needed to integrate n simultaneous first-order differential equations?

10-3. (Computer Problem) Repeat Problem 10-2 for the second-order Runge-Kutta procedure.

10-4. (Computer Problem) Solve

$$\dot{y} = (\cos t)y, \qquad y(0) = 1$$

using your computer programs of Problems 10-2 and 10-3. Try various step sizes. (Exact solution: $y = \epsilon^{\sin t}$.)

10-5. The voltage source in the network of Figure P10-5 is piecewise-constant,

$$u(t) = u(kT), \qquad kT < t \leq (k+1)T.$$

Find a difference equation for the capacitor voltage at sampling times, $v_C(kT)$. At $t = 0$, $v_C(0) = v_{c0}$.

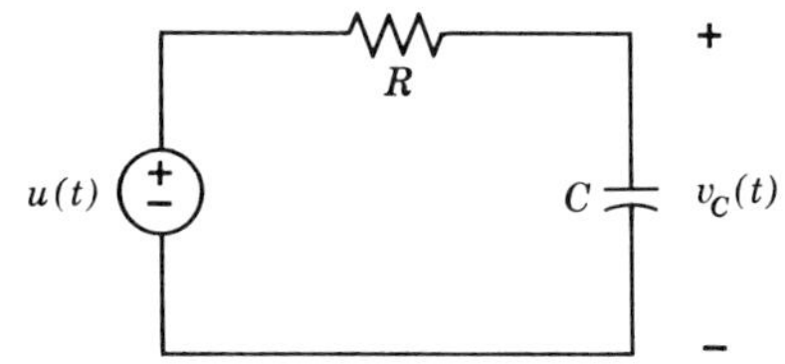

Figure P10-5

10-6. Construct a digital simulation diagram for the difference equation obtained in Problem 10-5.

10-7. (Computer Problem) Prepare a digital computer program from your simulation diagram of Problem 10-6. Obtain a computer solution for $v_c(kT)$ when $RC = 1$, $T = 0.1$, and $u(t)$ is a square wave, as shown in Figure P10-7.

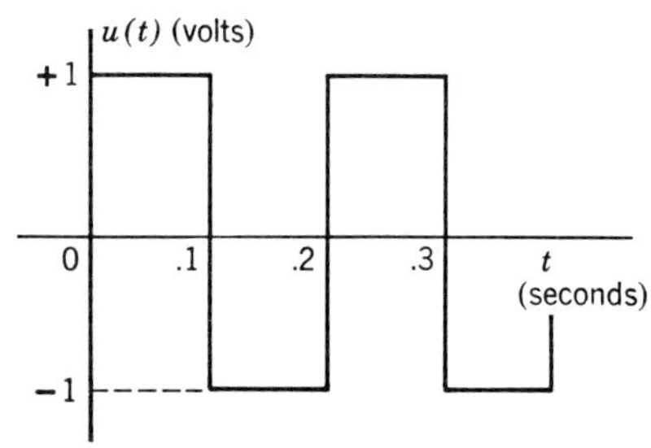

Figure P10-7

10-8. A simulation diagram of a discrete-time system is shown in Figure P10-8. Obtain the difference equation relating $c(kT)$ to $r(kT)$.

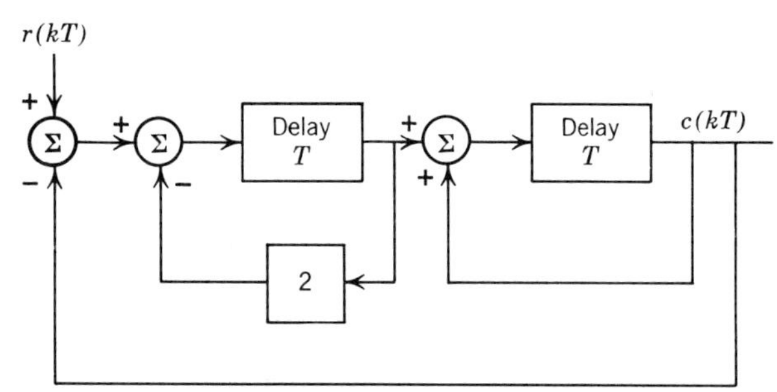

Figure P10-8

10-9. Repeat Problem 10-8 for the system in Figure P10-9.

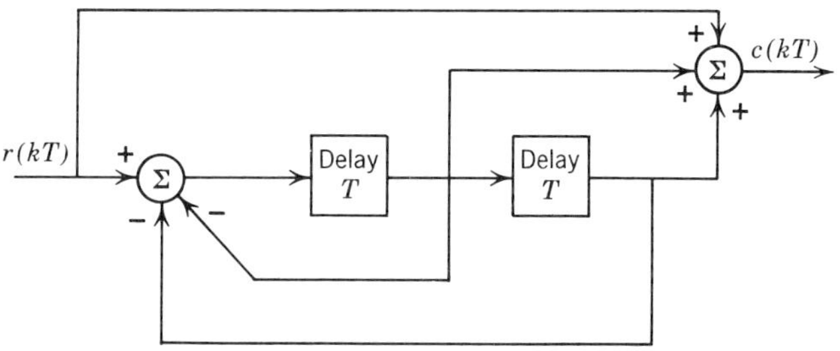

Figure P10-9

10-10. A system is modeled by the difference equations

$$x_1(n+1) = \tfrac{3}{2}x_1(n) + x_2(n) + u(n)$$
$$x_2(n+1) = -\tfrac{1}{2}x_1(n)$$
$$y(n) = x_1(n).$$

Construct a simulation block diagram for this system.

10-11. A system is described by

$$y(n+2) - y(n+1) = -\tfrac{2}{5}u(n) + \tfrac{3}{5}u(n+1).$$

Construct a simulation diagram for this system using the approach of Example 10-6.

10-12. Find z transforms of the following sequences. Express in closed form. In all cases, $f(n) = 0$ for $n < 0$.

(a) $f(n) = 1$.
(b) $f(n) = \sin n$.
(c) $f(n) = n^2$.

10-13. Invert the following transforms:

(a) $F(z) = \dfrac{z}{(z-1)(z-\frac{1}{2})}$

(b) $F(z) = \dfrac{z}{(z-\frac{1}{2})^2}$

(c) $F(z) = \dfrac{z^2+z^4}{(z-1)(z+1)(z-\frac{1}{2})^2}$.

10-14. Prove the initial value theorem:

$$f(0) = \lim_{z\to\infty} F(z).$$

10-15. Prove the final value theorem:

$$f(\infty) = \lim_{z\to 1} (1-z^{-1})F(z) \quad \text{if} \quad (1-z^{-1})F(z)$$

has no poles for $|z| \geq 1$.

10-16. Solve the difference equation of Problem 10-5 using z transforms.

10-17. Solve the difference equation of Problem 10-8 using z transforms when initial conditions are zero, and $r(kT) = kT$.

10-18. Repeat Problem 10-17 for $c(0) = 1$, $c(T) = 0$, $r(kT) = 1$.

10-19. Repeat Problem 10-17 for the system of Figure P10-9.

10-20. Repeat Problem 10-18 for the system of Figure P10-9.

10-21. Find $x_1(n)$ and $x_2(n)$ in Problem 10-10 when $x_1(0)$ and $x_2(0) = 0$, and

$$u(n) = \begin{cases} 1, & n = 0 \\ 0, & n \neq 0 \end{cases}$$

10-22. Repeat Problem 10-21 for $x_1(0) = x_{10}$, $x_2(0) = x_{20}$, $u(n) = 0$.

10-23. Find $Y(z)/U(z)$ for the system of Problem 10-10. Sketch location of poles and zeros in the z-plane.

10 24. Find $Y(z)/U(z)$ for the system of Problem 10-11. Sketch poles and zeros in the z-plane.

10-25. Consider the system of Figure P10-25. Find the state-transition matrix $\boldsymbol{\phi}(mT)$ using:

(a) $\boldsymbol{\Phi}(z) = (\mathbf{I} - z^{-1}\mathbf{F})^{-1}$.

(b) Transfer functions by Mason's rule applied to all-unit-delay diagram with initial conditions shown as inputs.

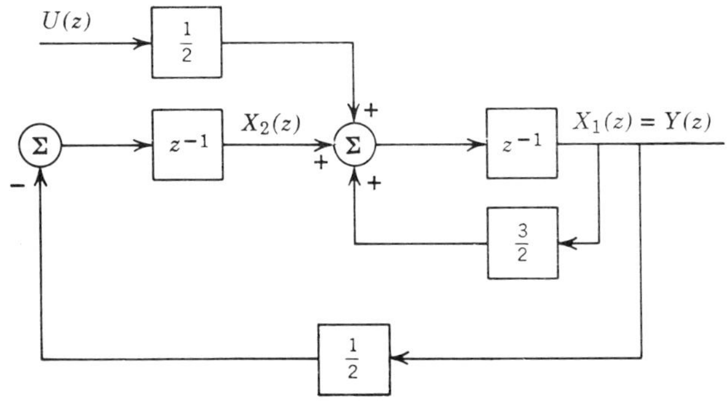

Figure P10-25

10-26. Find the state-transition matrix for the system of Problem 10-10.

10-27. For the system

$$\mathbf{x}_{k+1} = \mathbf{F}\mathbf{x}_k + \mathbf{G}\mathbf{u}_k$$
$$\mathbf{y}_k = \mathbf{C}\mathbf{x}_k + \mathbf{D}\mathbf{u}_k$$

it is shown in Equation (10.84) that the transfer function matrix $\mathbf{H}(z)$ is given by $[z^{-1}\mathbf{C}\mathbf{\Phi}(z)\mathbf{G}+\mathbf{D}]$. Suppose new state variables $\mathbf{w}_k$ are chosen that are related to the old state variables by

$$\mathbf{w}_k = \mathbf{T}\mathbf{x}_k,$$

where T is a constant nonsingular matrix.

(a) Find the state description for the $\mathbf{w}$ system in terms of $\mathbf{F}$, $\mathbf{G}$, $\mathbf{C}$, $\mathbf{D}$, and $\mathbf{T}$.

(b) Show that $\mathbf{H}(z)$ remains the same as before.

10-28. A linear *time-varying* discrete-time system is described by

$$\mathbf{x}(t_{k+1}) = \mathbf{F}(t_{k+1}, t_k)\mathbf{x}(t_k) + \mathbf{G}(t_{k+1}, t_k)\mathbf{u}(t_k),$$

with $\mathbf{x}(t_0) = \mathbf{x}_0$. Using recursion, find a time-domain solution to this system analogous to Equation 10.80.

10-29. Define and discuss the meaning of the characteristic equation for a linear time-invariant discrete-time system by analogy to the discussion of the continuous-time case in Chapter 8. Include a discussion of the relationship between z-plane location of roots of the characteristic equation and time response.

10-30. What is the discrete analog of the Nyquist stability criterion? (Suggestion: unit circle is boundary between stable and unstable z-plane regions.)

10-31. Determine if the roots are inside the unit circle for the following equations using the Jury stability table.

(a) $z^3 - \frac{1}{2}z^2 - \frac{9}{10}z + \frac{9}{20} = 0.$

(b) $z^6 - z^5 + \frac{1}{2}z^4 - z^2 + z - \frac{1}{2} = 0.$

10-32. Prove the following theorem: Consider the discrete-time system $\mathbf{x}_{k+1} = \mathbf{g}(\mathbf{x}_k)$ with $\mathbf{g}(\mathbf{0}) = \mathbf{0}$. Suppose that for *some* norm $\|\mathbf{g}(\mathbf{x})\| < \|\mathbf{x}\|$. Then the equilibrium state $\mathbf{x}_e = 0$ is asymptotically stable in the large, and $V(\mathbf{x}_k) = \|\mathbf{x}_k\|$ is one of its Liapunov functions.

10-33. In Problem 10-1 it was established that applying Newton's method to solve $x^2 - b = 0$ yields

$$x(t_{k+1}) = \frac{x(t_k)}{2} + \frac{b}{2x(t_k)}.$$

(a) Show that $x = \sqrt{b}$ is one equilibrium point for this system.

(b) Show that the change of variables

$$y(t_k) = x(t_k) - \sqrt{b}$$

produces a system with an equilibrium point $\mathbf{y}_e = \mathbf{0}$.

(c) Apply the theorem of Problem 10-32 to the system of part (b) to investigate the stability of $\mathbf{y}_e = \mathbf{0}$. (Suggestion: Use *magnitude* as a norm.)

(d) Discuss the stability of $x = -\sqrt{b}$.

(e) Interpret the results of this problem and comment on the use of Newton's method as a *computational* algorithm for finding square roots. (That is, does it converge? Does it converge to the correct answer?)

10-34. State and prove a discrete-time analog of Theorem 9.4.

10-35. Discuss extensions of the sensitivity function notions of previous chapters to the discrete-time case. In particular, discuss the obtaining of sensitivity functions from discrete-time simulation models or z-transform block diagrams.

10-36. A digital control system is shown in Figure P10-36.

(a) Write state equations for the continuous dynamic elements, CDE, using $x_1 = c$, $x_2 = \dot{c}$, and $u(t)$ as input. Draw all-integrator representation, including initial conditions.

(b)

$$D(z) = \frac{F(z)}{E(z)} = \frac{105.6 - 95.8z^{-1}}{1 + 0.472z^{-1}}.$$

Using the method discussed in Example 10.6, draw all-unit-delay diagram for the Digital Dynamic Elements, DDE, and write a state equation, calling the state x_3.

(c) Write (in matrix form) difference equations relating x_1, x_2, x_3 at $[(k+1)T]$ to x_1, x_2, x_3 at kT and $r(kT)$, for $T = 0.1$.

(d) Solve for $\mathbf{x}(mT)$ for $r(kT) = 1$, $m = 0, 1, 2, 3, 4$.

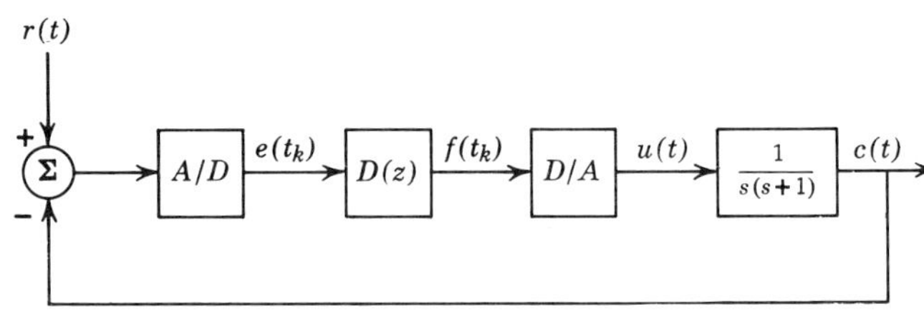

Figure P10-36

10-37. Consider the attitude control system of Example 10-16. Find the z-transform transfer function for the plant, $P(z) = \theta(z)/U(z)$, where $\theta(z) = \mathscr{Z}\{\theta(kT)\}$ and $U(z) = \mathscr{Z}\{u(kT)\} = \mathscr{Z}\{f(kT)\}$. Sketch the poles and zeros in the z-plane, and discuss and sketch the general nature of the vehicle response for step inputs [that is, piecewise-constant $u(t)$], for various values of T. (Suggestion: Find $x_1[(k+1)T] = \theta[(k+1)T]$ from Equation 10.108 by letting $t = (k+1)T$.)

10-38. Obtain the z-transform transfer function $\theta(z)/\theta_r(z) = H(z;\ K_1, K_2, T)$ for the attitude-control system of Example 10-16. Sketch a root locus showing the z-plane location of the poles of H for the following cases:

(a) K_1 variable, K_2 and T fixed.

(b) K_2 variable, K_1 and T fixed.

10-39. Repeat Problem 10-38 for K_1 and K_2 variable, T fixed. That is, draw "root contours" showing regions where the roots may be located.

PART III

Control and Optimization

Chapter 11. Feedback and Control

Chapter 12. Parameter Optimization I

Chapter 13. Parameter Optimization II

Control system design involves the determination of an input to a system (called the plant) to produce an acceptable performance. Performance may be an output time response or it may be the value of a scalar index such as cost. There may be several constraints on the determination of the plant input. The major constraints are: (1) plant output is related to plant input through a set of specified differential or difference equations, (2) plant input is limited by saturation and power considerations, (3) the subsystem (called the controller) whose output is the plant input may be structurally specified because of considerations such as ease of implementation and high reliability, (4) there may be uncertainty in the parameters of the plant equations, (5) there may be disturbance inputs, and (6) there may be a command input signal, which the system output is required to duplicate as much as possible. The essence of control is the manipulation of the plant input to achieve a desired performance in the presence of constraints.

The three chapters of Part III are concerned with control and optimization. Chapter 11, Feedback and Control, begins with a discussion of the general nature of control problems. The concepts of controllability and observability are discussed and illustrated in the context of our theme example. Some potential advantages of feedback structures are discussed. With this background on feedback and control, a feedback controller structure with adjustable parameters can be chosen. The next two chapters are concerned with the determination of the controller parameters to optimize a system performance index. In Chapter 12, Parameter Optimization I, the importance and relevance of a quadratic performance index in a variety of system problems are explained. Analytical procedures are developed. Also discussed is the concept of trade-off or compromise when two or more indices are relevant. In Chapter 13, Parameter Optimization II, two commonly used numerical optimization procedures are developed. The Newton-Raphson method is discussed first, followed by the gradient method. The sensitivity analyses described in previous chapters are used in conjunction with either the Newton-Raphson method or the gradient method for an automatic adjustment of the control parameters.

CHAPTER 11

Feedback and Control

In this chapter, we shall establish the control terminology that will be used. The general nature of control problems will be described and illustrated in terms of an attitude control problem.

The concepts of controllability and observability are defined and discussed in this chapter. These concepts are used to study the capability of reaction-wheel control for an attitude system. Starting from an arbitrary initial attitude and attitude rate, if the reaction wheel can generate a torque that will result in a specified attitude and attitude rate after a finite time duration, the attitude system is said to be controllable.

We consider a control-system configuration, consisting of a system to be controlled (called the plant) and state feedback. That is, the plant input is a linear combination of an external input and the plant state variables. For this feedback configuration, we shall show that if the plant is controllable, the feedback gains can always be chosen to achieve any pole locations of the transfer function of the composite system. For example, the feedback gains can always be chosen so that the composite system is stable even though the controllable plant is unstable. Furthermore, the transfer function of the feedback control system has exactly the same zeros as the transfer function of the plant alone. State feedback serves as one basis for design where specifications on an acceptable response can be translated into specifications on an acceptable transfer function. From a set of desirable pole locations, the state-feedback gains are computed.

Finally, the deviation in a system response due to a perturbation in system parameters, or due to the presence of disturbance inputs, is assessed by using the integral of the square of the response deviation. The deviation for a system employing feedback is compared to the deviation for a system with no feedback, using this measure. A comparison sensitivity function is introduced. Using the comparison sensitivity, a criterion for an insensitive feedback design is derived.

11.1 Nature of Control Problems

The purpose of this section is to state in a general way the nature of control problems and to establish some terminology for control-system analysis and design. A control problem is recognized as a problem of determining inputs to a given system to achieve desirable responses, in the presence of constraints.

The system to be controlled is defined as the *plant*. The plant is characterized by a set of inputs, a set of state variables, and a set of outputs. These variables are related by differential or difference equations characterizing the plant. These equations are said to impose structural constraints on the variables. A spacecraft whose outputs are the attitude angles and angular velocities and whose inputs are control torques and disturbance torques is an example of a plant. Figure 11.1*a* shows a block diagram for a spacecraft attitude control system, where the spacecraft is the plant. The control torque in Figure 11.1*a* is an example of a *plant control input* or, simply, *control input*. The disturbance input, such as the disturbance torque from micrometeorite impact and solar radiation, is not available for manipulation. The control input, on the other hand, can be manipulated by the system designer as an output of a subsystem called the *controller*, which is to be designed. In our attitude control system example, the control torque is generated by an actuator whose input comes from a control computer. The control computer may have two sets of inputs: desired attitude and measured vehicle orientation. The measured vehicle orientation is the set of outputs from sensors such as gyroscopes. The subsystem of actuator, control computer, and sensors is called the controller. The desired attitude is an example of an *external input* and the measured vehicle orientation is an example of *feedback*

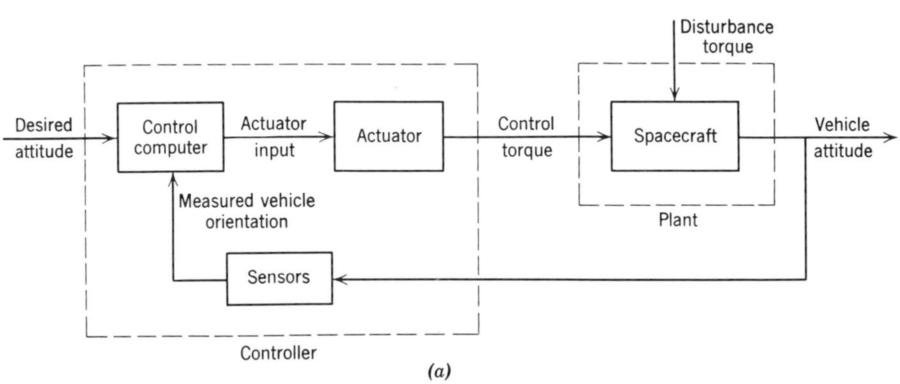

Figure 11.1(*a*)

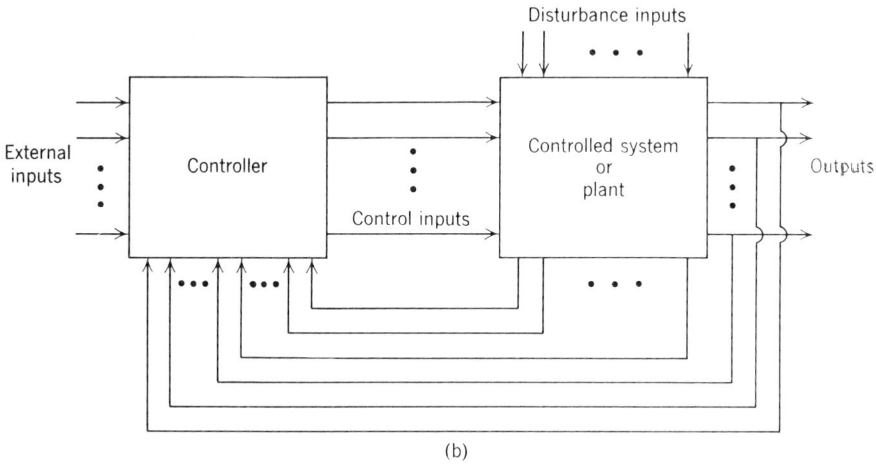

Figure 11.1(*b*) General configuration of a control system.

signals from the plant. With the understanding that sensors or transducers constitute part of the controller, we refer to the sensor inputs from the plant as the feedback signals to the controller.

The composite system, consisting of the plant and the controller, is called a *control system.* A general configuration is shown in Figure 11.1*b*. The inputs of the control system are the external inputs. The system outputs are the plant outputs. The system states are the plant states and the controller states.

The goals for which the controller is designed are varied. One goal is to achieve a specified time response. The specification may be in terms of overshoot, rise time, and settling time of the step response. The specifications may be converted to specifications of desirable pole-zero configurations of the transfer function for design using s-plane concepts.

Another type of goal for controller design is the minimization of an index of performance. For example, the performance index might be the time it takes the state to reach a specified value, starting from a specified initial state. Another index is the integral of the sum of the squares of state variables. The latter index is useful for designing regulators, where the ideal state is zero. The minimization of performance indices with respect to control parameters for a fixed controller structure is discussed in Chapters 12 and 13.

In addition to the plant structural constraints there may be other constraints. A common constraint is a limitation on the plant input. For example, consider our attitude control system where the plant input is the control torque. If the actuator is a gas-jet ejection system, the maximum torque is limited by the maximum rate of mass expulsion and the type of gas used. If the actuator is a reaction wheel, the maximum torque available is limited by the power rating of the motor. There may be constraints also on the state variables. For example, for our attitude system there may be a maximum allowable attitude rate. The foregoing constraints are examples of signal constraints.

The plant may contain some uncertainty, such as unknown coefficients in the differential or difference equations characterizing the plant. These unknown coefficients may result from uncertainties in sensor gains, spacecraft moments of inertia, reaction-wheel time constants, and other physical system parameters.

Two types of structure are defined. A controller that makes no use of the plant output or any signal from the plant is called an *open-loop controller*, and the resulting control system is called an *open-loop control system*. When plant signals are utilized, the controller is called a *feedback* or *closed-loop controller* and the resulting control system is called a *feedback control system*, or *closed-loop control system*.

When there are no disturbances or uncertainties in the plant, the type of structure plays a minor role in the behavior of the control system. But when disturbances or plant parameter uncertainties are present, the feedback structure is potentially superior. However, when feedback is employed, there is a possibility of additional noise or uncertainty in the measurement of the plant signals. This type of noise is called instrumentation or sensor noise. It does not arise in the open-loop case because no plant measurements are made. In assessing the overall merits of feedback, this instrumentation noise must be taken into account.

11.2 Controllability

As mentioned earlier, a class of problems in control-system design is the determination of a control input $\mathbf{u}(t)$ and the synthesis of the associated controller which generates $\mathbf{u}(t)$, such that a desired dynamic response is achieved. It is useful to know, before attempting to find $\mathbf{u}(t)$, whether such a control can be found. That is, it is useful to know whether the problem has a solution before trying to find one. In general, if the desired state $\mathbf{x}(t)$ is specified for all time, the requirements for the existence of a control input that will generate $\mathbf{x}(t)$ are very stringent.

A less ambitious but more realistic goal is to require only a partial specification of the state. One such partial specification, which has wide application, is forcing the state of a system to attain a specified value at some finite time in the future. That is, given an initial time t_0, an initial state $\mathbf{x}(t_0) = \mathbf{x}_0$, and a final state $\mathbf{x}_f$, find a control $\mathbf{u}(t)$, $t_0 \leq t \leq t+T$, for some finite T, such that $\mathbf{x}(t_0+T) = \mathbf{x}_f$. There may or may not be any control $\mathbf{u}(t)$ which can force the system to attain the state $\mathbf{x}_f$. Thereafter, it may be desirable to maintain the state at $\mathbf{x}_f$. If $\mathbf{x}_f$ is an equilibrium point and if the equilibrium point is stable in the sense of Lyapunov, the state will remain close to $\mathbf{x}_f$ for all future time. By a suitable choice of error coordinates such as state variables, the problem of reaching and maintaining a specified state is the problem of matching a desired dynamic response.

In this section, we discuss the conditions that the plant structure must satisfy in order to drive the state from an initial value $\mathbf{x}_0$ to another value $\mathbf{x}_f$, such as $\mathbf{0}$, in finite time. A system is said to be *completely state controllable* if, for any initial time t_0, it is possible to generate an unconstrained control vector $\mathbf{u}(t)$ that will take any given initial state $\mathbf{x}(t_0)$ to any final state $\mathbf{x}(t_f)$ in a finite time interval $t_0 \leq t \leq t_f$. For a linear system, the final state may be taken to be the origin without loss of generality. The concept of controllability and another closely related one called observability are due to Kalman.[1] The basic idea is presented in terms of a discrete-time system.

Consider a linear discrete-time system.

$$\mathbf{x}(k+1) = \mathbf{A}\mathbf{x}(k) + \mathbf{b}u(k) \tag{11.1}$$

where $\mathbf{x}(k)$ is a state n-vector, $u(k)$ is a *scc'lar* control sequence, $\mathbf{A}$ is a nonsingular $n \times n$ matrix, and $\mathbf{b}$ is an $n \times 1$ matrix. No bounds are placed on $u(k)$. Let us investigate under what conditions the state can be transferred from $\mathbf{x}(0)$ to $\mathbf{x}(1) = \mathbf{0}$ in one sampling period or one time unit of the discrete-time system. From (11.1) and if $\mathbf{x}(1) = \mathbf{0}$, $\mathbf{x}(0)$ must be

$$\mathbf{x}(0) = -\mathbf{A}^{-1}\mathbf{b}u(0). \tag{11.2}$$

Since $u(0)$ could have any value, the initial condition vector must lie on a line defined by the vector $-\mathbf{A}^{-1}\mathbf{b}$. If the system is second-order, the vector $-\mathbf{A}^{-1}\mathbf{b}$ is a line in the phase plane. Any initial condition that lies on the line as given in (11.2) can be taken to the origin in one sampling period. For a control over two sampling periods, if $\mathbf{x}(2) = \mathbf{0}$, (11.1) becomes

$$\mathbf{0} = \mathbf{A}\mathbf{x}(1) + \mathbf{b}u(1). \tag{11.3}$$

[1]R. E. Kalman, "On the General Theory of Control Systems," *Proc. First Congress of IFAC*, Moscow, USSR, 1960; Butterworths, London, **1**, 481–492 (1961).

Using (11.1) for expressing $\mathbf{x}(1)$ in terms of $\mathbf{x}(0)$, (11.3) can be further rewritten as

$$\mathbf{0} = \mathbf{A}[\mathbf{A}\mathbf{x}(0) + \mathbf{b}u(0)] + \mathbf{b}u(1).$$

Thus $\mathbf{x}(0)$ must be

$$\mathbf{x}(0) = -\mathbf{A}^{-1}\mathbf{b}u(0) - \mathbf{A}^{-2}\mathbf{b}u(1). \tag{11.4}$$

For a second-order system, $-\mathbf{A}^{-1}\mathbf{b}$ and $-\mathbf{A}^{-2}\mathbf{b}$ are lines on the phase plane passing through the origin. If these lines are distinct, that is, the lines are not coincident, any point $\mathbf{x}(0)$ in the phase plane can be expressed as a linear combination of the two lines.

Example 11-1. Consider a linear time-invariant second-order discrete-time system

$$x_1(k+1) = -\tfrac{1}{2}x_1(k) + \tfrac{1}{4}x_2(k) + u(k) \tag{11.5}$$

$$x_2(k+1) = \tfrac{1}{4}x_1(k) - \tfrac{3}{4}x_2(k) + 2u(k). \tag{11.6}$$

with initial conditions $x_1(0) = 4$, $x_2(0) = 6$. The lines $\mathbf{A}^{-1}\mathbf{b}$ and $\mathbf{A}^{-2}\mathbf{b}$ are

$$\mathbf{A}^{-1}\mathbf{b} = \begin{bmatrix} -\frac{1}{2} & \frac{1}{4} \\ \frac{1}{4} & -\frac{3}{4} \end{bmatrix}^{-1} \begin{bmatrix} 1 \\ 2 \end{bmatrix} = \begin{bmatrix} -4 \\ -4 \end{bmatrix} \tag{11.7}$$

$$\mathbf{A}^{-2}\mathbf{b} = \begin{bmatrix} -\frac{1}{2} & \frac{1}{4} \\ \frac{1}{4} & -\frac{3}{4} \end{bmatrix}^{-1} \begin{bmatrix} -4 \\ -4 \end{bmatrix} = \begin{bmatrix} \frac{64}{5} \\ \frac{48}{5} \end{bmatrix}. \tag{11.8}$$

The line segments are indicated in Figure 11.2 as OP_2 and OP_1, respectively. The initial condition $\mathbf{x}(0)$, represented by the vector

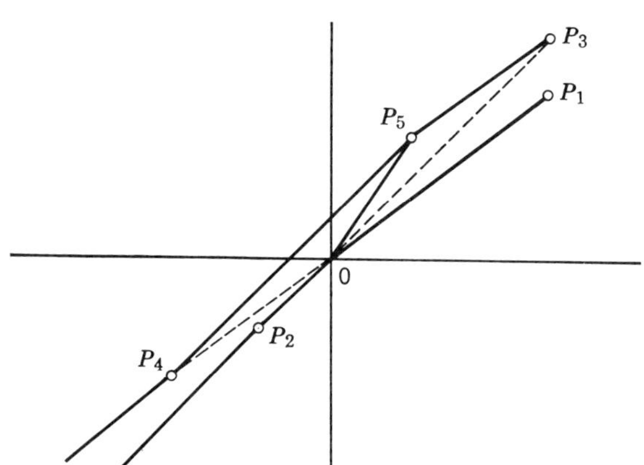

Figure 11.2 Phase plane showing the lines $\mathbf{A}^{-1}\mathbf{b}$ and $\mathbf{A}^{-2}\mathbf{b}$ as OP_2 and OP_1, respectively, for Example 11-1.

OP_5, may be expressed as a vector sum of OP_4 and OP_3, where OP_3 is colinear with OP_2 and OP_4 is colinear with OP_1. The points P_3 and P_4 are located by drawing lines parallel to OP_1 and OP_2 from P_5. Then, by (11.4), the values of $u(0)$ and $u(1)$ are determined because $-\mathbf{A}^{-1}\mathbf{b}u(0)$ is OP_3 and $-\mathbf{A}^{-2}\mathbf{b}u(1)$ is OP_4.

In this example, since only initial states colinear with OP_2 can be taken to the origin in one sampling period, it takes at least two sampling instants to take the initial state at P_5 to the origin. If $u(0)$ and $u(1)$ can be chosen so that (11.4) is satisfied, then the initial state can be taken to the origin in exactly two sampling periods. From (11.4), (11.7), (11.8), and P_5 in Figure 11.2, we have

$$\begin{bmatrix} 4 \\ 6 \end{bmatrix} = -u(0)\begin{bmatrix} -4 \\ -4 \end{bmatrix} - u(1)\begin{bmatrix} \frac{64}{5} \\ \frac{48}{5} \end{bmatrix} = \begin{bmatrix} 4 & -\frac{64}{5} \\ 4 & -\frac{48}{5} \end{bmatrix}\begin{bmatrix} u(0) \\ u(1) \end{bmatrix}$$

or

$$\begin{bmatrix} u(0) \\ u(1) \end{bmatrix} = \begin{bmatrix} 4 & -\frac{64}{5} \\ 4 & -\frac{48}{5} \end{bmatrix}^{-1}\begin{bmatrix} 4 \\ 6 \end{bmatrix} = \begin{bmatrix} 3 \\ \frac{5}{8} \end{bmatrix} \tag{11.9}$$

Thus, for $u(0) = 3$,

$$\mathbf{x}(1) = \begin{bmatrix} -\frac{1}{2} & \frac{1}{4} \\ \frac{1}{4} & -\frac{3}{4} \end{bmatrix}\begin{bmatrix} 4 \\ 6 \end{bmatrix} + \begin{bmatrix} 1 \\ 2 \end{bmatrix}3 = \begin{bmatrix} \frac{5}{2} \\ \frac{5}{2} \end{bmatrix} \tag{11.10}$$

which is colinear with OP_2. With $u(1) = \frac{5}{8}$,

$$\mathbf{x}(2) = \begin{bmatrix} -\frac{1}{2} & \frac{1}{4} \\ \frac{1}{4} & -\frac{3}{4} \end{bmatrix}\begin{bmatrix} \frac{5}{2} \\ \frac{5}{2} \end{bmatrix} + \begin{bmatrix} 1 \\ 2 \end{bmatrix} = \begin{bmatrix} 0 \\ 0 \end{bmatrix} \tag{11.11}$$

Thus the sequence of controls $u(0)$ and $u(1)$ in (11.9) takes the initial state $x_1(0) = 4$ and $x_2(0) = 6$ to the origin, in two sampling periods. ▲

If the initial state $\mathbf{x}(0)$ can be taken to the origin in m sampling periods, then (11.4) generalizes to

$$\begin{aligned} \mathbf{x}(0) &= -\mathbf{A}^{-1}\mathbf{b}u(0) - \mathbf{A}^{-2}\mathbf{b}u(1) - \cdots - \mathbf{A}^{-m}\mathbf{b}u(m-1) \\ &= -[\mathbf{A}^{-1}\mathbf{b} \mid \mathbf{A}^{-2}\mathbf{b} \mid \cdots \mid \mathbf{A}^{-m}\mathbf{b}]\begin{bmatrix} u(0) \\ u(1) \\ \vdots \\ u(m-1) \end{bmatrix}. \end{aligned} \tag{11.12}$$

But these linear algebraic equations (11.12) can have a solution for $u(0), u(1), \ldots, u(m-1)$ only if

$$\mathbf{Q}_1 = [\mathbf{A}^{-1}\mathbf{b} \mid \mathbf{A}^{-2}\mathbf{b} \mid \cdots \mid \mathbf{A}^{-m}\mathbf{b}]$$

and

$$\mathbf{Q}_2 = [\mathbf{A}^{-1}\mathbf{b} \mid \mathbf{A}^{-2}\mathbf{b} \mid \cdots \mid \mathbf{A}^{-m}\mathbf{b} \mid \mathbf{x}_0]$$

have the same rank.[2] If $\mathbf{x}(0)$ is an arbitrary n-dimensional vector, the above matrices can have the same rank if and only if

$$\text{rank of } \mathbf{Q}_1 = \text{rank of } [\mathbf{A}^{-1}\mathbf{b} \mid \mathbf{A}^{-2}\mathbf{b} \mid \cdots \mid \mathbf{A}^{-m}\mathbf{b}] = n. \quad (11.13)$$

It is possible to check the rank of $\mathbf{Q}_1$ without forming the inverse of $\mathbf{A}$. Since $\mathbf{A}$ is nonsingular and it has rank n, the rank of $\mathbf{Q}_1$ is equal to the rank of $\mathbf{A}^m\mathbf{Q}_1 = \mathbf{Q}_3$, so that (11.13) is equivalent to

$$\text{rank of } \mathbf{Q}_3 = \text{rank of } [\mathbf{A}^{m-1}\mathbf{b} \mid \mathbf{A}^{m-2}\mathbf{b} \mid \cdots \mid \mathbf{b}] = n.$$

The rank of a matrix is not changed if its columns are interchanged. Hence (11.13) is also equivalent to

$$\text{rank of } \mathbf{Q}_4 = \text{rank of } [\mathbf{b} \mid \mathbf{A}\mathbf{b} \mid \cdots \mid \mathbf{A}^{m-1}\mathbf{b}] = n, \quad (11.14)$$

where $\mathbf{Q}_4$ is formed from $\mathbf{Q}_3$ by reversing the order of the columns.

Note that $\mathbf{Q}_4$ has m columns. Thus it is necessary that $m \geqslant n$ in order that (11.14) be satisfied. If (11.14) is satisfied and $m = n$, (11.12) possesses a unique solution for $u(0), u(1), \ldots, u(n-1)$. If (11.14) is satisfied and $m > n$, (11.12) possesses infinitely many solutions.

If the rank of $\mathbf{Q}_4$ for a particular $m \geqslant n$ is less than n, it is not possible to drive the system to a zero final state in m sampling periods, starting from an arbitrary initial state. But can we wait and use a longer control sequence (more sampling periods) to achieve a zero final state? With the aid of the Cayley-Hamilton theorem in matrix theory, it will be seen that the answer is negative.

The Cayley-Hamilton theorem states that a square matrix satisfies its own characteristic equation. Thus, if the characteristic equation of a matrix $\mathbf{A}$ is

$$\lambda^n + a_{n-1}\lambda^{n-1} + \cdots + a_1\lambda + a_0 = 0,$$

$\mathbf{A}$ satisfies

$$\mathbf{A}^n + a_{n-1}\mathbf{A}^{n-1} + \cdots + a_1\mathbf{A} + a_0\mathbf{I} = \mathbf{0}.$$

[2] G. E. Shilov, *Theory of Linear Spaces*, Prentice Hall, Englewood Cliffs, 1916; See Chapter 3, Systems of Linear Equations.

It is seen that $\mathbf{A}^n$ can be expressed as a polynomial in $\mathbf{A}$ of degree $n-1$:

$$\mathbf{A}^n = -[a_{n-1}\mathbf{A}^{n-1} + \cdots + a_1\mathbf{A} + a_0\mathbf{I}]. \tag{11.15}$$

By a repeated application of (11.15), $\mathbf{A}^m$ for any $m \geqslant n$ can be expressed as

$$\mathbf{A}^m = k_0\mathbf{I} + k_1\mathbf{A} + \cdots + k_{n-1}\mathbf{A}^{n-1}.$$

where $k_0, \ldots, k_{n-1}$ are scalar constants. Moreover, for any $j \geqslant n$, $\mathbf{A}^j\mathbf{b}$ can be expressed as a linear combination of the column matrices $\mathbf{b}$, $\mathbf{Ab}, \ldots, \mathbf{A}^{n-1}\mathbf{b}$. Hence, for $m \geqslant n$,

$$\text{rank of } [\mathbf{b} \,\vdots\, \mathbf{Ab} \,\vdots\, \cdots \,\vdots\, \mathbf{A}^{m-1}\mathbf{b}] = \text{rank of } [\mathbf{b} \,\vdots\, \mathbf{Ab} \,\vdots\, \cdots \,\vdots\, \mathbf{A}^{n-1}\mathbf{b}]. \tag{11.16}$$

From (11.16), it is seen that if (11.14) is not satisfied for $m = n$, it cannot be satisfied for any m, no matter how large m is. We conclude that the system (11.1) is completely state controllable if and only if (11.14) is satisfied for $m = n$. Moreover, the reduction of the state to zero can be accomplished in n sampling periods.

If the system has a multidimensional control, described by

$$\mathbf{x}(k+1) = \mathbf{A}x(k) + \mathbf{Bu}(k) \tag{11.17}$$

where $\mathbf{x}(k)$ is the n-dimensional state vector, $\mathbf{u}(k)$ is the r-dimensional control vector, $\mathbf{A}$ is an $n \times n$ nonsingular constant matrix, and $\mathbf{B}$ is an $n \times r$ constant matrix, then the requirement that $\mathbf{x}(m) = \mathbf{0}$ leads to

$$\mathbf{x}(0) = -[\mathbf{A}^{-1}\mathbf{B} \,\vdots\, \mathbf{A}^{-2}\mathbf{B} \,\vdots\, \cdots \,\vdots\, \mathbf{A}^{-m}\mathbf{B}] \begin{bmatrix} \mathbf{u}(0) \\ \vdots \\ \mathbf{u}(m-1) \end{bmatrix}, \tag{11.18}$$

which is analogous to (11.12). For an arbitrary n-vector $\mathbf{x}(0)$, (11.18) can have a solution only if

$$\text{rank of } [\mathbf{A}^{-1}\mathbf{B} \,\vdots\, \mathbf{A}^{-2}\mathbf{B} \,\vdots\, \cdots \,\vdots\, \mathbf{A}^{-m}\mathbf{B}] = n.$$

By arguments similar to those leading to (11.16),

$$\text{rank of } [\mathbf{A}^{-1}\mathbf{B} \,\vdots\, \mathbf{A}^{-2}\mathbf{B} \,\vdots\, \cdots \,\vdots\, \mathbf{A}^{-m}\mathbf{B}] = \text{rank of } [\mathbf{B} \,\vdots\, \mathbf{AB} \,\vdots\, \cdots \,\vdots\, \mathbf{A}^{n-1}\mathbf{B}], \tag{11.19}$$

for $m \geqslant n$.

From the foregoing discussion, we have the following test for complete state controllability. A necessary and sufficient condition for the system in (11.17) to be completely state-controllable is that

$$\text{rank of } [\mathbf{B} \mid \mathbf{AB} \mid \cdots \mid \mathbf{A}^{n-1}\mathbf{B}] = n \tag{11.20}$$

must be satisfied. The matrix in (11.20) has nr columns. It should be noted that if it is not possible to take $\mathbf{x}(0)$ to the origin in n sampling periods, then it is not possible to take it to the origin at all for any finite number of sampling periods. On the other hand, if the system is completely state-controllable, the state can be reduced to zero in n sampling periods.

Example 11-2. Let us check the controllability of the system described by

$$\begin{bmatrix} x_1(k+1) \\ x_2(k+1) \end{bmatrix} = \begin{bmatrix} -2 & 0 \\ -1 & -1 \end{bmatrix} \begin{bmatrix} x_1(k) \\ x_2(k) \end{bmatrix} + \begin{bmatrix} 1 & -1 \\ 1 & 1 \end{bmatrix} \begin{bmatrix} u_1(k) \\ u_2(k) \end{bmatrix} \tag{11.21}$$

From (11.21),

$$\mathbf{B} = \begin{bmatrix} 1 & -1 \\ 1 & 1 \end{bmatrix},$$

$$\mathbf{AB} = \begin{bmatrix} -2 & 0 \\ -1 & -1 \end{bmatrix} \begin{bmatrix} 1 & -1 \\ 1 & 1 \end{bmatrix} = \begin{bmatrix} -2 & 2 \\ -2 & 0 \end{bmatrix}.$$

Form $\mathbf{Q} = [\mathbf{B} \mid \mathbf{AB}]$:

$$\mathbf{Q} = \begin{bmatrix} 1 & -1 & -2 & 2 \\ 1 & 1 & -2 & 0 \end{bmatrix}.$$

$\mathbf{Q}$ has exactly two linearly independent columns so that rank $\mathbf{Q} = 2$. Hence the system is completely state-controllable. ▲

For a linear, time-invariant continuous-time system

$$\dot{\mathbf{x}}(t) = \mathbf{Ax}(t) + \mathbf{Bu}(t) \tag{11.22}$$

where $\mathbf{x}(t)$ is n-dimensional and $\mathbf{u}(t)$ is r dimensional, a necessary and sufficient condition that the system be completely state-controllable is that (11.20) is satisfied. The derivation of this result is conceptually the same as that for the discrete-time case. A crucial result from the calculus of matrices that is needed is that exp $(\mathbf{A}t)$ may be expressed as a polynomial in $\mathbf{A}$ of degree $n-1$, where the scalar coefficients are functions of time. The formal derivation is somewhat abstract and is not included here.

An alternative definition of controllability to be stated below was presented by Gilbert in 1963.[3] He showed that his definition was equivalent to that of Kalman for the class of systems that he considered. Gilbert's definition gives additional physical insight into the nature of controllability and observability. For simplicity, only distinct eigenvalues are allowed for **A**.

Gilbert's definition is in terms of *normal coordinates*, which we describe first. As pointed out in previous chapters, there is no unique choice for state. For any given choice of a state vector **x**, we can select another vector **z**, which is related to **x** by a nonsingular transformation

$$\mathbf{x} = \mathbf{T}\mathbf{z} \tag{11.23}$$

where **T** is any nonsingular matrix. Substituting (11.23) in (11.22) we obtain

$$\dot{\mathbf{z}} = \mathbf{T}^{-1}\mathbf{A}T\mathbf{z} + \mathbf{T}^{-1}\mathbf{B}\mathbf{u}. \tag{11.24}$$

If **A** has distinct eigenvalues, **T** can always be chosen so that $\mathbf{T}^{-1}\mathbf{A}\mathbf{T}$ is diagonal. The diagonal elements of $\mathbf{T}^{-1}\mathbf{A}\mathbf{T}$ are the eigenvalues of **A**. Denote $\mathbf{T}^{-1}\mathbf{A}\mathbf{T}$ and $\mathbf{T}^{-1}\mathbf{B}$ by $\mathbf{\Lambda}$ and $\boldsymbol{\beta}$, respectively:

$$\mathbf{\Lambda} = \mathbf{T}^{-1}\mathbf{A}\mathbf{T} \tag{11.25}$$

$$\boldsymbol{\beta} = \mathbf{T}^{-1}\mathbf{B} \tag{11.26}$$

The components of the new state vector **z** are called the *normal coordinates* of the system. Equation 11.24 now becomes

$$\dot{\mathbf{z}} = \mathbf{\Lambda}\mathbf{z} + \boldsymbol{\beta}\mathbf{u}. \tag{11.27}$$

Gilbert defines the system described by (11.27) to be controllable if $\boldsymbol{\beta}$ has no rows that are zero. The coordinates z_i corresponding to the nonzero rows of β are called *controllable* coordinates or controllable modes. Coordinates corresponding to the zero rows of $\boldsymbol{\beta}$ are called *uncontrollable* coordinates or uncontrollable modes. Notice that the uncontrollable coordinates can not be affected by any control input **u**. That is, uncontrollable normal coordinates are not coupled to the control input. Gilbert proved that the above definition is equivalent to Kalman's definition given earlier. Thus, determining whether $\boldsymbol{\beta}$ has no nonzero rows is equivalent to determining whether the rank of $[\mathbf{B} \mid \mathbf{AB} \mid \cdots \mid \mathbf{A}^{n-1}\mathbf{B}]$ is equal to n.

[3]E. G. Gilbert, "Controllability and Observability in Multivariable Control Systems," *J.S.I.A.M. Control*, Ser. A, 2(1), 128–151.

The above controllability definition has a further interpretation in terms of transfer functions. From (11.27), the transfer function relating $\mathbf{U}(s)$ to $\mathbf{Z}(s)$ is

$$\mathbf{Z}(s) = (s\mathbf{I}-\boldsymbol{\Lambda})^{-1}\boldsymbol{\beta}\mathbf{U}(s). \tag{11.28}$$

But from (11.23) and (11.26), we also have

$$\begin{aligned}\mathbf{X}(s) &= \mathbf{T}(s\mathbf{I}-\boldsymbol{\Lambda})^{-1}\mathbf{T}^{-1}\mathbf{B}\mathbf{U}(s)\\ &= \boldsymbol{\Phi}(s)\mathbf{B}\mathbf{U}(s).\end{aligned} \tag{11.29}$$

Recall that we are assuming that $\mathbf{A}$ has distinct eigenvalues only. If a row of $\boldsymbol{\beta}$, such as the ith row, is zero, then λ_i does not appear as a pole of $(s\mathbf{I}-\boldsymbol{\Lambda})^{-1}\boldsymbol{\beta}$. Since $\mathbf{T}$ is a constant matrix, $\mathbf{T}(s\mathbf{I}-\boldsymbol{\Lambda})^{-1}\boldsymbol{\beta}=\boldsymbol{\Phi}(S)\mathbf{B}$ does not contain λ_i as a pole either. Similarly, since

$$(s\mathbf{I}-\boldsymbol{\Lambda})^{-1}\boldsymbol{\beta} = \mathbf{T}^{-1}\boldsymbol{\Phi}(s)\mathbf{B},$$

and if $\boldsymbol{\Phi}(s)\mathbf{B}$ does not contain λ_i as a pole, $(s\mathbf{I}-\boldsymbol{\Lambda})^{-1}\boldsymbol{\beta}$ cannot have λ_i as a pole either, and thus the ith row of $\boldsymbol{\beta}$ must be zero.

Denote the ith row jth column element of $\boldsymbol{\Phi}(s)\mathbf{B}$ by the transfer function

$$[\boldsymbol{\Phi}(s)\mathbf{B}]_{ij} = \frac{p_{ij}(s)}{\det(s\mathbf{I}-\mathbf{A})} = \frac{p_{ij}(s)}{\Delta(s)}$$

where the eigenvalues of $\mathbf{A}$ are the roots of $\det(s\mathbf{I}-\mathbf{A}) = 0$ and $p_{ij}(s)$ is a polynomial in s. From the foregoing discussion, we conclude that if $\mathbf{A}$ has distinct eigenvalues only, the system is controllable if and only if $p_{11}(s), \ldots, p_{1r}(s), \ldots, p_{n1}(s), \ldots, p_{nr}(s)$, and $\det(s\mathbf{I}-\mathbf{A})$ have no factor in common.[4]

Example 11-3. In the system of Figure 11.3, two first-order subsystems are connected in cascade. Clearly $\Phi(s)$ contains $s=-2$ and $s=-1$ as poles that are distinct eigenvalues of $\mathbf{A}$. The transfer functions from $U(s)$ to $X_1(s)$ and $X_2(s)$ are

$$\frac{X_1(s)}{U(s)} = \frac{p_{11}(s)}{\Delta(s)} = \frac{(s+1)}{(s+1)(s+2)}$$

$$\frac{X_2(s)}{U(s)} = \frac{p_{21}(s)}{\Delta(s)} = \frac{(s+1)}{(s+1)(s+2)}.$$

[4]This result also applies for the case when **A** has repeated eigenvalues and **B** is a column matrix (single-input system). See S. Butman and R. Sivan, "On Cancellations, Controllability and Observability," *IEEE Trans. on Automatic* Control, **AC-9**, 317–318 (July 1964).

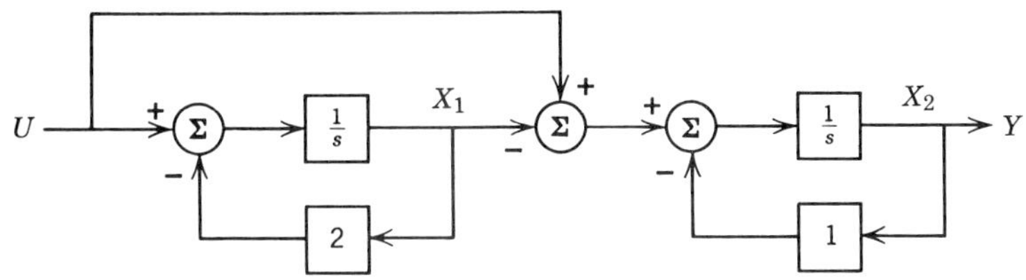

Figure 11.3 Uncontrollable system.

Since $p_{11}(s)$, $p_{21}(s)$, and $\Delta(s)$ have $(s+1)$ as a common factor, we conclude that the system is not controllable. ▲

Example 11-4. In Figure 11.4 two first-order subsystems are connected in cascade. These are the same subsystems as in Example 11-3 but the order is interchanged. By inspection we note that $s=-1$ and $s=-2$ are poles of $\mathbf{\Phi}(s)$ and distinct eigenvalues of $\mathbf{A}$. The transfer functions from U to the two states $X_1(s)$ and $X_2(s)$ are

$$\frac{X_1(s)}{U(s)}=\frac{p_{11}(s)}{\Delta(s)}=\frac{(s+2)}{(s+2)(s+1)}$$

$$\frac{X_2(s)}{U(s)}=\frac{p_{21}(s)}{\Delta(s)}=\frac{1}{(s+1)(s+2)}.$$

Since $p_{11}(s)$, $p_{21}(s)$, and $\Delta(s)$ do not have any factor in common, the system is controllable. Note, however, that $p_{11}(s)$ and $\Delta(s)$ have $(s+2)$ in common. ▲

Sometimes, a block diagram shows that the transfer function from every $U_j(s)$, $j=1,\ldots,r$ to a particular $X_i(s)$ is zero. In this case, $X_i(s)$ cannot be affected by any $U_j(s)$, and the system is not controllable.

Example 11-5. For the system in Figure 11.5, X_2 cannot be affected by the control U. Hence the system is uncontrollable. ▲

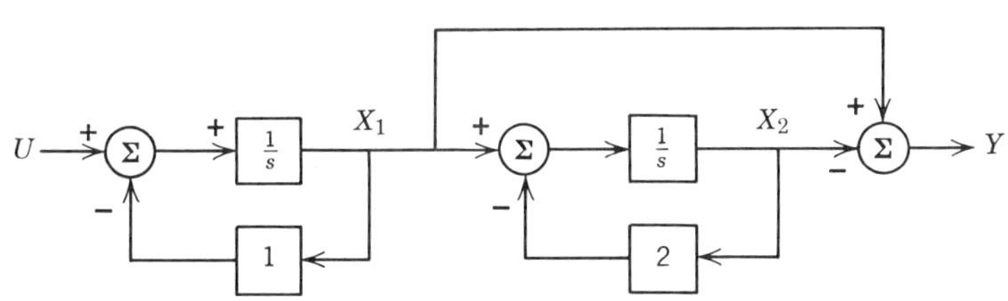

Figure 11.4 Controllable system.

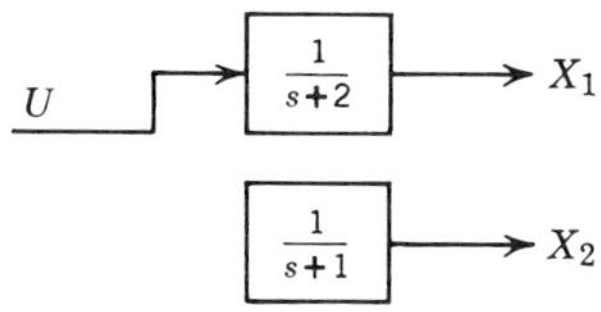

Figure 11.5 Uncontrollable system.

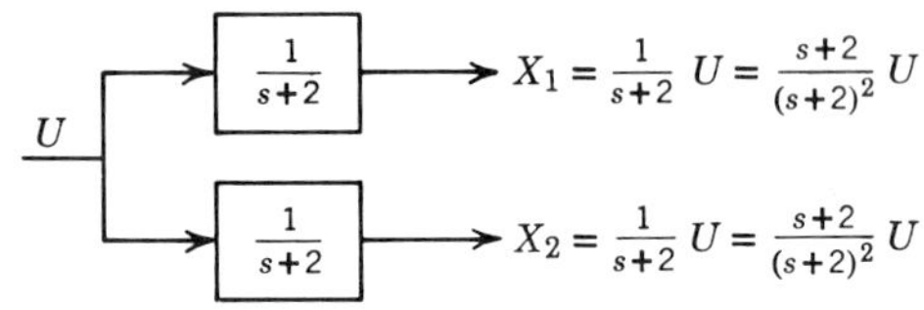

Figure 11.6 Uncontrollable system. Det($s\mathbf{I}-\mathbf{A}$) = $(s+2)^2$.

Example 11-6. In Figure 11.6 we have a second-order system where $s=-2$ is a repeated eigenvalue of the **A** matrix. For this simple single-input system, we notice that the transfer functions from U to X_1 and U to X_2 are equal. The numerators and det $(s\mathbf{I}-\mathbf{A})$ have a common factor. Hence the system is not controllable. Furthermore, the zero-input responses are $x_1(0)\epsilon^{-2t}$ and $x_2(0)\epsilon^{-2t}$, respectively. If $x_1(0)$ and $x_2(0)$ are not equal, it is impossible to find $u(t)$ such that $x_1(T)$ and $x_2(T)$ are zero at the same time, for a finite T. ▲

In case the state vector is not the output vector, it is useful to define a system to be *completely output-controllable* if it is possible to generate a control input $\mathbf{u}(t)$ so that, for any t_0, any initial output $\mathbf{y}(t_0)$ can be transferred to any final output $\mathbf{y}(t_f)$ in a finite time interval $t_0 \leqslant t \leqslant t_f$. For a linear system, there is no loss of generality in assuming that the final output $\mathbf{y}(t_f)$ is zero. For the linear time-invariant discrete-time system, with a p-dimensional output vector $\mathbf{y}(k)$.

$$\mathbf{y}(k)=\mathbf{C}x(k)+\mathbf{D}\mathbf{u}(k). \tag{11.30}$$

For the linear time-invariant continuous-time system, with a p-dimensional output vector $\mathbf{y}(t)$,

$$\mathbf{y}(t)=\mathbf{C}\mathbf{x}(t)+\mathbf{D}\mathbf{u}(t). \tag{11.31}$$

It is assumed that the $p\times n$ matrix **C** has p linearly independent columns, and $p\leqslant n$.

A necessary and sufficient condition for a linear time-invariant system to be completely output controllable[5] is that

$$\text{rank of } [\mathbf{D} \,\vdots\, \mathbf{CB} \,\vdots\, \mathbf{CAB} \,\vdots\, \cdots \,\vdots\, \mathbf{CA}^{n-1}\mathbf{B}] = p. \tag{11.32}$$

Notice that the matrix in (11.32) has $(n+1)r$ columns and p rows. The condition of (11.32) applies to both discrete-time and continuous-time

[5]E. Kreindler and P. E. Sarachik, "On the Concepts of Controllability and Observability of Linear Systems," *IEEE Trans. on Automatic Control*, **AC-9**, 129–137 (April 1964).

systems. Complete state controllability is not necessary, but it is sufficient for complete output controllability.

11.3 Observability

Observability[6] implies the determinability of the state from an observation of the output over a finite time interval, starting from the instant at which the state is desired. It is assumed that the input and the system state description are known.

Consider a linear time-invariant discrete-time system with scalar output described by

$$\mathbf{x}(k+1) = \mathbf{A}\mathbf{x}(k) + \mathbf{B}\mathbf{u}(k) \tag{11.33}$$

$$y(k) = \mathbf{c}'\mathbf{x}(k) + \mathbf{d}'\mathbf{u}(k). \tag{11.34}$$

From Chapter 10, the state $\mathbf{x}(k)$ is

$$\mathbf{x}(k) = \mathbf{A}^k x(0) + \sum_{i=0}^{k-1} \mathbf{A}^{k-i-1}\mathbf{B}\mathbf{u}(i). \tag{11.35}$$

Hence the output $y(k)$ is

$$y(k) = \mathbf{c}'\mathbf{A}^k\mathbf{x}(0) + \mathbf{c}' \sum_{i=0}^{k-1} \mathbf{A}^{k-i-1}\mathbf{B}\mathbf{u}(i) + \mathbf{d}'\mathbf{u}(k). \tag{11.36}$$

If the matrices $\mathbf{A}$, $\mathbf{B}$, $\mathbf{c}$, and $\mathbf{d}$ are known and if the input $\mathbf{u}(k)$ and output $y(k)$ are available for measurement, then the zero-input response can be computed as

$$y(k) - \mathbf{c}' \sum_{i=0}^{k-1} \mathbf{A}^{k-i-1}\mathbf{B}\mathbf{u}(i) - \mathbf{d}'\mathbf{u}(k) = \mathbf{c}'\mathbf{A}^k\mathbf{x}(0). \tag{11.37}$$

We can restrict the discussion to the case where the input sequence $\mathbf{u}(k)$ is zero because if it is not, the zero-state response can always be subtracted from the total response to obtain the zero-input response, as indicated in (11.37).

The system described by (11.33) and (11.34), with $\mathbf{u}(k) = 0$, is said to be *completely observable* if, for any arbitrary initial state $x(0)$, there is a finite m such that from measurements of $y(0), y(1), \ldots, y(m)$, the initial state $\mathbf{x}(0)$ can be computed.

[6]R. E. Kalman, "On the General Theory of Control Systems," *Proc. First Congress of IFAC*, Moscow, USSR, 1960; Butterworths, London, **1**, 481–492 (1961).

Thus, if $y(k)$ denotes the zero-input response,

$$y(k) = \mathbf{c}'\mathbf{A}^k\mathbf{x}(0). \tag{11.38}$$

For $k = 0, 1, 2, \ldots, (n-1)$, (11.38) becomes

$$\begin{aligned} y(0) &= \mathbf{c}'\mathbf{x}(0) \\ y(1) &= \mathbf{c}'\mathbf{A}\mathbf{x}(0) \\ &\vdots \\ y(n-1) &= \mathbf{c}'\mathbf{A}^{n-1}\mathbf{x}(0) \end{aligned}$$

or in matrix form,

$$\begin{bmatrix} y(0) \\ y(1) \\ \vdots \\ y(n-1) \end{bmatrix} = \begin{bmatrix} \mathbf{c}' \\ \mathbf{c}'\mathbf{A} \\ \vdots \\ \mathbf{c}'\mathbf{A}^{n-1} \end{bmatrix} \mathbf{x}(0). \tag{11.39}$$

If the $n \times n$ matrix

$$\begin{bmatrix} \mathbf{c}' \\ \mathbf{c}'\mathbf{A} \\ \vdots \\ \mathbf{c}'\mathbf{A}^{n-1} \end{bmatrix} \tag{11.40}$$

is nonsingular, then $\mathbf{x}(0)$ can be uniquely determined from (11.39). As discussed previously, for $k \geqslant n$, $\mathbf{A}^k$ can be expressed as a polynomial in $\mathbf{A}$ up to order $n-1$ so that $\mathbf{c}'\mathbf{A}^k$ for $k \geqslant n$ is linearly dependent on the rows $\mathbf{c}'$, $\mathbf{c}'\mathbf{A}, \ldots, \mathbf{c}'\mathbf{A}^{n-1}$. This means that if the matrix in (11.40) is singular, the rank of the matrix cannot be increased by adding more rows of the form $\mathbf{c}'\mathbf{A}^k$. Thus, no amount of additional measurement of $y(k)$ at additional sampling instants will enable us to recover $\mathbf{x}(0)$ if the matrix in (11.40) is singular. Hence the system is completely observable if and only if the $n \times n$ matrix in (11.40) is nonsingular.

For a vector output $\mathbf{y}(k)$ of dimension p, we replace $\mathbf{c}'$ in (11.34) by $\mathbf{C}$ which is $p \times n$ and the matrix in (11.40) becomes $pn \times n$.[7] A solution for $\mathbf{x}(0)$ exists if and only if

$$\text{rank of} \begin{bmatrix} \mathbf{C} \\ \mathbf{CA} \\ \vdots \\ \mathbf{CA}^{n-1} \end{bmatrix} = n. \tag{11.41}$$

[7]The reader is cautioned that $\mathbf{C}$ has p rows and n columns whereas $\mathbf{c}'$ has 1 row and n columns.

Alternatively, the transpose of the matrix in (11.41) must have rank n for a solution for $\mathbf{x}(0)$ to exist. A necessary and sufficient condition for complete observability is

$$\text{rank of } [\mathbf{C}' \,\vdots\, \mathbf{A}'\mathbf{C}' \,\vdots\, \cdots \,\vdots\, (\mathbf{A}')^{n-1}\mathbf{C}'] = n. \tag{11.42}$$

For a linear time-invariant continuous-time system, described by (11.22) and (11.31), the necessary and sufficient condition for complete observability is that (11.42) is satisfied. The derivation for the continuous time case is not presented here, since it is rather abstract.

As in controllability, there is an alternate definition of observability that provides additional physical insight. This alternate definition due to Gilbert is equivalent to that of Kalman.

Assuming distinct eigenvalues for $\mathbf{A}$, a nonsingular transformation $\mathbf{T}$ is used as in (11.23) such that $\mathbf{\Lambda}$ in (11.25) is diagonal. Equation 11.31 may be rewritten as

$$\mathbf{y} = \boldsymbol{\gamma}\mathbf{z} + \mathbf{D}\mathbf{u} \tag{11.43}$$

where

$$\boldsymbol{\gamma} = \mathbf{C}\mathbf{T}. \tag{11.44}$$

The system described by (11.27) and (11.43) is said to be completely observable if $\boldsymbol{\gamma}$ has no columns that are zero. Coordinates z_i corresponding to nonzero columns of $\boldsymbol{\gamma}$ are called *observable modes.* Coordinates z_i corresponding to zero columns of $\boldsymbol{\gamma}$ are called *unobservable modes.* Thus, unobservable coordinates or modes do not appear in any of the output components. Since Gilbert's definition is equivalent to Kalman's, when $\mathbf{A}$ has only distinct eigenvalues, determining whether $\boldsymbol{\gamma}$ has nonzero columns is equivalent to determining whether $[\mathbf{C}' \,\vdots\, \mathbf{A}'\mathbf{C}' \,\vdots\, \cdots \,\vdots\, (\mathbf{A}')^{n-1}\mathbf{C}']$ has rank n.

For the case when $\mathbf{A}$ has distinct eigenvalues only, observability has a transfer-function interpretation. From (11.27) and (11.43), and for $\mathbf{u}(t) \equiv \mathbf{0}$, the transform of the zero-input response is

$$\mathbf{Y}(s) = \boldsymbol{\gamma}(s\mathbf{I} - \mathbf{\Lambda})^{-1}\mathbf{z}(0) \tag{11.45}$$

If the ith column of $\boldsymbol{\gamma}$ is zero, then the ith eigenvalue λ_i does not appear as a pole in the transfer function

$$\boldsymbol{\gamma}(s\mathbf{I} - \mathbf{\Lambda})^{-1}.$$

Recall that $\mathbf{T}$ is a constant nonsingular matrix. Hence

$$\boldsymbol{\gamma}(s\mathbf{I} - \mathbf{\Lambda})^{-1}\mathbf{T}^{-1}$$

does not have a pole at λ_i either. But from (11.23),

$$\mathbf{z}(0) = \mathbf{T}^{-1}\mathbf{x}(0)$$

so that (11.45) becomes

$$\begin{aligned}\mathbf{Y}(s) &= \boldsymbol{\gamma}(s\mathbf{I}-\boldsymbol{\Lambda})^{-1}\mathbf{T}^{-1}\mathbf{x}(0)\\ &= \mathbf{CT}(s\mathbf{I}-\boldsymbol{\Lambda})^{-1}\mathbf{T}^{-1}\mathbf{x}(0)\\ &= \mathbf{C}\boldsymbol{\Phi}(s)\mathbf{x}(0). \end{aligned} \tag{11.46}$$

That is, if λ_i is not a pole of $\boldsymbol{\gamma}(s\mathbf{I}-\boldsymbol{\Lambda})^{-1}$, it is not a pole of $\mathbf{C}\boldsymbol{\Phi}(s) = \boldsymbol{\gamma}(s\mathbf{I}-\boldsymbol{\Lambda})^{-1}\mathbf{T}^{-1}$ either. From (11.46),

$$\boldsymbol{\gamma}(s\mathbf{I}-\boldsymbol{\Lambda})^{-1} = \mathbf{C}\boldsymbol{\Phi}(s)\mathbf{T}. \tag{11.47}$$

Since $\mathbf{T}$ is a constant matrix, if λ_i is a pole of $\boldsymbol{\gamma}(s\mathbf{I}-\boldsymbol{\Lambda})^{-1}$, it must be a pole of $\mathbf{C}\boldsymbol{\Phi}(s)$ also. Thus λ_i is a pole of $\boldsymbol{\gamma}(s\mathbf{I}-\boldsymbol{\Lambda})^{-1}$ if and only if it is a pole of $\mathbf{C}\boldsymbol{\Phi}(s)$.

Denote the ith row jth column element of $\mathbf{C}\boldsymbol{\Phi}(s)$ by $q_{ij}(s)/\det(s\mathbf{I}-\mathbf{A})$. From the foregoing discussion, we conclude that if $\mathbf{A}$ has distinct eigenvalues only, the system is completely observable if and only if $q_{11}(s), \ldots q_{1n}(s), \ldots, q_{p1}(s), \ldots, q_{pn}(s)$ and $\det(s\mathbf{I}-\mathbf{A}) = \Delta(s)$ have no factor in common.

Example 11-7. The uncontrollable system of Figure 11.3 is redrawn in Figure 11.7 where $U=0$ and the initial conditions are indicated. To determine whether the system is observable, we determine

$$\frac{Y(s)}{x_1(0)} = \frac{q_{11}(s)}{\Delta(s)} = \frac{1}{(s+1)(s+2)}$$

$$\frac{Y(s)}{x_2(0)} = \frac{q_{21}(s)}{\Delta(s)} = \frac{(s+2)}{(s+1)(s+2)}$$

Since $q_{11}(s)$, $q_{21}(s)$, and $\Delta(s)$ do not have a common factor the system is observable.

The controllable system of Figure 11.4 is redrawn in Figure 11.8 where $U=0$ and the initial conditions are indicated. To determine

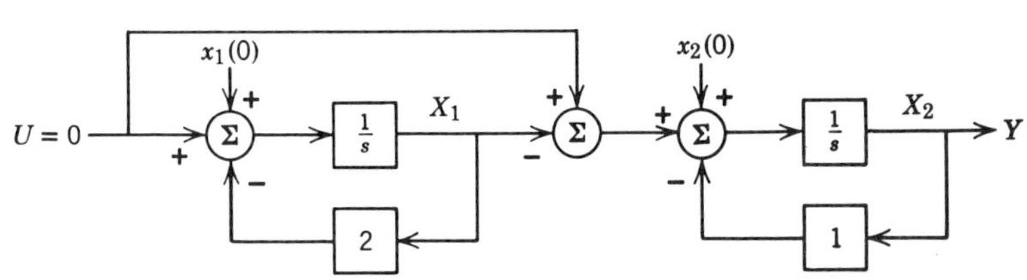

Figure 11.7 Observable system.

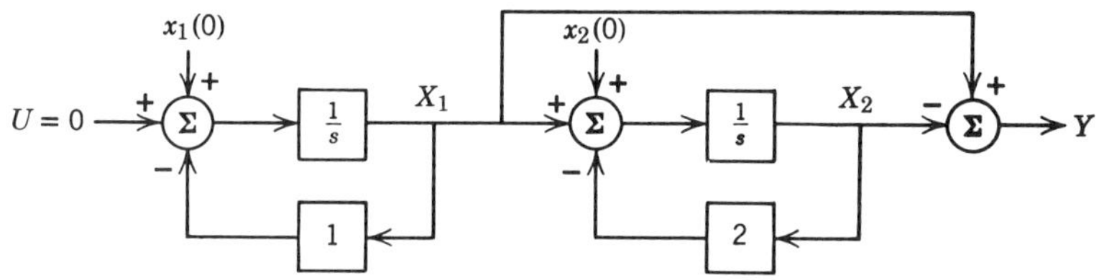

Figure 11.8 Unobservable system.

whether the system is observable we determine

$$\frac{Y(s)}{x_1(0)} = \frac{q_{11}(s)}{\Delta(s)} = \frac{(s+1)}{(s+1)(s+2)}$$

$$\frac{Y(s)}{x_2(0)} = \frac{q_{21}(s)}{\Delta(s)} = \frac{(s+1)}{(s+1)(s+2)}.$$

Since $q_{11}(s)$, $q_{21}(s)$, and $\Delta(s)$ have $(s+1)$ as a common factor the system is unobservable. Thus the system in Figure 11.3 is uncontrollable but observable. If the subsystems are interchanged, maintaining the same transfer function from $U(s)$ to $Y(s)$, as in Figure 11.4, the resulting system is controllable but unobservable. ▲

In some control schemes, it is necessary to determine the state of the system in order to generate a control signal. Observation of the output of a completely observable system, over a finite time interval, yields sufficient information to determine the state of the system at the beginning of the interval. In some cases, it is possible to determine the current state. However, more generally, the state that can be computed is for a past time instant. If the present state is desired, in general only an estimate can be made.

11.4 Controllability and Observability of Single-Axis Reaction-Wheel Attitude Control Systems

The linearized single-axis reaction-wheel attitude control system, studied in previous chapters, will be examined for controllability and observability. A block diagram is given in Figure 11.9. The state equations are:

$$\begin{bmatrix} \dot{x}_1 \\ \dot{x}_2 \\ \dot{x}_3 \end{bmatrix} = \begin{bmatrix} 0 & 1 & 0 \\ 0 & 0 & a_{23} \\ 0 & 0 & a_{33} \end{bmatrix} \begin{bmatrix} x_1 \\ x_2 \\ x_{33} \end{bmatrix} + \begin{bmatrix} 0 \\ b_2 \\ b_3 \end{bmatrix} u \qquad (11.48)$$

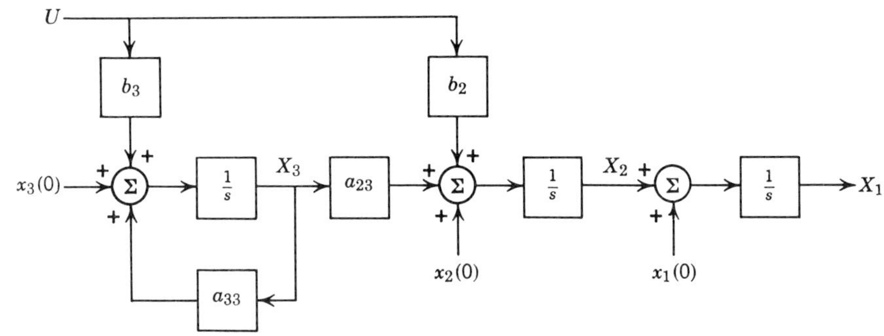

Figure 11.9 Block diagram of linearized single-axis reaction-wheel attitude control system.

where x_1 is the attitude angle, x_2 is the attitude rate, and x_3 is the wheel speed relative to the spacecraft. The elements of the **A** and **B** matrices are given by

$$a_{23} = \frac{1}{J_v}\left(D + \frac{k^2}{R_a}\right)$$

$$a_{33} = -\left(D + \frac{k^2}{R_a}\right)\left(\frac{1}{J_v} + \frac{1}{J_w}\right) \tag{11.49}$$

$$b_2 = -\frac{k}{R_a J_v}$$

$$b_3 = \frac{k}{R_a}\left(\frac{1}{J_v} + \frac{1}{J_w}\right)$$

where J_v, J_w, D, R_a, and k are system constants, as discussed in previous chapters. To check complete state controllability, form

$$[\mathbf{B} \mid \mathbf{AB} \mid \mathbf{A}^2\mathbf{B}] = \begin{bmatrix} 0 & b_2 & a_{23}b_3 \\ b_2 & a_{23}b_3 & a_{23}a_{33}b_3 \\ b_3 & a_{33}b_3 & a_{33}^2 b_3 \end{bmatrix}. \tag{11.50}$$

But

$$\det \begin{bmatrix} 0 & b_2 & a_{23}b_3 \\ b_2 & a_{23}b_3 & a_{23}a_{33}b_3 \\ b_3 & a_{33}b_3 & a_{33}^2 b_3 \end{bmatrix} = 2a_{23}a_{33}b_2b_3{}^2 - b_2{}^2b_3a_{33}^3 - b_3{}^3a_{23}^2 \tag{11.51}$$

Substituting the system constants, (11.51) becomes

$$\frac{2(D + k^2/R_2)^2(1/J_v + 1/J_w)^3(k)^3(1-1)}{J_v{}^2R_a{}^2} = 0$$

Hence the rank of the matrix in (11.50) is less than 3, and the system is not completely state-controllable. It is not possible to find a control armature voltage applied for a finite time duration so that, starting from an arbitrary initial attitude, initial attitude rate, initial wheel speed, the final attitude, final attitude rate, and final speed are zero.

The above conclusion can also be reached by examining the transfer-function block diagram. Clearly, the state-transition matrix transform $\Phi(s)$ contains poles at $s = a_{33}$, $s = 0$, and $s = 0$. In order to apply the controllability test of Section 11.2 for distinct eigenvalues, consider a reduced system involving x_3 and x_2 only, where the integrator for x_1 is excluded. A simplified diagram is shown in Figure 11.10. The formulas in (11.49) have been used to simplify the transfer function from U to X_2, which is $X_2/U = p_{21}(s)/\Delta(s) = b_2 s/[s(s-a_{33})]$. From Figure 11.9, $X_1/U = p_{11}(s)/\Delta(s) = b_3 s/[s(s-a_{33})]$. Since $p_{11}(s)$, $p_{21}(s)$, and $\Delta(s)$ have s as a common factor, the reduced system is uncontrollable. Since x_2 and x_3 cannot be reduced to zero at the same time, neither can x_1, x_2 and x_3. Hence the system in Figure 11.9 is not completely state-controllable.

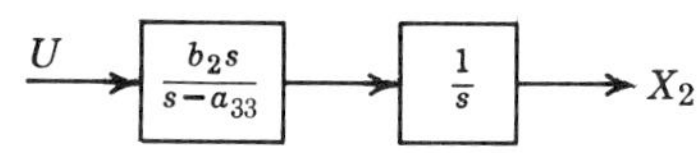

Figure 11.10 Reduced system of Figure 11.9 where X_1 is excluded.

Although the attitude system is not completely state-controllable, it may be possible to drive some (but not all) of the final state variables to zero in a finite time. That is, some of the state variables may be controllable. A convenient method to check whether some of the state variables are controllable is to assign the state variables in question as outputs. Then the system is examined for output controllability. For example, let us investigate whether the attitude and attitude rate are controllable. Consider the output $y(t)$ as

$$\mathbf{y}(t) = \begin{bmatrix} 1 & 0 & 0 \\ 0 & 1 & 0 \end{bmatrix} \begin{bmatrix} x_1 \\ x_2 \\ x_3 \end{bmatrix} \tag{11.52}$$

Let us check whether the system is completely output-controllable. Form

$$[\mathbf{CB} \mid \mathbf{CAB}] = \begin{bmatrix} 0 & b_2 \\ b_2 & a_{23}b_3 \end{bmatrix}. \tag{11.53}$$

Then

$$\det \begin{bmatrix} 0 & b_2 \\ b_2 & a_{23}b_3 \end{bmatrix} = -b_2^2 = -\frac{K_a^2 k^2}{R_a^2 J_v^2} \neq 0.$$

Hence the matrix in (11.53) has rank 2, (11.32) is satisfied, and the system is completely output-controllable. That is, starting from an arbitrary $y_1(t_0)$ and $y_2(t_0)$, there exists a control such that, for a finite t_f, we can force $y_1(t_f)$ and $y_2(t_f)$ to be zero. Since $y_1 = x_1$ and $y_2 = x_2$, these two state variables can be reduced to zero at $t = t_f$.

However, x_3, the reaction-wheel speed relative to the spacecraft, cannot be reduced to zero at the same time that the attitude and attitude rate are zero.

If we wish to maintain x_1 and x_2 to be zero for $t \geqslant t_f$, then $\dot{x}_1$ and $\dot{x}_2$ must be zero for $t \geqslant t_f$. From (11.48),

$$\dot{x}_1 = x_2$$

$$\dot{x}_2 = a_{23}x_3 + b_2u$$

$$\dot{x}_3 = a_{33}x_3 + b_3u.$$

If the control is chosen to satisfy

$$a_{23}x_3 + b_2u = 0 \tag{11.54}$$

for $t \geqslant t_f$, then the attitude and attitude rate will remain zero. However, because of this choice of control, the wheel speed x_3 must satisfy

$$\dot{x}_3 = a_{33}x_3 + b_3u = a_{33}x_3 + b_3\left(-\frac{a_{23}}{b_2}\right)x_3$$
$$= \left(a_{33} - \frac{a_{23}b_3}{b_2}\right)x_3.$$

From (11.49), the above equation becomes

$$\dot{x}_3 = -\left(D + \frac{k^2}{R_a}\right)\left(\frac{1}{J_v} + \frac{1}{J_w}\right) + \left(D + \frac{k^2}{R_a}\right)\left(\frac{1}{J_v} + \frac{1}{J_w}\right)$$

Hence

$$x_3(t) = x_3(t_f)$$

for $t \geqslant t_f$. The required control is

$$u = -\frac{a_{23}}{b_2}x_3 = \frac{R_a}{k}\left(D + \frac{k^2}{R_a}\right)x_3(t_f)$$

for $t \geqslant t_f$. That is, to maintain attitude and attitude rate to be zero for $t \geqslant t_f$, a constant armature voltage must be applied continuously for $t \geqslant t_f$. The wheel speed remains constant at a value $x_3(t_f)$ for all $t \geqslant t_f$.

Depending on the mission and number of times the reaction wheel has to drive the attitude and attitude rate to zero because of disturbance impulsive torques, the reaction wheel momentum buildup may be excessive, and the wheel speed may exceed safe values. The reaction wheel must have some means of interacting with the outside world to "dump" the excess momentum. Typically, this is done by an auxiliary reaction jet system, or magnetic torquing against the earth's magnetic field.

Let us check whether the single-axis reaction-wheel attitude-control system is completely observable. From (11.52), form

$$\mathbf{C}' = \begin{bmatrix} 1 & 0 \\ 0 & 1 \\ 0 & 0 \end{bmatrix} \tag{11.55}$$

and from (11.48) and (11.55),

$$\mathbf{A}'\mathbf{C}' = \begin{bmatrix} 0 & 0 & 0 \\ 1 & 0 & 0 \\ 0 & a_{23} & a_{33} \end{bmatrix} \begin{bmatrix} 1 & 0 \\ 0 & 1 \\ 0 & 0 \end{bmatrix} = \begin{bmatrix} 0 & 0 \\ 1 & 0 \\ 0 & a_{23} \end{bmatrix}. \tag{11.56}$$

In order to check the rank of the matrix in (11.42), for a third-order system $(\mathbf{A}')^2\mathbf{C}'$ has to be formed in general. However, it is seen from (11.55) and (11.56) that it is possible to form a 3×3 nonsingular submatrix from the columns of $\mathbf{C}'$ and $\mathbf{A}'\mathbf{C}'$, namely a matrix consisting of the columns of $\mathbf{C}'$ and the second column of $\mathbf{A}'\mathbf{C}'$:

$$\begin{bmatrix} 1 & 0 & 0 \\ 0 & 1 & 0 \\ 0 & 0 & a_{23} \end{bmatrix}$$

The determinant of this matrix is equal to a_{23} which is

$$a_{23} = \frac{1}{J_v}\left(D + \frac{k^2}{R_a}\right)$$

from (11.49). We conclude that our third-order single-axis reaction-wheel attitude control system is completely observable.

In summary, we conclude that the single-axis linearized reaction-wheel attitude control system depicted in Figure 11.9 with parameters given in (11.49) and state equations given by (11.48 and (11.52) is not completely state-controllable. However, it is completely output-controllable. Moreover, it is completely observable.

11.5★ Controllability and Observability of Composite Systems

In this section we study the controllability and observability properties of interconnections of subsystems. Of course, it is possible to write the state equations for the entire system and then apply the tests described in the previous sections. However, additional insight is obtained if we can express the controllability and observability of the system in terms of those of the subsystems.

Following Gilbert,[8] we examine the controllability and observability of three basic interconnections of subsystems: parallel, cascade, and feedback. Controllability and observability of more complex composite systems consisting of parallel, cascade, and feedback connections of subsystems may be studied by repeated application of the results for the basic interconnections.

Figure 11.11 shows a parallel connection of linear time-invariant subsystems S^a and S^b, which have a common input **u** and whose outputs $\mathbf{y}^a$ and $\mathbf{y}^b$ add to produce the system output **y**. It is assumed that the eigen values for S^a and S^b are distinct. If S^a is of order n_a and S^b is of order n_b, then the composite system S is of order n_a+n_b, and the eigenvalues for S are the eigenvalues for S^a and the eigenvalues for S^b. Furthermore, a necessary and sufficient condition that S be completely controllable (or observable) is that both S^a and S^b are completely controllable (or observable). By stacking the state equations for S^a and S^b, the state equations for the composite systems S is clearly of order n_a+n_b, and that the eigenvalues of S are those of S^a and S^b. If either S^a or S^b is not controllable (or observable), then the composite system S is also not controllable (or observable) because the state variables of S are those of S^a and S^b. Conversely if S^a and S_b are controllable (or observable), then S is controllable (or observable).

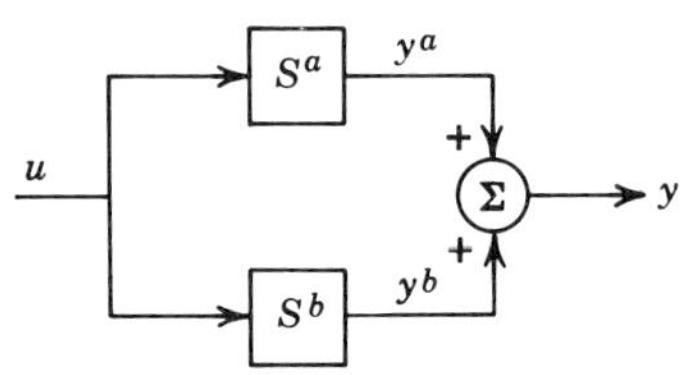

Figure 11.11 Parallel connection of subsystems.

Next we consider two linear time-invariant systems in cascade as shown in Figure 11.12. Again, it is assumed that the eigenvalues for S^a and for S^b are distinct. If n_a and n_b are the orders of S^a and S^b respectively, then the order of S is n_a+n_b. The eigenvalues for S are those for S^a and S^b. The state variables for S^a and S^b may be chosen as the state variables for S. Clearly, if S is controllable (or observable), then S^a must be controllable (or observable). Likewise, if S is controllable

[8]E. G. Gilbert, *op. cit.*

(or observable) then S^b is also controllable (or observable). That is, a necessary condition for S to be controllable (or observable) is that both S^a and S^b be controllable (or observable). However, the controllability (or observability) of S^a and S^b are not enough to guarantee that S be controllable (or observable). By inspection of Figure 11.12, it is clear that if S^a and S^b are both controllable, then the uncontrollable state variables of S must belong to S^b. Similarly, if S^a and S^b are both observable, then the unobservable state variables of S must belong to S^a.

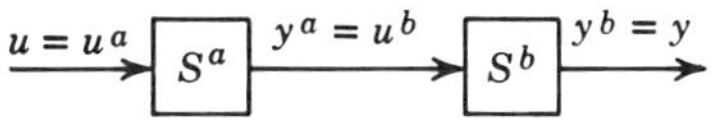

Figure 11.12 Systems in cascade.

Example 11-8. The systems S^a and S^b in Figure 11.13 are clearly controllable and observable. Let us analyze the composite system for controllability and observability. Choosing x_{1a} as x_1 and x_{1b} as x_2, the composite state equations are

$$\begin{bmatrix} x_1(k+1) \\ x_2(k+1) \end{bmatrix} = \begin{bmatrix} -\frac{1}{2} & 0 \\ -\frac{1}{4} & -\frac{1}{4} \end{bmatrix} \begin{bmatrix} x_1(k) \\ x_2(k) \end{bmatrix} + \begin{bmatrix} 1 \\ 1 \end{bmatrix} u(k) \tag{11.57}$$

$$y(k) = [1 \quad -1] \begin{bmatrix} x_1(k) \\ x_2(k) \end{bmatrix} - 4u(k) \tag{11.58}$$

To determine controllability, form

$$[\mathbf{B} \mid \mathbf{AB}] = \begin{bmatrix} 1 & -\frac{1}{2} \\ 1 & -\frac{1}{2} \end{bmatrix}. \tag{11.59}$$

The matrix in (11.59) has rank 1. Since the system is second-order, the composite system is not controllable. To determine observability, form

$$\begin{bmatrix} \mathbf{C} \\ \mathbf{CA} \end{bmatrix} = \begin{bmatrix} 1 & -1 \\ -\frac{1}{4} & \frac{1}{4} \end{bmatrix} \tag{11.60}$$

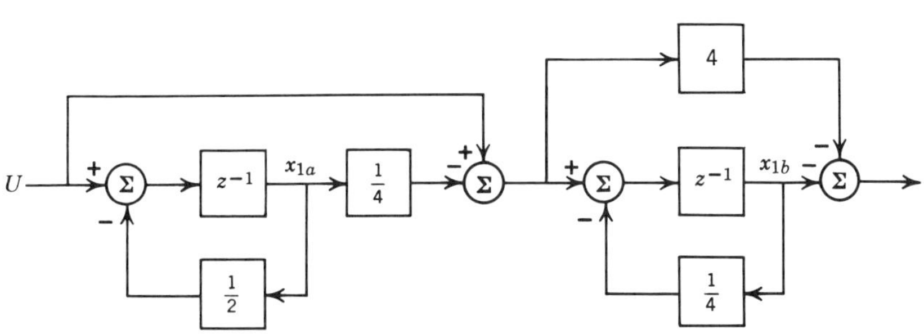

Figure 11.13 Block diagram for Example 11-8.

The matrix of (11.60) has rank 1 so that the composite system is not observable. Hence the composite system is neither controllable nor observable.

The above conclusions may be obtained also by examining the transfer functions $\mathbf{\Phi}(z)\mathbf{B}$ and $\mathbf{C\Phi}(z)$. For complete controllability, all the eigenvalues of $\mathbf{A}$ must appear as poles of $\mathbf{\Phi}(z)\mathbf{B}$. By inspection of the block diagram of Figure 11.13, it is seen that

$$\frac{X_1}{U}=\frac{p_{11}(z)}{\Delta(z)}=\frac{(z+\frac{1}{4})}{(z+\frac{1}{2})(z+\frac{1}{4})}$$

$$\frac{X_2}{U}=\left[1-\frac{\frac{1}{4}}{z+\frac{1}{2}}\right]\left[\frac{1}{z+\frac{1}{4}}\right]=\frac{p_{21}(z)}{\Delta(z)}=\frac{(z+\frac{1}{4})}{(z+\frac{1}{2})(z+\frac{1}{4})}$$

Since $(z+\frac{1}{4})$ appears as a common factor of $p_{11}(z)$, $p_{21}(z)$ and $\Delta(z)$, the eigenvalue $z=-\frac{1}{4}$ does not appear as a pole in any element of $\mathbf{\Phi}(z)\mathbf{B}$. Thus the system is not controllable.

Similarly, for complete observability, all the eigenvalues of $\mathbf{A}$ must appear in $\mathbf{C\Phi}(z)$. By inspection of Figure 11.13,

$$\frac{Y}{x_{10}}=\frac{q_{11}(z)}{\Delta(z)}$$
$$=\frac{-z(z+\frac{1}{2})}{(z+\frac{1}{4})(z+\frac{1}{2})}$$

and

$$\frac{Y}{x_{20}}=\frac{q_{21}(z)}{\Delta(z)}=\frac{-z(z+\frac{1}{2})}{(z+\frac{1}{4})(z+\frac{1}{2})}$$

The factor $(z+\frac{1}{2})$ is common to $q_{11}(z)$, $q_{21}(z)$, and $\Delta(z)$ so that $z=-\frac{1}{2}$ does not appear as a pole of any of the elements of $\mathbf{C\Phi}(z)$. Hence the system is not observable. ▲

Another result due to Gilbert is on the controllability and observability of the linear time-invariant feedback system of Figure 11.14. It is assumed that the eigenvalues for S^a and S^b are distinct. It is also assumed that $(\mathbf{I}+\mathbf{D}_a\mathbf{D}_b)$ is nonsingular. As in the parallel and cascade composite systems, the order of the composite system is n_a+n_b where n_a and n_b are the orders of S^a and S^b, respectively. A necessary and sufficient condition for the controllability of the feedback system of Figure 11.14 is that the open-loop system

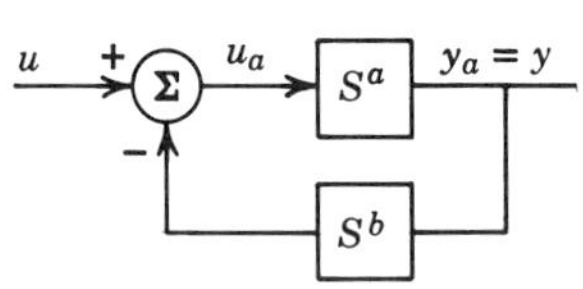

Figure 11.14 Feedback composite system.

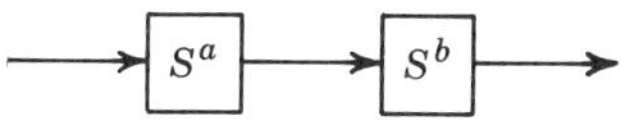

Figure 11.15 Feedback system in Figure 11.14 is controllable if and only if open-loop system in Figure 11.15 is controllable.

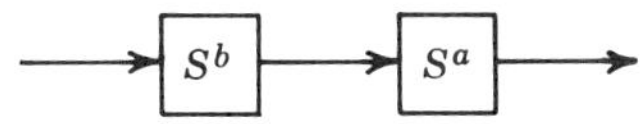

Figure 11.16 Feedback system in Figure 11.14 is observable if and only if open-loop system of Figure 11.16 is observable.

in Figure 11.15 be controllable. The necessary part follows immediately by the method of contradiction. Gilbert proves the sufficiency also by contradiction. Similarly, a necessary and sufficient condition for the observability of the feedback system in Figure 11.14 is that the open-loop system in Figure 11.16 is observable. In general, the open-loop systems are easier to analyze for controllability and observability.

From results stated earlier, a necessary but not sufficient condition for the controllability (or observability) of the feedback system is the controllability (or observability) of both S^a and S^b. Thus, if S^a is not controllable, there is no system S^b that can make the feedback system controllable. Similarly, if S^a is not observable, there is no S^b that can make the feedback system observable.

11.6 State Feedback for Controllable Systems

Consider a plant that is linear time-invariant single-input single-output and completely state controllable represented by the state equations

$$\dot{\mathbf{x}} = \mathbf{A}\mathbf{x} + \mathbf{b}u \tag{11.61}$$

$$y = \mathbf{c}'\mathbf{x} + du \tag{11.62}$$

where u is the scalar control input, y is the scalar output, and $\mathbf{x}$ is an n-vector representing the state. Also consider a controller whose output is a linear combination of an external input r and the state variables:

$$u = r - \mathbf{k}'\mathbf{x}, \tag{11.63}$$

where $\mathbf{k}'$ is a $1 \times n$ row matrix of constants. The control system is shown in block diagram form in Figure 11.17.

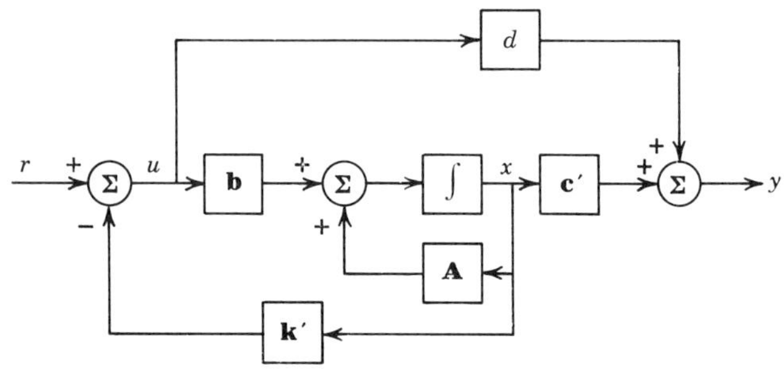

Figure 11.17 Single-input single-output system with state feedback.

The state equations of the composite system become

$$\dot{\mathbf{x}} = (\mathbf{A} - \mathbf{b}\mathbf{k}')\mathbf{x} + \mathbf{b}r \tag{11.64}$$

and

$$y = (\mathbf{c}' - d\mathbf{k}')\mathbf{x} + dr \tag{11.65}$$

Hence the transfer function of the feedback system is

$$\frac{Y(s)}{R(s)} = (\mathbf{c}' - d\mathbf{k}')(s\mathbf{I} - \mathbf{A} + \mathbf{b}\mathbf{k}')^{-1}\mathbf{b} + d$$

$$= \frac{(\mathbf{c}' - d\mathbf{k}')\,\mathrm{adj}\,(s\mathbf{I} - \mathbf{A} + \mathbf{b}k')\mathbf{b} + d[\det(s\mathbf{I} - \mathbf{A} + \mathbf{b}\mathbf{k}')]}{\det(s\mathbf{I} - \mathbf{A} + \mathbf{b}\mathbf{k}')}. \tag{11.66}$$

Denote the numerator polynomial in (11.66) by $p(s)$:

$$p(s) = (c' - dk')\,\mathrm{adj}\,(s\mathbf{I} - \mathbf{A} + \mathbf{b}\mathbf{k}')b + d[\det(s\mathbf{I} - \mathbf{A} + \mathbf{b}\mathbf{k}')] \tag{11.67}$$

so that

$$\frac{Y(s)}{R(s)} = \frac{p(s)}{\det(s\mathbf{I} - \mathbf{A} + \mathbf{b}\mathbf{k}')}. \tag{11.68}$$

In this section, we discuss the following two results due to Morgan:[9]

1. For any given $\mathbf{A}$, a feedback matrix $\mathbf{k}'$ can always be computed such that $\mathbf{A} - \mathbf{b}\mathbf{k}'$ can have any configuration of n eigenvalues. Of course, since we are only considering real matrices, any complex

[9]B. S. Morgan, Jr., "The Synthesis of Single Variable Systems by State Variable Feedback," *Proc. First Allerton Conf. on Circuit and System Theory*, University of Illinois, Urbana, Illinois, 1963, pp. 509–520. See also, R. W. Brockett, "Poles, Zeros, and Feedback: State Space Interpretation," *IEEE Trans. on Automatic Control*, **AC-10**, 129–135 (April 1965).

eigenvalue must appear in conjugate pairs. Thus, for any given set of n numbers where conjugates of complex numbers are also in the set, there is a choice of a real $\mathbf{k}'$ such that these numbers are the roots of

$$\det(s\mathbf{I}-\mathbf{A}+\mathbf{b}\mathbf{k}')=0.$$

2. The polynomial $p(s)$ is independent of $\mathbf{k}'$. A change in k' does not cause a change in $p(s)$. In particular, we may set $\mathbf{k}'=\mathbf{0}'$ in (11.67) and $p(s)$ can be written as

$$p(s)=\mathbf{c}'\,\mathrm{adj}\,(s\mathbf{I}-\mathbf{A})\mathbf{b}+d[\det(s\mathbf{I}-\mathbf{A})] \tag{11.69}$$

Before we derive the above results, let us discuss their design implications. Using state feedback as a controller for a given completely state-controllable plant, the closed-loop transfer function is given in (11.68) where $p(s)$ is fixed. However, by an appropriate choice of $\mathbf{k}'$, the roots of $\det(s\mathbf{I}-\mathbf{A}+\mathbf{b}\mathbf{k}')=0$ can be placed anywhere. If desired, some of the factors of $p(s)$ may be cancelled by some of the factors of $\det(s\mathbf{I}-\mathbf{A}+\mathbf{b}\mathbf{k}')$. From Chapters 6 and 7, we know that the unit impulse response is the inverse Laplace transform of the transfer function. The unit step response is also obtainable as the inverse Laplace transform of the transfer function divided by s. Thus, time response is related to the pole-zero configuration of the transfer function.

One design approach is to determine a pole configuration to be associated with a fixed set of zeros, which yields an acceptable time response. From the desired pole locations $\mathbf{k}'$ can then be computed. For example, for a second-order transfer function with no zeros, we know the relationship between overshoot, rise time, and settling time of a step response, and the damping ζ and undamped natural frequency ω_n corresponding to the poles of the transfer function. For a second-order transfer function with one finite zero, we can obtain a similar relationship. For the same set of two poles, a transfer function with a zero tends to have a step response with more overshoot than one without a finite zero. Similarly, for the same set of two complex poles, a transfer function with an additional pole tends to have a more sluggish step response than one without an extra pole. Based on experience on low-order transfer functions, the pole configuration for a low-order transfer function may be chosen to correspond with an acceptable step response.

We shall derive the above results first for the case when the state variables are related to each other by $\dot{x}_{i+1}=x_i$, $i=1,\ldots,n-1$. These state variables are called *canonic phase variables*. A system state representation is said to be in *canonic phase variable form* if the state equations are

$$\dot{\mathbf{z}} = \begin{bmatrix} 0 & 1 & 0 & 0 & \cdots & 0 \\ 0 & 0 & 1 & 0 & \cdots & 0 \\ 0 & 0 & 1 & 0 & \cdots & 0 \\ \vdots & & \ddots & & & \\ & & & & 1 & 0 \\ 0 & \cdots & & & 0 & 1 \\ -a_1 & -a_2 & \cdots & & -a_{n-1} & -a_n \end{bmatrix} \mathbf{z} + \begin{bmatrix} 0 \\ 0 \\ 0 \\ \vdots \\ 0 \\ 1 \end{bmatrix} u \qquad (11.70)$$

where the components of the state vector $\mathbf{z}$ are called the canonic phase variables. In terms of $\mathbf{z}$ and u, the output equation is

$$y = [\hat{c}_1 \hat{c}_2 \cdots \hat{c}_n]\mathbf{z} + du = \hat{\mathbf{c}}'\mathbf{z} + du \qquad (11.71)$$

Figure 11.18 shows a block diagram representation of the system. Equation 11.70 is easily verified as the state equation. From the

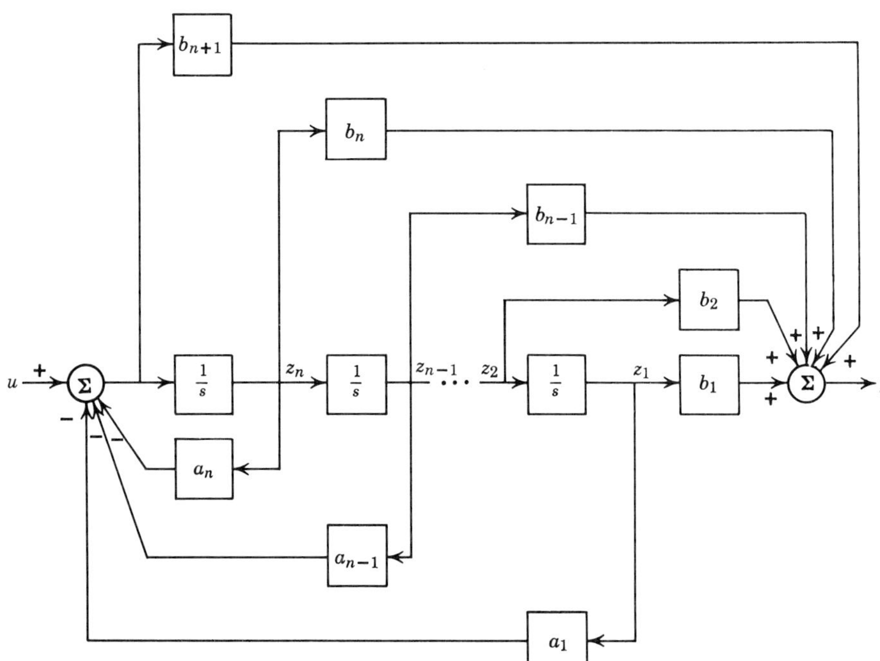

Figure 11.18 Canonic phase-variable state representation.

diagram, the output is

$$
\begin{aligned}
y &= b_1 z_1 + b_2 z_2 + \cdots + b_n z_n + b_{n+1} \dot{z}_n \\
&= b_1 z_1 + b_2 z_2 + \cdots + b_n z_n - b_{n+1} a_1 z_1 - b_{n+1} a_2 z_2 \\
&\qquad - \cdots - b_{n+1} a_n z_n + b_{n+1} u
\end{aligned}
\tag{11.72}
$$

Comparing with (11.71), it is seen that

$$b_{n+1} = d \tag{11.73}$$

$$b_k = \hat{c}_k + b_{n+1} a_k. \tag{11.74}$$

By Mason's rule, the transfer function $Y(s)/U(s)$ is

$$
\begin{aligned}
\frac{Y(s)}{U(s)} &= \frac{b_{n+1} + \dfrac{1}{s} b_n + \dfrac{1}{s^2} b_{n-1} + \cdots + \dfrac{b_1}{s^n}}{1 + \dfrac{a_n}{s} + \dfrac{a_{n-1}}{s^2} + \cdots + \dfrac{a_1}{s^n}} \\
&= \frac{b_{n+1} s^n + b_n s^{n-1} + b_{n-1} s^{n-2} + \cdots + b_1}{s^n + a_n s^{n-1} + \cdots + a_1}
\end{aligned}
\tag{11.75}
$$

Let us now consider a phase variable feedback

$$u = r - \hat{k}' z = r - \hat{k}_1 z_1 - \cdots - \hat{k}_n z_n \tag{11.76}$$

The composite system is shown in Figure 11.19.
By Mason's rule, the transfer function $Y(s)/R(s)$ of the composite feedback system is

$$
\begin{aligned}
\frac{Y(s)}{R(s)} &= \frac{b_{n+1} + \dfrac{1}{s} b_n + \cdots + \dfrac{b_1}{s^n}}{1 + \dfrac{a_n + \hat{k}_n}{s} + \dfrac{a_{n-1} + \hat{k}_{n-1}}{s^2} + \cdots + \dfrac{a_1 + \hat{k}_1}{s^n}} \\
&= \frac{b_{n+1} s^n + b_n s^{n-1} + \cdots + b_1}{s^n + (a_n + \hat{k}_n) s^{n-1} + \cdots + (a_1 + \hat{k}_1)}
\end{aligned}
\tag{11.77}
$$

Since the denominator polynomial is nth degree,

$$\det(s\mathbf{I} - \mathbf{A} + \hat{\mathbf{b}}\hat{\mathbf{k}}') = s^n + (a_n + \hat{k}_n) s^{n-1} + \cdots + (a_1 + \hat{k}_1) = 0 \tag{11.78}$$

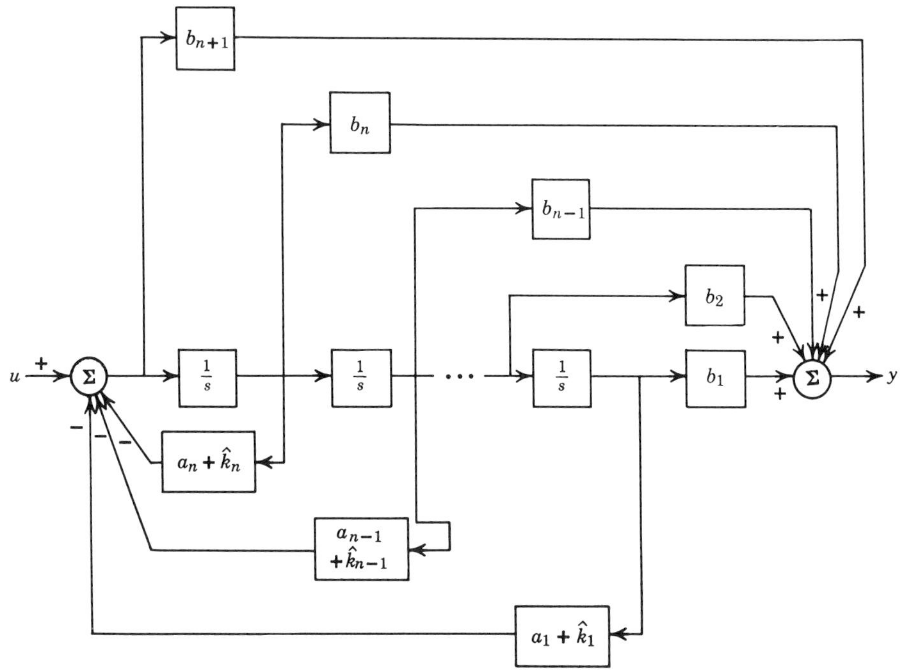

Figure 11.19 Canonic phase-variable state representation with phase-variable feedback added.

gives the n eigenvalues of $\hat{\mathbf{A}}-\hat{\mathbf{b}}\hat{\mathbf{k}}'$, where $\hat{\mathbf{A}}$ and $\hat{\mathbf{b}}$ have the special form

$$\hat{\mathbf{A}} = \begin{bmatrix} 0 & 1 & 0 & 0 & \cdots & & 0 \\ 0 & 0 & 1 & 0 & \cdots & & 0 \\ 0 & 0 & 0 & 1 & \cdots & & 0 \\ \cdot & & & \ddots & & & \vdots \\ & \cdot & & & & & \\ & & & & & 1 & 0 \\ 0 & \cdots & & & & 0 & 1 \\ -a_1 & -a_2 & \cdots & & & & -a_n \end{bmatrix} \tag{11.79}$$

$$\hat{\mathbf{b}} = \begin{bmatrix} 0 \\ 0 \\ \vdots \\ 0 \\ 1 \end{bmatrix} \tag{11.80}$$

For any given set of $a_1, \ldots, a_n$ and any given set of coefficients of the nth degree equation in (11.78), we can determine the corresponding set of $\hat{k}_1, \hat{k}_2, \ldots, \hat{k}_n$. Thus (11.78) can have any set of n roots by an appropriate choice of $\hat{k}_1, \hat{k}_2, \ldots, \hat{k}_n$. This proves Result 1 for the canonic phase variable representation.

From (11.73), (11.74), and (11.76), we see that the numerator polynomial of the feedback-system transfer function is independent of $\hat{\mathbf{k}}$ and

$$p(s) = b_{n+1}s^n + b_n s^{n-1} + \cdots + b_1. \tag{11.81}$$

This proves Result 2 for the canonic phase variable representation.

To proceed with the proof of the general case, we recall from Chapter 7 that if $\mathbf{T}$ is a nonsingular $n \times n$ matrix, relating the state vector $\mathbf{x}$ to another state vector $\mathbf{z}$ by

$$\mathbf{x} = \mathbf{T}\mathbf{z},$$

(11.64) and (11.65) may be written in terms of $\mathbf{z}$ as

$$\dot{\mathbf{z}} = \mathbf{T}^{-1}(\mathbf{A} - \mathbf{b}\mathbf{k}')\mathbf{T}\mathbf{z} + \mathbf{T}^{-1}\mathbf{b}r \tag{11.82}$$

$$y = (\mathbf{c}' - \mathbf{d}\mathbf{k}')\mathbf{T}\mathbf{z} + dr. \tag{11.83}$$

Based on the above equations, the transfer function $Y(s)/R(s)$ is

$$\frac{Y(s)}{R(s)} = (\mathbf{c}' - \mathbf{d}\mathbf{k}')\mathbf{T}(s\mathbf{I} - \mathbf{T}^{-1}\mathbf{A}\mathbf{T} + \mathbf{T}^{-1}\mathbf{b}\mathbf{k}'\mathbf{T})^{-1}\mathbf{T}^{-1}\mathbf{b} + d. \tag{11.84}$$

Since

$$\begin{aligned}(s\mathbf{I} - \mathbf{T}^{-1}\mathbf{A}\mathbf{T} + \mathbf{T}^{-1}\mathbf{b}\mathbf{k}'\mathbf{T})^{-1}\mathbf{T}^{-1} &= [\mathbf{T}(s\mathbf{I} - \mathbf{T}^{-1}\mathbf{A}\mathbf{T} + \mathbf{T}^{-1}\mathbf{b}\mathbf{k}'\mathbf{T})]^{-1} \\ &= (s\mathbf{T} - \mathbf{A}\mathbf{T} + \mathbf{b}\mathbf{k}'\mathbf{T})^{-1},\end{aligned}$$

and since

$$\mathbf{T}(s\mathbf{T} - \mathbf{A}\mathbf{T} + \mathbf{b}\mathbf{k}'\mathbf{T})^{-1} = [(s\mathbf{T} - \mathbf{A}\mathbf{T} + \mathbf{b}\mathbf{k}'\mathbf{T})\mathbf{T}^{-1}]^{-1} = (s\mathbf{I} - \mathbf{A} + \mathbf{b}\mathbf{k}')^{-1},$$

(11.84) can be reduced to

$$\frac{Y(s)}{R(s)} = (\mathbf{c}' - \mathbf{d}\mathbf{k}')(s\mathbf{I} - \mathbf{A} + \mathbf{b}\mathbf{k}')^{-1}b + d \tag{11.85}$$

Comparing (11.85) and (11.66), we conclude that the transfer function is independent of the nonsingular transformation $\mathbf{T}$.

Kalman[10] proved that if the single-input system of (11.61) is completely state-controllable, there exists a nonsingular transformation $\mathbf{T}$, which relates $\mathbf{x}$ to $\mathbf{z}$ by $\mathbf{x} = \mathbf{Tz}$, so that

$$\mathbf{T}^{-1}\mathbf{AT} = \hat{\mathbf{A}} \tag{11.86}$$

$$\mathbf{T}^{-1}\mathbf{b} = \hat{\mathbf{b}} \tag{11.87}$$

are equal to the matrices in (11.79) and (11.80), respectively. That is, there exists a nonsingular matrix $\mathbf{T}$ that will transform the state to canonic phase variable form. Using such a transformation, (11.82) and (11.83) become

$$\dot{\mathbf{z}} = (\hat{\mathbf{A}} - \hat{\mathbf{b}}\mathbf{k}'\mathbf{T})\mathbf{z} + \mathbf{b}r \tag{11.88}$$

$$y = (\mathbf{c}'\mathbf{T} - d\mathbf{k}'\mathbf{T})\mathbf{z} + dr. \tag{11.89}$$

Denoting $\mathbf{k}'\mathbf{T}$ by $\hat{k}'$ and $\mathbf{c}'\mathbf{T}$ by $\hat{\mathbf{c}}'$, we see that (11.88) and (11.89) correspond to the block diagram of Figure 11.19. Since we have already shown that Results 1 and 2 hold for the canonic phase variable form of (11.88) and (11.89), and since the transfer function $Y(s)/R(s)$ for the canonic phase variable form is equal to the transfer function $Y(s)/R(s)$ using the original state representation in (11.64) and (11.65), we conclude that Results 1 and 2 hold as well for the original state representation.

An algorithm for constructing the matrix $\mathbf{T}$, which transforms a given state representation as in (11.61) to a canonic phase variable form as in (11.70), was presented by Wonham and Johnson.[11] It is given below.

Starting from

$$\mathbf{t}_n = \mathbf{b}, \tag{11.90}$$

solve the recursion equation

$$\mathbf{t}_{i-1} = \mathbf{A}\mathbf{t}_i + a_i\mathbf{t}_n \tag{11.91}$$

[10]R. E. Kalman, "Mathematical Description of Linear Dynamic Systems," *J.S.I.A.M. Control*, Ser. A, **1**(**2**), 152–192(1963).

[11]W. M. Wonham and C. D. Johnson, "Optimal Bang-bang Control with Quadratic Performance Index," *J. Basic Engineering, Trans. ASME*, **86**, 107–115 (March 1964); C. D. Johnson and W. M. Wonham, "Another Note on the Transformation to Canonical (Phase Variable) Form," *IEEE Trans. on Automatic Control*, **AC-11**, 609–610 (July 1966), see also B.S. Morgan, *op. cit.*

for $i = n, n-1, \ldots, 2$, where $a_2, \ldots, a_n$ are the coefficients in the characteristic equations. Then

$$\mathbf{T} = [\mathbf{t}_1 \,\vdots\, \mathbf{t}_2 \,\vdots\, \cdots \,\vdots\, \mathbf{t}_n] \tag{11.92}$$

is the desired transformation to canonic phase variable form.

Example 11-9. Let us transform

$$\begin{aligned} \dot{x}_1 &= -x_1 + 2x_2 + u \\ \dot{x}_2 &= x_1 - 3x_2 + u \end{aligned} \tag{11.93}$$

to canonic phase variable form. First let us check whether the system is completely state-controllable. Form

$$\mathbf{Ab} = \begin{bmatrix} -1 & 2 \\ 1 & -3 \end{bmatrix} \begin{bmatrix} 1 \\ 1 \end{bmatrix} = \begin{bmatrix} 1 \\ -2 \end{bmatrix}.$$

$$\text{rank}\,[\mathbf{b} \,\vdots\, \mathbf{AB}] = \text{rank} \begin{bmatrix} 1 & 1 \\ 1 & -2 \end{bmatrix} = 2$$

since

$$\det \begin{bmatrix} 1 & 1 \\ 1 & -2 \end{bmatrix} \neq 0.$$

Hence the system is completely state-controllable. Next form the characteristic equation of $\mathbf{A}$:

$$\det(s\mathbf{I} - \mathbf{A}) = \det \begin{pmatrix} s+1 & -2 \\ -1 & s+3 \end{pmatrix} = 0$$

or

$$(s+1)(s+3) - 2 = s^2 + 4s + 1 = 0. \tag{11.94}$$

Thus $a_2 = 4$, $a_1 = 1$. To construct $\mathbf{T}$, form $\mathbf{t}_2$ from (11.90)

$$\mathbf{t}_2 = \begin{bmatrix} 1 \\ 1 \end{bmatrix}$$

and $\mathbf{t}_1$ from (11.91):

$$\mathbf{t}_1 = \begin{bmatrix} -1 & 2 \\ 1 & -3 \end{bmatrix} \begin{bmatrix} 1 \\ 1 \end{bmatrix} + 4 \begin{bmatrix} 1 \\ 1 \end{bmatrix} = \begin{bmatrix} 5 \\ 2 \end{bmatrix}.$$

Hence

$$\mathbf{T} = \begin{bmatrix} 5 & 1 \\ 2 & 1 \end{bmatrix} \tag{11.95}$$

is the desired transformation:

$$\mathbf{x} = \begin{bmatrix} 5 & 1 \\ 2 & 1 \end{bmatrix} \mathbf{z}.$$

To verify that indeed $\mathbf{T}^{-1}\mathbf{AT}$ has the form in (11.79), determine

$$\mathbf{T}^{-1} = \begin{bmatrix} 1 & -1 \\ -2 & 5 \end{bmatrix}.$$

$$\begin{aligned} \mathbf{T}^{-1}\mathbf{AT} &= \tfrac{1}{3}\begin{bmatrix} 1 & -1 \\ -2 & 5 \end{bmatrix}\begin{bmatrix} -1 & 2 \\ 1 & -3 \end{bmatrix}\begin{bmatrix} 5 & 1 \\ 2 & 1 \end{bmatrix} \\ &= \begin{bmatrix} 0 & 1 \\ -1 & -4 \end{bmatrix}. \end{aligned} \tag{11.96}$$

We also verify that the characteristic equation of the matrix $\mathbf{T}^{-1}\mathbf{AT}$ in (11.96) is

$$\det\begin{pmatrix} s & -1 \\ 1 & s+4 \end{pmatrix} = s(s+4)+1 = s^2+4s+1 = 0$$

which is the same as (11.94). Thus the transformation $\mathbf{T}$ does not affect the eigenvalues. Finally, we verify that $\mathbf{T}^{-1}\mathbf{b}$ has the form in (11.80):

$$\mathbf{T}^{-1}\mathbf{b} = \tfrac{1}{3}\begin{bmatrix} 1 & -1 \\ -2 & 5 \end{bmatrix}\begin{bmatrix} 1 \\ 1 \end{bmatrix} = \begin{bmatrix} 0 \\ 1 \end{bmatrix}.$$

Thus, with the transformation $\mathbf{T}$ in (11.95), (11.93) transform to the canonic phase variable form

$$\dot{\mathbf{z}} = \begin{bmatrix} 0 & 1 \\ -1 & -4 \end{bmatrix}\mathbf{z} + \begin{bmatrix} 0 \\ 1 \end{bmatrix}u. \blacktriangle \tag{11.97}$$

Example 11-10. It is desired to apply state feedback for the system in Example 11-9 so that the new eigenvalues are at $s=-5$ and $s=-1$. The desired characteristic equation is

$$(s+5)(s+1) = s^2+6s+5 = 0. \tag{11.98}$$

Using a control

$$u = r - \hat{k}_1 z_1 - \hat{k}_2 z_2, \tag{11.99}$$

(11.97) becomes

$$\dot{\mathbf{z}} = \begin{bmatrix} 0 & 1 \\ -(1+\hat{k}_1) & -(4+\hat{k}_2) \end{bmatrix}\mathbf{z} + \begin{bmatrix} 0 \\ 1 \end{bmatrix} r. \tag{11.100}$$

The new characteristic equation is

$$s^2 + (4+\hat{k}_2)s + (1+\hat{k}_1) = 0. \tag{11.101}$$

Comparing (11.101) with (11.98), we have

$$\begin{aligned} 4+\hat{k}_2 &= 6 \\ 1+\hat{k}_1 &= 5 \end{aligned}$$

or

$$\begin{aligned} \hat{k}_1 &= 4 \\ k_2 &= 2. \end{aligned}$$

Hence

$$u = r - [4 \quad 2]\begin{bmatrix} z_1 \\ z_2 \end{bmatrix}$$

will produce a characteristic equation specified by (11.98). Finally, the state feedback in terms of **x** is

$$\begin{aligned} u &= r - [4 \quad 2]\mathbf{T}^{-1}\mathbf{x} \\ &= r - [4 \quad 2]\tfrac{1}{3}\begin{bmatrix} 1 & -1 \\ -2 & 5 \end{bmatrix}\mathbf{x} \\ &= r - [0 \quad 2]\mathbf{x} \\ &= r - 2x_2. \end{aligned} \tag{11.102}$$

Substituting u of (11.102) in (11.93), we have

$$\begin{aligned} \dot{x}_1 &= -x_1 + r \\ \dot{x}_2 &= x_1 - 5x_2 + r. \end{aligned} \tag{11.103}$$

The eigenvalues for the new system are easily verified from

$$\begin{aligned} &\det\begin{pmatrix} s+1 & 0 \\ -1 & s+5 \end{pmatrix} = 0 \\ &= (s+5)(s+1) = 0 \end{aligned} \tag{11.104}$$

which checks with (11.98). ▲

Example 11-11. Suppose that the output equation for a system whose state equation is given in (11.93) is

$$y = x_1 + x_2. \tag{11.105}$$

Consider the system as a plant. State variable feedback is to be applied as shown in Figure 11.20.

Let us determine the expression for the transfer function $Y(s)/R(s)$. By Mason's rule we have

$$\frac{Y(s)}{R(s)} = \frac{\dfrac{2}{s^2} + \dfrac{1}{s}\left(1 + \dfrac{1}{s}\right) + \dfrac{1}{s}\left(1 + \dfrac{3}{s}\right)}{1 + \left(\dfrac{3}{s} + \dfrac{1}{s} - \dfrac{2}{s^2} + \dfrac{k_2}{s} + \dfrac{2k_1}{s^2} + \dfrac{k_1}{s} + \dfrac{k_2}{s^2}\right) - \left(\dfrac{3}{s^2} + \dfrac{k_2}{s^2}\right)}$$

$$= \frac{2s + 6}{s^2 + (4 + k_2 + k_1)s + (2k_1 - 5)}.$$

Notice that the numerator polynomial is invariant with k_1 and k_2. Furthermore, the poles of the transfer function may be located anywhere. The poles are the roots of

$$s^2 + (4 + k_2 + k_1)s + (2k_1 - 5) = 0.$$

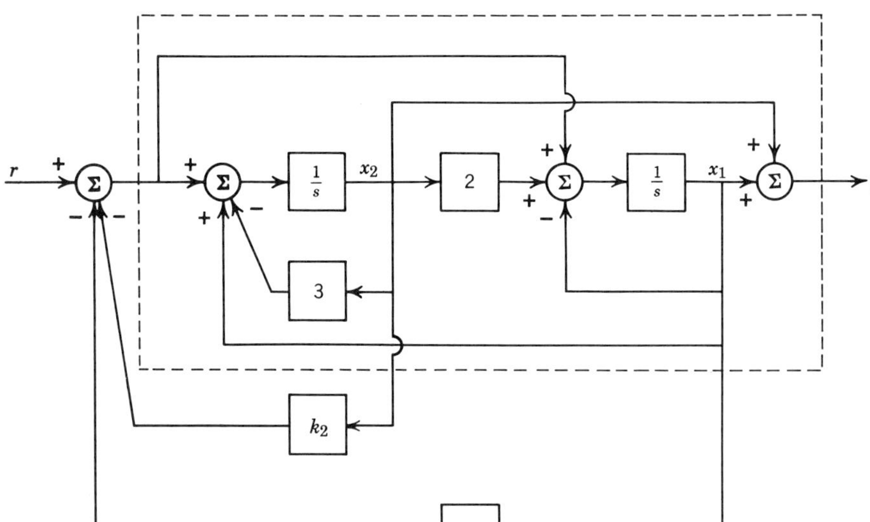

Figure 11.20 Control system for Example 11-11.

Thus k_1 may be chosen for any specified value of $(2k_1-5)$. With k_1 thus determined, k_2 may be chosen for any specified value of $(4+k_2+k_1)$. ▲

Example 11-12. Consider again the linear time-invariant second-order discrete-time single-input system of Example 11-1. The state equations are reproduced below:

$$x_1(k+1) = -\tfrac{1}{2}x_1(k) + \tfrac{1}{4}x_2(k) + u(k)$$
$$x_2(k+1) = \tfrac{1}{4}x_1(k) - \tfrac{3}{4}x_2(k) + 2u(k)$$

or

$$\mathbf{x}(k+1) = \begin{bmatrix} -\frac{1}{2} & \frac{1}{4} \\ \frac{1}{4} & -\frac{3}{4} \end{bmatrix}\mathbf{x}(k) + \begin{bmatrix} 1 \\ 2 \end{bmatrix}u(k). \tag{11.106}$$

In Example 11-1, it was shown that the system is completely state-controllable and that a control sequence $u(0)$ and $u(1)$ can be chosen so that the state is zero at the second sampling instant and for all sampling instants thereafter. The control obtained is open-loop.

It is now desired to obtain a feedback control $u(k)$ that is a linear combination of the state variables, which will drive the state to the origin in two sampling periods. Let the feedback coefficients be denoted by $-K_1$ and $-K_2$ so that

$$u(k) = -K_1x_1(k) - K_2x_2(k) = -[K_1 \quad K_2]\mathbf{x}(k) = -\mathbf{K}'\mathbf{x}(k). \tag{11.107}$$

The proposed control block diagram is shown in Figure 11.21. With the feedback structure of $u(k)$ in (11.107), (11.106) becomes

$$\mathbf{x}(k+1) = \begin{bmatrix} -(\frac{1}{2}+K_1) & +(\frac{1}{4}-K_2) \\ (\frac{1}{4}-2K_1) & -(\frac{3}{4}+2K_2) \end{bmatrix}\mathbf{x}(k)$$
$$= \hat{\mathbf{A}}\mathbf{x}(k). \tag{11.108}$$

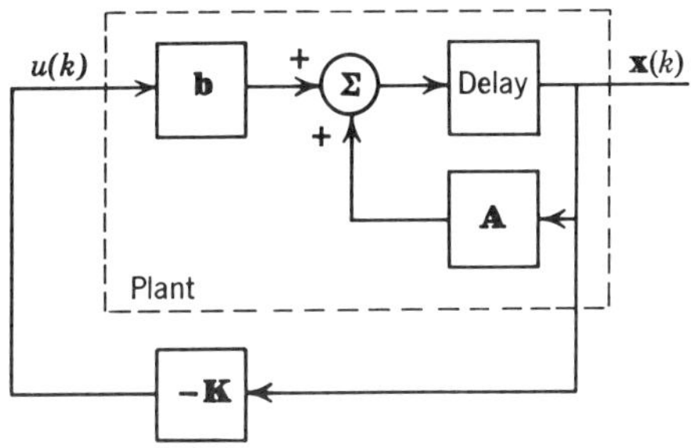

Figure 11.21 Feedback control for Example 11-12.

We shall demonstrate that if the eigenvalues of $\hat{\mathbf{A}}$ are both chosen to be at the origin, by an appropriate choice of K_1 and K_2, then $\mathbf{x}(k) = \mathbf{0}$ for $k \geqslant 2$ no matter what the value of $\mathbf{x}(0)$ is. From our previous discussion, since the system is completely state-controllable, the closed-loop eigenvalues can be placed anywhere.

The z-transform of (11.108) is

$$z\mathbf{X}(z) - z\mathbf{x}(0) = \hat{\mathbf{A}}\mathbf{X}(z).$$

Thus

$$\mathbf{X}(z) = [z\mathbf{I} - \hat{\mathbf{A}}]^{-1} z\mathbf{x}(0).$$

Recall from the definition of z-transforms that

$$\mathbf{X}(z) = \sum_{k=0}^{\infty} \mathbf{x}(k) z^{-k}.$$

Hence, if it is desired that $\mathbf{x}(k) = 0$ for $k \geqslant 2$, the form of $\mathbf{X}(z)$ must be

$$\mathbf{X}(z) = \mathbf{x}(0) + \frac{1}{z}\mathbf{x}(1). \tag{11.109}$$

We deduce from (11.109) that the eigenvalues of $\hat{\mathbf{A}}$ should be at the origin. Hence the problem becomes the choice of K_1 and K_2 so that

$$\det(z\mathbf{I} - \hat{\mathbf{A}}) = 0 \tag{11.110}$$

has a double root at the origin. Substituting the expression for $\hat{\mathbf{A}}$ from (11.108) in (11.110) we have

$$\det \begin{bmatrix} z + (\frac{1}{2} + K_1) & -(\frac{1}{4} - K_2) \\ -(\frac{1}{4} - 2K_1) & z + (\frac{3}{4} + 2K_2) \end{bmatrix}$$
$$= [z + (\tfrac{1}{2} + K_1)][z + (\tfrac{3}{4} + 2K_2)] - (\tfrac{1}{4} - K_2)(\tfrac{1}{4} - 2K_1) = 0$$

or

$$z^2 + (\tfrac{5}{4} + K_1 + 2K_2)z + (\tfrac{5}{16} + \tfrac{5}{4}K_1 + \tfrac{5}{4}K_2) = 0. \tag{11.111}$$

Equation 11.111 has a double root at the origin if and only if

$$\tfrac{5}{4} + K_1 + 2K_2 = 0 \tag{11.112}$$

and

$$\tfrac{5}{16} + \tfrac{5}{4}K_1 + \tfrac{5}{4}K_2 = 0. \tag{11.113}$$

The unique values of K_1 and K_2, which satisfy (11.112) and (11.113), are

$$K_1 = \tfrac{3}{4}$$

and

$$K_2 = 1.$$

For these values of K_1 and K_2, (11.108) becomes

$$\mathbf{x}(k+1) = \begin{bmatrix} -\frac{5}{4} & \frac{5}{4} \\ -\frac{5}{4} & \frac{5}{4} \end{bmatrix} \mathbf{x}(k).$$

Hence

$$\mathbf{x}(2) = \hat{\mathbf{A}}\hat{\mathbf{A}}\mathbf{x}(0) = \begin{bmatrix} -\frac{5}{4} & \frac{5}{4} \\ -\frac{5}{4} & \frac{5}{4} \end{bmatrix} \begin{bmatrix} -\frac{5}{4} & \frac{5}{4} \\ -\frac{5}{4} & \frac{5}{4} \end{bmatrix} \mathbf{x}(0)$$

$$= \begin{bmatrix} 0 & 0 \\ 0 & 0 \end{bmatrix} \mathbf{x}(0) = \mathbf{0}$$

and

$$\mathbf{x}(k) = \mathbf{0}.$$

for $k \geqslant 2$, and any arbitrary value of $\mathbf{x}(0)$.

For the initial condition of Example 11-1, the resulting control $u(k)$ should be exactly the same as before. For example, since

$$u(0) = -K_1 x_1(0) - K_2 x_2(0),$$
$$x_1(0) = 4, \text{ and } x_2(0) = 6,$$
$$u(0) = -\tfrac{3}{4}(4) - (-1)(6) = 3$$

which agrees with (11.9). Similarly,

$$u(1) = -K_1 x_1(1) - K_2 x_2(1).$$

But

$$\begin{bmatrix} x_1(1) \\ x_2(1) \end{bmatrix} = \begin{bmatrix} -\frac{5}{4} & \frac{5}{4} \\ -\frac{5}{4} & \frac{5}{4} \end{bmatrix} \begin{bmatrix} x_1(0) \\ x_2(0) \end{bmatrix} = \begin{bmatrix} \frac{5}{2} \\ \frac{5}{2} \end{bmatrix}$$

Hence

$$u(1) = -\tfrac{3}{4}(\tfrac{5}{2}) - (-1)(\tfrac{5}{2}) = \tfrac{5}{8}$$

which agrees with (11.9) also. The main advantage of the feedback control in this example is that K_1 and K_2 are constants whose values do not depend on the initial state $\mathbf{x}(0)$. Once computed and implemented in the structure of Figure 11.21, any initial state $\mathbf{x}(0)$ will result in a zero state after two sampling intervals. ▲

For a completely state-controllable nth-order single-input linear time-discrete system, linear state feedback can always be applied so that all the eigenvalues of the closed-loop $\mathbf{A}$ matrix are at the origin. Analogous to (11.109), the z-transform of the state becomes

$$\mathbf{X}(z) = \mathbf{x}(0) + \frac{1}{z}\mathbf{x}(1) + \cdots + \frac{1}{z^{n-1}}\mathbf{x}(n-1) \tag{11.114}$$

when all the eigenvalues of $\hat{\mathbf{A}}$ are at the origin. This guarantees that for an arbitrary initial state $\mathbf{x}(0), \mathbf{x}(k) = 0$ for $k \geqslant n$. The feedback coefficients may be computed by first computing the corresponding feedback coefficients for the canonic phase variable form, which will make all the eigenvalues equal to zero.

11.7 Comparison of Open-Loop and Closed-Loop Structure when Plant Is Perturbed

In this section, we shall compare the output errors induced by parameter changes in the plant for a control structure with no feedback, as shown in Figure 11.22, and a control structure with feedback, as shown in Figure 11.23. For simplicity in the discussion, consider single-input single-output linear time-invariant systems only. The plant is represented by a transfer function $P(s)$. Because of uncertainties, such as inaccuracies in the modeling, aging of components and simplification in the analysis, the actual plant transfer function may differ from P. Consequently, the plant output may differ from that when the plant transfer function is P. For a nominal plant transfer function P, if G_1, G, and H are chosen so that $u_c = u_0$, the structures are said to be *nominally equivalent*. The outputs y_0 and y_c of nominally equivalent structures, when the plant transfer function is equal to the nominal P, are equal. In general, if there is any change in the

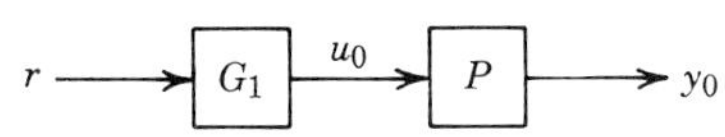

Figure 11.22 Open-loop control.

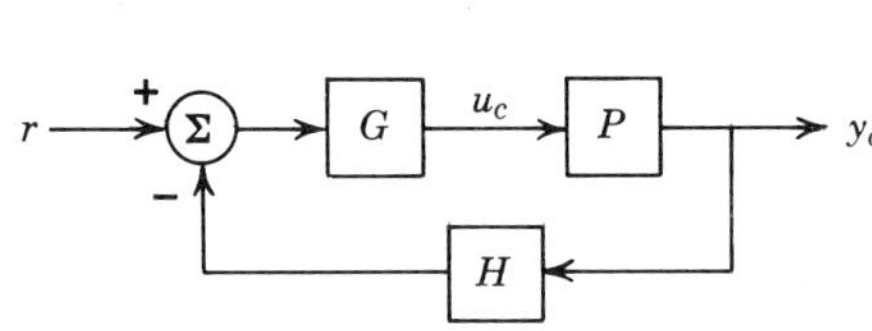

Figure 11.23 Feedback control.

plant parameters, the resulting change in the open-loop output will not be equal to the resulting change in the closed-loop output. Let us compare the two structures when only the plant is subject to parameter variations and G_1, G, and H are accurately known. For differential changes in the parameters, denote the open-loop output deviation by dy_0 and the closed-loop output deviation by dy_c.

The output deviations are generally functions of time that depend on the command input r and on the structure. A convenient way of comparing the errors or deviations is to compare

$$W_0 = \int_0^\infty (dy_0)^2 \, dt \tag{11.115}$$

to

$$W_c = \int_0^\infty (dy_c)^2 \, dt, \tag{11.116}$$

if the integrals exist. Both positive and negative output deviations yield a positive contribution to the integral. This measure will be discussed further in Chapter 12. If

$$W_0 \geqslant W_c \tag{11.117}$$

for all inputs r, we say that the feedback structure is *less sensitive* to plant variation than the open-loop structure.

Let us express W_0 and W_c in terms of $dY_0(j\omega)$ and $dY_c(j\omega)$, the Fourier transforms of $dy_0(t)$ and $dy_c(t)$. Using the inverse transform integral for $dy_0(t)$, (11.107) can be written as

$$\int_0^\infty [dy_0(t)]^2 dt = \int_0^\infty dy_0(t) \left[\frac{1}{2\pi} \int_{-\infty}^\infty dY_0(j\omega) \epsilon^{j\omega t} d\omega\right] dt. \tag{11.118}$$

Assuming that the order of integration can be interchanged, (11.118) can be rewritten as

$$\int_0^\infty [dy_0(t)]^2 dt = \frac{1}{2\pi} \int_{-\infty}^\infty dY_0(j\omega) \left[\int_0^\infty dy_0(t) \epsilon^{j\omega t} dt\right] d\omega. \tag{11.119}$$

But

$$\int_0^\infty dy_0(t) \epsilon^{j\omega t} \, dt = dY_0(-j\omega)$$

from the definition of the Fourier transform. Hence (11.119) becomes

$$\int_0^\infty [dy_0(t)]^2 dt = \frac{1}{2\pi} \int_{-\infty}^\infty dY_0(j\omega) dY_0(-j\omega) \, d\omega = \frac{1}{2\pi} \int_{-\infty}^\infty |dY_0(j\omega)|^2 \, d\omega. \tag{11.120}$$

The last integral in (11.120) is obtained by noting that for a real valued $dy_0(t)$, $dY_0(-j\omega)$ is the conjugate of $dY_0(j\omega)$. The identity in (11.120) is known as *Parseval's theorem*. Similarly,

$$\int_0^\infty [dy_c(t)]^2\, dt = \frac{1}{2\pi} \int_{-\infty}^{\infty} |dY_c(j\omega)|^2\, d\omega.$$

Hence, for a feedback structure that is less sensitive than an open loop structure,

$$\int_{-\infty}^{\infty} |dY_0(j\omega)|^2 d\omega \geqslant \int_{-\infty}^{\infty} |dY_c(j\omega)|^2 d\omega. \tag{11.121}$$

To carry on the comparison further, let us derive a relationship between $dY_0(j\omega)$ and $dY_c(j\omega)$. From Figure 11.23,

$$Y_c = PU_c \tag{11.122}$$

and

$$U_c = GR - GHY_c. \tag{11.123}$$

Taking differentials, we have

$$dY_c = dPU_c + P\, dU_c \tag{11.124}$$

and

$$dU_c = -GH\, dY_c, \tag{11.125}$$

where G and H are fixed but P and Y_c are subject to variation. Substituting (11.125) in (11.124), we obtain

$$dY_c = \frac{1}{1+PGH} dPU_c. \tag{11.126}$$

From Figure 11.22,

$$Y_0 = PU_0$$

Since U_0 is fixed for fixed G_1,

$$dY_0 = dPU_0. \tag{11.127}$$

For nominally equivalent structures, $U_0 = U_c$ so that

$$dPU_c = dPU_0 = dY_0 \tag{11.128}$$

Hence, (11.126) becomes

$$dY_c = \frac{1}{1+PGH} dY_0. \tag{11.129}$$

The relationship between dY_0 and dY_c is of the form

$$dY_c = S\, dY_0. \tag{11.130}$$

S is called the *comparison sensitivity.** It relates the open-loop output deviation to the closed-loop output deviation. From (11.129), the comparison sensitivity is

$$S(j\omega) = \frac{1}{1+P(j\omega)G(j\omega)H(j\omega)}. \tag{11.131}$$

Using (11.130), (11.121) can be written as

$$\int_{-\infty}^{\infty} |dY_0(j\omega)|^2\, d\omega \geq \int_{-\infty}^{\infty} |S(j\omega)\; dY_0(j\omega)|^2\, d\omega$$

or

$$\int_{-\infty}^{\infty} |dY_0(j\omega)|^2\, d\omega \geq \int_{-\infty}^{\infty} |S(j\omega)|^2 |dY_0(j\omega)|^2\, d\omega. \tag{11.132}$$

The relation in (11.132) can be rewritten further as

$$\int_{-\infty}^{\infty} (1-|S(j\omega)|^2)\,|dY_0(j\omega)|^2\, d\omega \geq 0. \tag{11.133}$$

Thus a feedback system is less sensitive than an open-loop system, according to the defining criterion of (11.117), if and only if (11.133) is satisfied.

Since $|dY_0(j\omega)|^2 \geq 0$, (11.133) will be satisfied if

$$1-|S(j\omega)|^2 \geq 0 \tag{11.134}$$

for all real ω. That is, the condition in (11.134) is *sufficient* to guarantee that

$$\int_0^{\infty} [dy_0(t)]^2\, dt \geq \int_0^{\infty} [dy_c(t)]^2\, dt \tag{11.135}$$

will be satisfied for all inputs r for which the integrals exist. From (11.131), (11.134) becomes

$$\left|\frac{1}{1+P(j\omega)G(j\omega)H(j\omega)}\right| \leq 1 \tag{11.136}$$

or

$$|1+P(j\omega)G(j\omega)H(j\omega)| \geq 1 \tag{11.137}$$

*This comparison was first introduced in J. B. Cruz, Jr. and W. R. Perkins, "A New Approach to the Sensitivity Problem in Multivariable Feedback System Design," *IEEE Trans. on Automatic Control*, **AC-9**, 216–223 (July, 1964).

for all real ω. If $r(t)$ is bandlimited, that is, $|R(j\omega)|$ is neglible outside a finite range of ω, then y_c and y_0 are bandlimited also. For such a class of bandlimited inputs, it is sufficient to satisfy (11.137) over the frequency band of the signals in order for (11.135) to hold.

From a sensitivity standpoint, it is desirable to make the magnitude of $P(j\omega)G(j\omega)H(j\omega)$ large. However, as noted in Chapter 8, such a requirement might make the system unstable. Hence, G and H should be chosen so that $|1+P(j\omega)G(j\omega)H(j\omega)|$ is as large as possible without making the system unstable.

In addition to the comparison criterion of (11.135) where the error for all time is of importance, and the steady-state error must be zero if the integrals exist, we can compare the parameter-induced steady-state errors for $t \to \infty$. From the final value theorem of Appendix B

$$\lim_{t\to\infty} e(t) = \lim_{s\to 0} sE(s).$$

Hence

$$\lim_{t\to\infty} dy_c(t) = \lim_{s\to 0} s\, dY_c(s) \tag{11.138}$$

and

$$\lim_{t\to\infty} dy_0(t) = \lim_{s\to 0} s\, dY_0(s). \tag{11.139}$$

Thus

$$\lim_{s\to 0} s\, dY_c(s) = \lim_{s\to 0} [sS(s)\, dY_0(s)].$$

If

$$\lim_{s\to 0} S(s)$$

exists, and if the limits in (11.138) and (11.139) exist,

$$\lim_{t\to\infty} dy_c(t) = [\lim_{s\to 0} S(s)][\lim_{s\to\infty} dy_0(t)]. \tag{11.140}$$

Hence

$$|\lim_{s\to 0} S(s)| < 1 \tag{11.141}$$

is sufficient to guarantee that the magnitude of the closed-loop steady-state error is less than the magnitude of the open-loop steady-state error.

Example 11-13. The open loop and closed loop systems in Figure 11.24 are nominally equivalent for $K = 1$.

For a step input of amplitude R, the steady-state output deviation of the open-loop system due to a differential change in K from $K = 1$ is

$$\lim_{t\to\infty} dy_0(t) = R\, dK.$$

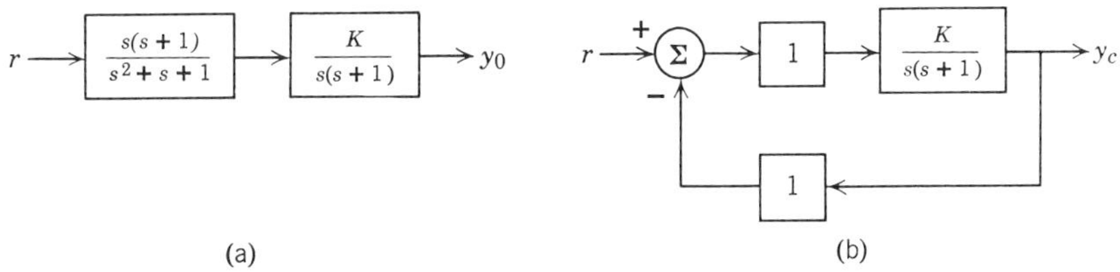

Figure 11.24 Open-loop and closed-loop systems which are nominally equivalent for $K = 1$.

The comparison sensitivity for the structure in Figure 11.24*b* is

$$S(s) = \frac{1}{1+PGH} + \frac{1}{1+1/s(s+1)} \tag{11.142}$$

It is seen from (11.142) that

$$\lim_{s\to 0} S(s) = 0$$

Hence

$$\lim_{t\to\infty} dy_c(t) = 0.$$

This means that differential changes in K have no effect on the steady state value of $y_c(t)$. For this example, let us show that even for large changes in K, the steady-state value of $y_c(t)$ does not change. The transfer function for the system in Figure 11.24*b* is

$$\frac{Y_c(s)}{R(s)} = \frac{K}{s(s+1)+K}.$$

Hence for $R(s) = R/s$,

$$\lim_{t\to\infty} y_c(t) = \lim_{s\to 0}\left[s\frac{R}{s}\frac{K}{s(s+1)+K}\right] = R \tag{11.143}$$

which is independent of K. Hence changes in K do not affect the steady state value of $y_c(t)$. ▲

Example 11-14. For the structure in Figure 11.24*b*, the comparison sensitivity for a nominal value of $K = 1$ is

$$S(s) = \frac{s(s+1)}{s^2+s+1}.$$

A plot of $20 \log |S(j\omega)|$ together with its piecewise linear approximation is shown in Figure 11.25.

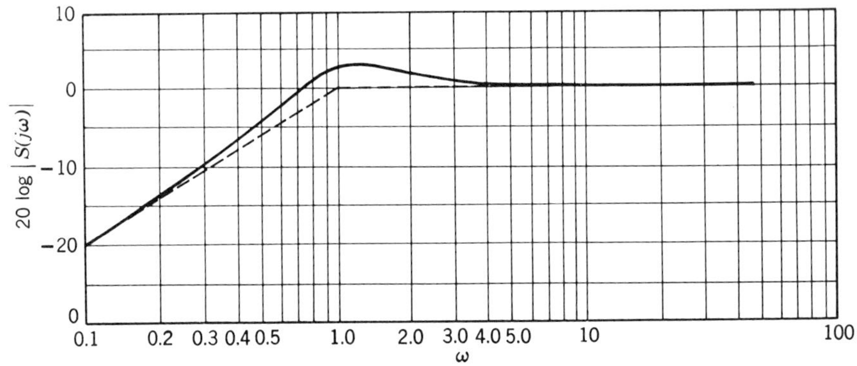

Figure 11.25 Plot of 20 log $S(j\omega)$ versus ω for the system in Example 11-14.

It is seen that $|S(j\omega)|$ is not less than unity for all ω. However, if the input is bandlimited containing only low frequencies (for example, for $\omega < 0.5$ rad/sec) the feedback structure is less sensitive than the open-loop structure. ▲

11.8 Comparison of Open-Loop and Closed-Loop Structures when Plant Has Disturbance Input

In Figure 11.26*a* and 11.26*b*, we have open-loop and closed-loop control for a plant with a disturbance input. For a linear time-invariant plant, there is no loss of generality by assuming that the disturbance

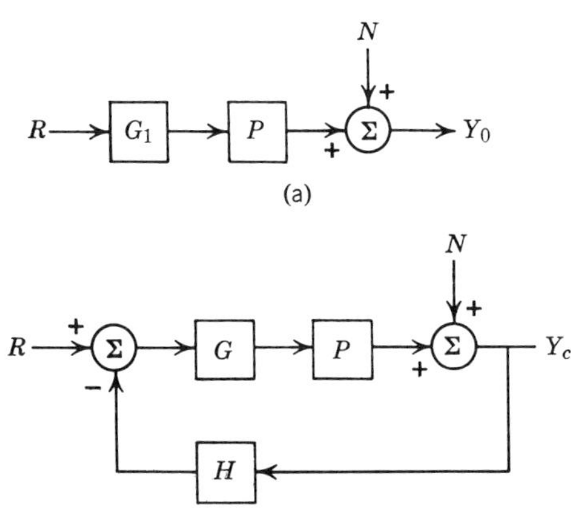

Figure 11.26 Open-loop and closed-loop structures with plant-disturbance input.

adds to the output. The disturbance may also include the effects of initial conditions of the plant. We wish to compare the output components due to the disturbance N. For the open-loop system, the output component due to N is

$$\Delta Y_0 = N. \tag{11.144}$$

For the closed-loop system in Figure 11.26*b*, the component of the output due to N is

$$\Delta Y_c = \frac{1}{1+PGH}\Delta Y_0. \tag{11.145}$$

The coefficient which relates ΔY_0 to ΔY_c is the comparison sensitivity derived in the previous section. Thus the same sufficient condition as in (11.137) will guarantee that the integral of the square of the output component due to the disturbance is less for the closed-loop structure as compared to that for the open-loop structure.

In the practical application of the feedback control scheme, an additional consideration must be examined. In measuring the output for feedback, there is usually some inaccuracy. This inaccuracy may be represented by an additional disturbance, as shown in Figure 11.27. The output contribution due to the feedback disturbance N_F is

$$\Delta Y_c = \frac{-PGH}{1+PGH}N_F. \tag{11.146}$$

Recall that for low sensitivity due to plant parameter variations or plant disturbance N, $|P(j\omega)G(j\omega)H(j\omega)|$ must be large. But if this is done, (11.146) indicates that $|\Delta Y_c| \approx |N_F|$. Thus the feedback disturbance will propagate to the output with practically no attenuation. Since the requirement for a low output contribution from N_F is in conflict with the requirement for low sensitivity to plant parameter

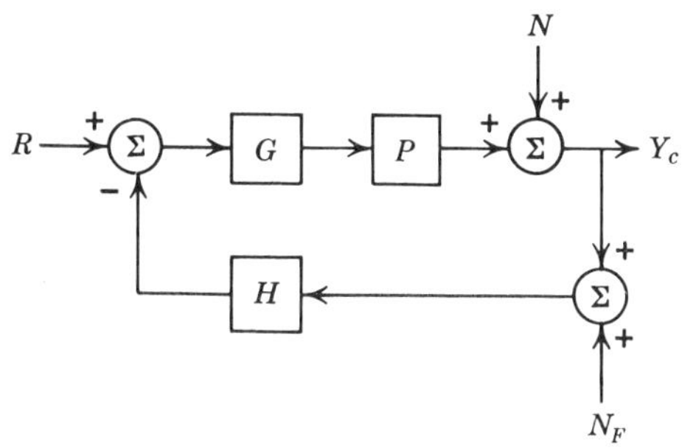

Figure 11.27 Feedback system with feedback disturbance.

variations and low output contribution due to plant disturbance N, the design of G and H must be a compromise. In particular, for frequencies outside the frequency band of the input, the magnitude $|P(j\omega)G(j\omega)H(j\omega)|$ must be made as small as possible. This insures that the high-frequency components of N_F do not propagate to the output.

Problems

11-1. (a) Verify that the system in Figure P11-1 is completely state-controllable.

(b) It is proposed to apply a control

$$\begin{aligned} u(t) &= K_1, \quad 0 \leqslant t < 1 \\ &= K_2, \quad 1 \leqslant t < 2 \\ &= 0, \quad t \geqslant 2. \end{aligned}$$

Suppose that $x_1(0) = 3$, $x_2(0) = -5$. Determine K_1 and K_2 so that $x_1(t) = 0$ for $t \geqslant 2$ and $x_2(t) = 0$ for $t \geqslant 2$.

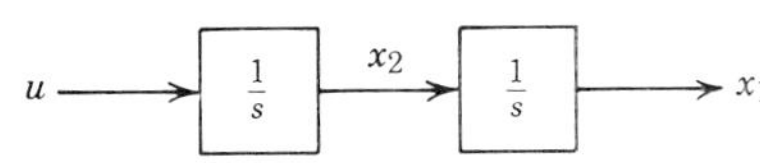

Figure P11-1

11-2. Suppose that for the system in Figure P11-1, the proposed control is

$$\begin{aligned} u(t) &= K_1\epsilon^{-t} + K_2\epsilon^{-2t}, \quad 0 \leqslant t \leqslant 1 \\ &= 0 \quad \text{for} \quad t > 1. \end{aligned}$$

Determine K_1 and K_2 as functions of $x_1(0)$ and $x_2(0)$ so that $x_1(t) = x_2(t) = 0$ for $t \geqslant 0$.

11-3. (a) Using v_{C_1} and v_{C_2} as state variables and v as input, verify that the network in Figure P11.3 is completely state-controllable provided that $R_1C_1 \neq R_2C_2$.

(b) Suppose that $R_1 = R_2 = 1000\Omega$, $C_1 = .01$ microfarad, and $C_2 = 0.02$ microfarad. A proposed voltage pulse is

$$\begin{aligned} v(t) &= K \sin(2\pi \times 10^5 t) + K_2 \sin(4\pi \times 10^5 t), \quad t \leqslant t \leqslant 10^{-5} \text{ seconds} \\ &= 0 \quad \text{for} \quad t > 10^{-5} \text{ seconds}. \end{aligned}$$

Compute K_1 and K_2 as functions of $v_{C1}(0)$ and $v_{C2}(0)$ so that

$$v_{C1}(t) = v_{C2}(t) = 0 \quad \text{for} \quad t \geqslant 10^{-5} \text{ seconds}.$$

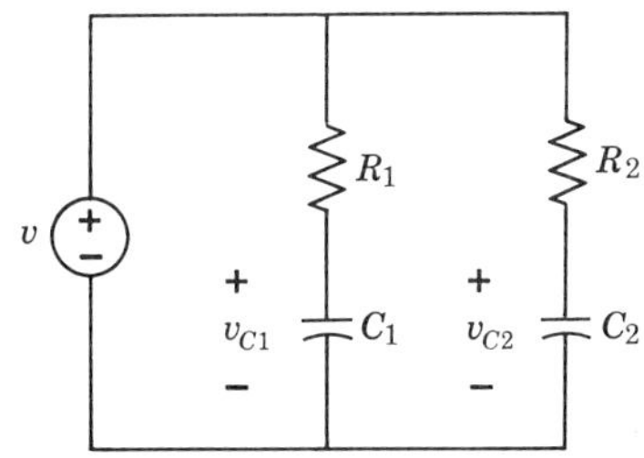

Figure P11-3

11-4. For the network in Figure P11-3, suppose that $R_1 = 10{,}000\Omega$, $R_2 = 20{,}000\Omega$, $C_1 = 0.001$ microfarad, $C_2 = 0.001$ microfarad, and

$$\begin{aligned} v(t) &= K_1, \quad 0 \leqslant t < 10^{-5} \text{ seconds} \\ &= K_2, \quad 10^{-5} \leqslant t < 2 \times 10^{-5} \text{ seconds} \\ &= 0 \text{ elsewhere.} \end{aligned}$$

Determine K_1 and K_2 such that

$$v_{C1}(t) = v_{C2}(t) = 0 \quad \text{for} \quad t > 2 \times 10^{-5} \text{ second.}$$

11-5. Figure P11-5 shows a block diagram for a position control system using a 2-phase servo motor. The transfer function for the linearized model of the motor is

$$\frac{10}{s(0.1s+1)}.$$

where the input is control phase voltage and the output is motor shaft position θ_m.

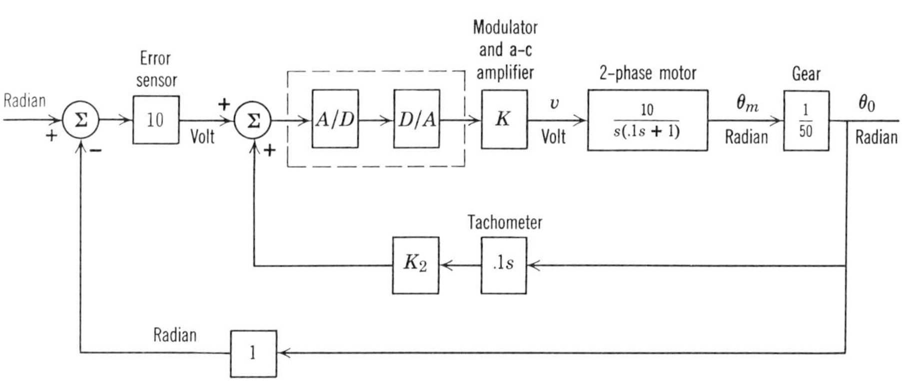

Figure P11-5

(a) Determine a state representation for the motor portion using the motor shaft position θ_m and motor shaft angular velocity $\dot{\theta}_m$ as state variables.

(b) The sampling period of the analog-to-digital and digital-to-analog converters is 0.1 second. It is desired to obtain a time-discrete state representation for the motor at the sampling instants. Assuming that the control phase voltage input is piecewise constant, obtain the difference equations for the motor.

(c) Determine if the discrete-time system in (b) is completely state-controllable.

(d) The complete system in Figure P11-5 may be considered to employ state feedback where the coefficient for θ_m is $(1/50)(10)K$ and the coefficient for $\dot{\theta}_m$ is $(1/50)(0.1)(K_2)(K)$. For a zero input, determine the gain constants K and K_2 so that the state at the end of the second sampling period is zero.

(e) Suppose the input $r(t)$ is a step function. Instead of θ_m, choose the output of the error sensor as a state variable and $\dot{\theta}_m$ as the other state variable. Determine the discrete-time state representation of the feedback system, using the new state variables. Determine K and K_2 so that the new state vector is zero at the end of the second sampling period. Thus the output θ_0 will follow the step input after two sampling periods.

11-6. (a) By Mason's rule determine the transfer function Y/R (Figure P11-6).

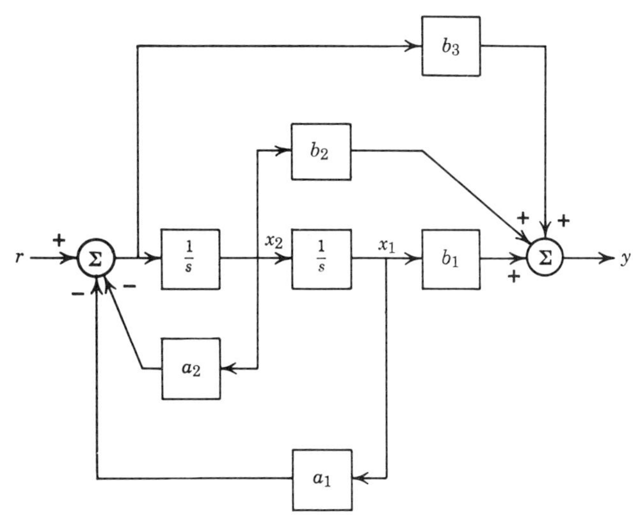

Figure P11-6

(b) Using x_1 and x_2 as state variables, r as input and y as output, determine the matrices **A**, **B**, **C** and **D** in the state representation.

(c) Determine the complete state controllability of the system by examining the rank of the matrix $[\mathbf{B} \vdots \mathbf{AB}]$.

(d) Determine the output controllability of the system.

(e) Determine the complete observability of the system by checking the rank of the matrix $[\mathbf{C}' \vdots \mathbf{C}'\mathbf{A}']$.

11-7. For the system in Figure P11-6,

(a) Determine the complete state controllability by checking whether in $\mathbf{\Phi}(s)\mathbf{B}$ the numerator polynomials and the characteristic polynomial have any factor in common. Use Mason's rule for computing the transfer functions (elements of $\mathbf{\Phi}(s)\mathbf{B}$).

(b) Determine the complete observability of the system by checking whether the numerator polynomials of $\mathbf{C\Phi}(s)$ and the characteristic polynomial have any factors in common. Use Mason's rule for computing the transfer functions [elements of $\mathbf{C\Phi}(s)$].

11-8. Repeat Problem 11-6 for the system in Figure P11-8.

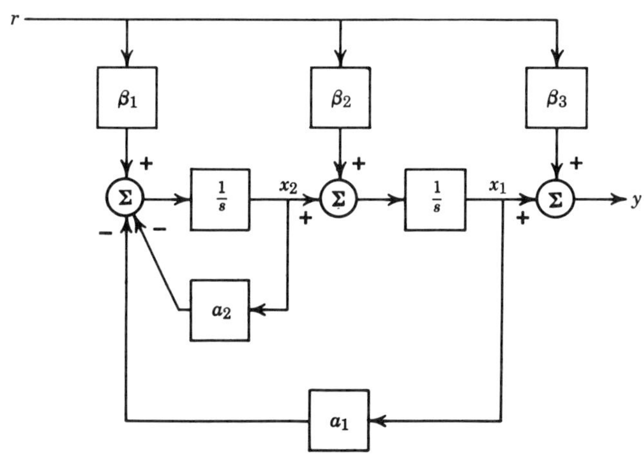

Figure P11-8

11-9. Repeat Problem 11-7 for the system in Figure P11-8.

11-10. Repeat Problem 11-6 for the nth-order system in Figure P11-10.

11-11. Repeat Problem 11-7 for the nth-order system in Figure P11-10.

11-12. Repeat Problem 7-6 for the nth-order system in Figure P11-12.

11-13. Repeat Problem 7-7 for the nth-order system in Figure P11-12.

11-14. For the single-axis reaction-wheel attitude control system considered in Section 11.4,

(a) Suppose that the wheel speed is the output. Is the system output-controllable?

(b) Suppose that the wheel speed and attitude position are the outputs. Is the system output-controllable?

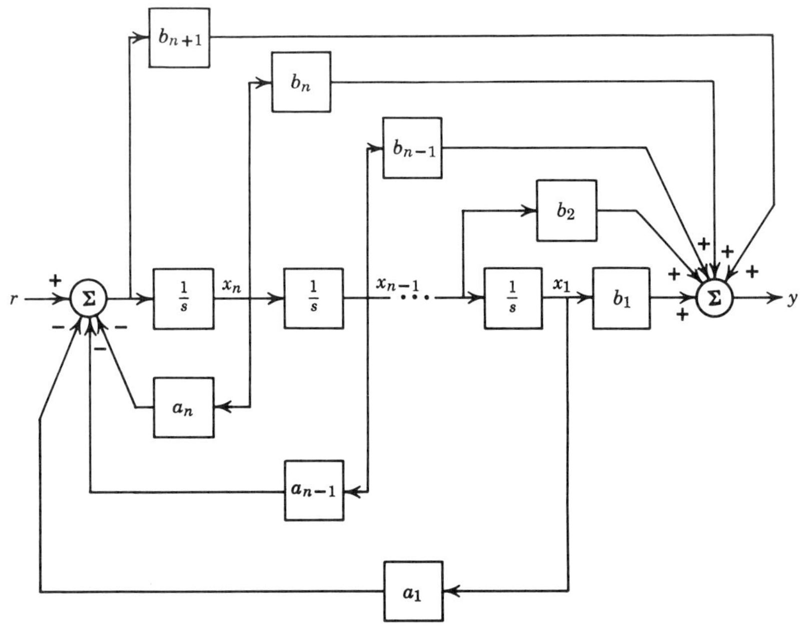

Figure P11-10

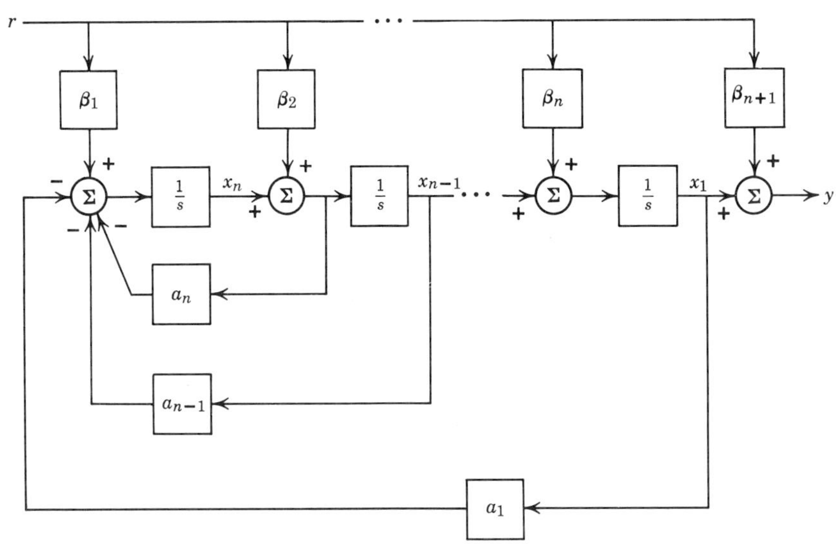

Figure P11-12

11-15. (a) The state representation for a second-order system is

$$\dot{x}_1 = -x_1 + u$$
$$\dot{x}_2 = x_1 - 2x_2 + u.$$

Verify that the system is not completely state-controllable.

(b) It is proposed to apply state feedback control to the system in (a) by choosing u as

$$u = -k_1 x_1 - k_2 x_2 + r$$

where r is the new external input and k_1 and k_2 are feedback coefficients. Show that there exists no choice for k_1 and k_2 that will make the feedback system completely state-controllable.

11-16. For the system in Problem 11-15(a), which is not completely state controllable

(a) If x_1 is the output, is the system output-controllable?

(b) If x_2 is the output, is the system output-controllable?

11-17. (a) For the system in Problem 11-15(a), suppose that the control u is piecewise-constant, being the output of a sample-and-hold device. Assuming a uniform sampling period of 1 second, determine the discrete-time state equations for the state at sampling instants.

(b) Verify that the discrete-time system is not completely state-controllable.

(c) Determine the set of initial states that can be "controlled" in the sense that a control sequence $u(k)$ can be determined to drive the state to the origin in one sampling period. *Hint.* The set is a straight line in the state plane, passing through the origin.

11-18. A second-order system has the following state description:

$$\dot{x}_1 = -2x_1 + x_2$$
$$\dot{x}_2 = -x_2 + u$$
$$y = x_1 + x_2$$

(a) Verify that the system is completely state-controllable.

(b) Suppose u is chosen as

$$u = -k_1 x_1 - k_2 x_2 + r$$

where r is the new external input and k_1 and k_2 are state feedback coefficients. Show that the feedback system remains completely state-controllable for any choice of k_1 and k_2.

(c) Verify that the zeros of the transfer function are independent of k_1 and k_2.

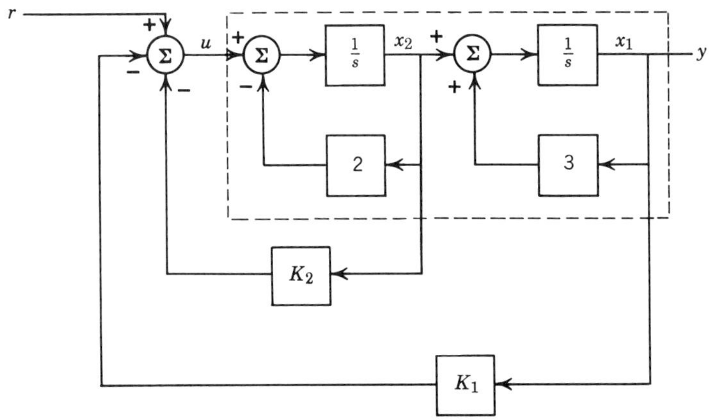

Figure P11-19

11-19. (a) Verify that the plant inside the dotted box in Figure P11-19 is completely state-controllable. Notice that is is unstable.

(b) For what choices of the feedback coefficients K_1 and K_2 will the feedback system be stable?

(c) Choose K_1 and K_2 so that the characteristic equations is

$$s^2+\sqrt{2}s+1=0.$$

11-20. (a) Draw a sensitivity block diagram (see Chapter 6) for determining $\partial Y_c/\partial K$ for nominal values $K=1, \tau=1$ (Figure P11-20).

(b) Draw a nominally equivalent open-loop system for $K=1$, $\tau=1$. Show that $\partial Y_0/\partial K=(\partial P/\partial K)U_c$.

(c) From the results of (a) and (b), verify that

$$\frac{\partial Y_c}{\partial K}=S\frac{\partial Y_0}{\partial K}$$

where

$$S=\frac{1}{1+P},$$

$$P=\frac{K}{s(\tau s+1)}$$

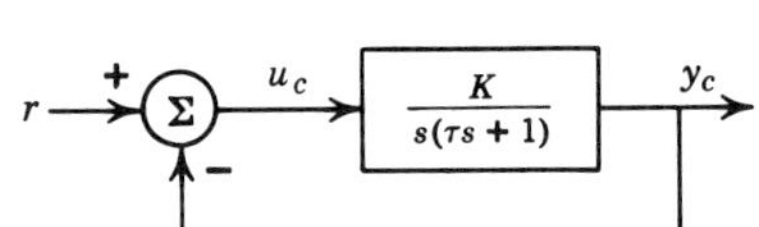

Figure P11-20

and $K = 1, \tau = 1$. If K is the only parameter subject to variation,

$$dY_c = \frac{\partial Y_c}{\partial K} dK$$

and

$$dY_c = \frac{\partial Y_0}{\partial K} dK \cdot$$

These output deviations may be computed from the block diagram of (a) and the parameter perturbation dK.

11-21. Repeat Problem 11-20 with K fixed at 1 and τ subject to variation from nominal value of 1. That is, consider $\partial Y_c/\partial\tau$ and $\partial Y_0/\partial\tau$ instead of $\partial Y_c/\partial K$ and $\partial Y_0/\partial K$. Notice that the S that relates $\partial Y_0/\partial\tau$ to $\partial Y_c/\partial\tau$ is the same S that relates $\partial Y_0/\partial K$ to $\partial Y_c/\partial K$.

11-22. Denote the transfer function $P(s)G(s)H(s)$ by $L(s)$. The sufficient condition for a feedback structure to be less sensitive than an open-loop system is

$$|1 + L(j\omega)| > 1$$

for all real ω. Determine a geometric interpretation of this inequality in terms of the complex $L(j\omega)$ plane.

11-23. Let the plant transfer function P be a function of a parameter α. G and H do not depend on α (see Figure P11-23). Denote the closed-loop transfer function by T. The Bode sensitivity function is defined as

$$S_P{}^T = \frac{\dfrac{\partial T}{\partial \alpha} P}{T \dfrac{\partial P}{\partial \alpha}}$$

(H. W. Bode's original definition is the reciprocal of this, but the above form is commonly used at present). Note that

$$dT = \frac{\partial T}{\partial \alpha} d\alpha$$

and

$$dP = \frac{\partial P}{\partial \alpha} d\alpha$$

Hence, $S_P{}^T$ can be written as

$$S_P{}^T = \frac{dT/T}{dP/P} = \frac{d \ln T}{d \ln P}$$

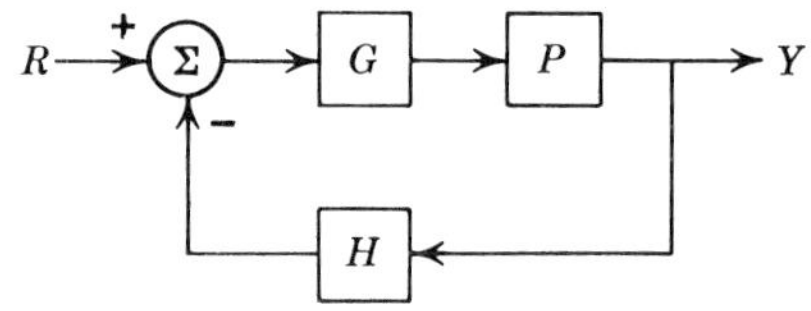

Figure P11-23

For real valued T and P, this has the physical interpretation of a ratio of the percent change in T to the percent change in P. Hence it is desirable to have $S_P{}^T$ less than one. For complex T and complex P, a physical interpretation is not readily evident. Show that

$$S_P{}^T = \frac{1}{1+PGH}.$$

Note that $S_P{}^T$ is equal to the comparison sensitivity described in the chapter. Hence, $S_P{}^T$ has the additional physical interpretation that it relates dY_0 to dY_c.

11-24. For the feedback structure in Figure P11-23, verify that

$$S = 1 - TH.$$

Verify that G and H can be expressed in terms of P, S, and T:

$$H = \frac{1-S}{T}$$

$$G = \frac{T}{PS}.$$

11-25. The block diagram for a fourth-order system is shown in Figure P11-25. Using x_1, x_2, x_3, and x_4 as state variables, write the state equations for the system. Determine whether or not the system is completely state-controllable.

11-26. A system has the following state representation:

$$\dot{x}_1 = -x_1 + x_2$$

$$\dot{x}_2 = -x_2 + x_3$$

$$\dot{x}_3 = -x_3 + u$$

(a) Verify that the system is completely state-controllable.
(b) Transform the representation to canonic phase variable form.
(c) Apply state variable feedback to the original representation in (a) so that the eigenvalues are $s = -5$, $s = -1 \pm j1$. Compute the

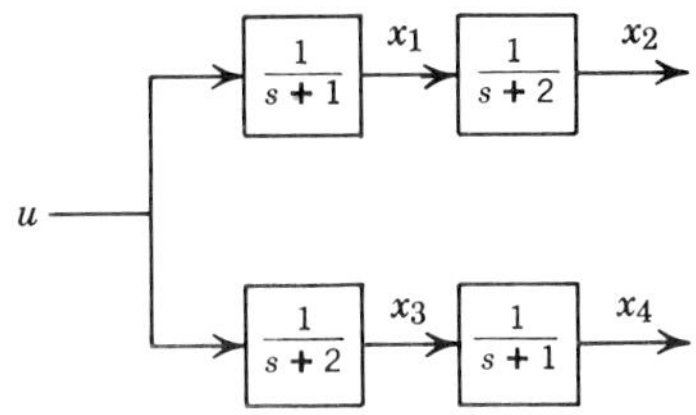

Figure P11-25

feedback coefficients by computing the phase variable feedback coefficients first.

11-27. Verify that the system in (a) is completely state controllable if and only if the system in (b) is completely observable (Figure P11-27).

Figure P11-27

11-28. A controllable discrete-time system is described by

$$x_1(k+1) = -\tfrac{1}{2}x_1(k) + x_2(k)$$
$$x_2(k+1) = -\tfrac{1}{2}x_2(k) + x_3(k)$$
$$x_3(k+1) = -\tfrac{1}{2}x_3(k) + u(k)$$

The control $u(k)$ is constructed as

$$u(k) = r(k) - k_1 x_1(k) - k_2 x_2(k) - k_3 x_3(k)$$

where $r(k)$ is an external input. It is desired to choose k_1, k_2, and k_3 so that for a step input $r(k)$, $x_1(k) = r(k)$ for $k \geqslant 3$. This may be accomplished by placing all the poles of $X_1(z) - R(z)$ at the origin. Determine k_1, k_2, and k_3.

CHAPTER 12

Parameter Optimization I

The engineer finds that there are alternate ways of accomplishing the design of most engineering systems. Sometimes there are a very large number of alternatives, so large that they cannot all be investigated individually. Yet it is the engineer's job to select the "best" of these alternatives. This implies that all designs can be evaluated unambiguously in terms of a desired set of system characteristics. The design that is "best" with regard to these desired system characteristics is termed "optimum."

Engineers have always been interested in producing the best designs they could, of course. Thus the idea of optimum design is not new. In the past, however, the desired system characteristics often have been expressed as "rules of thumb" found by experience to produce acceptable designs. Some remarkably effective designs have resulted from the application of such an approach by experienced and ingenious engineers. However, the synthesis of very complex large-scale systems, involving the consideration of a large number of alternative designs and demanding high performance, requires a different approach. The design procedure must be made as organized as possible. In particular, the desired system characteristics must be stated precisely, in terms of mathematical expressions, if possible.

Since about 1950 there has been very active study of the problem of optimum system design. A large body of theory now exists, and significant engineering applications are beginning to be made. We shall consider here only one important class of system optimization problems—parameter optimization. In this problem, the structure of the system is supposed to have been fixed initially. For example, we recall from Chapter 7 that we can form a closed-loop system having eigenvalues anywhere by employing state feedback for a controllable plant. Thus, state feedback is a useful system structure and will be used in this chapter for optimization. Other configurations will be discussed also.

The discussion of the selection of the subsystem parameters will proceed in two steps:

(1) Selection of a performance criterion.
(2) Calculation of optimum parameters.

Step (1) will be discussed in this chapter. A beginning of a discussion of Step (2) will also be made, with this discussion continuing in the following chapter.

12.1 Quadratic Performance Criteria

There are many criteria by which system performance might be judged. Steady-state or transient response, reliability, cost, energy consumption, and weight are all possible criteria. For analytic design, we must be able to relate the performance criterion mathematically to the design parameters to be selected. A popular performance index involves the integral of the sum of squares of several system variables. Such indices are called quadratic performance criteria. One attractive feature of these criteria is that they are mathematically tractable. In addition, they have a reasonable engineering interpretation in some (but not all, of course) real problems. Let us consider an example, a simplified version of our single-axis attitude control problem.

In previous chapters, it has been shown that under suitable assumptions a single-axis reaction-wheel attitude control system may be modeled by

$$\begin{aligned}\dot{\theta} &= \omega \\ \dot{\omega} &= \frac{k^2}{J_v R_a}\Omega - \frac{k}{J_v R_a}u_1 \\ \dot{\Omega} &= -\left(\frac{1}{J_v} + \frac{1}{J_w}\right)\frac{k^2}{R_a}\Omega + \left(\frac{1}{J_v} + \frac{1}{J_w}\right)\frac{k}{R_a}u_1\end{aligned} \tag{12.1}$$

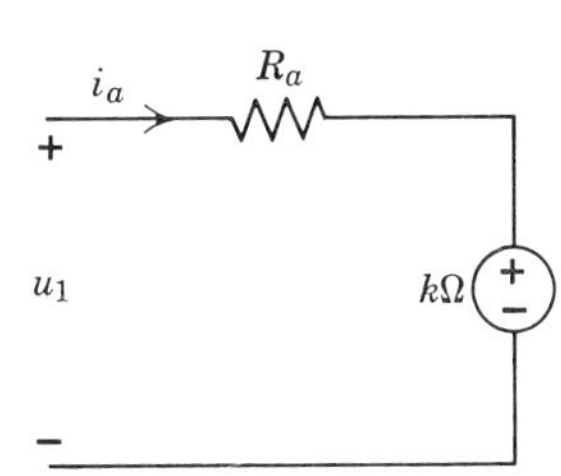

Figure 12.1 Armature circuit of the d-c motor.

An additional assumption will be made here to further simplify the system representation. Recall from Chapter 3 that the armature circuit of the reaction-wheel motor can be represented by the network diagram of Figure 12.1. It is often the case that the "back emf" $k\Omega$ is much smaller than the applied armature voltage u. If $k\Omega$

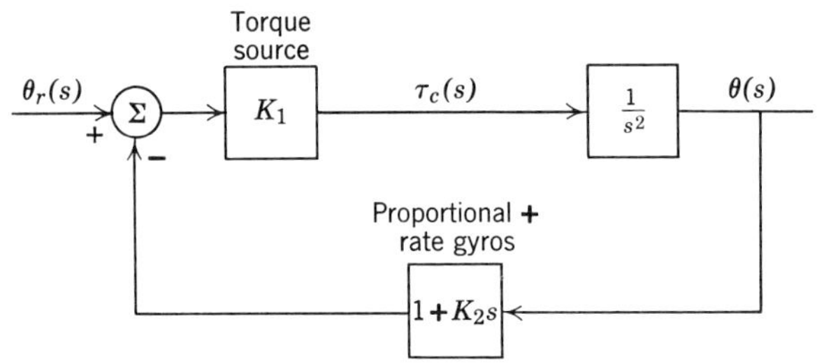

Figure 12.2 Single-axis attitude-control system.

is neglected in comparison with u_1 in (12.1), then

$$\dot{\omega} = -\frac{k}{J_v R_a} u_1, \tag{12.2}$$

which is *decoupled* completely from wheel speed. Thus the system may be modeled by Figure 12.2.

Notice that wheel speed is not shown in Figure 12.2, although (12.1) yields

$$\dot{\Omega} = \left(\frac{1}{J_v} + \frac{1}{J_w}\right)\frac{k}{R_a} u_1, \tag{12.3}$$

neglecting back emf. As was discussed in Chapter 11, the wheel speed is an uncontrollable output, if either θ or $\dot{\theta}$ or both are also outputs. Moreover, if back emf is neglected, the wheel speed is decoupled from the θ and ω responses. Finally, state feedback cannot make output-controllable a plant that is not originally output-controllable. Thus, there is no point in feeding back wheel speed for u, and hence the wheel speed is omitted from Figure 12.2. In practice, Ω may build up, and "momentum dumping" must be employed (using gas jets, for example) to prevent dangerously large wheel speeds from occurring, as was discussed in Chapter 11.

We wish to select the design parameters K_1 and K_2 so that the spacecraft attitude θ is controlled in a desirable manner by the reference attitude input θ_r. The quantity $e = \theta_r - \theta$ is defined as the *system error.* It is a measure of how well the actual attitude is following the desired (reference) attitude. Ideally, perhaps, we would like $e(t)$ to be zero at *all time instants*. This would mean that the spacecraft *always* maintains the desired orientation. This utopian design is generally not possible, however. Instantaneous response at the output requires very large input signals to the plant (doublets, in this case). Such large inputs are not only difficult to realize physically but generally result in saturation and other nonlinear effects, thus voiding our linear model. It is therefore unreasonable to try to make $e(t)$ zero (or even just "very small") at all time instants.

Moreover, making $e(t)$ small at only some time instants does not seem appropriate either. Which instants are most important? Both transient (small times) and steady-state (large time) are important, in general. It appears that what we need is a performance criterion that would avoid large errors at any time and would average effects of errors at all times to produce a single numerical indication of system quality. This suggests, perhaps, using an integral of the error as a criterion.

Consider as a performance criterion

$$I_1 = \int_0^\infty e(t)\, dt. \tag{12.4}$$

Such a criterion permits cancellation of contributions to I_1 due to large negative errors by contributions to I_1 due to large positive errors. Thus I_1 could be small, yet $e(t)$ could be large for some time interval. This cancellation difficulty could be overcome by using

$$I_2 = \int_0^\infty |e(t)|\, dt. \tag{12.5}$$

This criterion is actually used in practice. Unfortunately, it is difficult to treat analytically in closed form. A mathematically more attractive criterion is

$$I_3 = \int_0^\infty e^2(t)\, dt. \tag{12.6}$$

Notice that this criterion produces a large penalty for large errors, whether positive or negative. Moreover, it includes both transient and steady-state effects. In fact, if $e(t)$ does not approach zero as $t \to \infty$, this integral diverges.

Let us proceed with our example, using I_3 as the performance index. To evaluate the integral (as a function of K_1 and K_2), $\theta_r(t)$ must be specified. This is a drawback of that type of performance index. It produces a design that is "tuned" to a particular input. In applications, therefore, we must examine the criterion for all inputs that we reasonably expect will occur.

Suppose, in the case at hand, θ_r is the unit step function. We can then evaluate $I_3\,(K_1,K_2)$. For this case,

$$E(s) = \frac{s + K_1K_2}{s^2 + K_2K_1s + K_1} \tag{12.7}$$

By inversion of the transform, we could find $e(t)$. Squaring this function and integrating, we would obtain $I_3\ (K_1,K_2)$. This is a lengthy and tedious procedure. Some shorter methods exist for finding I_3. These will be discussed later in this chapter. Using these methods, we find that

$$I_3(K_1, K_2) = \frac{1}{2K_1K_2} + \frac{K_2}{2} \tag{12.8}$$

The values of K_1 and K_2 that minimize I_3 are found by setting the partial derivatives to zero:

$$\frac{\partial I_3}{\partial K_1} = -\frac{1}{2K_1{}^2K_2} = 0$$

$$\frac{\partial I_3}{\partial K_2} = -\frac{1}{2K_1K_2{}^2} + \tfrac{1}{2} = 0. \tag{12.9}$$

These equations cannot be solved simultaneously for *finite* values of K_1 and K_2. If $K_1 \to \infty$ and $K_2 \to 0$ such that $K_1K_2{}^2 = 1$, the equations are solved.

It appears we have come full circle, returning to our "infinite gain" solution with which we began this discussion. What is the source of the difficulty? We are, in effect, requiring infinite control torque τ_c by requiring infinite gains K_1 and K_2. Thus constraining the torque permitted should result in a more acceptable design. An appropriate way to constrain the torque when using the integral square error as an index is to append a term proportional to the integral square value of the torque. This term penalizes us heavily for large torques. Suppose we consider

$$I(K_1, K_2) = \int_0^\infty (e^2 + \rho\tau_c{}^2)\, dt, \tag{12.10}$$

where ρ is a constant. The parameter ρ represents a weighting of the relative importance of the integral square error and integral square control. For ρ much smaller than unity, errors will be penalized more than control effort, leading to large values of τ_c producing small values of e. This leads to the "infinite gain" case discussed above for $\rho \to 0$. Conversely, if $\rho >> 1$ the designer is penalizing control effort more heavily than system error. In practice, the designer chooses some compromise, based perhaps on the nature of the time response $\theta(t)$, the error $e(t)$, and the torque as a function of time, $\tau_c(t)$. More will be said about the selection of ρ later in this chapter. Evaluating I

(by methods to be discussed later), we obtain

$$I = \frac{1}{2K_1K_2} + \frac{K_2}{2} + \rho\,\frac{K_1}{2K_2} \tag{12.11}$$

Minimizing,

$$\frac{\partial I}{\partial K_1} = -\frac{1}{2K_1{}^2K_2} + \frac{\rho}{2K_2} = 0 \qquad \frac{\partial I}{\partial K_2} = -\frac{1}{2K_1K_2{}^2} + \frac{1}{2} - \frac{\rho K_1}{2K_2{}^2} = 0.$$

Solving these equations, we obtain

$$K_1 = \frac{1}{\sqrt{\rho}}, \quad K_2 = \sqrt{2\sqrt{\rho}} \tag{12.12}$$

(Note that as $\rho \to 0$, $K_1 \to \infty$ and $K_2 \to 0$, as discussed above.)

Thus K_1 and K_2 are determined if ρ is known.

12.2 Optimization Using Parseval's Theorem

In Section 12.1 we introduced the quadratic performance index. The index as presented in that section has the general form

$$I = \int_0^\infty \sum_{i=1}^{k} f_i^2(t)\,dt, \tag{12.13}$$

where the $f_i(t)$ are various signals of importance. In the example discussed above, $f_1 = e$ and $f_2 = \sqrt{\rho}\tau_c$. In this section, we discuss a method of finding I in terms of the system design parameters. The method involves Parseval's theorem and a table.

Consider a term in the quadratic index,

$$I_j = \int_0^\infty f_j^2(t)\,dt. \tag{12.14}$$

Assuming $f_j(t)$ to be Laplace-transformable, we may write

$$I_j = \int_0^\infty f_j(t)\cdot f_j(t)\,dt$$

$$= \int_0^\infty f_j(t)\left[\frac{1}{2\pi j}\int_{c-j\infty}^{c+j\infty} F_j(s)\epsilon^{st}\,ds\right]dt. \tag{12.15}$$

Let us suppose that it is possible to interchange the order of performing the two integrations in (12.15):

$$I_j = \frac{1}{2\pi j}\int_{c-j\infty}^{c+j\infty} F_j(s)\left[\int_0^{\infty} f_j(t)\epsilon^{st}\,dt\right] ds$$

$$= \frac{1}{2\pi j}\int_{c-j\infty}^{c+j\infty} F_j(s)F_j(-s)\,ds. \tag{12.16}$$

Hence

$$I_j = \int_0^{\infty} f_j^2(t)\,dt = \frac{1}{2\pi j}\int_{c-j\infty}^{c+j\infty} F_j(s)F_j(-s)\,ds. \tag{12.17}$$

This result is known as Parseval's theorem. A table[1] has been obtained for evaluating the frequency-domain integral on the right-hand side of (12.17). For linear, lumped, time-invariant systems having inputs with rational transforms, $F_j(s)$ will be a ratio of polynomials in s:

$$F_j(s) = \frac{c(s)}{d(s)} = \frac{c_{n-1}s^{n-1} + \cdots + c_0}{d_n s^n + \cdots + d_0}, \tag{12.18}$$

where n is the order of the denominator polynomial. A few integrals for $F_j(s)$ of this form are tabulated in Table 12.1. The results are given in terms of the coefficients of the numerator and denominator polynomials. The results given in Table 12.1 assume that the system is stable. Thus, stability of the final design must be checked in each application.

Example 12-1: For the attitude-control example of Figure 12.2, (12.7) states that from $\theta_r = 1/s$:

$$E(s) = \frac{s + K_1K_2}{s^2 + K_1K_2 s + K_1}. \tag{12.19}$$

Noting that

$$\int_0^{\infty} e^2(t)\,dt = \frac{1}{2\pi j}\int_{c-j\infty}^{c+j\infty} E(s)E(-s)\,ds, \tag{12.20}$$

we may apply Table 12.1 to the evaluation of the integral squared

[1]See G. C. Newton, Jr., L. A. Gould, and J. F. Kaiser, *Analytic Design of Linear Feedback Controls*, Wiley, New York, 1957, pp. 372–381.

error by noting that, for $E(s)$,

$$\begin{aligned} c_0 &= K_1K_2 \qquad & d_0 &= K_1 \\ c_1 &= 1 & d_1 &= K_1K_2 \\ & & d_2 &= 1 \end{aligned} \tag{12.21}$$

Since $n = 2$, we look up I_2 in Table 12.1:

$$I_2 = \int_0^\infty e^2\,dt = \frac{K_1 + K_1^2K_2^2}{2K_1^2K_2} = \frac{1}{2K_1K_2} + \frac{K_2}{2}, \tag{12.22}$$

as was given in (12.8). Equation 12.11 is obtained in a similar manner. ▲

Table 12.1 Evaluation of an Integral

$$I_n = \frac{1}{2\pi j}\int_{-j\infty}^{j\infty} \frac{c(s)c(-s)}{d(s)d(-s)}\,ds$$

$$c(s) = c_{n-1}s^{n-1} + \cdots + c_0$$
$$d(s) = d_ns^n + \cdots + d_0$$

$$I_1 = \frac{c_0^2}{2d_0d_1}$$

$$I_2 = \frac{c_1^2d_0 + c_0^2d_2}{2d_0d_1d_2}$$

$$I_3 = \frac{c_2^2d_0d_1 + (c_1^2 - 2c_0c_2)d_0d_3 + c_0^2d_2d_3}{2d_0d_3(-d_0d_3 + d_1d_2)}$$

$$I_4 = \frac{c_3^2(-d_0^2d_3 + d_0d_1d_2) + (c_2^2 - 2c_1c_3)d_0d_1d_4}{\Delta} + \frac{(c_1^2 - 2c_0c_2)d_0d_3d_4 + c_0^2(-d_1d_4^2 + d_2d_3d_4)}{\Delta}$$

where

$$\Delta = 2d_0d_4(-d_0d_3^2 - d_1^2d_4 + d_1d_2d_3)$$

In principle, these tables can be used to express any quadratic performance index in terms of the design parameters. The optimum parameters could then be obtained by setting to zero the partial derivatives of the index with respect to the parameters and by solving the resulting equations simultaneously. This procedure quickly becomes unwieldy as the order of the system and the number of design parameters is increased. Thus, we naturally seek computation-oriented procedures for accomplishing parameter optimization.

One possibility is discussed in the next section. Chapter 13 presents further discussion of this problem.

12.3 The Optimum State Regulator

In this section, we consider a special case of the general quadratic performance index of (12.13) in which the $f_i(t)$ are either weighted plant state variables $x_i(t)$ or weighted plant inputs $u_j(t)$:

$$I = \int_0^{\infty} \left(\sum_{i=1}^{n} q_i x_i^2(t) + \sum_{j=1}^{m} r_j u_j^2(t) \right) dt. \tag{12.23}$$

Using the matrix representation for state variables and inputs,

$$\mathbf{x} = \begin{bmatrix} x_1(t) \\ x_2(t) \\ \cdot \\ \cdot \\ \cdot \\ x_n(t) \end{bmatrix}, \qquad \mathbf{u} = \begin{bmatrix} u_1(t) \\ u_2(t) \\ \cdot \\ \cdot \\ \cdot \\ u_m(t) \end{bmatrix}, \tag{12.24}$$

(12.23) can be written as

$$I = \int_0^{\infty} (\mathbf{x}^T\mathbf{Q}\mathbf{x} + \mathbf{u}^T\mathbf{R}\mathbf{u})\, dt, \tag{12.25}$$

where

$$\mathbf{Q} = \begin{bmatrix} q_1 & & & & 0 \\ & q_2 & & & \\ & & \cdot & & \\ & & & \cdot & \\ & & & & \cdot \\ 0 & & & & q_n \end{bmatrix}, \qquad \mathbf{R} = \begin{bmatrix} r_1 & & & & 0 \\ & r_2 & & & \\ & & \cdot & & \\ & & & \cdot & \\ & & & & \cdot \\ 0 & & & & r_n \end{bmatrix}. \tag{12.26}$$

A more general index can be considered where $\mathbf{Q}$ and $\mathbf{R}$ have off-diagonal entries. However, the diagonal case is usually easier to understand physically, and thus is most often used in practice. For the index to be physically meaningful, the contribution of any trajectory to $\mathbf{x}^T\mathbf{Q}\mathbf{x}$ must be non-negative. Thus we require $q_i \geqslant 0$ for all i. Furthermore, the cost of any control should be strictly positive, and thus we require $r_i > 0$ for all i.

With q_i only non-negative, the possibility exists that the optimum system could be unstable. This can happen if an unstable mode does not affect the performance index. Since $\mathbf{x}^T\mathbf{Q}\mathbf{x}$ with $\mathbf{Q}$ diagonal is a sum of squares, an unstable mode will not affect I if and only if that mode is not observable from the state variables appearing in $\mathbf{x}^T\mathbf{Q}\mathbf{x}$, considering those state variables as system outputs. To avoid these difficulties, we shall require that all state variables are observable from those x_i's in $q_i x_i^2$ corresponding to $q_i \neq 0$. Furthermore, we shall require the system to be completely state controllable. If this were not true, unstable uncontrollable modes may appear in the resulting optimum system. Clearly this is undesirable physically. It can be shown that all the above conditions guarantee that the resulting optimal system is asymptotically stable.

The plant is described in state form by

$$\dot{\mathbf{x}} = \mathbf{A}\mathbf{x} + \mathbf{B}\mathbf{u}, \quad \mathbf{x}(0) = \mathbf{x}_0. \tag{12.27}$$

We shall further restrict the problem by requiring $\mathbf{u}$, the control input to the plant, to be constructed from the state variables by linear, time-invariant feedback,

$$\mathbf{u} = \mathbf{K}\mathbf{x}, \tag{12.28}$$

where $\mathbf{K}$ is an $m \times n$ matrix of constants. The problem then is one of parameter optimization: Choose $\mathbf{K}$ to minimize the quadratic performance index I of (12.25). We see from the performance index that we are penalized for both large deviations in the state variables from zero and for the use of large control inputs. A system designed in this manner is called an optimum state regulator.

The derivation of the optimal $\mathbf{K}$ is straightforward, although it involves some matrix calculus. The details are deferred until Section 12.4. The result is the following: the optimal matrix $\mathbf{K}$ minimizing the index of (12.25) is given by

$$\mathbf{K} = -\mathbf{R}^{-1}\mathbf{B}^T\mathbf{P}, \tag{12.29}$$

where $\mathbf{R}$ is given in the performance index (12.25) (notice that $\mathbf{R}^{-1}$ always exists), and $\mathbf{B}$ is given in the plant model (12.27). The matrix $\mathbf{P}$ is the *positive definite, symmetric* solution to the *matrix Riccati equation*

$$\mathbf{A}^T\mathbf{P} + \mathbf{P}\mathbf{A} - \mathbf{P}\mathbf{B}\mathbf{R}^{-1}\mathbf{B}^T\mathbf{P} + \mathbf{Q} = \mathbf{0}, \tag{12.30}$$

with $\mathbf{A}$ from (12.27) and $\mathbf{Q}$ from (12.25). Since $\mathbf{P}$ is an $n \times n$ matrix, (12.30) represents n^2 simultaneous, nonlinear (quadratic), algebraic, equations for the n^2 elements of **P**. However, since **P** is symmetric, not all of these equations are different. In fact, since there are n^2 elements in **P**, and n elements on the diagonal, there are (n^2-n) off-diagonal elements, of which half are distinct. Thus, there are

$$\frac{(n^2-n)}{2}+n=\frac{n(n+1)}{2} \tag{12.31}$$

distinct elements of the P matrix to be found from (12.30). Since the equations are quadratic, many different sets of roots occur. It can be shown that there is a unique *positive definite* solution, and that this **P** determines the optimum **K**. The direct solution of these nonlinear equations is difficult, even with machine computation. The isolation of the desired positive definite solution is especially troublesome. Several indirect methods for machine computation have been suggested. See Section 12.4.

It is interesting to observe that the optimum parameter matrix **K** is independent of the initial conditions $\mathbf{x}_0$. Thus **K** can be computed once and for all, and will be optimum no matter what the initial conditions are. A block diagram of the optimum regulator is given in Figure 12.3.

As noted above, the optimum state regulator can be shown to be asymptotically stable, that is, the eigenvalues of $\mathbf{A}-\mathbf{B}\mathbf{R}^{-1}\mathbf{B}^T\mathbf{P}$ are always in the left-half s-plane.

Example 12-2. We again consider the single-axis attitude control example of Figure 12.2, with the performance of (12.10), involving both squared error and squared control. It may appear at first glance that

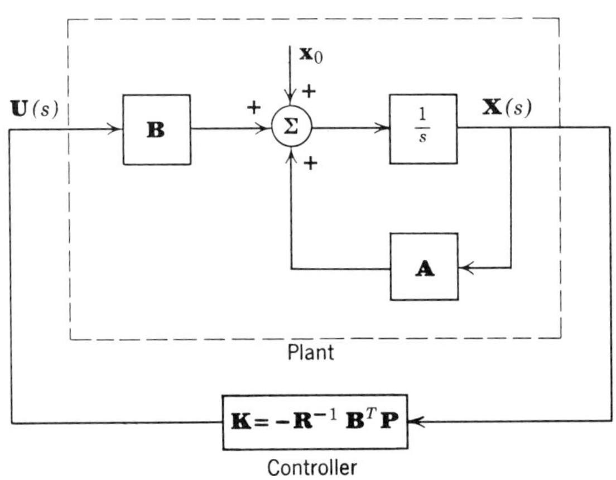

Figure 12.3 Block diagram of optimum state-regulator system.

this is not a state regulator problem at all, since there is an external system input and no initial conditions, whereas the state regulator requires initial conditions but no external input. However, by an appropriate choice of state variables, this problem can be put into the state regulator framework. (It should be strongly emphasized that such a transformation is not always possible for all problems!) Suppose we choose as state variables

$$\begin{aligned} x_1 &= e = \theta_r - \theta, \\ x_2 &= \dot{e} = -\dot{\theta}, \end{aligned} \tag{12.32}$$

assuming θ_r is a unit step function. The state equations become

$$\begin{aligned} \dot{x}_1 &= x_2 \\ \dot{x}_2 &= u, \end{aligned} \tag{12.33}$$

where $u = -\tau_c$. The performance index of (12.10) becomes, taking $\rho = \frac{1}{4}$ for concreteness,

$$I = \int_0^\infty (x_1^2 + \tfrac{1}{4}u^2)\, dt. \tag{12.34}$$

This index has the standard form of (12.25) with

$$\mathbf{Q} = \begin{pmatrix} 1 & 0 \\ 0 & 0 \end{pmatrix}, \qquad \mathbf{R} = \tfrac{1}{4}. \tag{12.35}$$

We see that the index (12.34) controls only the state variable x_1, with no penalty provided for deviations of x_2 from zero. $\mathbf{R}$ is 1×1 since u is a scalar.

From (12.33),

$$\mathbf{A} = \begin{pmatrix} 0 & 1 \\ 0 & 0 \end{pmatrix}, \qquad \mathbf{b} = \begin{pmatrix} 0 \\ 1 \end{pmatrix}. \tag{12.36}$$

We first solve for $\mathbf{P}$ in the Riccati equation, (12.30),

$$\begin{pmatrix} 0 & 0 \\ 1 & 0 \end{pmatrix}\begin{pmatrix} p_{11} & p_{12} \\ p_{12} & p_{22} \end{pmatrix} + \begin{pmatrix} p_{11} & p_{12} \\ p_{12} & p_{22} \end{pmatrix}\begin{pmatrix} 0 & 1 \\ 0 & 0 \end{pmatrix}$$
$$-4\begin{pmatrix} p_{11} & p_{12} \\ p_{12} & p_{22} \end{pmatrix}\begin{pmatrix} 0 \\ 1 \end{pmatrix}(0 \quad 1)\begin{pmatrix} p_{11} & p_{12} \\ p_{12} & p_{22} \end{pmatrix} + \begin{pmatrix} 1 & 0 \\ 0 & 0 \end{pmatrix} = \begin{pmatrix} 0 & 0 \\ 0 & 0 \end{pmatrix}. \tag{12.37}$$

Carrying out the indicated matrix operations, we obtain

$$\begin{aligned} 1-4p_{12}{}^2&=0\\ p_{11}-4p_{12}p_{22}&=0\\ 2p_{12}-4p_{22}{}^2&=0. \end{aligned} \tag{12.38}$$

These are solved rather easily in this case for the positive definite matrix

$$\mathbf{P}=\begin{bmatrix}1 & \frac{1}{2}\\ \frac{1}{2} & \frac{1}{2}\end{bmatrix}. \tag{12.39}$$

From (12.29),

$$\mathbf{K}=-\mathbf{R}^{-1}\mathbf{B}^T\mathbf{P}=-4(0 \quad 1)\begin{pmatrix}1 & \frac{1}{2}\\ \frac{1}{2} & \frac{1}{2}\end{pmatrix} \tag{12.40}$$

Thus

$$u=-2x_1-2x_2. \tag{12.41}$$

Translating this result to the system of Figure 12.2,

$$\begin{aligned} -\tau_c&=-2e-2\dot{e}\\ &=-2(\theta_r-\theta)-2(-\dot{\theta}). \end{aligned} \tag{12.42}$$

From Figure 12.1,

$$\tau_c=K_1(\theta_r-\theta)+K_1K_2(-\dot{\theta}). \tag{12.43}$$

Thus, comparing (12.42) and (12.43):

$$K_1=2, \qquad K_2=1. \tag{12.44}$$

This is exactly the result obtained previously, (12.12), with $\rho=\frac{1}{4}$. ▲

12.4★ A Derivation of the Riccati Equation

The optimum state regulator problem has been studied thoroughly by R. E. Kalman.[2] The results in this section were obtained first by

[2]See, for example, R. E. Kalman, "Contributions to the Theory of Optimal Control," *Bol. Soc. Mat. Mex.*, **5**, 102–119 (1960).

him in the context of a general optimization of $\mathbf{u}$. It turns out that a state feedback solution $\mathbf{u} = \mathbf{Kx}$ is obtained even if a general $\mathbf{u}$ is assumed. Since then, many alternate methods of deriving the Riccati equation have been proposed. We shall follow a procedure suggested by Luenberger.[3]

Given a plant described by

$$\dot{\mathbf{x}} = \mathbf{Ax} + \mathbf{Bu}, \qquad \mathbf{x}(0) = \mathbf{x}_0 \tag{12.45}$$

and a performance index

$$I = \int_0^\infty (\mathbf{x}^T\mathbf{Qx} + \mathbf{u}^T\mathbf{Ru})\, dt, \tag{12.46}$$

where $\mathbf{Q}$ is symmetric and positive semidefinite, $\mathbf{R}$ is symmetric and positive definite.[4] Furthermore, with $\mathbf{Q}$ written as $\mathbf{C}^T\mathbf{C}$, we require (12.45) and $\mathbf{y} = \mathbf{Cx}$ to be a completely observable system.[5] Moreover, we require (12.45) to be a completely state controllable system. Finally, we restrict $\mathbf{u}$ to the form

$$\mathbf{u} = \mathbf{Kx}. \tag{12.47}$$

The problem, then, is to find $\mathbf{K}$ to minimize the index (12.46) subject to the above conditions.

We solve this problem by an elementary application of *calculus of variations*. We observe that $\mathbf{u}$ can be written as the optimum $\mathbf{u}$, denoted by $\mathbf{u}^0$, plus another function $\eta\mathbf{u}'$, where η is a scalar parameter, and $\mathbf{u}'$ is an arbitrary vector time function:

$$\mathbf{u} = \mathbf{u}^0 + \eta\mathbf{u}' = \mathbf{Kx} + \eta\mathbf{u}'. \tag{12.48}$$

If this $\mathbf{u}$ is substituted into the index I, it is clear that, as a function of η, I has a minimum for $\eta = 0$ (this is because $\mathbf{u}^0$ minimizes I). Thus

$$\frac{dI(\eta)}{d\eta} = 0 \quad \text{at} \quad \eta = 0. \tag{12.49}$$

[3]D. G. Luenberger, "A New Derivation of the Quadratic Loss Control Equation," *IEEE Trans. on Automatic Control*, pp. 202–203, April 1965.

[4]The choice of $\mathbf{Q}$ and $\mathbf{R}$ to be diagonal with $q_i \geqslant 0$ and $r_i > 0$ as in Sec. 12.3 is a special case of $\mathbf{Q}$ positive semidefinite and $\mathbf{R}$ positive definite.

[5]A positive semidefinite matrix can always be factored as $\mathbf{C}^T\mathbf{C}$. See A. S. Householder, *The Theory of Matrices in Numerical Analysis*, Blaisdell Publishing Company, Waltham, Mass., 1964, p. 13.

Our method of solution is to substitute into I the $\mathbf{u}$ given in (12.48). We then differentiate as in (12.49). In performing this derivative, we must be aware that $\mathbf{x}$ is a function of η as well as t because of the form selected for $\mathbf{u}$:

$$\frac{d\mathbf{x}}{dt} = \dot{\mathbf{x}} = \mathbf{Ax} + \mathbf{Bu} = \mathbf{Ax} + \mathbf{BKx} + \mathbf{B}\eta\mathbf{u}'$$

$$\dot{\mathbf{x}} = (\mathbf{A} + \mathbf{BK})\mathbf{x} + \eta\mathbf{Bu}' \tag{12.50}$$

or

$$\dot{\mathbf{x}} = \bar{\mathbf{A}}\mathbf{x} + \eta\mathbf{Bu}', \tag{12.51}$$

where

$$\bar{\mathbf{A}} = \mathbf{A} + \mathbf{BK}. \tag{12.52}$$

The solution of (12.51) is $\mathbf{x}(t, \eta)$, a function of both t and η.

We now proceed with the differentiation:

$$\frac{dI}{d\eta} = \int_0^\infty \frac{d}{d\eta}[\mathbf{x}^T\mathbf{Qx} + (\mathbf{x}^T\mathbf{K}^T + \eta\mathbf{u}'^T)\mathbf{R}(\mathbf{Kx} + \eta\mathbf{u}')]\,dt. \tag{12.53}$$

Some terms can be combined in (12.53), yielding

$$\frac{dI}{d\eta} = \int_0^\infty \frac{d}{d\eta}[\mathbf{x}^T\bar{\mathbf{Q}}\mathbf{x} + 2\eta\mathbf{x}^T\mathbf{K}^T\mathbf{Ru}' + \eta^2\mathbf{u}'^T\mathbf{Ru}']\,dt, \tag{12.54}$$

where

$$\bar{\mathbf{Q}} = \mathbf{Q} + \mathbf{K}^T\mathbf{RK}. \tag{12.55}$$

We consider the derivatives term by term.

$$\frac{d}{d\eta}(\mathbf{x}^T\bar{\mathbf{Q}}\mathbf{x}) = \frac{d}{d\eta}\left(\sum_{j=1}^n\sum_{i=1}^n \bar{q}_{ij}x_ix_j\right) = \sum_{j=1}^n\sum_{i=1}^n \bar{q}_{ij}\left(x_i\frac{\partial x_j}{\partial\eta} + \frac{\partial x_i}{\partial\eta}x_j\right)\cdot \tag{12.56}$$

But

$$\sum_{j=1}^n\sum_{i=1}^n \bar{q}_{ij}x_i\frac{\partial x_j}{\partial\eta} = \sum_{j=1}^n\sum_{i=1}^n \bar{q}_{ij}\frac{\partial x_i}{\partial\eta}x_j, \tag{12.57}$$

since, on interchanging the dummy indices i and j in the second summation, and using the fact that $\bar{q}_{ij} = \bar{q}_{ji}$ (note from (12.55) that $\bar{Q}$ is symmetric), the two sums are seen to be equal. Thus, in matrix form,

$$\frac{d}{d\eta}(\mathbf{x}^T\bar{\mathbf{Q}}\mathbf{x}) = 2\mathbf{x}^T\bar{\mathbf{Q}}\frac{\partial\mathbf{x}}{\partial\eta}\cdot \tag{12.58}$$

In a similar manner it can be shown that, since $\mathbf{u}'$ is independent of η,

$$\frac{d}{d\eta}\,(2\eta\mathbf{x}^T\mathbf{K}^T\mathbf{R}\mathbf{u}') = 2\mathbf{x}^T\mathbf{K}^T\mathbf{R}\mathbf{u}' + 2\eta\frac{\partial\mathbf{x}^T}{\partial\eta}\,\mathbf{K}^T\mathbf{R}\mathbf{u}'. \tag{12.59}$$

Furthermore,

$$\frac{d}{d\eta}\,(\eta^2\mathbf{u}'^T\mathbf{R}\mathbf{u}') = 2\eta\mathbf{u}'^T\mathbf{R}\mathbf{u}'. \tag{12.60}$$

We recall from (12.49) that these derivatives are to be evaluated at $\eta = 0$. Thus the second term on the right-hand side of (12.59) vanishes, as does the whole derivative given in (12.60). Equation 12.54 then becomes

$$\left.\frac{dI}{d\eta}\right|_{\eta=0} = \int_0^\infty \left[2\mathbf{x}^T\bar{\mathbf{Q}}\,\frac{\partial\mathbf{x}}{\partial\eta} + 2\mathbf{x}^T\mathbf{K}^T\mathbf{R}\mathbf{u}'\right]_{\eta=0} dt = 0. \tag{12.61}$$

The partial derivative of $\mathbf{x}$ with respect to η can be found from the solution of (12.51). Using the results established in Chapter 7, the solution to (12.51) is

$$\mathbf{x} = \boldsymbol{\epsilon}^{\bar{\mathbf{A}}t}\,\mathbf{x}_0 + \int_0^t \boldsymbol{\epsilon}^{\bar{\mathbf{A}}(t-\tau)}\mathbf{B}\eta\mathbf{u}'(\tau)\,d\tau. \tag{12.62}$$

Forming the partial derivative at $\eta = 0$ yields

$$\left.\frac{\partial\mathbf{x}}{\partial\eta}\right|_{\eta=0} = \int_0^t \boldsymbol{\epsilon}^{\bar{\mathbf{A}}(t-\tau)}\,\mathbf{B}\mathbf{u}'(\tau)\,d\tau. \tag{12.63}$$

Also $\mathbf{x}$ at $\eta = 0$ is obtained from (12.62):

$$\mathbf{x}|_{\eta=0} = \boldsymbol{\epsilon}^{\bar{A}t}\mathbf{x}_0. \tag{12.64}$$

Substituting (12.63) and (12.64) into (12.61) yields

$$\int_0^\infty \left[\mathbf{x}_0{}^T\boldsymbol{\epsilon}^{\bar{\mathbf{A}}^Tt}\bar{\mathbf{Q}}\int_0^t \boldsymbol{\epsilon}^{\bar{\mathbf{A}}(t-\tau)}\mathbf{B}\mathbf{u}'(\tau)\,d\tau + \mathbf{x}_0{}^T\boldsymbol{\epsilon}^{\bar{\mathbf{A}}^Tt}\,\mathbf{K}^T\mathbf{R}\mathbf{u}'(t)\right] dt = 0. \tag{12.65}$$

We next change the order of integration in the first term of (12.64),

$$\mathbf{x}_0{}^T\int_0^\infty \left[\int_\tau^\infty \boldsymbol{\epsilon}^{\bar{\mathbf{A}}^Tt}\bar{\mathbf{Q}}\boldsymbol{\epsilon}^{\bar{\mathbf{A}}(t-\tau)}\mathbf{B}\,dt\right]\mathbf{u}'(\tau)\,d\tau + \mathbf{x}_0{}^T\int^\infty \boldsymbol{\epsilon}^{\bar{\mathbf{A}}^T\tau}\,\mathbf{K}^T\mathbf{R}\mathbf{u}'(\tau)\,d\tau = 0. \tag{12.66}$$

We then replace $t - \tau$ by σ in the t integral in the first term, yielding

$$\mathbf{x}_0{}^T\int_0^\infty \boldsymbol{\epsilon}^{\bar{\mathbf{A}}^T\tau}\left(\int_0^\infty \boldsymbol{\epsilon}^{\bar{\mathbf{A}}^T\sigma}\,\bar{\mathbf{Q}}\boldsymbol{\epsilon}^{\bar{\mathbf{A}}\sigma}\mathbf{B}\,d\sigma + \mathbf{K}^T\mathbf{R}\right)\mathbf{u}'(\tau)\,d\tau = 0. \tag{12.67}$$

This expression must equal zero for *all* $\mathbf{x}_0^T$ and $\mathbf{u}'(\tau)$. Clearly, a sufficient condition is that the expression in the parentheses be zero. It can be shown that this is necessary also. Thus

$$\mathbf{RK} + \mathbf{B}^T \int_0^\infty \boldsymbol{\epsilon}^{\bar{\mathbf{A}}^T\sigma} \overline{\mathbf{Q}} \boldsymbol{\epsilon}^{\bar{\mathbf{A}}\sigma} d\sigma = 0. \tag{12.68}$$

This is a complicated integral equation for **K** (recall that **K** appears in $\bar{\mathbf{A}}$ and $\overline{\mathbf{Q}}$). A simpler expression can be obtained by means of a trick. Consider the matrix differential equation

$$\frac{d\mathbf{Z}}{dt} = \bar{\mathbf{A}}^T\mathbf{Z} + \mathbf{Z}\bar{\mathbf{A}}, \; \mathbf{Z}(0) = \overline{\mathbf{Q}}. \tag{12.69}$$

Direct substitution verifies that

$$\mathbf{Z} = \boldsymbol{\epsilon}^{\bar{\mathbf{A}}^T t} \overline{\mathbf{Q}} \boldsymbol{\epsilon}^{\bar{\mathbf{A}}t} \tag{12.70}$$

is the solution to (12.69). However, integration of both sides of (12.69) yields

$$\mathbf{Z}(\infty) - \overline{\mathbf{Q}} = \bar{\mathbf{A}}^T \left(\int_0^\infty \mathbf{Z}\, dt \right) + \left(\int_0^\infty \mathbf{Z}\, dt \right) \bar{\mathbf{A}}. \tag{12.71}$$

We shall suppose the system of (12.69) to be asymptotically stable, so that $\mathbf{Z}(\infty) = \mathbf{0}$. Furthermore, by (12.70):

$$\int_0^\infty \mathbf{Z}\, dt = \int_0^\infty \boldsymbol{\epsilon}^{\bar{\mathbf{A}}^T t} \overline{\mathbf{Q}} \boldsymbol{\epsilon}^{\bar{\mathbf{A}}t}\, dt. \tag{12.72}$$

Thus, (12.71) becomes

$$-\overline{\mathbf{Q}} = \bar{\mathbf{A}}^T\mathbf{P} + \mathbf{P}\bar{\mathbf{A}}, \tag{12.73}$$

where, by definition,

$$\mathbf{P} = \int_0^\infty \mathbf{Z}\, dt. \tag{12.74}$$

Thus, (12.68) yields

$$\mathbf{K} = -\mathbf{R}^{-1}\mathbf{B}^T\mathbf{P}, \tag{12.75}$$

where

$$-\overline{\mathbf{Q}} = \bar{\mathbf{A}}^T\mathbf{P} + \mathbf{P}\bar{\mathbf{A}}. \tag{12.76}$$

Substitution for $\bar{\mathbf{A}}$, $\overline{\mathbf{Q}}$, and **K** in (12.76) reveals this equation to be the matrix Riccati equation,

$$\mathbf{A}^T\mathbf{P} + \mathbf{P}\mathbf{A} - \mathbf{P}\mathbf{B}\mathbf{R}^{-1}\mathbf{B}^T\mathbf{P} + \mathbf{Q} = \mathbf{0}. \tag{12.77}$$

It is clear from (12.72) and (12.74) that $\mathbf{P}$ is symmetric. Furthermore, it may be shown that a positive definite solution of (12.71) always exists provided the conditions listed at the beginning of this section are met. Furthermore, the optimum system is stable, that is, the eigenvalues of $\bar{\mathbf{A}}$ have negative real parts.

Unfortunately, solution of (12.77) is not easy in practical applications. Kalman has suggested the following approach. Consider the matrix differential equation

$$\mathbf{A}^T\mathbf{P}(t)+\mathbf{P}(t)\mathbf{A}-\mathbf{P}(t)\mathbf{B}\mathbf{R}^{-1}\mathbf{B}^T\mathbf{P}(t)+\mathbf{Q}=-\dot{\mathbf{P}}(t), \qquad (12.78)$$

with the boundary condition

$$\mathbf{P}(T)=\mathbf{0}. \qquad (12.79)$$

Equation 12.78 is solved *backward* in time, starting with $\mathbf{P}(T)=\mathbf{0}$. It can be shown that the solution of (12.78) approaches a constant steady-state value $\mathbf{P}$, which is the desired positive-definite solution of (12.77). Thus, as the starting time $T\to\infty$, the transient due to $\mathbf{P}(T)$ dies out, and we approach $\mathbf{P}(0)=\mathbf{P}$, the desired Riccati solution.

12.5 Design Considerations Using Parameter Optimization and Quadratic Indices

We now have some experience in the optimization of linear systems with quadratic indices. In particular, we have studied the optimum state regulator problem. We shall now consider optimum system design using quadratic indices.

There are two major difficulties involved in design of systems via minimization of quadratic indices:

1 How are the weighting terms $\mathbf{Q}$ and $\mathbf{R}$ selected?

2 How is the optimum control computed for complex, high-order systems?

We have touched briefly on the second problem above. More will be said in Chapter 13. We shall concentrate our discussion here on the first problem, the selection of the weighting terms. For additional discussion, see the Problems.

It is difficult to give general guidelines for the selection of weighting matrices outside the context of a specific problem. The physical meaning of the variables in the index, and hence their relative importance, can be discussed only within the context of particular examples. The diagonal entries in Q, q_{ii}, indicate the relative importance to be assigned to the "energies" represented by the integrals of the squares of the state variables. The diagonal entries of $\mathbf{R}$ indicate the relative importance of control energies to be supplied by the various inputs.

Another related approach to choosing the weighting factors is to observe that minimizing the index of (12.10) is equivalent to minimizing the integral squared error

$$\hat{I}(K_1, K_1) = \int_0^\infty e^2\,dt, \tag{12.80}$$

subject to an *equality constraint* on the integral of the square of control torque:

$$\int_0^\infty \tau_c{}^2\,dt = M, \tag{12.81}$$

where M is a *given* constant. (In many practical problems, the integral of the square of the control variable is proportional to the control energy. Thus a constraint of this type is called a *control energy constraint*.) This equivalence is established by noting that (12.81) establishes a constraint between K_1 and K_2:

$$\frac{K_1}{2K_2} = M. \tag{12.82}$$

Table 12.1 can be used to evaluate $\int_0^\infty \tau_c{}^2\,dt$. Thus the optimization of $\hat{I}$ is a constrained optimization:

$$\hat{I} = \int_0^\infty e^2\,dt = \left[\frac{1}{2K_1K_2} + \frac{K_2}{2}\right]K_1 = 2K_2M \tag{12.83}$$

Therefore,

$$\hat{I}(K_2) = \frac{1}{4MK_2{}^2} + \frac{K_2}{2}, \tag{12.84}$$

and the minimum is given by

$$\frac{d\hat{I}}{dK_2} = -\frac{1}{2MK_2{}^3} + \tfrac{1}{2} = 0, \tag{12.85}$$

or

$$K_2 = M^{-1/3},$$

$$K_1 = 2M^{2/3}. \tag{12.86}$$

The relationship between M and ρ is found by comparing (12.86) with (12.12):

$$\rho = \tfrac{1}{4}M^{-4/3}. \tag{12.87}$$

Thus the minimization of $\hat{I}$ with the control energy constraint of (12.81) is equivalent to minimizing I with $\rho = \frac{1}{4}M^{-4/3}$.

The choice of weighting parameters can also be interpreted in terms of the time responses of the outputs or, equivalently, in terms of locations of system characteristic roots. For example, the system of Figure 12.2 has as characteristic equation

$$s^2 + K_1K_2s + K_1 = 0. \tag{12.88}$$

Thus, for this system,

$$\omega_n = \sqrt{K_1} \tag{12.89}$$

$$\zeta = \tfrac{1}{2}K_2\sqrt{K_1}. \tag{12.90}$$

With optimum parameters, as in (12.12), we have

$$\omega_n = \rho^{-1/4} \tag{12.91}$$

$$\zeta = \frac{\sqrt{2}}{2}. \tag{12.92}$$

Notice that the damping ratio ζ does not change with ρ, while the system natural frequency varies as the inverse fourth root of ρ. Similar statements can be made concerning the effects of M by utilizing (12.87).

The following example illustrates some of these ideas.

Example 12-3. Consider an attitude-control system modelled by (12.1) with back emf neglected. Choosing as state variables $x_1 = \theta$, $x_2 = \dot{\theta} = \omega$, and suppressing $\dot{\Omega}$,

$$\dot{x}_1 = x_2$$

$$\dot{x}_2 = u = -\frac{k}{J_vR_a}u_1\,. \tag{12.93}$$

As shown in Fig. 12.4, we employ state feedback,

$$u = k_1x_1 + k_2x_2$$

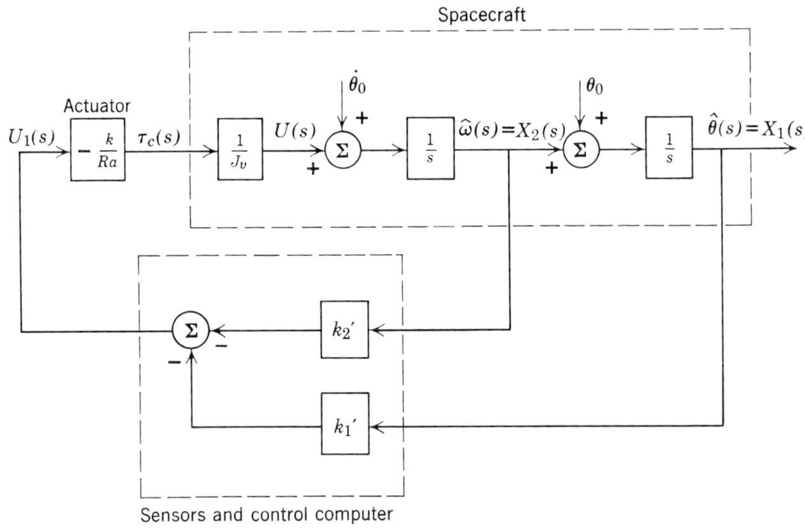

Figure 12.4 Spacecraft with linear feedback attitude control.

where

$$k_1 = \frac{k}{J_v R_a} k_1' ,$$

$$k_2 = \frac{k}{J_v R_a} k_2' . \tag{12.94}$$

In system design many goals are of interest, as has been noted above. The time responses $\theta(t)$, $\omega(t)$, and $u_1(t)$ are generally of paramount interest in attitude control, as specifications are often given in terms of acceptable maxima and decay times for θ and ω for given command inputs or disturbance inputs, and a specified maximum value of $u_1(t)$ to avoid motor saturation or the exceeding of motor ratings.

Also, since back emf has been neglected, $u_1 = i_a R_a$, and thus motor input power, $u_1 i_a$, is proportional to $u_1{}^2$. Hence

$$\frac{1}{R_a}\int_0^\infty u_1{}^2\,dt = E$$

is the actual electrical energy required for attitude control. This energy may be important in spacecraft design.

We shall design the system by minimizing a quadratic performance index

$$I = I_2 + \lambda I_1,$$

where

$$I_2 = \int_0^\infty (q_1 x_1^2 + q_2 x_2^2)\, dt$$

and

$$I_1 = \int_0^\infty u^2\, dt.$$

As discussed above, fixing λ is equivalent to fixing

$$\int_0^\infty u^2\, dt = \frac{k^2}{J_v^2 R_a} \int_0^\infty \frac{u_1^2}{R_a}\, dt. \tag{12.95}$$

Since

$$\int_0^\infty u^2\, dt = \frac{k^2}{J_v^2 R_a} E, \tag{12.96}$$

fixing λ is equivalent to fixing E. Plots of I_1 and I_2 for various values of q_1 and q_2, together with time responses $\theta(t)$, $\omega(t)$, and $u(t)$ can be used by the designer to clarify the "trade-offs" or compromises involved.

We find the optimum values of k_1 and k_2 (as functions of λ, q_1 and q_2) to minimize I, using the Riccati equation. For this problem,

$$\mathbf{A} = \begin{bmatrix} 0 & 1 \\ 0 & 0 \end{bmatrix}, \quad \mathbf{b} = \begin{bmatrix} 0 \\ 1 \end{bmatrix}$$
$$\mathbf{Q} = \begin{bmatrix} q_1 & 0 \\ 0 & q_2 \end{bmatrix}, \qquad \mathbf{R} = \lambda. \tag{12.97}$$

Substituting the matrices in (12.97) into (12.30), we obtain

$$\begin{bmatrix} 0 & 0 \\ 1 & 0 \end{bmatrix}\begin{bmatrix} p_{11} & p_{12} \\ p_{12} & p_{22} \end{bmatrix} + \begin{bmatrix} p_{11} & p_{12} \\ p_{12} & p_{22} \end{bmatrix}\begin{bmatrix} 0 & 1 \\ 0 & 0 \end{bmatrix}$$
$$-\frac{1}{\lambda}\begin{bmatrix} p_{11} & p_{12} \\ p_{12} & p_{22} \end{bmatrix}\begin{bmatrix} 0 \\ 1 \end{bmatrix}[0 \quad 1]\begin{bmatrix} p_{11} & p_{12} \\ p_{21} & p_{22} \end{bmatrix} + \begin{bmatrix} q_1 & 0 \\ 0 & q_2 \end{bmatrix}$$
$$= \begin{bmatrix} 0 & 0 \\ 0 & 0 \end{bmatrix}. \tag{12.98}$$

Solution of this set of equations for the positive definite P matrix yields

$$\mathbf{P} = \begin{bmatrix} \sqrt{q_1(2\sqrt{q_1\lambda} + q_2)} & \sqrt{q_1\lambda} \\ \sqrt{q_1\lambda} & \sqrt{2\lambda\sqrt{q_1\lambda} + q_2\lambda} \end{bmatrix} \tag{12.99}$$

The control parameters are obtained from (12.29):

$$\mathbf{K} = -\mathbf{R}^{-1}\mathbf{B}^T\mathbf{P}.$$

For this case,

$$k_1 = -\sqrt{q_1/\lambda}$$
$$k_2 = -\sqrt{1\sqrt{q_1/\lambda} + q_2/\lambda} \tag{12.100}$$

We have now obtained the optimum design in terms of the weighting parameters q_1, q_2, and λ. Notice that there are in effect only two design parameters, q_1/λ and q_2/λ.

Insight into the effect of q_1/λ and q_2/λ on the time responses can be obtained by relating these parameters to the damping ratio ζ and natural frequency ω_n appearing in the standard form of a second-order characteristic equation:

$$s^2 + 2\zeta\omega_n s + \omega_n^2 = 0.$$

For our system of Figure 12.4, the characteristic equation is

$$s^2 - k_2 s - k_1 = 0. \tag{12.101}$$

Thus

$$\omega_n = \sqrt{-k_1}$$

$$\zeta = \frac{-k_2}{2\sqrt{-k_1}}. \tag{12.102}$$

Solving (12.100) and (12.102) simultaneously leads to relations between ζ, ω_n and q_1/λ, q_2/λ:

$$\frac{q_1}{\lambda} = \omega_n^2$$

$$\frac{q_2}{\lambda} = 2\omega_n^2(2\zeta^2 - 1). \tag{12.103}$$

For this second-order case, time responses for various values of q_1/λ and q_2/λ can be obtained from plots in Appendix B. In more complex systems, simulation together with root-locus plots can be used to study the effects on the time response of different weighting parameters.

The integrals I_1 and I_2 can be found using Table 12.1. For simplicity of presentation, we consider here only the case $q_2 = 0$. In an actual design, other values of q_2 would be studied. With $q_2 = 0$, we obtain

$$I_1 = \int_0^\infty u^2\, dt = \frac{-k_1^2 x_{10}^2 - 2k_1 k_2 x_{20} x_{10} + (k_1 - k_2^2) x_{20}^2}{2k_2} \tag{12.104}$$

and

$$I_2 = \int_0^\infty x_1^2\, dt = \frac{-k_1 x_{10}^2 + [x_{20} - k_2 x_{10}]^2}{2k_1 k_2}. \tag{12.105}$$

These expressions are functions of initial conditions, of course, as well as k_1 and k_2. The data in Figure 12.5 was computed from (12.100), (12.104), and (12.105), using $q_1 = 1$, $q_2 = 0$, and values for λ as shown. (Recall that fixing λ in effect fixes I_1.)

Figure 12.6 shows a plot of I_1 vs I_2 for several fixed values of the initial condition ratio $\alpha = x_{20}/x_{10}$, and with λ as a parameter for each curve. This plot can be used for design. Suppose, for example, that we investigate the consequences of constraining the integral square attitude by the relation

$$\frac{I_2}{x_{10}^2} \leqslant 0.3 \times 10^4,$$

for expected initial condition ratios in the range $\alpha \leqslant 0.035$. Then we see from Figure 12.6 that the smallest value of I_1 meeting the above specifications occurs at

$$I_1 = 0.935 \times 10^{-5} x_{10}^2. \tag{12.106}$$

The corresponding optimum values of the feedback coefficients from Figure 12.5 are

$$\begin{aligned} k_1 &= -0.375 \times 10^{-4} \\ k_2 &= -0.866 \times 10^{-2}. \end{aligned} \tag{12.107}$$

From (12.103), we find that these optimum coefficients lead to a closed-loop system characterized by

$$\begin{aligned} \zeta &= 0.707 \\ \omega_n &= 6.12 \times 10^{-3}\ \text{sec}^{-1}. \end{aligned} \tag{12.108}$$

The settling time corresponding to the values of ζ and ω_n of (12.108) is about 900 seconds. If a smaller settling time is desired, ω_n must be increased [ζ is always 0.707 for $q_2 = 0$. See (12.103)], which means increasing k_1. This can be done by decreasing the value of I_2. As a penalty, I_1 will increase, of course. ▲

QX20/X10	Q	LAMBDA	I1/X10SQUARED	I2/X10SQUARED	K1	K2
.0000E+00	.0000E+00	.1000E+08	.1988E-05	.5965E+02	-.3162E-03	-.2515E-01
.0000E+00	.0000E+00	.1100E+09	.3292E-06	.1086E+03	-.9535E-04	-.1381E-01
.0000E+00	.0000E+00	.2100E+09	.2027E-06	.1277E+03	-.6901E-04	-.1175E-01
.0000E+00	.0000E+00	.3100E+09	.1513E-06	.1407E+03	-.5680E-04	-.1066E-01
.0000E+00	.0000E+00	.4100E+09	.1227E-06	.1509E+03	-.4939E-04	-.9938E-02
.0000E+00	.0000E+00	.5100E+09	.1042E-06	.1594E+03	-.4428E-04	-.9411E-02
.0000E+00	.0000E+00	.6100E+09	.9109E-07	.1667E+03	-.4049E-04	-.8999E-02
.0000E+00	.0000E+00	.7100E+09	.8129E-07	.1731E+03	-.3753E-04	-.8664E-02
.0000E+00	.0000E+00	.8100E+09	.7364E-07	.1789E+03	-.3514E-04	-.8383E-02
.0000E+00	.0000E+00	.9100E+09	.6748E-07	.1842E+03	-.3315E-04	-.8142E-02
.5000E-02	.0000E+00	.1000E+08	.4041E-05	.7703E+02	-.3162E-03	-.2515E-01
.5000E-02	.0000E+00	.1100E+09	.1065E-05	.1706E+03	-.9535E-04	-.1381E-01
.5000E-02	.0000E+00	.2100E+09	.7680E-06	.2156E+03	-.6901E-04	-.1175E-01
.5000E-02	.0000E+00	.3100E+09	.6352E-06	.2494E+03	-.5680E-04	-.1066E-01
.5000E-02	.0000E+00	.4100E+09	.5560E-06	.2776E+03	-.4939E-04	-.9938E-02
.5000E-02	.0000E+00	.5100E+09	.5020E-06	.3023E+03	-.4428E-04	-.9411E-02
.5000E-02	.0000E+00	.6100E+09	.4623E-06	.3245E+03	-.4049E-04	-.8999E-02
.5000E-02	.0000E+00	.7100E+09	.4314E-06	.3448E+03	-.3753E-04	-.8664E-02
.5000E-02	.0000E+00	.8100E+09	.4065E-06	.3637E+03	-.3514E-04	-.8383E-02
.5000E-02	.0000E+00	.9100E+09	.3859E-06	.3814E+03	-.3315E-04	-.8142E-02
.1000E-01	.0000E+00	.1000E+08	.7037E-05	.9756E+02	-.3162E-03	-.2515E-01
.1000E-01	.0000E+00	.1100E+09	.2318E-05	.2515E+03	-.9535E-04	-.1381E-01
.1000E-01	.0000E+00	.2100E+09	.1774E-05	.3343E+03	-.6901E-04	-.1175E-01
.1000E-01	.0000E+00	.3100E+09	.1519E-05	.3994E+03	-.5680E-04	-.1066E-01
.1000E-01	.0000E+00	.4100E+09	.1362E-05	.4553E+03	-.4939E-04	-.9938E-02
.1000E-01	.0000E+00	.5100E+09	.1253E-05	.5052E+03	-.4428E-04	-.9411E-02
.1000E-01	.0000E+00	.6100E+09	.1171E-05	.5509E+03	-.4049E-04	-.8999E-02
.1000E-01	.0000E+00	.7100E+09	.1106E-05	.5934E+03	-.3753E-04	-.8664E-02
.1000E-01	.0000E+00	.8100E+09	.1054E-05	.6333E+03	-.3514E-04	-.8383E-02
.1000E-01	.0000E+00	.9100E+09	.1010E-05	.6711E+03	-.3315E-04	-.8142E-02
.1500E-01	.0000E+00	.1000E+08	.1098E-04	.1212E+03	-.3162E-03	-.2515E-01
.1500E-01	.0000E+00	.1100E+09	.4090E-05	.3514E+03	-.9535E-04	-.1381E-01
.1500E-01	.0000E+00	.2100E+09	.3220E-05	.4838E+03	-.6901E-04	-.1175E-01
.1500E-01	.0000E+00	.3100E+09	.2802E-05	.5907E+03	-.5680E-04	-.1066E-01
.1500E-01	.0000E+00	.4100E+09	.2541E-05	.6839E+03	-.4939E-04	-.9938E-02
.1500E-01	.0000E+00	.5100E+09	.2356E-05	.7681E+03	-.4428E-04	-.9411E-02
.1500E-01	.0000E+00	.6100E+09	.2217E-05	.8459E+03	-.4049E-04	-.8999E-02
.1500E-01	.0000E+00	.7100E+09	.2106E-05	.9188E+03	-.3753E-04	-.8664E-02
.1500E-01	.0000E+00	.8100E+09	.2015E-05	.9878E+03	-.3514E-04	-.8383E-02
.1500E-01	.0000E+00	.9100E+09	.1939E-05	.1054E+04	-.3315E-04	-.8142E-02

DX20/X1	Q	LAMBDA	I1/X10SQUARED	I2/X10SQUARED	K1	K2
.2000E-01	.0000E+00	.1000E+08	.1586E-04	.1480E+03	-.3162E-03	-.2515E-01
.2000E-01	.0000E+00	.1100E+09	.6379E-05	.4703E+03	-.9535E-04	-.1381E-01
.2000E-01	.0000E+00	.2100E+09	.5107E-05	.6642E+03	-.6901E-04	-.1175E-01
.2000E-01	.0000E+00	.3100E+09	.4485E-05	.8233E+03	-.5680E-04	-.1066E-01
.2000E-01	.0000E+00	.4100E+09	.4092E-05	.9634E+03	-.4939E-04	-.9938E-02
.2000E-01	.0000E+00	.5100E+09	.3813E-05	.1091E+04	-.4428E-04	-.9411E-02
.2000E-01	.0000E+00	.6100E+09	.3600E-05	.1210E+04	-.4049E-04	-.8999E-02
.2000E-01	.0000E+00	.7100E+09	.3431E-05	.1321E+04	-.3753E-04	-.8664E-02
.2000E-01	.0000E+00	.8100E+09	.3291E-05	.1427E+04	-.3514E-04	-.8383E-02
.2000E-01	.0000E+00	.9100E+09	.3173E-05	.1529E+04	-.3315E-04	-.8142E-02
.2500E-01	.0000E+00	.1000E+08	.2168E-04	.1780E+03	-.3162E-03	-.2515E-01
.2500E-01	.0000E+00	.1100E+09	.9186E-05	.6082E+03	-.9535E-04	-.1381E-01
.2500E-01	.0000E+00	.2100E+09	.7435E-05	.8754E+03	-.6901E-04	-.1175E-01
.2500E-01	.0000E+00	.3100E+09	.6567E-05	.1097E+04	-.5680E-04	-.1066E-01
.2500E-01	.0000E+00	.4100E+09	.6016E-05	.1294E+04	-.4939E-04	-.9938E-02
.2500E-01	.0000E+00	.5100E+09	.5622E-05	.1474E+04	-.4428E-04	-.9411E-02
.2500E-01	.0000E+00	.6100E+09	.5321E-05	.1642E+04	-.4049E-04	-.8999E-02
.2500E-01	.0000E+00	.7100E+09	.5081E-05	.1800E+04	-.3753E-04	-.8664E-02
.2500E-01	.0000E+00	.8100E+09	.4882E-05	.1951E+04	-.3514E-04	-.8383E-02
.2500E-01	.0000E+00	.9100E+09	.4713E-05	.2096E+04	-.3315E-04	-.8142E-02
.3000E-01	.0000E+00	.1000E+08	.2845E-04	.2111E+03	-.3162E-03	-.2515E-01
.3000E-01	.0000E+00	.1100E+09	.1251E-04	.7650E+03	-.9535E-04	-.1381E-01
.3000E-01	.0000E+00	.2100E+09	.1020E-04	.1118E+04	-.6901E-04	-.1175E-01
.3000E-01	.0000E+00	.3100E+09	.9049E-05	.1412E+04	-.5680E-04	-.1066E-01
.3000E-01	.0000E+00	.4100E+09	.8313E-05	.1675E+04	-.4939E-04	-.9938E-02
.3000E-01	.0000E+00	.5100E+09	.7785E-05	.1917E+04	-.4428E-04	-.9411E-02
.3000E-01	.0000E+00	.6100E+09	.7380E-05	.2143E+04	-.4049E-04	-.8999E-02
.3000E-01	.0000E+00	.7100E+09	.7055E-05	.2357E+04	-.3753E-04	-.8664E-02
.3000E-01	.0000E+00	.8100E+09	.6786E-05	.2561E+04	-.3514E-04	-.8383E-02
.3000E-01	.0000E+00	.9100E+09	.6558E-05	.2756E+04	-.3315E-04	-.8142E-02
.3500E-01	.0000E+00	.1000E+08	.3616E-04	.2473E+03	-.3162E-03	-.2515E-01
.3500E-01	.0000E+00	.1100E+09	.1635E-04	.9409E+03	-.9535E-04	-.1381E-01
.3500E-01	.0000E+00	.2100E+09	.1341E-04	.1390E+04	-.6901E-04	-.1175E-01
.3500E-01	.0000E+00	.3100E+09	.1193E-04	.1769E+04	-.5680E-04	-.1066E-01
.3500E-01	.0000E+00	.4100E+09	.1098E-04	.2108E+04	-.4939E-04	-.9938E-02
.3500E-01	.0000E+00	.5100E+09	.1030E-04	.2420E+04	-.4428E-04	-.9411E-02
.3500E-01	.0000E+00	.6100E+09	.9776E-05	.2712E+04	-.4049E-04	-.8999E-02
.3500E-01	.0000E+00	.7100E+09	.9355E-05	.2990E+04	-.3753E-04	-.8664E-02
.3500E-01	.0000E+00	.8100E+09	.9005E-05	.3255E+04	-.3514E-04	-.8383E-02
.3500E-01	.0000E+00	.9100E+09	.8709E-05	.3509E+04	-.3315E-04	-.8142E-02

Figure 12.5 Optimum values of k_1 and k_2, with resulting I_1 and I_2.

.4000E-01	.0000E+00	.1000E+08	.4482E-04	.2867E+03	-.3162E-03	-.2515E-01
.4000E-01	.0000E+00	.1100E+09	.2071E-04	.1136E+04	-.9535E-04	-.1381E-01
.4000E-01	.0000E+00	.2100E+09	.1706E-04	.1694E+04	-.6901E-04	-.1175E-01
.4000E-01	.0000E+00	.3100E+09	.1521E-04	.2167E+04	-.5680E-04	-.1066E-01
.4000E-01	.0000E+00	.4100E+09	.1402E-04	.2591E+04	-.4939E-04	-.9938E-02
.4000E-01	.0000E+00	.5100E+09	.1317E-04	.2983E+04	-.4428E-04	-.9411E-02
.4000E-01	.0000E+00	.6100E+09	.1251E-04	.3350E+04	-.4049E-04	-.8999E-02
.4000E-01	.0000E+00	.7100E+09	.1198E-04	.3699E+04	-.3753E-04	-.8664E-02
.4000E-01	.0000E+00	.8100E+09	.1154E-04	.4033E+04	-.3514E-04	-.8383E-02
.4000E-01	.0000E+00	.9100E+09	.1116E-04	.4355E+04	-.3315E-04	-.8142E-02
.4500E-01	.0000E+00	.1000E+08	.5441E-04	.3293E+03	-.3162E-03	-.2515E-01
.4500E-01	.0000E+00	.1100E+09	.2559E-04	.1350E+04	-.9535E-04	-.1381E-01
.4500E-01	.0000E+00	.2100E+09	.2115E-04	.2029E+04	-.6901E-04	-.1175E-01
.4500E-01	.0000E+00	.3100E+09	.1889E-04	.2606E+04	-.5680E-04	-.1066E-01
.4500E-01	.0000E+00	.4100E+09	.1744E-04	.3125E+04	-.4939E-04	-.9938E-02
.4500E-01	.0000E+00	.5100E+09	.1639E-04	.3605E+04	-.4428E-04	-.9411E-02
.4500E-01	.0000E+00	.6100E+09	.1558E-04	.4057E+04	-.4049E-04	-.8999E-02
.4500E-01	.0000E+00	.7100E+09	.1493E-04	.4486E+04	-.3753E-04	-.8664E-02
.4500E-01	.0000E+00	.8100E+09	.1439E-04	.4897E+04	-.3514E-04	-.8383E-02
.4500E-01	.0000E+00	.9100E+09	.1393E-04	.5293E+04	-.3315E-04	-.8142E-02
.5000E-01	.0000E+00	.1000E+08	.6495E-04	.3749E+03	-.3162E-03	-.2515E-01
.5000E-01	.0000E+00	.1100E+09	.3099E-04	.1582E+04	-.9535E-04	-.1381E-01
.5000E-01	.0000E+00	.2100E+09	.2568E-04	.2394E+04	-.6901E-04	-.1175E-01
.5000E-01	.0000E+00	.3100E+09	.2297E-04	.3086E+04	-.5680E-04	-.1066E-01
.5000E-01	.0000E+00	.4100E+09	.2123E-04	.3710E+04	-.4939E-04	-.9938E-02
.5000E-01	.0000E+00	.5100E+09	.1996E-04	.4288E+04	-.4428E-04	-.9411E-02
.5000E-01	.0000E+00	.6100E+09	.1899E-04	.4832E+04	-.4049E-04	-.8999E-02
.5000E-01	.0000E+00	.7100E+09	.1820E-04	.5350E+04	-.3753E-04	-.8664E-02
.5000E-01	.0000E+00	.8100E+09	.1755E-04	.5846E+04	-.3514E-04	-.8383E-02
.5000E-01	.0000E+00	.9100E+09	.1699E-04	.6324E+04	-.3315E-04	-.8142E-02

Figure 12.5 Optimum values of k_1 and k_2, with resulting I_1 and I_2.

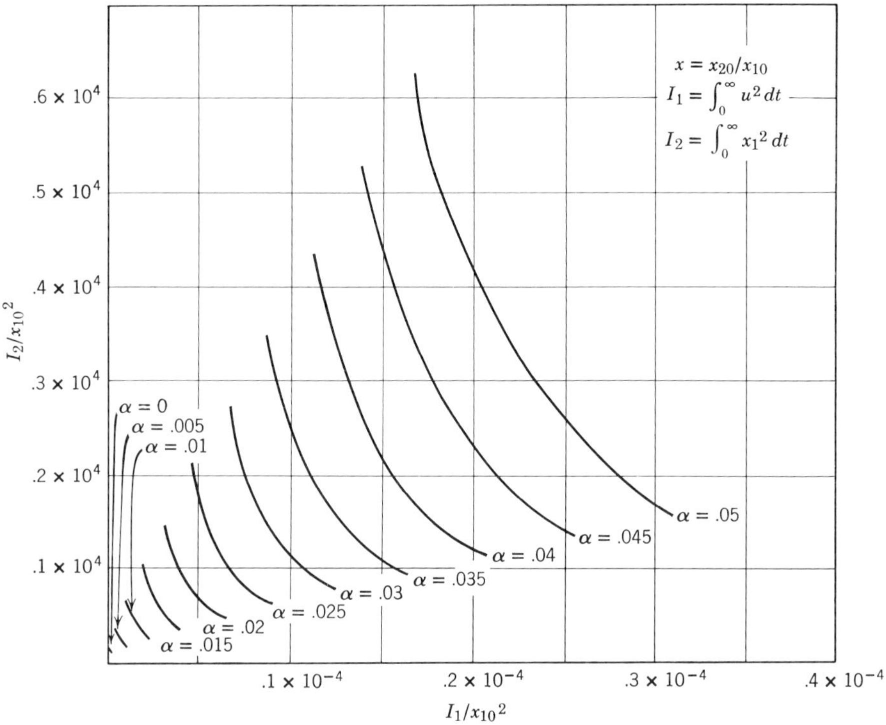

Figure 12.6

Yet another approach to determining an appropriate quadratic index is to try to force the system response to follow the response of another "ideal" system. This is basically the idea behind much of the classical trial-and-error design—to make the system behave like a second-order system having desirable characteristics. The application of this concept to the optimum state regulator problem is due to Kalman. He suggests representing an "ideal" or desired system by the mathematical model

$$\dot{\boldsymbol{\eta}} = \mathbf{L}\boldsymbol{\eta}. \tag{12.109}$$

The system matrix L is chosen so that the time-domain response of the system of (12.109) has the properties desired, that is, a specified percent overshoot, a desired transfer-function matrix, a given rise time, or whatever is important in the application at hand. We then consider the system to be designed. The plant is represented by

$$\dot{\mathbf{x}} = \mathbf{A}\mathbf{x} + \mathbf{B}u, \tag{12.110}$$

with output relation

$$\mathbf{y} = \mathbf{C}\mathbf{x}, \tag{12.111}$$

and the control input constrained to the form

$$\mathbf{u} = \mathbf{K}\mathbf{x}. \tag{12.112}$$

We choose as the performance index

$$I = \int_0^\infty \{[\mathbf{y} - \mathbf{L}\mathbf{y}]^T \mathbf{Q} [\mathbf{y} - \mathbf{L}\mathbf{y}] + \mathbf{u}^T \mathbf{R} \mathbf{u}\}\, dt. \tag{12.113}$$

It is seen that if the designed system behaves exactly like the desired model, the first term in the index will be zero. As before, the diagonal elements of $\mathbf{Q}$ indicate the relative importance of matching the corresponding components of $\mathbf{y}$ to the model, and the diagonal elements of $\mathbf{R}$ show control energy weightings. Generally, if the elements of $\mathbf{Q}$ are large and those of $\mathbf{R}$ are relatively smaller, the designed system will approximately match the model, and this will be accomplished via large control inputs.

It can be shown that this "model-following" problem can be converted into an optimum state regulator problem, and is solved by the Riccati equation of the form of (12.30):

$$\hat{\mathbf{A}}^T \mathbf{P} + \mathbf{P}\hat{\mathbf{A}} - \mathbf{P}\mathbf{B}\hat{\mathbf{R}}^{-1}\mathbf{B}^T\mathbf{P} + \hat{\mathbf{Q}} = \mathbf{0}, \tag{12.114}$$

where

$$\begin{aligned}
&\hat{\mathbf{A}} = \mathbf{A} - \mathbf{B}\hat{\mathbf{R}}^{-1}\mathbf{N}^T\mathbf{Q}\mathbf{M},\\
&\hat{\mathbf{Q}} = \mathbf{Q} - \mathbf{Q}\mathbf{N}\hat{\mathbf{R}}^{-1}\mathbf{N}^T\mathbf{Q},\\
&\hat{\mathbf{R}} = \mathbf{N}^T\mathbf{Q}\mathbf{N} + \mathbf{R}, \quad \mathbf{N} = \mathbf{C}\mathbf{B},\\
&\mathbf{M} = \mathbf{C}\mathbf{A} - \mathbf{L}\mathbf{C}.
\end{aligned}$$

The optimum $\mathbf{K}$ is given by

$$\mathbf{K} = -\hat{\mathbf{R}}^{-1}(\mathbf{B}^T\mathbf{P} + \mathbf{N}^T\mathbf{Q}\mathbf{M}). \tag{12.115}$$

12.6 Parameter Optimization of Digital Systems

All techniques discussed in this chapter for parameter optimization of continuous-time systems can be extended readily to the optimization of discrete-time systems. For example, corresponding to the continuous-time index (12.13), we have

$$I = \sum_{n=0}^{\infty} \sum_{i=1}^{k} f_i^2(nT) \tag{12.116}$$

for linear time-invariant discrete-time systems. Consider a quadratic term in (12.116),

$$I_j = \sum_{n=0}^{\infty} f_j^2(nT). \tag{12.117}$$

Using z-transforms,

$$I_j = \sum_{n=0}^{\infty} f_j(nT)\left[\frac{1}{2\pi j}\oint_{\Gamma} F(z)z^{n-1}\,dz\right]$$

$$= \frac{1}{2\pi j}\oint_{\Gamma} F(z)z^{-1}\left[\sum_{n=0}^{\infty} f_j(nT)z^{n}\right]dz$$

$$I_j = \frac{1}{2\pi j}\oint_{\Gamma} F(z)F(z^{-1})z^{-1}\,dz. \tag{12.118}$$

This integral can be evaluated using Table 12.2, which is the Z-transform counterpart of Table 12.1. (This table is adapted from Table III of *Theory and Application of the Z-Transform Method*, by E. I. Jury. See Jury, pp. 297–299, for additional entries.)

Table 12.2 Evaluation of an Integral

$$I_n = \frac{1}{2\pi j}\oint_{\Gamma} F(z)F(z^{-1})z^{-1}dz.$$

1. $F(z) = \dfrac{b_0z + b_1}{a_0z + a_1}$

$$I_1 = \frac{(b_0^2 + b_1^2)a_0 - 2b_0b_1a_1}{a_0(a_0^2 - a_1^2)}$$

2. $F(z) = \dfrac{b_0z^2 + b_1z + b_2}{a_0z^2 + a_1z + a_2}$

$$I_2 = \frac{B_0a_0e_1 - B_1a_0a_1 + B_2(a_1^2 - a_2e_1)}{a_0[(a_0^2 - a_2^2)e_1 - (a_0a_1 - a_1a_2)a_1]}$$

where

$$B_0 = b_0^2 + b_1^2 + b_2^2$$

$$B_1 = 2(b_0b_1 + b_1b_2)$$

$$B_2 = 2b_0b_2$$

$$e_1 = a_0 + a_2$$

The linear, time-invariant discrete-time optimum state regulator problem is formulated as follows. The plant is described by the vector difference equation

$$\mathbf{x}_{k+1} = \mathbf{F}\mathbf{x}_k + \mathbf{G}\mathbf{u}_k, \tag{12.119}$$

where the "shorthand" notation of Chapter 10 is used. The initial condition is $\mathbf{x}_0$. The quadratic performance index is

$$I = \sum_{k=1}^{N} [\mathbf{x}_k{}^T\mathbf{Q}\mathbf{x}_k + \mathbf{u}_{k-1}^T\mathbf{R}\mathbf{u}_{k-1}]. \tag{12.120}$$

It can be shown that (see, for example, J. T. Tou, *Optimum Design of Digital Control Systems*, Academic Press, 1963, pp. 42-53) the solution to this problem is

$$\mathbf{u}_k = -\mathbf{K}_k\mathbf{x}_k, \tag{12.121}$$

where the feedback-gain matrix $\mathbf{K}_k$ is obtained from

$$\mathbf{K}_k = [\mathbf{G}^T\mathbf{P}_k\mathbf{G} + \mathbf{R}]^{-1}\mathbf{G}^T\mathbf{P}_k\mathbf{F}, \tag{12.122}$$

$\mathbf{P}_k$ being the solution of the *sampled-data Riccati matrix difference* equation

$$\mathbf{P}_k = \mathbf{Q} + \mathbf{F}^T\mathbf{P}_{k+1}[\mathbf{F} - \mathbf{G}\mathbf{K}_{k+1}] \tag{12.123}$$

with boundary condition

$$\mathbf{P}_N = \mathbf{Q}. \tag{12.124}$$

Discussion concerning choice of weighting matrices $\mathbf{Q}$ and $\mathbf{R}$ parallels that of Section 12.5.

Problems

12-1. Shown in Figure P12-1 is a simplified single-axis attitude-control system. During a portion of the mission, the spacecraft is to revolve slowly on its axis ("barbecue roll") to prevent excessive temperature buildup. Thus the reference input is a *ramp*, $\theta(t) = t$, $t \geqslant 0$. We choose as a performance index

$$I(k, a) = \int_0^\infty (\theta_r - \theta)^2\, dt + \rho \int_0^\infty u^2\, dt.$$

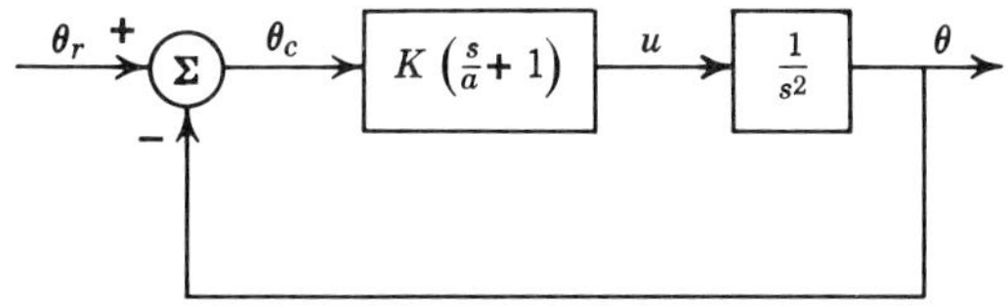

Figure P12-1

Find the optimum values of K and a as functions of ρ, using Parseval's theorem and Table 12.1.

12-2. Repeat Problem 12-1 by converting the problem to an equivalent state regulator problem and using the Riccati equation. (Suggestion: $x_1 = \theta - \theta_r$, $x_2 = \dot{\theta} - \dot{\theta}_r$.)

12-3. Select ρ in Problem 12-1 so that the system satisfies the control energy constraint

$$\int_0^\infty u^2\,dt = \tfrac{3}{2}.$$

What are the resulting values of K and a?

12-4. Consider the attitude-control system of Figure 12.2 of the text. Sometimes the feedback of attitude rate cannot be done satisfactorily because of hardware problems, such as noise in rate gyros. Approximate differentiation can be accomplished using a lead network in the forward loop, as shown in Figure P12-4. Find K_1, K_2, and α to minimize

$$J = \int_0^\infty \left(e^2 + \tfrac{1}{4}\tau_c^{\,2}\right) dt,$$

where $e(t) = \theta_r(t) - \theta(t)$, with $\theta_r(t) =$ unit step function. Discuss and interpret any notable aspects of the result.

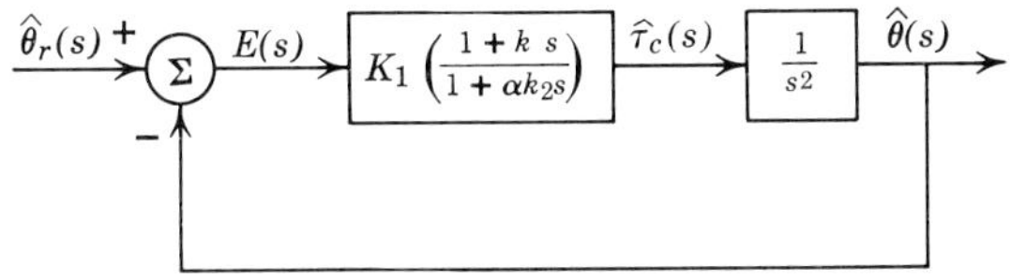

Figure P12-4

12-5. Some indices other than squared error indices can be optimized, using Parseval's theorem and Table 12.1. Consider the following:

(a) Show $\mathscr{L}[te(t)] = -\dfrac{dE(s)}{ds}$, and thus

$$\int_0^\infty t^2 e^2\,dt = \frac{1}{2\pi j}\int_{-j\infty}^{j\infty}\left[-\frac{dE(s)}{ds}\right]\left[-\frac{dE(-s)}{ds}\right]ds.$$

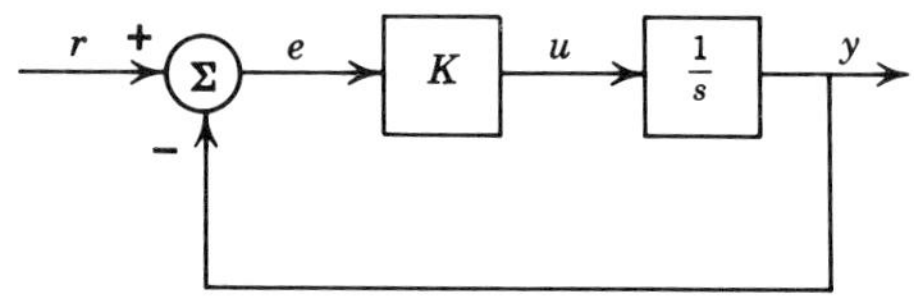

Figure P12-5

(b) Find K in Figure P12-5 to minimize

$$\int_0^\infty t^2 e^2\,dt \quad \text{subject to} \quad \int_0^\infty u^2\,dt = \tfrac{1}{2}.$$

(c) Discuss the significance of the index

$$\int_0^\infty t^2 e^2\,dt.$$

12-6. (Computer Problem) Write a computer program (using FORTRAN or your favorite language) to solve the Riccati equation following the suggestion at the end of Section 12.4.

12-7. The attitude control example of Chapter 12 was simple enough so that the time response could be correlated easily to the choice of weighting parameter ρ through expressions for ω_n and ζ, Equations 12.91 and 12.92. For more complex situations, such closed-form equations for root locations cannot be obtained. One approach to these more complex cases is to draw a root locus with respect to the parameter ρ or, more generally, to draw families root loci with respect to the entries in the weighting matrices $\mathbf{Q}$ and $\mathbf{R}$. Such plots provide the designer with guidelines and insight concerning the time response of the optimal system resulting from any selection of $\mathbf{Q}$ and $\mathbf{R}$.

(a) Show that the characteristic equation of the system of Figure 12.2 with optimum K_1 and K_2 as a function of ρ is

$$\Delta(s) = s^2 + \sqrt{2}\,\rho^{-1/4}s + \rho^{-1/2} = 0.$$

We would like to sketch the root locations as a function of ρ. However, the characteristic equation cannot be put into "standard" root-locus form, $1+\rho G(s) = 0$. It has been suggested (see S. S. L. Chang, *Synthesis of Optimum Control Systems*, McGraw-Hill, 1961) that the "root-square locus," $\Delta(s)\Delta(-s)$, be plotted instead. For this case, $\Delta(s)\Delta(-s)$ is

$$s^4 + \frac{1}{\rho} = 0.$$

(b) Sketch a root locus for $s^4 + (1/\rho) = 0$. Show that the root locations correspond to Equations 12.91 and 12.92.

(c) Discuss and sketch several time responses $\theta(t)$ and the corresponding control inputs $\tau_c(t)$ for several values of ρ when $\theta_r(t) =$ unit step. In particular, show θ and τ_c vs t for $\rho = \frac{1}{4}$.

12.8. Sketch and comment on a root-square locus (see Problem 12-7) as a function of ρ for the system of Problem 12-1.

12-9. Consider the digital attitude control system of Figure 10.20. Select K_1 and K_2 to minimize

$$I = \sum_{n=0}^{\infty} [e^2(kT) + \rho u^2(kT)],$$

where

$$e(kT) = \theta_r(kT) - \theta(kT),$$

when $\theta_r(0) = 1$, and $\theta_r(kT) = 0$, $k \geqslant 1$. (Suggestion: Using Table 12.2, obtain optimum K_1 and K_2 as functions of ρ and T.)

12-10. Discuss various time responses that can be obtained for various choices of ρ and T in Problem 12-9. Sketch time responses for $\theta(kT)$ and $u(kT)$.

CHAPTER 13

Parameter Optimization II

The classical calculus provides a fundamental tool in the analytic formulation and solution of parameter optimization problems. The ultimate success of the application of calculus depends on the complexity of the analytic form of the performance index, granting that an analytic form can be obtained in the first place. The optimum parameters are solutions of algebraic equations, which are obtained by setting the partial derivatives of the performance index equal to zero. Chapter 12 deals with a class of problems that can be successfully solved in such a manner. With the widespread availability of electronic computers, numerical experimentation is a feasible alternative that must be considered. For complex system optimization, it may be the only recourse available.

In this chapter, we describe a few guidelines in carrying out numerical experimentation. Basically, numerical experimentation is a trial and error procedure that involves guessing the parameter values, computing the values of the performance index in the neighborhood of the guessed values of the parameters, and making subsequent new guesses for the parameters that are, hopefully, better than the previous guesses. A strategy or plan for making the guesses is defined as a *search technique.* The first technique that will be described is Newton's method for determining the roots of an equation (such as the partial derivative of the performance index set equal to zero). The other methods are primarily for the minimization or maximization of an algebraic function. We shall restrict our discussion to real functions of real parameters.

13.1 The Newton-Raphson Method

Consider an equation

$$f(x) = 0 \tag{13.1}$$

where f is a function of a single variable x. The function f may represent the derivative with respect to x of a performance index that we wish to maximize or minimize. The variable x may represent some control parameter that we are to choose in order to maximize the performance index. The problem is to determine $x = x^*$ which satisfies (13.1), that is, $f(x^*) = 0$. A class of numerical methods useful for solving such an equation depends on converting (13.1) to another equation

$$x = F(x) \tag{13.2}$$

such that the solution of (13.2) satisfies (13.1) as well. There are infinitely many ways of choosing $F(x)$, and the success of the method depends on how $F(x)$ is chosen. For example, (13.1) may be written as

$$x = x + f(x) \equiv F(x). \tag{13.3}$$

Clearly, the solution of (13.3) satisfies (13.1). To solve (13.3), the *method of successive approximations* is used. Briefly, this method of successive approximations consists of constructing a sequence $x_0, x_1, \ldots, x_n \ldots$ such that

$$x_{n+1} = F(x_n), \text{ for } n = 0, 1, \ldots \tag{13.4}$$

where the first number of the sequence, x_0, is arbitrary. Thus, from x_0, (13.4) is used to compute x_1. Then, from x_1, (13.4) yields x_2. Hopefully, the sequence may converge to the solution x^* as $n \to \infty$. Unfortunately, not any choice of $F(x)$ for which (13.1) and (13.2) have the same solutions leads to a convergent series as constructed by means of (13.4).

We now describe a particular choice for $F(x)$, and for constructing the sequence, which leads to the *Newton-Raphson* method. Before stating the method as well as the conditions under which the series will converge, let us explain the motivation for the method. Let us assume that $f(x)$ is continuous and differentiable at a point x_0. Thus, if we expand $f(x)$ in Taylor's series and *approximate* $f(x)$ only by its linear terms, we have

$$f(x) \approx f(x_0) + (x - x_0) f'(x_0). \tag{13.5}$$

Instead of (13.1), we solve the linear equation

$$f(x_0) + (x - x_0) f'(x_0) = 0. \tag{13.6}$$

Denoting the solution of (13.6) by $x = x_1$, we immediately obtain

$$x_1 = x_0 - \frac{f(x_0)}{f'(x_0)}. \tag{13.7}$$

Equation 13.7 provides a first approximation for the solution of (13.1). The process may be continued by expanding $f(x)$ in Taylor's series at x_1 and approximating $f(x)$ by its linear terms only, yielding

$$f(x) \approx f(x_1) + (x - x_1) f'(x_1). \tag{13.8}$$

The solution of

$$f(x_1) + (x - x_1) f'(x_1) = 0, \tag{13.9}$$

which is a linear approximation of (13.1), is denoted by $x = x_2$, and

$$x_2 = x_1 - \frac{f(x_1)}{f'(x_1)}. \tag{13.10}$$

A third approximation may be obtained by expanding $f(x)$ in Taylor's series at x_2 and retaining only the linear terms. This procedure suggests constructing a sequence $x_0, x_1, \ldots, x_n, \ldots$, using the algorithm

$$x_{n+1} = x_n - \frac{f(x_n)}{f'(x_n)}. \tag{13.11}$$

Comparing (13.11) to (13.4), we observe that

$$F(x_n) = x_n - \frac{f(x_n)}{f'(x_n)}. \tag{13.12}$$

Hence, this suggests writing (13.2) as

$$x = F(x) = x - \frac{f(x)}{f'(x)}. \tag{13.13}$$

It is clear that if x^* satisfies (13.13), it satisfies (13.1) also. The algorithm of (13.11) is known as the Newton-Raphson method.

Example 13-1. Let us apply the Newton-Raphson algorithm for determining a root of

$$f(x) = x\epsilon^{-x} - 0.2 = 0.$$

First, form
$$f'(x) = \epsilon^{-x} - x\epsilon^{-x}.$$

From (13.11), we have

$$x_{n+1} = x_n - \frac{x_n\epsilon^{-x_n} - 0.2}{\epsilon^{-x_n} - x_n\epsilon^{-x_n}} = x_n - \frac{x_n - 0.2\epsilon^{x_n}}{1 - x_n}.$$

Starting with $x_0 = 0.5$,

$$x_1 = 0.5 - \frac{0.5 - 0.2\epsilon^{0.5}}{0.5} = 0.16$$

$$x_2 = 0.16 - \frac{0.16 - 0.2\epsilon^{0.16}}{0.84} = 0.25$$

$$x_3 = 0.25 - \frac{0.25 - 0.2\epsilon^{0.25}}{0.75} = 0.259$$

$$x_3 = 0.259 - \frac{0.259 - 0.2\epsilon^{0.259}}{0.741} = 0.259.$$

Hence

$$x^* \approx 0.259 \text{ is a root. } \blacktriangle$$

In the application of the Newton-Raphson method, it may not be computationally feasible to determine whether the sequence $x_0, x_1, \ldots, x_n, \ldots$ will converge. A common practice is to use the algorithm to generate a sequence. The process is stopped when the magnitude of the difference between x_n and x_{n-1} is less than a predetermined arbitrarily chosen small number, or when the magnitude of $f(x_n)$ is less than an arbitarily chosen small number. These stopping criteria will determine the accuracy of the results. If the values $f(x_n)$ progressively depart farther away from zero, or if the number of iterations exceeds a predetermined number, the process is stopped.

If certain conditions are met, the sequence is guaranteed to converge. A set of sufficient conditions due to Ostrowski-Kantorovich,[1] which guarantees convergence, is listed below:

(a) $f'(x_0) \neq 0$,
(b) $f(x)$ has a continuous second derivative for all x in $|x - x_0| \leq r$,

[1]T. L. Saaty and J. Bram, *Nonlinear Mathematics*, McGraw–Hill, New York, 1964.

(c) $h = \alpha\beta \leqslant \frac{1}{2}$ where $\alpha \equiv |f(x_0)/f'(x_0)|$ and $\left|\frac{f''(x)}{f'(x_0)}\right| \leqslant \beta$, for $|x - x_0| \leqslant r$,

(d) $\left(\frac{1-\sqrt{1-2h}}{h}\right)\alpha \leqslant r$.

If in addition to the above conditions, the following condition,

(e) $r < \left(\frac{1+\sqrt{1-2h}}{h}\right)\alpha$ for $h < \frac{1}{2}$ and $r \leqslant \left(\frac{1+\sqrt{1-2h}}{h}\right)\alpha$ for $h \leqslant \frac{1}{2}$,

is satisfied also, the solution x^* is unique within the interval $|x - x_0| \leqslant r$.

The first two conditions guarantee the existence of α and β in (c). Condition (d) defines the smallest value for the radius r. The computation of β requires finding a bound on the value of $f''(x)$ within the interval $|x - x_0| \leqslant r$. In complex cases, the amount of computation to obtain this may be as much as that for constructing the sequence. Hence, it may be practical to proceed with the sequence construction even without a guarantee of convergence.

A modification of the algorithm in (13.11) into

$$x_{n+1} = x_n - \frac{f(x_n)}{f'(x_0)} \tag{13.14}$$

is known as the *modified Newton-Raphson method*. To distinguish this method from the original algorithm, the earlier algorithm in (13.11) is called the *basic Newton-Raphson method*.

Conditions (a) to (e) above are also sufficient to guarantee convergence of the modified method. Note that the modified method is simpler for each iteration since $f'(x_n)$ does not have to be computed for each n, as in the basic method. However, the rate of convergence is slower. Nevertheless, the total computing time might be less compared to the basic method. The rate of convergence for the basic method is indicated by

$$|x^* - x_n| \leqslant \frac{1}{2^n}(2h)^{2^n}\left|\frac{f'(x_0)}{f''(x_0)}\right|. \tag{13.15}$$

For the modified method, the rate of convergence is indicated by

$$|x^* - x_n| \leqslant (1-\sqrt{1-2h})^{n+1}\left|\frac{f'(x_0)}{f''(x_0)}\right|. \tag{13.16}$$

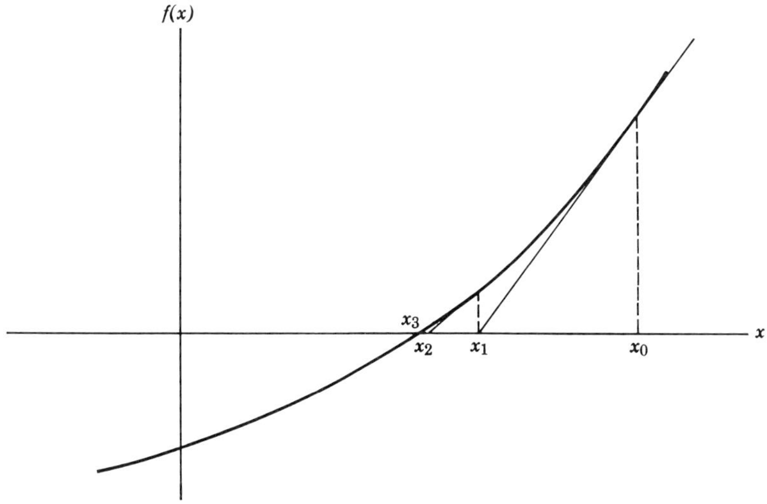

Figure 13.1 Illustration for the basic Newton-Raphson method.

Figure 13.1 shows an example of how the basic Newton-Raphson method generates a sequence $x_0, x_1, \ldots,$ which converges to the solution x^*. Notice that the slope of each straight line approximation is equal to the slope of the curve at the nth iterate x_n. Figure 13.2 shows an example of how the modified Newton-Raphson algorithm

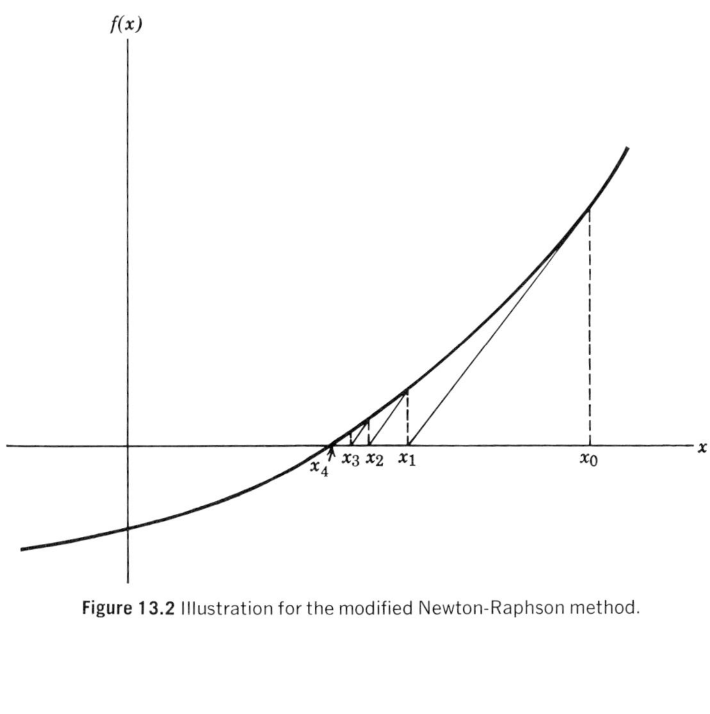

Figure 13.2 Illustration for the modified Newton-Raphson method.

generates the sequence that converges to x^*. In contrast to the variable slope straight-line approximations in Figure 13.1, the straight-line approximations in Figure 13.2 have the same slope, obtained from the slope of the curve at the initial guess x_0. Although there are many more iterations relative to the basic algorithm in Figure 13.1, the slopes at the various iterates x_n do not have to be computed.

Flow Charts 13.1 and 13.2 show the essential computation steps for

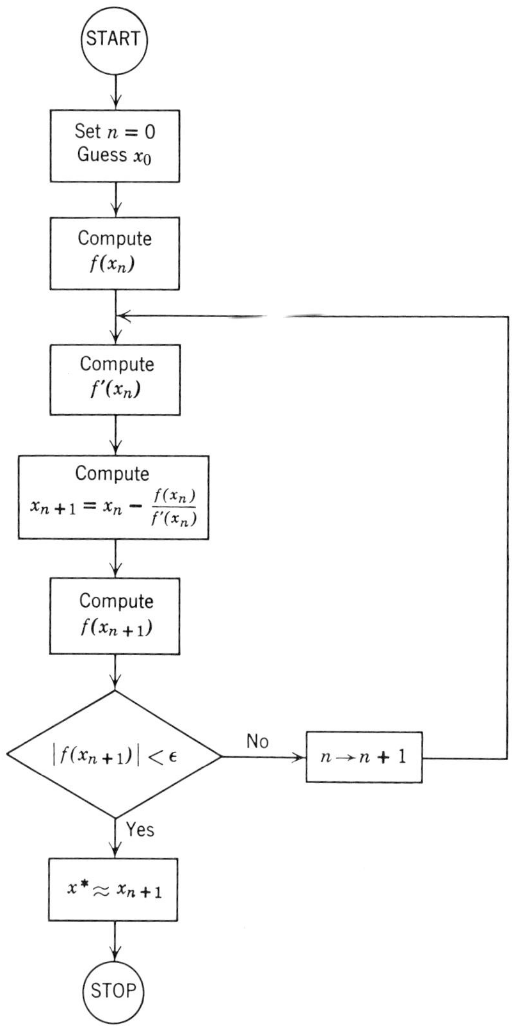

Flow Chart 13.1 Basic Newton-Raphson method.

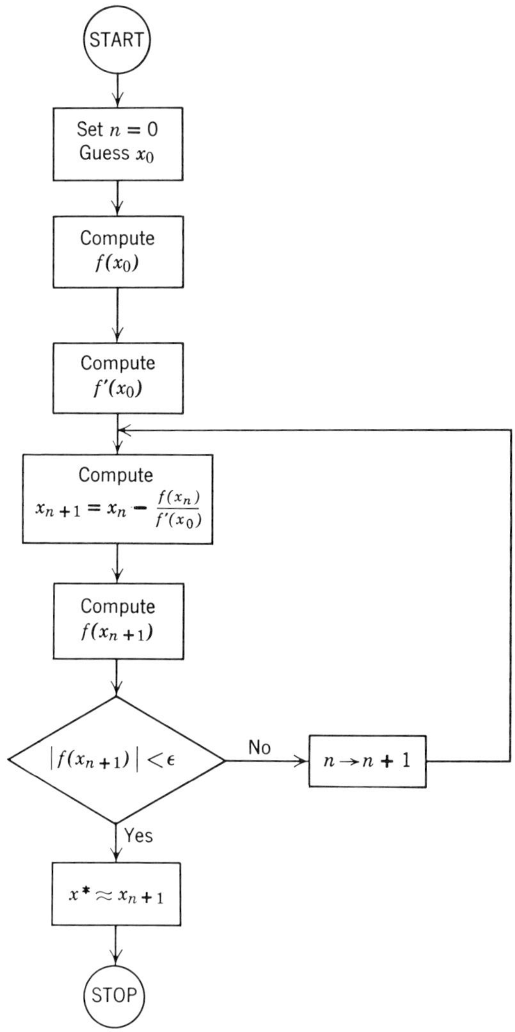

Flow Chart 13.2 Modified Newton-Raphson method.

the basic Newton-Raphson and the modified Newton-Raphson algorithms. From the flow charts, it is clear that the modified Newton-Raphson scheme requires the computation of the derivative of $f(x)$ only at x_0, which is done once. If $f(x)$ is not known analytically, $f'(x)$ must be computed numerically. For such cases, the basic Newton-Raphson algorithm may be slower because of the need to compute $f'(x_n)$ for each new x_n.

It is clear that the closer $f(x)$ is to a straight line, the faster is the convergence. In some problems, it may be possible to change variables so that the resulting function is more nearly a straight line in terms of the new coordinates. That is, $f(x)=0$ might be replaced by $f[g(y)]=0$ where $x=g(y)$ so that the new function of y, $w(y)\equiv f[g(y)]$ is closer to a straight line. Of course, it may not be easy to think of a transformation $x=g(y)$ such that this is the case. Nevertheless, the complexity or simplicity of the problem is influenced by the choice of coordinates, so that, in any given situation, some consideration might be given to choices for coordinates.

The simple algorithm above may be extended to several variables. For example, suppose that we have a system of two equations in two variables,

$$f_1(x_1, x_2)=0 \tag{13.17}$$

$$f_2(x_1, x_2)=0. \tag{13.18}$$

To solve for the roots of these equations by iteration, we expand each function in Taylor's series in the two variables x_1, x_2, evaluated at the nth iteration. Retaining only the linear terms and denoting the next approximation by $(x_1)_{n+1}$ and $(x_2)_{n+1}$ we have

$$f_1[(x_1)_{n+1}, (x_2)_{n+1}]=f_1[(x_1)_n, (x_2)_n]+\left(\frac{\partial f_1}{\partial x_1}\right)_n[(x_1)_{n+1}-(x_1)_n] + \left(\frac{\partial f_1}{\partial x_2}\right)_n[(x_2)_{n+1}-(x_2)_n]=0 \tag{13.19}$$

and

$$f_2[(x_1)_{n+1}, (x_2)_{n+1}]=f_2[(x_1)_n, (x_2)_n]+\left(\frac{\partial f_2}{\partial x_1}\right)[(x_1)_{n+1}-(x_1)_n] + \left(\frac{\partial f_2}{\partial x_2}\right)_n[(x_2)_{n+1}-(x_2)_n]=0 \tag{13.20}$$

where the partial derivatives are evaluated at $(x_1)_n$ and $(x_2)_n$, the nth guess for x_1 and x_2 after the very first choice $(x_1)_0$ and $(x_2)_0$. We now have a set of two linear equations in two unknowns. Denote $(x_1)_{n+1}-(x_1)_n$ by $(\Delta x_1)_n$ and $(x_2)_{n+1}-(x_2)_n$ by $(\Delta x_2)_n$. Then we may write (13.19) and (13.20) in matrix form as

$$\begin{bmatrix} \left(\frac{\partial f_1}{\partial x_1}\right)_n & \left(\frac{\partial f_1}{\partial x_2}\right)_n \\ \left(\frac{\partial f_2}{\partial x_1}\right)_n & \left(\frac{\partial f_2}{\partial x_2}\right)_n \end{bmatrix} \begin{bmatrix} (\Delta x_1)_n \\ (\Delta x_2)_n \end{bmatrix} = \begin{bmatrix} -(f_1)_n \\ -(f_2)_n \end{bmatrix} \tag{13.21}$$

If the matrix premultiplying $(\Delta x)_n$ is nonsingular, we may solve for $(\Delta x_1)_n$ and $(\Delta x_2)_n$ as

$$\begin{bmatrix} (\Delta x_1)_n \\ \\ (\Delta x_2)_n \end{bmatrix} = \begin{bmatrix} \left(\frac{\partial f_1}{\partial x_1}\right)_n & \left(\frac{\partial f_1}{\partial x_2}\right)_n \\ \\ \left(\frac{\partial f_2}{\partial x_1}\right)_n & \left(\frac{\partial f_2}{\partial x_2}\right) \end{bmatrix}^{-1} \begin{bmatrix} -(f_1)_n \\ \\ -(f_2)_n \end{bmatrix} \tag{13.22}$$

Equation 13.22 provides the Newton-Raphson basic algorithm for generating $(\Delta x_1)_n$ and $(\Delta x_2)_n$. The modified algorithm differs slightly from the above in that the partial derivatives of f_i are not needed for each stage of the iteration. Thus

$$\begin{bmatrix} (\Delta x_1)_n \\ (\Delta x_2)_n \end{bmatrix} = \begin{bmatrix} \left(\frac{\partial f_1}{\partial x_1}\right)_0 & \left(\frac{\partial f_1}{\partial x_2}\right)_0 \\ \left(\frac{\partial f_2}{\partial x_1}\right)_0 & \left(\frac{\partial f_2}{\partial x_2}\right)_0 \end{bmatrix}^{-1} \begin{bmatrix} -(f_1)_n \\ -(f_2)_n \end{bmatrix} \tag{13.23}$$

These algorithms are generalized to the nth-order case in a straightforward manner. For the basic algorithm, we have

$$\begin{bmatrix} (\Delta x_1)_n \\ (\Delta x_2)_n \\ \vdots \\ (\Delta x_p)_n \end{bmatrix} = \begin{bmatrix} \left(\frac{\partial f_1}{\partial x_1}\right)_n & \left(\frac{\partial f_1}{\partial x_2}\right)_n & \cdots & \left(\frac{\partial f_1}{\partial x_p}\right)_n \\ \vdots & & & \\ \left(\frac{\partial f_p}{\partial x_1}\right)_n & \left(\frac{\partial f_p}{\partial x_2}\right)_n & \cdots & \left(\frac{\partial f_p}{\partial x_p}\right)_n \end{bmatrix}^{-1} \begin{bmatrix} -(f_1)_n \\ -(f_2)_n \\ \vdots \\ -(f_p)_n \end{bmatrix} \tag{13.24}$$

The modified algorithm is generated by using only the partial derivatives at the first guess x_0. It is given by

$$\begin{bmatrix} (\Delta x_1)_n \\ (\Delta x_2)_n \\ \vdots \\ (\Delta x_p)_n \end{bmatrix} = \begin{bmatrix} \left(\frac{\partial f_1}{\partial x_1}\right)_0 & \cdots & \left(\frac{\partial f_1}{\partial x_p}\right)_0 \\ \vdots & & \\ \left(\frac{\partial f_p}{\partial x_1}\right)_0 & \cdots & \left(\frac{\partial f_p}{\partial x_p}\right)_0 \end{bmatrix}^{-1} \begin{bmatrix} -(f_1)_n \\ \vdots \\ -(f_p)_n \end{bmatrix} \tag{13.25}$$

Just as in the single variable problem, the sequence may not converge. However, if the following conditions are met, convergence is guaranteed.

(a) The matrix whose (i, j) element is $(\partial f_i/\partial x_j)_0$ evaluated at the initial guess x_0 is nonsingular and

$$\frac{1}{D}\max_k \sum |A_{jk}| \leqslant B, k = 1, \ldots, p \tag{13.26}$$

where D is the determinant of the matrix and A_{ij} is the cofactor of the (i, j) element of the matrix.

(b) $\max[|\Delta x_1|_0, |\Delta x_2|_0, \ldots, |\Delta x_p|_0] \leqslant \alpha.$ (13.27)

(c) $$\left|\frac{\partial^2 f_i}{\partial x_j \partial x_k}\right| \leqslant C, (i, j, k = 1, 2, \ldots, p). \tag{13.28}$$

in the region $|x_i - (x_i)_0| \leqslant r$.

(d) $h = \alpha B C p^2 \leqslant \frac{1}{2}$. (13.29)

(e) $$\left(\frac{1-\sqrt{1-2h}}{h}\right)\alpha \leqslant r. \tag{13.30}$$

These conditions are sufficient for convergence for both the basic and modified algorithm. Furthermore, if

$$r < \frac{1+\sqrt{1-2h}}{h}\alpha \tag{13.31}$$

then the solution is unique in the interval $|x - x_0| \leqslant r$.

In the application of Newton's method, some preliminary searching may be needed so that the starting point x_0 is close enough to the solution. For continuous functions whose second partial derivatives exist, the above conditions for convergence will be satisfied if x_0 is close enough to the true solution.

The left-hand sides in conditions (a) and (b) above are computable in a straightforward manner. The quantities $(\Delta x_i)_0$, $i = 1, 2, \ldots, p$ are obtained from (13.24). Thus the two bounds α and B are readily obtained. For the third condition (c), we have to examine the second partial derivative of f_i for x_i in the region $|x_i - (x_i)_0| \leqslant r$ for all $i = 1, 2, \ldots, p$. The quantity r specifies the neighborhood of $(x_i)_0$ over which the second partial derivative is to be bounded, and its value is left to our choice, except that it must satisfy condition (3) also. It is desirable to choose as small a value of r as possible, since for a given initial guess $(x_i)_0$, the maximum of the magnitude of the second partial derivative of f_i, in the region defined by r, increases as r is increased, necessitating a larger value of C which may not satisfy condition (d).

A convenient choice for r is 2α. For h between 0 and $\frac{1}{2}$, $0 < h \leqslant \frac{1}{2}$, it can be shown that

$$\frac{1-\sqrt{1-2h}}{h} \leqslant 2. \tag{13.32}$$

In fact the quantity $[(1-\sqrt{1-2h})/h]$ increases monotonically from 1 to 2 as h increased from 0 to $\frac{1}{2}$.

An estimate of the rate of convergence of the series is provided by the inequality

$$\max_{\substack{i \\ i=1,2,\ldots,p}} [|x_i^* - (x_i)_n|] \leqslant \frac{1}{2^n}(2h)^{2^n}\frac{1}{BCp^2}, \quad (n=0, 1, \ldots) \tag{13.33}$$

for the basic algorithm, and

$$\max_{\substack{i \\ i=1,\ldots,p}} [|x_i^* - (x_i)_n|] \leqslant (1-\sqrt{1-2h})^{n+1}\frac{1}{BCp^2}, \quad (n=0, 1, \ldots,) \tag{13.34}$$

for the modified algorithm.

Example 13-2. Consider the system of equations

$$x_1^2 + 3x_1x_2^2 - 1 = 0 \tag{13.35}$$

$$x_1^3x_2 + x_2^2 - 1 = 0. \tag{13.36}$$

From a graphical plot of the two equations it is known that the solution is in the neighborhood of $(x_1)_0 = 0.32$ and $(x_2)_0 = 0.98$. Let us construct on algorithm for improving on this approximation. First, we compute $(\Delta x_1)_0$ and $(\Delta x_2)_0$. Equation 13.21, for $n = 0$, evaluated at $(x_1)_0 = 0.32$ and $(x_2)_0 = 0.98$ yields

$$3.52(\Delta x_1)_0 + 1.88(\Delta x_2)_0 = -0.0245 \tag{13.37}$$

$$0.301(\Delta x_1)_0 + 3.798(\Delta x_2)_0 = 0.046. \tag{13.38}$$

Solving for $(\Delta x_1)_0$ and $(\Delta x_2)_0$, we obtain

$$\begin{bmatrix}(\Delta x_1)_0\\(\Delta x_2)_0\end{bmatrix}=\frac{1}{13.03}\begin{bmatrix}3.798 & -1.88\\-0.301 & 3.52\end{bmatrix}\begin{bmatrix}-0.0245\\0.046\end{bmatrix}=\begin{bmatrix}-0.0138\\0.013\end{bmatrix} \quad (13.39)$$

Hence

$$\alpha \geqslant \max\,[|-0.0138|,|0.013|]$$

so we may choose

$$\alpha = 0.014. \quad (13.40)$$

To choose B, first we compute

$$\frac{1}{13.03}[3.52+0.301]=0.293$$

and

$$\frac{1}{13.03}[1.88+3.798]=0.436.$$

Thus

$$\max\,[0.293, 0.436]=0.436 \leqslant B.$$

Choose

$$B=0.437. \quad (13.41)$$

Next we have to compute C. We examine the second partial derivatives for $r=2\alpha=0.028$. This means that

$$|x_1-0.32| \leqslant 0.028$$

or

$$0.292 \leqslant x_1 \leqslant 0.348 \quad (11.42)$$

and

$$|x_2-0.98| \leqslant 0.028$$

or

$$0.952 \leqslant x_2 \leqslant 1.008. \quad (13.43)$$

From (13.35) and (13.36), the second partial derivatives are

$$\frac{\partial^2 f_1}{\partial x_1 \partial x_1}=2; \qquad \frac{\partial^2 f_1}{\partial x_1 \partial x_2}=6x_2; \qquad \frac{\partial^2 f_1}{\partial x_2 \partial x_2}=6x_1$$

$$\frac{\partial^2 f_2}{\partial x_1 \partial x_1}=6x_1x_2; \qquad \frac{\partial^2 f_2}{\partial x_1 \partial x_2}=3x_1^2; \qquad \frac{\partial^2 f_2}{\partial x_2 \partial x_2}=12x_2^2.$$

Clearly,

$$\max\left|\frac{\partial^2 f_1}{\partial x_i \partial x_j}\right| = 12(1.008)^2 = 12.17 \leqslant C$$

and we choose

$$C = 12.2. \tag{13.44}$$

Now compute

$$h = \alpha B C p^2 = (0.014)(0.437)(12.2)(4) = 0.298 < \tfrac{1}{2}.$$

Hence all the conditions for convergence are satisfied. The modified Newton-Raphson algorithm is

$$\begin{bmatrix}(\Delta x_1)_n\\(\Delta x_2)_n\end{bmatrix} = \begin{bmatrix}(x_1)_{n+1} & -(x_1)_n\\(x_2)_{n+1} & -(x_2)_n\end{bmatrix} = \frac{1}{13.03}\begin{bmatrix}3.798 & -1.88\\-0.301 & 3.52\end{bmatrix}\begin{bmatrix}-(f_1)_n\\-(f_2)_n\end{bmatrix} \tag{13.45}$$

where the starting point is $(x_1)_0 = 0.32$ and $(x_2)_0 = 0.98$. The next approximations are

$$(x_1)_1 = (x_1)_0 + (\Delta x_1)_0 = 0.32 + (-0.0138) = 0.3062$$
$$(x_2)_1 = (x_2)_0 + (\Delta x_2)_0 = 0.98 + (0.013) = 0.993.$$

From (13.33), the error between $(x_1)_1$ and x_1^*, and $(x_2)_1$ and x_2^* is less than

$$|\text{error}| \leqslant \tfrac{1}{2}(2h)^2\frac{B}{Cp^2} = \tfrac{1}{2}(0.596)^2\frac{(0.436)}{(12.2)(4)} \leqslant 0.00158. \blacktriangle \tag{13.46}$$

For a more complicated example, the main difficulty in checking the conditions for convergence is checking the second partial derivative. In a practical case, the algorithm is used even without checking the conditions for convergence, since such a check may be as difficult as the original problem itself. Of course there is no guarantee of convergence in this case.

13.2 The Gradient Method

Consider a function of two variables $f(x_1, x_2)$, which we wish to maximize with respect to a choice of x_1 and x_2, and imagine a plot of f with respect to x_1 and x_2 in three dimensions. Suppose we find the intersection of planes parallel to the (x_1, x_2) plane with the surface $f(x_1, x_2)$. Then these intersections or contours represent loci of x_1 and x_2 which yield a specific constant value for f. Figure 13.3 illustrates

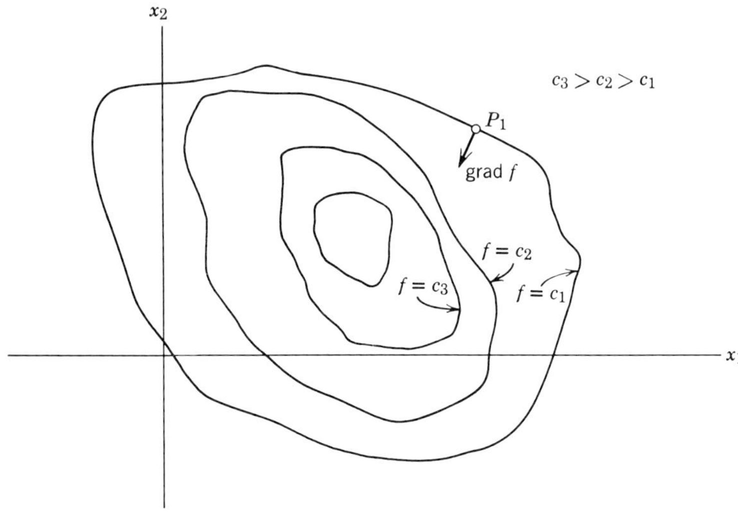

Figure 13.3 Contours of constant f for a function of two variables.

possible contours for a function of two variables. Suppose that the point P_1 on the $x_1 - x_2$ plane is given and it is desired to investigate how we should move to another point P_2 of fixed distance from P_1 such that the new value of f at P_2 results in the greatest increase of f from its value at P_1. The point P_1 is shown as lying on the contour for which $f = c_1$. If P_2 is chosen to be on this $f = c_1$ contour, then there will be no increase in the value of f as we move from P_1 to P_2. If the distance $P_2 - P_1$ is differentially small, then the line segment $P_2 - P_1$, if P_2 is also on $f = c_1$, will be approximately tangent to the contour $f = c_1$ at the point P_1. Alternatively, if $|P_2 - P_1|$ is differentially small and if P_2 is chosen to lie on a line tangent to the $f = c_1$ contour at the point P_1, then P_2 will be approximately on the contour $f = c_1$ also. To a linear approximation, moving to a point P_2 on a line tangent to $f = c_1$ at P_1 results in no increase in f. This observation leads to a conjecture that to a linear approximation, moving to a point P_2 on a line perpendicular to the tangent to $f = c_1$ at P_1 yields the greatest increase in f, for a specified but small value for $|P_2 - P_1|$. That this is indeed the case will be shown below.

Consider passing a plane through the point P_1 of the f surface, which is perpendicular to the (x_1, x_2) plane. The surface and the plane will intersect in a curve. Such an intersection is illustrated in Figure 13.4. We now wish to examine all possible angles for the plane so that the slope of the curve at P_1 will be the largest. The surface may be imagined as some kind of a hill or mountain and the side view of

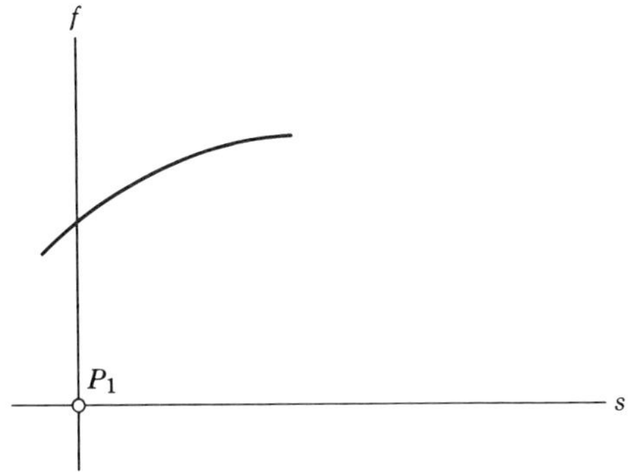

Figure 13.4 Side view of intersection of the f surface and a plane through the point P_1 and perpendicular to the (x_1, x_2) plane.

Figure 13.4 as a cross-section view of the hill. If the plane is made tangent to $f = c_1$ at P_1, then the intersection curve of Figure 13.4 will have a zero slope at P_1. As discussed earlier, this simply means that, to a linear approximation, f will not be increased if we move to a point P_2 such that the line segment $P_2 - P_1$ is tangent to the curve $f = c_1$ and $|P_2 - P_1|$ is differentially small.

Let the new direction be represented by s. We wish to move to a point P_2 of fixed distance ds from P_1 such that if a linear approximation is used, the change in f is greatest. The change in f, to a linear approximation, is represented by the differential df which is

$$df = \frac{\partial f}{\partial x_1} dx_1 + \frac{\partial f}{\partial x_2} dx_2. \tag{13.47}$$

Equation 13.47 is the scalar product of two vectors whose matrix representations are

$$\begin{bmatrix} \dfrac{\partial f}{\partial x_1} \\ \dfrac{\partial f}{\partial x_2} \end{bmatrix} \quad \text{and} \quad \begin{bmatrix} dx_1 \\ dx_2 \end{bmatrix}.$$

Thus

$$df = \begin{bmatrix} \dfrac{\partial f}{\partial x_1} \\ \dfrac{\partial f}{\partial x_2} \end{bmatrix}^T \begin{bmatrix} dx_1 \\ dx_2 \end{bmatrix}. \tag{13.48}$$

The first vector is defined as the gradient of f with respect to x_1 and x_2 and denoted by grad f or $\nabla_x f$.

$$\text{Matrix representation of } \nabla_x f = \begin{bmatrix} \dfrac{\partial f}{\partial x_1} \\ \dfrac{\partial f}{\partial x_2} \end{bmatrix}. \tag{13.49}$$

The gradient is computed at the point P_1. If the vector dx, whose matrix representation is

$$\begin{bmatrix} dx_1 \\ dx_2 \end{bmatrix}, \tag{13.50}$$

has length ds,

$$(dx_1)^2 + (dx_2)^2 = (ds)^2$$

or

$$\left(\frac{dx_1}{ds}\right)^2 + \left(\frac{dx_2}{ds}\right)^2 = 1. \tag{13.51}$$

Since the magnitude of the vector in (13.50) is ds which is fixed, the only adjustable quantity is its angle. To maximize the change df, we should choose the vector to be colinear with the gradient vector $\overrightarrow{\nabla_x f}$, that is, we choose the angle between the two vectors to be zero. Hence we can choose dx/ds as

$$\frac{\overrightarrow{\mathbf{dx}}}{ds} = \frac{\overrightarrow{\nabla_x f}}{\sqrt{\left(\dfrac{\partial f}{\partial x_1}\right)^2 + \left(\dfrac{\partial f}{\partial x_2}\right)^2}} \tag{13.52}$$

This choice makes the left-hand side of (13.52) have the same direction as the gradient and the magnitude is unity. The differentials for changing x_1 and x_2 to $x_1 + dx_1$ and $x_2 + dx_2$ are given by

$$dx_1 = \frac{\dfrac{\partial f}{\partial x_1} ds}{\sqrt{\left(\dfrac{\partial f}{\partial x_1}\right)^2 + \left(\dfrac{\partial f}{\partial x_2}\right)^2}} \tag{13.53}$$

and

$$dx_2 = \frac{\dfrac{\partial f}{\partial x_2} ds}{\sqrt{\left(\dfrac{\partial f}{\partial x_1}\right)^2 + \left(\dfrac{\partial f}{\partial x_2}\right)^2}} \tag{13.54}$$

If a minimization instead of a maximization is desired, then instead of making the angle between the two vectors in (13.48) zero, we make it 180° so that

$$\begin{bmatrix} \dfrac{dx_1}{ds} \\ \\ \dfrac{dx_2}{ds} \end{bmatrix} = \frac{-1}{\sqrt{\left(\dfrac{\partial f}{\partial x_1}\right)^2 + \left(\dfrac{\partial f}{\partial x_2}\right)^2}} \begin{bmatrix} \dfrac{\partial f}{\partial x_1} \\ \\ \dfrac{\partial f}{\partial x_2} \end{bmatrix}. \tag{13.55}$$

The gradient of f at point P_1 is a vector that is perpendicular to the contour $f = c_1$ at the point P_1. Hence the direction in moving from P_1 to P_2 represented by (13.52), which is the direction of the gradient of f, is perpendicular to the contour $f = c_1$ at P_1, as stated earlier. The direction represented by $\overrightarrow{\nabla_x f}$ is the direction of *steepest ascent* and the direction $-\overrightarrow{\nabla_x f}$ is the direction of *steepest descent*.

The actual increase or decrease in f, when taking a finite step ds which is not small, is of course not necessarily largest. The closeness of the approximation depends on the closeness of $f(x_1, x_2)$ to a linear function in the neighborhood of the point under examination (such as P_1). This closeness to linearity, in turn, depends on the choice of coordinates. That is, if x_1 and x_2 can be transformed by $x_1 = g_1(y_1)$ and $x_2 = g_2(y_2)$ so that

$$f(x, y) = f[g_1(y_1), g_2(y_2)] \equiv h(y_1, y_2), \tag{13.56}$$

the new function $h(y_1, y_2)$ may be closer to a linear function and hence the gradient procedure may be more effective for h.

The gradient procedure may be generalized to functions of more than two variables such as $f(x_1, \ldots, x_n)$. Subject to the restriction that

$$ds = \sqrt{(dx_1)^2 + (dx_2)^2 + \cdots + (dx_n)^2} \tag{13.57}$$

is a fixed number, the corrections in moving from x_1 to $x_1 + dx_1$, x_2 to $x_2 + dx_2, \ldots$, and x_n to $x_n + dx_n$, for a maximization problem, are given by

$$\begin{aligned} dx_1 &= \frac{\dfrac{\partial f}{\partial x_1} ds}{\sqrt{\left(\dfrac{\partial f}{\partial x_1}\right)^2 + \cdots + \left(\dfrac{\partial f}{\partial x_n}\right)^2}} \\ &\;\vdots \\ dx_n &= \frac{\dfrac{\partial f}{\partial x_n} ds}{\sqrt{\left(\dfrac{\partial f}{\partial x_1}\right)^2 + \cdots + \left(\dfrac{\partial f}{\partial x_n}\right)^2}} \end{aligned} \tag{13.58}$$

For a minimization problem, we multiply the right-hand sides in (13.58) by (-1). The algorithm for generating the $(i+1)$th approximation for $(x_k)_{i+1}$ from $(x_k)_i$ is provided by (13.58), where all the partial derivatives on the right-hand side of (13.58) are evaluated at $(x_1)_i$, $(x_2)_i, \ldots, (x_n)_i$ and the left-hand side is replaced by $(x_k)_{i+1}-(x_k)_i$. That is,

$$
\begin{aligned}
(x_1)_{i+1}-(x_1)_i &= \frac{\left(\dfrac{\partial f}{\partial x_1}\right)_i ds}{\sqrt{\left(\dfrac{\partial f}{\partial x_1}\right)_i^2+\cdots+\left(\dfrac{\partial f}{\partial x_n}\right)_i^2}} \\
&\vdots \\
(x_n)_{i+1}-(x_n)_i &= \frac{\left(\dfrac{\partial f}{\partial x_n}\right)_i ds}{\sqrt{\left(\dfrac{\partial f}{\partial x_1}\right)_i^2+\cdots+\left(\dfrac{\partial f}{\partial x_n}\right)^2}}
\end{aligned}
\tag{13.59}
$$

The step-size ds is undetermined so far, and its choice depends a great deal on the particular function, so that some experimentation with several values of ds is desirable in any given problem. As the sequence tends to converge to some limit, it may be advisable to reduce ds so that the searching will be confined to a smaller neighborhood of the nth iterate. The procedure may be terminated when the partial derivatives have magnitudes that are less than some predetermined small quantity close to zero. Note that, at a maximum, the partial derivatives are all zero. The algorithm is shown in Flow Chart 13.3.

At a maximum or minimum, the gradient of f must be zero:

$$\vec{\nabla}_x f = 0. \tag{13.60}$$

For the one-dimensional function $f(x)$ shown in Figure 13.5, zero gradient simply means that the slope is zero. The points x_1, x_3, and x_5 are called *local maxima.* The points x_2, x_4, and x_6 are called *local minima.* x_7 is called an inflection point. At these local minima, local maxima, and inflection point, the gradient is zero.

In the immediate vicinity of a local minimum, such as x_2, the value $f(x_2)$ is not greater than $f(x)$ for any x in the vicinity of x_2. However, if

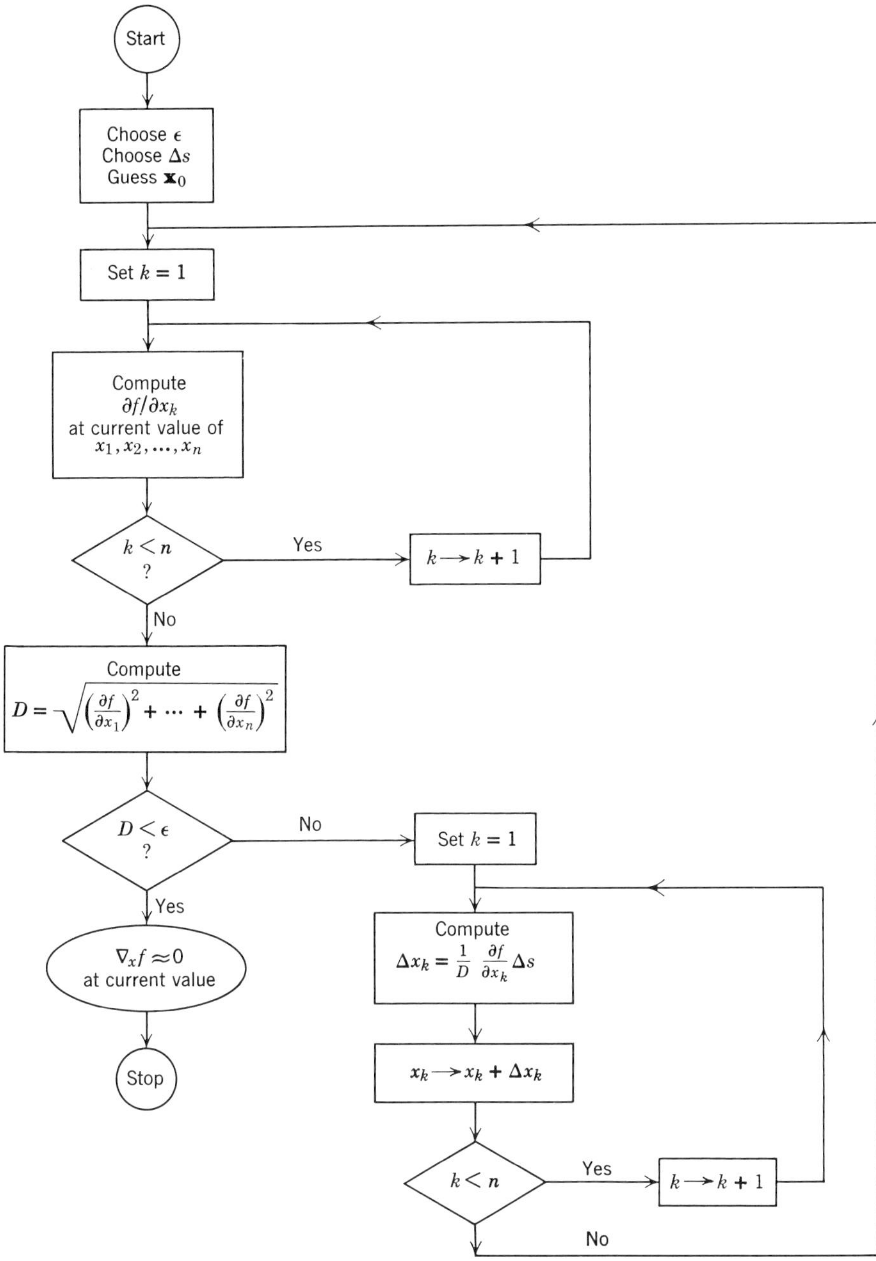

Flow Chart 13.3 Steepest ascent procedure. For steepest descent, instead of $x_k \rightarrow x_k + x_k + \Delta x_k$, use $x_k \rightarrow x_k - \Delta x_k$.

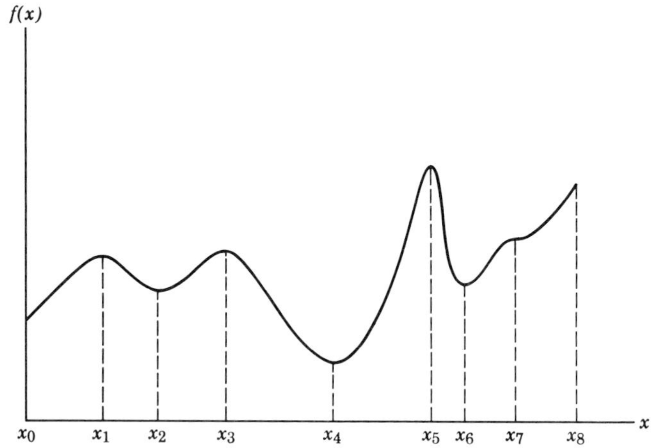

Figure 13.5 A function with several local minima and maxima.

x is not in the immediate vicinity of $x_2, f(x)$ may be lower than $f(x_2)$. If the allowed range of x is from x_0 to x_8, $f(x)$ attains a minimum at x_4. The point x_4 is said to be a *global minimum*. It has the property that $f(x_4) \leq f(x)$ for any x between x_0 and x_8. In the immediate vicinity of a local maximum, such as x_3, $f(x_3) \geq f(x)$, provided that x is close enough to x_3. However in the complete interval from x_0 to x_8 where x is allowed to have a value, there are points for which $f(x) > f(x_3)$. A point which yields the largest value of f over the entire interval is called a *global maximum*. The point x_8 is such a point. Notice that the slope of f at x_8 is not zero.

The gradient method of following the direction of steepest ascent will yield a point with zero gradient, which may be a local maximum or an inflection point. There is no guarantee that the point is a global maximum. Other local maxima may be obtained by using several different initial guesses x_0. The value of f is also computed on the boundary.

Similarly, the gradient method of following the direction of steepest descent will yield a point with zero gradient, which may be a local minimum or an inflection point. Several different guesses of initial points x_0 may lead to different local minima. From the local minima and boundary point, the one that yields the smallest value of f is the best numerical answer for the minimization. However, there is no guarantee that it is the global minimum. For a general function, there is no known numerical method that guarantees a global minimum or maximum.

The gradient method may be used to solve for the roots of a system of equations such as

$$
\begin{aligned}
f_1(x_1, x_2, \ldots, x_p) &= 0 \\
f_2(x_1, x_2, \ldots, x_p) &= 0 \\
&\vdots \\
f_p(x_1, x_2, \ldots, x_p) &= 0.
\end{aligned}
\tag{13.61}
$$

Form $F(x_1, \ldots, x_p)$,

$$F(x_1, x_2, \ldots, x_p) = \sum_{i=1}^{p} f_i^2(x_1, x_2, \ldots, x_p). \tag{13.62}$$

The new quantity $F(x_1, \ldots, x_p)$ may be minimized, using the gradient method. Note that F is minimum when it has zero value. This happens only when all the f_i are zero, and hence (13.61) is satisfied.

The Newton-Raphson method may be used to solve a minimization or maximization problem. Thus, if $f(x)$ is to be minimized or maximized, then a local minimum or local maximum must satisfy (13.60). The Newton-Raphson method is employed to obtain values of x that satisfy (13.60). Of course, if $\nabla_x f = 0$ has several solutions, there is no systematic way of insuring that all the solutions will be obtained. The gradient method and the Newton-Raphson method are complementary methods which may be used for solving the same problem. As a general guide, the gradient method gives fairly good improvements, starting from a rough initial guess to some vicinity of the solution. The Newton-Raphson method, on the other hand, works nicely, with convergence realized, whenever the starting point is close to the solution. Thus a reasonable strategy is to employ the gradient method for a starter, switching to the Newton-Raphson method as the improvements in the iteration become relatively small.

13.3 Use of Sensitivity Functions in Newton-Raphson and Gradient Methods

In Sections 13.1 and 13.2, we assumed that the functions to be minimized, or the equations whose roots are to be determined, are given explicitly in terms of the unknown parameters. For example, in Chapter 12, using the performance index

$$I = \int_0^\infty e^2(t)\, dt \tag{13.63}$$

for a linear time-invariant system where $e(t)$ is some error signal, if the order of the stable dynamic system is small, the algebraic expression for I may be obtained from Table 12.1. The algebraic expression for I is in terms of the numerator and denominator coefficients of the Laplace transform of $e(t)$. In turn, the coefficients are readily expressed, in terms of the control parameters which are to be optimized, by a routine analysis of the system block diagram or signal flow graph. Thus, if $x_1, x_2, \ldots, x_p$ are the control parameters, the performance index in (13.63) is equal to a function of $x_1, x_2, \ldots, x_p$, denoted by $I(x_1, x_2, \ldots, x_p)$. The function I is then minimized (or maximized, depending on the problem) by the gradient method, or the gradient of I is formed and equated to zero, for solution by the Newton-Raphson method.

Unfortunately, when the upper limit of the integral in (13.63) is not infinity, or if the integrand is not a quadratic, or if the system is not linear, or if the system is of high order, or if a combination of these conditions prevail, I is not easily expressed in terms of the parameters explicitly. However, as we shall see shortly, the gradient of I may be written in terms of partial derivatives of signals, such as e, with respect to the control parameters. Hence we may apply the Newton-Raphson method or the gradient method by machine minimization of I. In order to avoid confusion of control parameters with our usual notation for state variables, let us denote the parameters by $\alpha_1, \alpha_2, \ldots, \alpha_p$. The signals in the system are dependent not only on the time variable t but also on these parameters. Since we shall be taking partial derivatives with respect to these parameters, let us indicate the dependence of the signals on these parameters by writing the signals with $t, \alpha_1, \ldots, \alpha_p$ as arguments. Thus, instead of $e(t)$, let us write $e(t, \alpha_1, \alpha_2, \ldots, \alpha_p)$.

For example, suppose that the performance index to be minimized is

$$I = \int_0^T e^4(t, \alpha_1, \alpha_2, \ldots, \alpha_p)\, dt. \tag{13.64}$$

Then, using the chain rule for partial differentiation and assuming that T is not dependent on $\alpha_1, \alpha_2, \ldots, \alpha_p$, we obtain

$$\begin{aligned}
\frac{\partial I}{\partial \alpha_1} &= \int_0^T 4e^3(t, \alpha_1, \alpha_2, \ldots, \alpha_p)\,\frac{\partial e}{\partial \alpha_1}(t, \alpha_1, \ldots, \alpha_p)\, dt \\
\frac{\partial I}{\partial \alpha_2} &= \int_0^T 4e^3(t, \alpha_1, \alpha_2, \ldots, \alpha_p)\,\frac{\partial e}{\partial \alpha_2}(t, \alpha_1, \ldots, \alpha_p)\, dt \\
&\;\;\vdots \\
\frac{\partial I}{\partial \alpha_p} &= \int_0^T 4e^3(t, \alpha_1, \alpha_2, \ldots, \alpha_p)\,\frac{\partial e}{\partial \alpha_p}(t, \alpha_1, \alpha_2, \ldots, \alpha_p)\, dt.
\end{aligned} \tag{13.65}$$

The left-hand side of the set of equations above represent the gradient of I with respect to $\alpha_1, \alpha_2, \ldots, \alpha_p$. The quantities $\partial e/\partial\alpha_1$ and $\partial e/\partial\alpha_p$ are called *sensitivity functions*. It was demonstrated in Chapters 6 and 7 that the sensitivity functions may be obtained as signals from models of the system being optimized, where the inputs of the sensitivity models come from the system itself. Hence the expressions on the right-hand side of the set of equations in (13.64) may be generated from a computer *simulation*, using an analog, hybrid, or digital computer.

Consider the system to be described in state form and for simplicity assume that the differential equations are linear and time-invariant. Let the control parameters to be chosen be denoted by $\alpha_1, \alpha_2, \ldots, \alpha_p$, and let us display the possible dependence of signals on these parameters by explicitly indicating these as arguments of the functions. Denote the output vector by $\mathbf{y}(t, \alpha_1, \ldots, \alpha_p)$, the state vector by $\mathbf{x}(t, \alpha_1, \alpha_2, \ldots, \alpha_p)$, and the input vector by $\mathbf{r}(t)$. We write the system equations as

$$\frac{\partial \mathbf{x}}{\partial t}(t, \alpha_1, \ldots, \alpha_p) = \mathbf{A}(\alpha_1, \ldots, \alpha_p)\mathbf{x}(t, \alpha_1, \ldots, \alpha_p) + \mathbf{B}(\alpha_1, \ldots, \alpha_p)\mathbf{r}(t)$$

$$\mathbf{y}(t, \alpha_1, \ldots, \alpha_p) = \mathbf{C}(\alpha_1, \ldots, \alpha_p)\mathbf{x}(t, \alpha_1, \ldots, \alpha_p) + \mathbf{D}(\alpha_1, \ldots, \alpha_p)\mathbf{r}(t) \tag{13.66}$$

The matrices $\mathbf{A}$, $\mathbf{B}$, $\mathbf{C}$, and $\mathbf{D}$ are time-invariant, but they depend on the parameters $\alpha_1, \alpha_2, \ldots, \alpha_p$. Let us assume that the partial derivatives of the matrices with respect to the parameters exist and that these are continuous functions with respect to these parameters. Then the signals are continuous and differentiable functions of these parameters. Let us assume also that all second-order partial derivatives are continuous. Then if we take the partial derivative of the vector equations with respect to α_i, and since $\partial^2 x/\partial t\,\partial\alpha_i = \partial^2 x/\partial\alpha_i\partial t$, we have

$$\begin{aligned} \frac{\partial}{\partial t}\frac{\partial \mathbf{x}}{\partial \alpha_i} &= \mathbf{A}\frac{\partial \mathbf{x}}{\partial \alpha_i} + \frac{\partial \mathbf{A}}{\partial \alpha_i}\mathbf{x} + \frac{\partial \mathbf{B}}{\partial \alpha_i}\mathbf{r} \\ \frac{\partial \mathbf{y}}{\partial \alpha_i} &= \mathbf{C}\frac{\partial \mathbf{x}}{\partial \alpha_i} + \frac{\partial \mathbf{C}}{\partial \alpha_i}\mathbf{x} + \frac{\partial \mathbf{D}}{\partial \alpha_i}\mathbf{r}. \end{aligned} \tag{13.67}$$

In (13.67), the partial derivatives with respect to α_i are evaluated for fixed α_i, as is required by either the Newton-Raphson or the gradient procedure. Thus, (13.67) is an *ordinary* differential equation for the *time functions* $\partial\mathbf{x}/\partial\alpha_i$ and $\partial\mathbf{y}/\partial\alpha_i$. These equations are similar to the

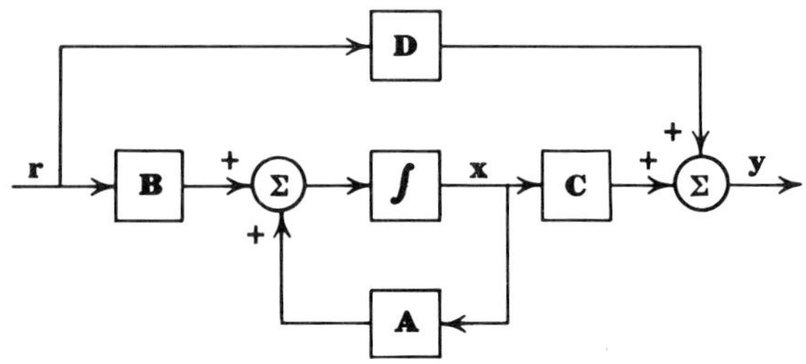

Figure 13.6 Block diagram for Equation 13.66.

ones in (13.66) except that instead of $\mathbf{x}$ as a state vector and $\mathbf{y}$ as the output vector, we have $\partial\mathbf{x}/\partial\alpha_i$ and $\partial\mathbf{y}/\partial\alpha_i$, respectively.

Figure 13.6 shows a block-diagram representation for the state equations in (13.66). The sensitivity function vectors $\partial\mathbf{x}/\partial\alpha_i$ and $\partial\mathbf{y}/\partial\alpha_i$ may be generated by simulation, using the block diagram of Figure 13.7. Notice the similarities between portions of the block diagrams. The inputs to the sensitivity-system model consists of $\mathbf{r}$ and $\mathbf{x}$, where $\mathbf{r}$ is the input to the original system and $\mathbf{x}$ is the state vector of the original system. Thus $\mathbf{r}$ and $\mathbf{x}$ may be obtained as signals generated from the block diagram of Figure 13.6. The boxes corresponding to the matrices $\partial\mathbf{A}/\partial\alpha_i$, $\partial\mathbf{B}/\partial\alpha_i$, $\partial\mathbf{C}/\partial\alpha_i$, and $\partial\mathbf{D}/\partial\alpha_i$ are simply constant gain amplifiers on an analog simulation. In the special case when α_i is an element of one of the matrices, then all the partial derivatives are zero except for one entry in one matrix, which will be unity. For example, if $\alpha_i = b_{kj}$, then all the matrices $\partial\mathbf{A}/\partial\alpha_i$, $\partial\mathbf{C}/\partial\alpha_i$, and $\partial\mathbf{D}/\partial\alpha_i$ will be zero and the matrix $\partial\mathbf{B}/\partial\alpha_i$ consists of zero elements except the (k, j)th element, which is unity. Hence the amplifiers are eliminated except for one with unit gain (or just a direct connection).

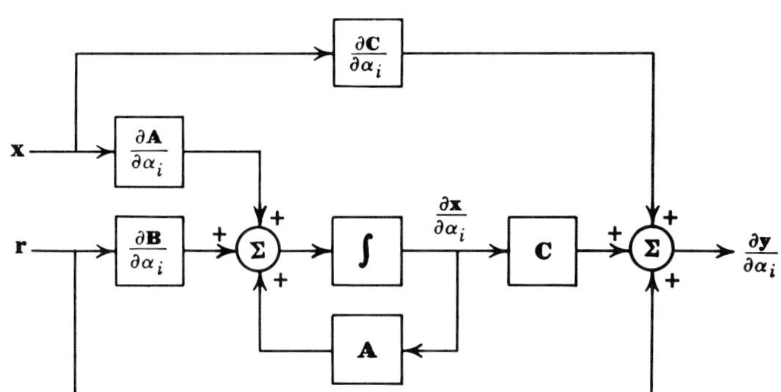

Figure 13.7 Block diagram for Equation 13.67.

The sensitivity-system model simulations are useful for generating the sensitivity functions that are needed in the computation of the gradients for the gradient-optimization procedure. For example, the signals in the integrands of (13.65) may be considered to be outputs of simulation models, and the integrals may be computed using an analog or hybrid computer. With these gradients, the new choices for the parameter settings $\alpha_1 + d\alpha_1, \alpha_2 + d\alpha_2, \ldots, \alpha_p + d\alpha_p$ may be obtained using the scheme in Section 13.2. With the parameters thus adjusted, the process may be repeated for another adjustment on the parameters until the performance index is minimized (or maximized). The following example illustrates the procedure.

Example 13-3. Consider the simplified single-axis reaction-wheel attitude control system with vehicle-orientation feedback, where the back emf of the reaction-wheel motor is neglected. The block diagram is shown in Figure 13.8.

The system equations are

$$\dot{\theta} = \omega \tag{13.68}$$

$$\dot{\omega} = -\frac{1}{J} K_m K_a K_s \theta - \frac{1}{J} K_m K_a K_r \omega + \frac{1}{J} K_m K_a K_s \theta_i . \tag{13.69}$$

θ and ω are the attitude position and attitude rate, respectively. The control τ is the reaction torque applied by the motor. The constants K_s, K_m, and J are known, and the constants K_a and K_r are to be chosen to minimize the performance index

$$I = \int_0^\infty [(\theta - \theta_i)^2 + \beta \tau^2] \, dt \tag{13.70}$$

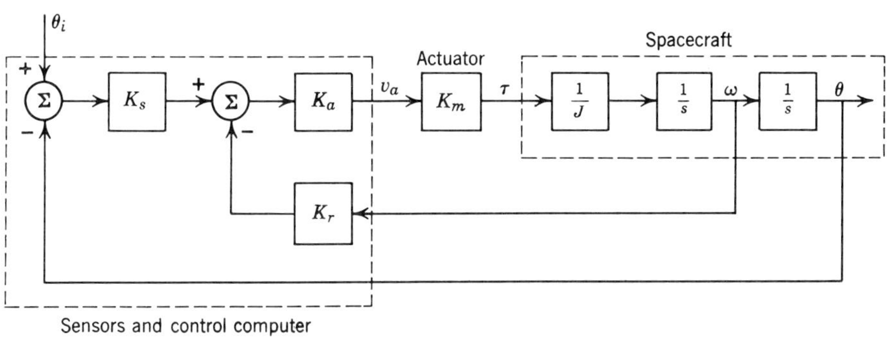

Figure 13.8 Simplified single-axis reaction-wheel attitude control system.

for a given weighting constant β. As discussed in Chapter 12, the inclusion of the term $\beta\tau^2$ limits the energy associated with the torque τ. The command input θ_i is a step function.

Let us choose $\theta - \theta_i$ and ω as state variables x_1 and x_2, respectively. Since θ_i is a step function, $\dot{\theta}_i = 0$ for $t > 0$ and the system equations become

$$\dot{x}_1 = x_2 \tag{13.71}$$

$$\dot{x}_2 = -\frac{1}{J} K_m K_a K_s x_1 - \frac{1}{J} K_m K_a K_r x_2. \tag{13.72}$$

Denote the constants $-\frac{1}{J}K_m K_a K_s$ and $-\frac{1}{J}K_m K_a K_r$ by K_1 and K_2, respectively:

$$K_1 = -\frac{1}{J} K_m K_a K_s \tag{13.73}$$

$$K_2 = -\frac{1}{J} K_m K_a K_r. \tag{13.74}$$

Equation 13.72 becomes

$$\dot{x}_2 = K_1 x_1 + K_2 x_2. \tag{13.75}$$

The performance index in (13.70) becomes

$$I = \int_0^\infty (x_1^2 + \beta\tau^2)\, dt. \tag{13.76}$$

The problem is now reduced to the minimization of I in (13.76) with respect to the parameters K_1 and K_2. Once these parameters are determined, K_a and K_r can be computed from (13.73) and (13.74), since J, K_m, and K_s are given. This problem was solved analytically in Chapter 12. Here K_1 and K_2 are obtained numerically by a gradient procedure. The gradient of I with respect to K_1 and K_2 is

$$\frac{\partial I}{\partial K_1} = \int_0^\infty 2\left(x_1 \frac{\partial x_1}{\partial K_1} + 2\beta\tau \frac{\partial \tau}{\partial K_1}\right) dt \tag{13.77}$$

$$\frac{\partial I}{\partial K_2} = \int_0^\infty 2\left(x_1 \frac{\partial x_1}{\partial K_2} + 2\beta\tau \frac{\partial \tau}{\partial K_2}\right) dt. \tag{13.78}$$

The sensitivity functions $\partial x_1/\partial K_1$ and $\partial x_1/\partial K_2$ are obtained from the sensitivity model. Thus, from (13.71) and (13.75),

$$\frac{\partial \dot{x}_1}{\partial K_1} = \frac{\partial x_2}{\partial K_1} \tag{13.79}$$

$$\frac{\partial \dot{x}_2}{\partial K_1} = K_1 \frac{\partial x_1}{\partial K_1} + K_2 \frac{\partial x_2}{\partial K_1} + x_1 \tag{13.80}$$

$$\frac{\partial \dot{x}_1}{\partial K_2} = \frac{\partial x_2}{\partial K_2} \tag{13.81}$$

$$\frac{\partial \dot{x}_2}{\partial K_2} = K_1 \frac{\partial x_1}{\partial K_2} + K_2 \frac{\partial x_2}{\partial K_2} + x_2. \tag{13.82}$$

From Figure 13.7, we note that

$$\tau = J\dot{\omega} = J\dot{x}_2 = J(K_1x_1 + K_2x_2). \tag{13.83}$$

Hence

$$\frac{\partial \tau}{\partial K_1} = J\frac{\partial \dot{x}_2}{\partial K_1} = J\left(K_1 \frac{\partial x_1}{\partial K_1} + K_2 \frac{\partial x_2}{\partial K_1} + x_1\right) \tag{13.84}$$

$$\frac{\partial \tau}{\partial K_2} = J\frac{\partial \dot{x}_2}{\partial K_2} = J\left(K_1 \frac{\partial x_1}{\partial K_2} + K_2 \frac{\partial x_2}{\partial K_2} + x_2\right). \tag{13.85}$$

For a known set of initial conditions, $x_1(0)$ and $x_2(0)$, and a specific choice for K_1 and K_2, (13.71) and (13.75) are integrated to yield $x_1(t)$ and $x_2(t)$. The torque $\tau(t)$ is computed from (13.83). The sensitivity functions are obtained by integrating (13.79), (13.80), (13.81), and (13.82). Note that these are ordinary differential equations in the variables $\partial x_1/\partial K_1$, $\partial x_2/\partial K_1$, $\partial x_1/\partial K_2$, and $\partial x_2/\partial K_2$. The initial conditions $\partial x_1(0)/\partial K_1$, $\partial x_2(0)/\partial K_1$, $\partial x_2(0)/\partial K_1$, and $\partial x_2(0)/\partial K_2$ are zero, since $x_1(0)$ and $x_2(0)$ are not functions of K_1 and K_2. The torque sensitivities are then computed from (13.84) and (13.85). The quantities in the integrands of (13.77) and (13.78) are now known, and $\partial I/\partial K_1$ and $\partial I/\partial K_2$ are computed from (13.77) and (13.78).

The next choice for K_1 and K_2 are determined from

$$(K_1)_{i+1} = (K_1)_i - h\frac{\left(\dfrac{\partial I}{\partial K_1}\right)_i}{\sqrt{\left(\dfrac{\partial I}{\partial K_1}\right)_i^2 + \left(\dfrac{\partial I}{\partial K_2}\right)_i^2}} \tag{13.86}$$

$$(K_2)_{i+1} = (K_2)_i - h\frac{\left(\dfrac{\partial I}{\partial K_2}\right)_i}{\sqrt{\left(\dfrac{\partial I}{\partial K_1}\right)_i^2 + \left(\dfrac{\partial I}{\partial K_2}\right)_i^2}}. \tag{13.87}$$

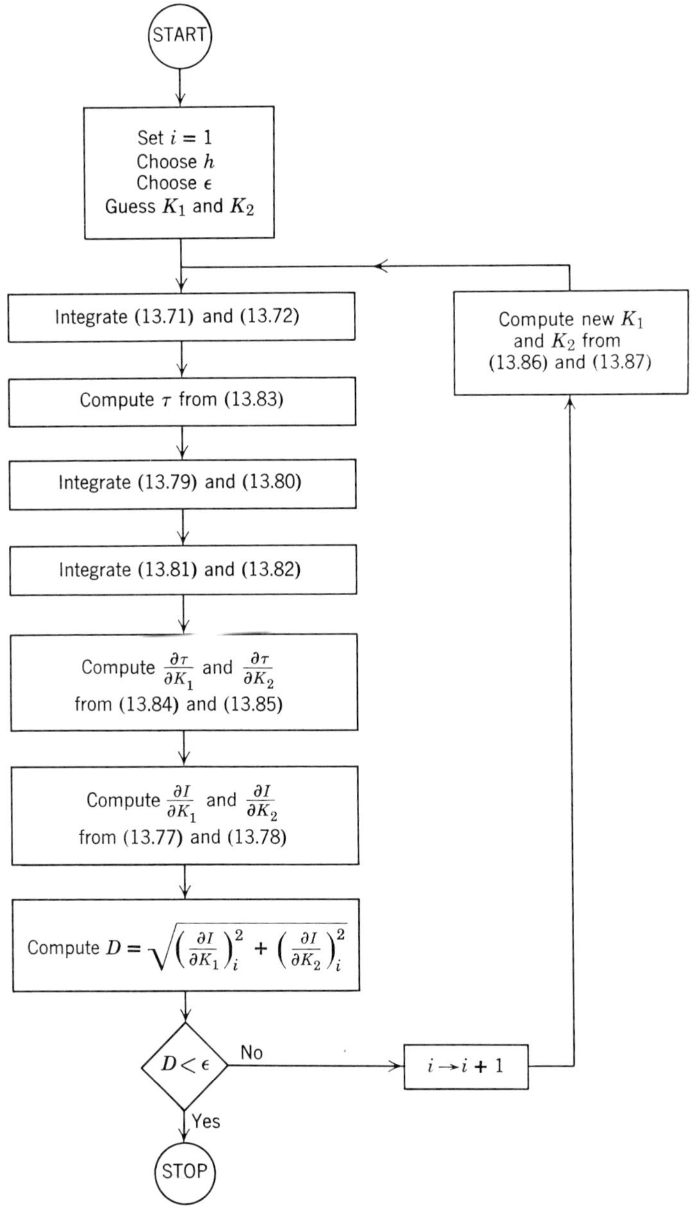

Flow Chart 13.4 Summary of calculations for Example 13-3.

These equations are based on the gradient method of Section 13.2. The step-size h is experimentally chosen. The calculations are summarized in Flow Chart 13.4. For $\beta J^2 = \frac{1}{4}$, the result for a computer gradient minimization[2] is shown in Figure 13.9 for different starting values of K_1 and K_2. The number of iterations required to reach the optimal solution $K_1 = -2$, $K_2 = -2$ is indicated beside the curves. Figure 13.10 shows plots of $x_1(t)$ for five iterations on K_1 and K_2 starting from $K_1 = -4$ and $K_2 = 0$, with initial conditions $x_1(0) = 1$ and $x_2(0) = 0$. Figure 13.11 shows a plot of $x_1(t)$ for the optimal values $K_1 = -2$ and $K_2 = -2$. ▲

Sensitivity functions obtained by simulation are useful also in connection with the use of the Newton-Raphson method when the

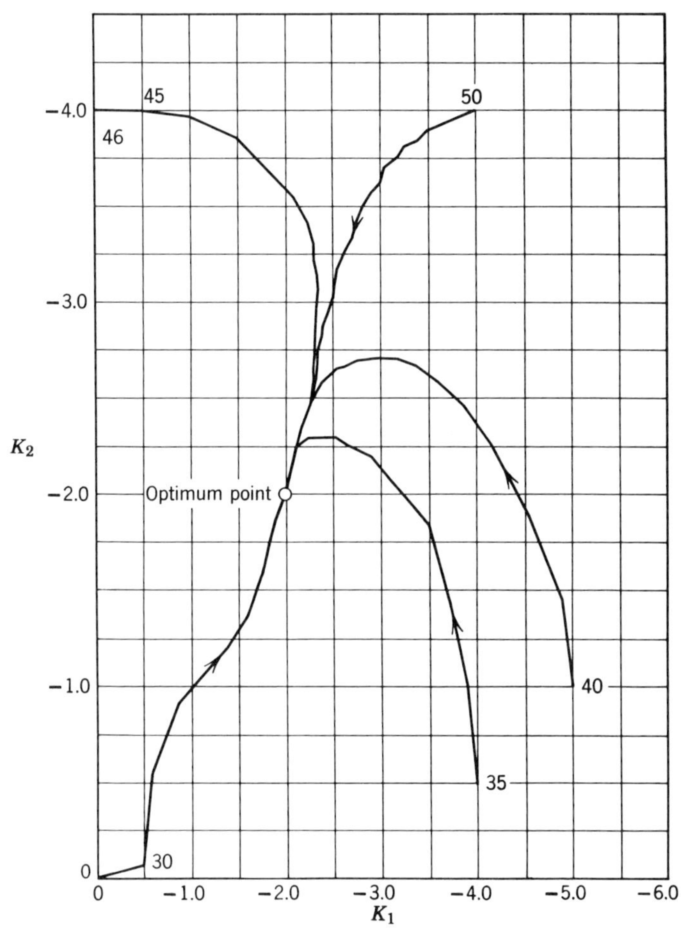

Figure 13.9 Using five different starting values for K_1 and K_2, the gradient procedure yields $K_1 = -2$ and $K_2 = -2$ as the optimum values for Example 13-3.

[2]Dr. J. E. Heller obtained the numerical results in this example.

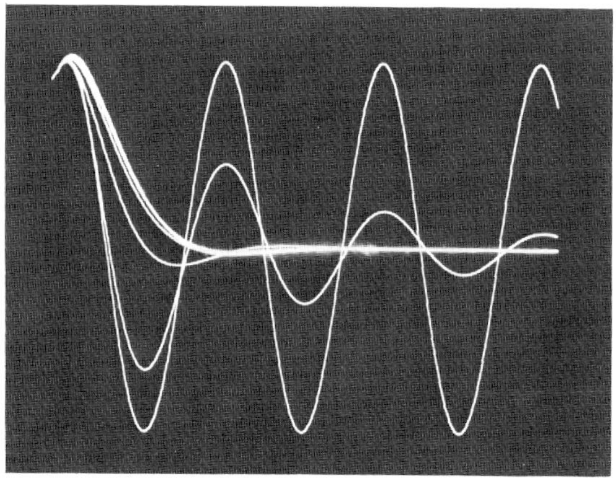

Figure 13.10

equations are not given in terms of explicit functions of the parameters. For example, consider (13.64) again, with each expression equated to zero. The resulting equations have to be solved for the parameters that make each integral equal to zero.

To apply the Newton-Raphson method, we expand the left-hand

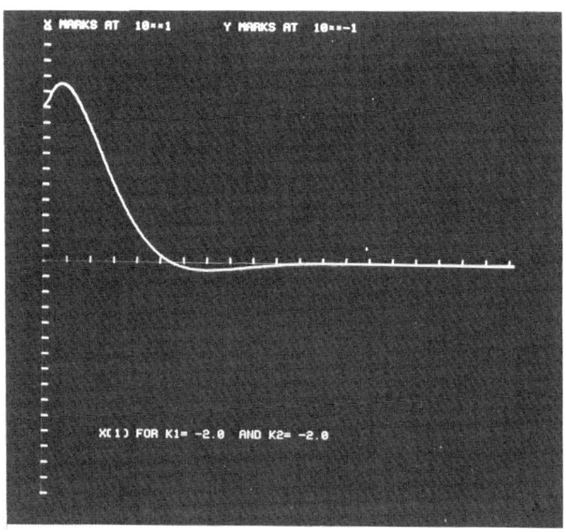

Figure 13.11

side of (13.65) in Taylor's series and retain only the linear terms. We have

$$\left.\frac{\partial I}{\partial \alpha_1}\right|_{i+1} = \left.\frac{\partial I}{\partial \alpha_1}\right|_i + \sum_{k=1}^{p} \left.\frac{\partial^2 I}{\partial \alpha_k \partial \alpha_1}\right|_i [(\alpha_k)_{i+1} - (\alpha_k)_i]$$

$$\left.\frac{\partial I}{\partial \alpha_2}\right|_{i+1} = \left.\frac{\partial I}{\partial \alpha_2}\right|_i + \sum_{k=1}^{p} \left.\frac{\partial^2 I}{\partial \alpha_k \partial \alpha_2}\right|_i [(\alpha_k)_{i+1} - (\alpha_k)_i] \tag{13.88}$$

$$\vdots$$

$$\left.\frac{\partial I}{\partial \alpha_p}\right|_{i+1} = \left.\frac{\partial I}{\partial \alpha_p}\right|_i + \sum_{k=1}^{p} \left.\frac{\partial^2 I}{\partial \alpha_k \partial \alpha_p}\right|_i [(\alpha_k)_{i+1} - (\alpha_k)_i].$$

The subscripts i and $i+1$ denote ith and $(i+1)$th iteration, respectively, anticipating that these equations will be used for an iterative scheme or algorithm. If each equation is now set equal to zero, we may solve the linear equations for the differences $[(\alpha_k)_{i+1} - (\alpha_k)_i]$. The coefficients of the algebraic equations

$$\frac{\partial^2 I}{\partial \alpha_k \partial \alpha_j}$$

are computed with the aid of sensitivity functions. For example, if the performance index I is as given in (13.64), then

$$\frac{\partial I}{\partial \alpha_k} = \int_0^T 4e^3(t, \alpha_1, \ldots, \alpha_p) \frac{\partial e}{\partial \alpha_k} (t, \alpha_1, \ldots, \alpha_p)\, dt. \tag{13.89}$$

Now, if we take the partial derivative of (13.89) with respect to α_j, we have

$$\frac{\partial^2 I}{\partial \alpha_k \partial \alpha_j} = \int_0^T \left[12e^2 \frac{\partial e}{\partial \alpha_j} \frac{\partial e}{\partial \alpha_k} + 4e^3 \frac{\partial^2 e}{\partial \alpha_j \partial \alpha_k}\right] dt. \tag{13.90}$$

Equation 13.90 is the formula for the coefficients for all i,j from 1 to p if I is given by (13.64). The integrand contains signals e and $\partial e/\partial \alpha_k$ $(k = 1, \ldots, p)$ which are obtainable by simulation, and it also contains the second partial derivatives, $\partial^2 e/\partial \alpha_k \partial \alpha_j$, which are called *second-order sensitivity functions*. These second-order sensitivity functions may be obtained by simulation also. This is easily shown by taking the partial derivative of (13.67) with respect to α_k and substituting j for i. If this differentiation is carried out, the result is

$$\frac{\partial}{\partial t}\frac{\partial \mathbf{x}}{\partial\alpha_k\partial\alpha_j} = \mathbf{A}\frac{\partial^2\mathbf{x}}{\partial\alpha_k\partial\alpha_j} + \frac{\partial\mathbf{A}}{\partial\alpha_k}\frac{\partial\mathbf{x}}{\partial\alpha_j} + \frac{\partial\mathbf{A}}{\partial\alpha_j}\frac{\partial\mathbf{x}}{\partial\alpha_k} + \frac{\partial^2\mathbf{A}}{\partial\alpha_k\partial\alpha_j}\mathbf{x} + \frac{\partial^2\mathbf{B}}{\partial\alpha_k\partial\alpha_j}\mathbf{r}$$

$$\frac{\partial^2\mathbf{y}}{\partial\alpha_k\partial\alpha_j} = \mathbf{C}\frac{\partial^2\mathbf{x}}{\partial\alpha_k\partial\alpha_j} + \frac{\partial\mathbf{C}}{\partial\alpha_k}\frac{\partial\mathbf{x}}{\partial\alpha_j} + \frac{\partial\mathbf{C}}{\partial\alpha_j}\frac{\partial\mathbf{x}}{\partial\alpha_k} + \frac{\partial^2\mathbf{C}}{\partial\alpha_k\partial\alpha_j}\mathbf{x} + \frac{\partial^2\mathbf{D}}{\partial\alpha_k\partial\alpha_j}\mathbf{r}. \tag{13.91}$$

Equation 13.91 has some resemblance to (13.66) except that the system variables are now the second-order sensitivity functions, and the inputs are the old system inputs and the first-order sensitivity functions.

In block-diagram form, (13.91) corresponds to the structure shown in Figure 13.12. Note that the inputs to the model can be generated

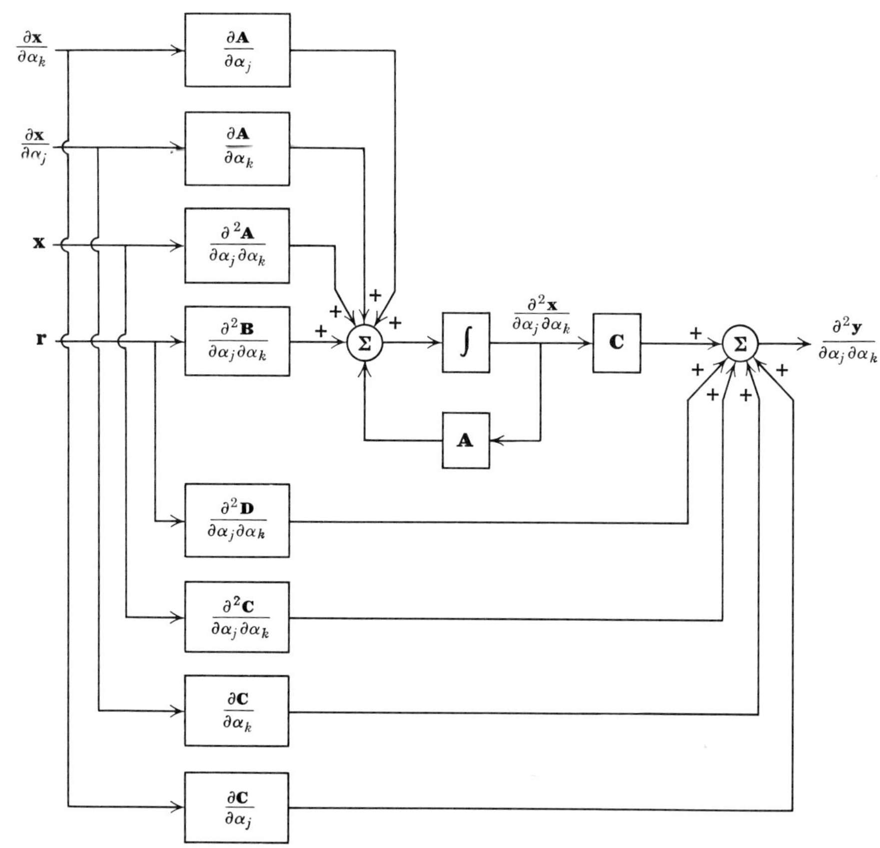

Figure 13.12 Simulation model for second-order sensitivity functions.

from the original system model and the first-order sensitivity model. Hence, all the factors in the integrand of (13.90) may be obtained by simulation. This method of computing the second-order partial derivatives of I then yields the system of equations in (13.88) so that we may proceed with the solution of $[(\alpha_k)_{i+1}-(\alpha_k)_i]$. The process is repeated and a sequence of parameter values is generated. In a practical case, it is difficult to verify whether the sequence will converge before the sequence is constructed. Usually, the time spent in verification may be as much as constructing the sequence, so that one usually proceeds with the construction of the sequence without absolute assurance of convergence. As mentioned before, if the initial guess is close enough to the solution, the sequence will converge.

The sensitivity models developed in Chapters 6 and 7 can be used to generate the partial derivatives of signals with respect to control parameters which are to be chosen to minimize or maximize a performance index. The generation of those partial derivatives may be performed by computer simulation. The algorithms of the previous sections, Newton's method and the gradient method, may be employed to adjust the control parameters. The complete optimization procedure may be automated.

Problems

13-1. It is desired to find a root of

$$f(x) = x^3 - 6x^2 + 3x + 10 = 0$$

by Newton's method. Using the basic Newton-Raphson algorithm, compute a sequence of 5 numbers, starting with $x_0 = 10$, that is, determine x_1, x_2, x_3, x_4, and x_5.

13-2. Repeat Problem 13-1 starting with $x_0 = -5$.

13-3. Repeat Problem 13-1 starting with $x_0 = +1$.

13-4. Repeat Problem 13-1 using the modified Newton-Raphson algorithm.

13-5. Repeat Problem 13-1 using the modified Newton-Raphson algotithm but for $x_0 = -5$.

13-6. Repeat Problem 13-1 using the modified Newton-Raphson algorithm but for $x_0 = +1$.

13-7. Write a FORTRAN program for solving the roots of $f(x) = 0$ in Problem 13-1 using the basic Newton-Raphson algorithm. Do 20 iterations, starting with $x_0 = 10$. Repeat for $x_0 = -5$ and $x_0 = +1$.

13-8. Write a FORTRAN program for solving the roots of $f(x) = 0$ in Problem 13-1 using the modified Newton-Raphson algorithm. Do 40 iterations, using $x_0 = 10$. Repeat for $x_0 = -5$ and $x_0 = +1$.

13-9. (a) Sketch a root-locus of the poles of the closed loop transfer function as a function of K for $K > 0$ (see Figure P13-9).

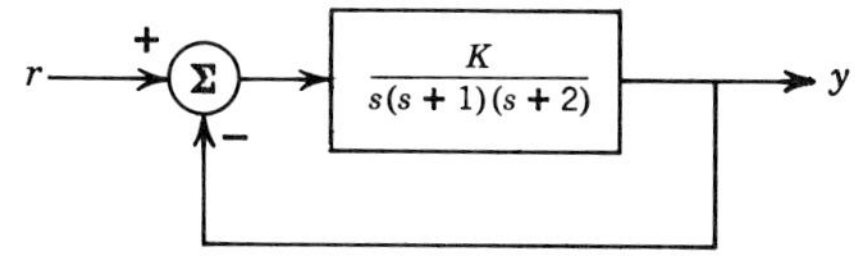

Figure P13-9

(b) It is desired to obtain the value of K for which a pair of complex poles of the transfer function has negative real parts and a damping ratio of $\zeta = 1/\sqrt{2}$. Verify that the complex roots are of the form $s = -\sigma \pm j\sigma$ where σ is an unknown real number of be determined. Substituting $s = -\sigma + j\sigma$ in the characteristic equation, obtain a new equation in terms of K and σ. Note that the equation has a real and an imaginary part. Are the real solutions for K and σ of this equation the values that will make the complex poles have a damping ratio of $1/\sqrt{2}$? Determine two real equations equivalent to the single complex equation.

(c) Set up the algorithm for computing K and σ, using the basic Newton-Raphson algorithm.

(d) Set up the algorithm for computing K and σ, using the modified Newton-Raphson algorithm.

(e) Using an initial guess of σ and K based on the root locus (draw a 45° line intersecting the root locus), determine an improvement of the guessed solution, using one iteration of the basic Newton-Raphson algorithm.

13-10. A simplified single-axis attitude control system is shown in Figure P13-10, where θ is the attitude angle and θ_r is the desired attitude angle which is a unit step function. K_1 and K_2 are control parameters to be chosen. The performance index to be minimized is

$$I = \int_0^\infty (e^2 + \hat{\rho}\tau_c^{\,2})\, dt$$

where $\hat{\rho}$ is a given weighting parameter. Using $K_1 = \hat{K}_1/J$ and $\hat{\rho} = J^2\rho$, verify that

$$I = \frac{1}{2K_1K_2} + \frac{K_2}{2} + \rho\frac{K_1}{2K_2}.$$

[See Figure 12.2 and (12.11)]. Using the gradient method, draw a

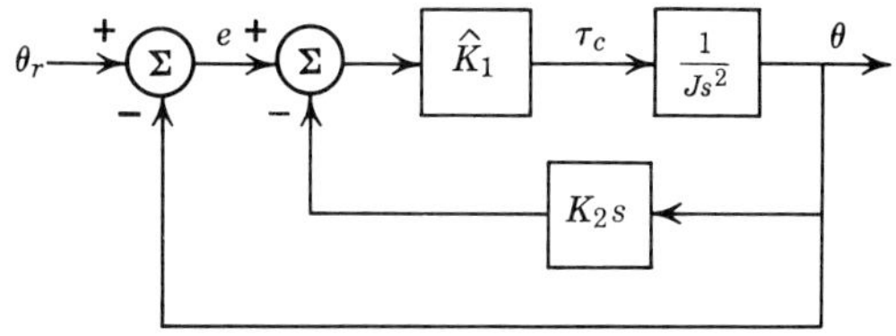

Figure P13-10

flow chart for computing the values of K_1 and K_2 that would minimize I. Write a FORTRAN program based on your flow chart.

13-11. For the system in Figure P13-10 and the performance index in Problem 13-10,

(a) Draw a block diagram for computing $\partial I/\partial \hat{K}_1$ and $\partial I/\partial K_2$.

(b) Based on the diagram in (a) and the gradient method, suggest an automatic experimental procedure for adjusting $\hat{K}_1$ and K_2 so as to minimize I.

13-12. Obtain an algorithm for minimizing $f^2(x)$ by the gradient method, where $f(x)$ is given in Problem 13-1. Note that if x is a solution of $f(x) = 0$, it makes the gradient of $f^2(x)$ equal to zero also.

13-13. Obtain an algorithm for minimizing the index I in Problem 13-10 with respect to K_1 and K_2, using the basic Newton-Raphson method.

13-14. Repeat Problem 13-13 using the modified Newton-Raphson method.

13-15. For the system in Figure P13-10, and the performance index in Problem 13-10,

(a) Draw a block diagram for computing $\partial^2 I/\partial K_1\, \partial K_2$, $\partial^2 I/\partial K_1{}^2$, and $\partial^2 I/\partial K_2{}^2$.

(b) Suggest an automatic experimental procedure for adjusting K_1 and K_2, using the diagram in (a) and Newton's method. The scheme may involve a hybrid simulation where some computation may be by digital means and others by analog means.

APPENDIX A

Basic Matrix Algebra

A brief survey of basic matrix algebra is presented in this Appendix.

A *matrix* is a rectangular array of numbers or functions. For example,

$$\mathbf{C} = \begin{bmatrix} c_{11} & c_{12} \\ c_{21} & c_{22} \\ c_{31} & c_{32} \end{bmatrix}$$

$$\mathbf{A}(t) = \begin{bmatrix} a_{11}(t) & a_{12}(t) & a_{13}(t) \\ a_{21}(t) & a_{22}(t) & a_{23}(t) \\ a_{31}(t) & a_{32}(t) & a_{33}(t) \end{bmatrix}$$

$$\mathbf{Z}(s) = \begin{bmatrix} z_{11}(s) & z_{12}(s) \\ z_{21}(s) & z_{22}(s) \end{bmatrix}$$

$$\mathbf{d} = \begin{bmatrix} d_1 \\ d_2 \\ d_3 \end{bmatrix} \quad \text{and} \quad \mathbf{h} = \begin{bmatrix} 1 \\ 5 \\ -2 \end{bmatrix}$$

are matrices. In general, we denote the *elements* of the matrix **A** by a_{ij}, where the first subscript indicates the *row* in the array in which the element appears and the second subscript indicates the *column*. The *dimension* of a matrix is denoted by $m \times n$, where m is the number of rows and n the number of columns. Thus the matrix **C** above is 3×2, $\mathbf{A}(t)$ is 3×3, $\mathbf{Z}(s)$ is 2×2, **d** is 3×1 and **h** is 3×1. If $m = n$, the matrix is said to be *square*. A square matrix is *diagonal* if $a_{ij} = 0$ whenever $i \neq j$. A square matrix is *symmetrical* if $a_{ij} = a_{ji}$. The *identity matrix* is a diagonal matrix having all diagonal elements equal to unity. For example,

$$\mathbf{I} = \begin{bmatrix} 1 & 0 \\ 0 & 1 \end{bmatrix}$$

is the 2×2 identity matrix. Sometimes a subscript is appended to the identity matrix symbol I to indicate the dimension. $\mathbf{I}_n$ is the $n \times n$ identity matrix.

A useful algebra of matrices is available. Some of the basic operations are the following:

Equality: $\mathbf{A} = \mathbf{B}$ means that the corresponding elements of $\mathbf{A}$ and $\mathbf{B}$ are equal, that is, $a_{ij} = b_{ij}$ for all i, j. $\mathbf{A}$ and $\mathbf{B}$ must have the same dimensions.

Addition: $\mathbf{C} = \mathbf{A} + \mathbf{B}$ means that $c_{ij} = a_{ij} + b_{ij}$ for all i, j. Notice that addition is commutative, $\mathbf{A} + \mathbf{B} = \mathbf{B} + \mathbf{A}$, and also associative, $(\mathbf{A} + \mathbf{B}) + \mathbf{C} = \mathbf{A} + (\mathbf{B} + \mathbf{C})$. Notice also that $\mathbf{A}$ and $\mathbf{B}$ must have the same dimensions.

Multiplication by a scalar: $\mathbf{C} = k\mathbf{A}$ means $c_{ij} = ka_{ij}$ for all i, j.

Transpose: $\mathbf{C} = \mathbf{A}^T$ means $c_{ij} = a_{ji}$. Note: If $\mathbf{A}$ is symmetric, then $\mathbf{A}^T = \mathbf{A}$.

Matrix multiplication is somewhat more complicated. The rule for matrix multiplication is set up to be especially useful in representing simultaneous equations.

Multiplication: $\mathbf{AB} = \mathbf{C}$ means $c_{ij} = \sum_{k=1}^{r} a_{ik}b_{kj}$ where $\mathbf{A}$ is $m \times r$, $\mathbf{B}$ is $r \times n$, and $\mathbf{C} = m \times n$. Note that if the number of columns of $\mathbf{A}$ is not equal to the number of rows of $\mathbf{B}$, then $\mathbf{AB}$ is not defined. It should also be noted that $\mathbf{AB} \neq \mathbf{BA}$, in general.

Example:

$$\begin{bmatrix} 3 & 1 \\ 5 & 2 \end{bmatrix}\begin{bmatrix} 2 & 0 \\ 1 & 1 \end{bmatrix} = \begin{bmatrix} (3)(2) + (1)(1) & (3)(0) + (1)(1) \\ (5)(2) + (2)(1) & (5)(0) + (2)(1) \end{bmatrix}$$

$$= \begin{bmatrix} 7 & 1 \\ 12 & 2 \end{bmatrix}.$$

Although matrix multiplication is not in general *commutative* ($\mathbf{AB} \neq \mathbf{BA}$), it is *associative*,

$$(\mathbf{AB})\mathbf{C} = \mathbf{A}(\mathbf{BC}),$$

and *distributive*,

$$\mathbf{A}(\mathbf{B} + \mathbf{C}) = \mathbf{AB} + \mathbf{AC}.$$

Matrix Inverse: The inverse of a matrix, denoted by $\mathbf{A}^{-1}$, is defined by

$$\mathbf{AA}^{-1} = \mathbf{A}^{-1}\mathbf{A} = \mathbf{I},$$

where $\mathbf{I}$ is the identity matrix. Not all matrices have inverses. Clearly, $\mathbf{A}$ must be square. Furthermore, it can be shown that $\mathbf{A}^{-1}$ exists if and only if the determinant of $\mathbf{A}$, det $\mathbf{A}$, is not identically zero.

Several methods exist for hand and machine computation of $\mathbf{A}^{-1}$. For example,

$$\mathbf{A}^{-1} = \frac{\text{adj } \mathbf{A}}{\det \mathbf{A}},$$

where adj $\mathbf{A}$ denotes the *adjoint matrix* of $\mathbf{A}$ and is formed by the following steps:

1. Replace each element of **A** by its cofactor.
2. Transpose the resulting matrix of cofactors.

Example:

$$A = \begin{bmatrix} s & -1 \\ 1 & s+2 \end{bmatrix}$$

$$\begin{aligned} \det A &= s(s+2)+1 \\ &= s^2+2s+1 = (s+1)^2 \end{aligned}$$

$$\text{adj } A = \begin{bmatrix} s+2 & 1 \\ -1 & s \end{bmatrix}$$

So

$$A^{-1} = \frac{1}{(s+1)^2}\begin{bmatrix} s+2 & 1 \\ -1 & s \end{bmatrix}$$

Partial check:

$$\begin{aligned} AA^{-1} &= \frac{1}{(s+1)^2}\begin{bmatrix} s & -1 \\ 1 & s+2 \end{bmatrix}\begin{bmatrix} s+2 & 1 \\ -1 & s \end{bmatrix} \\ &= \frac{1}{(s+1)^2}\begin{bmatrix} s^2+2s+1 & 0 \\ 0 & 1+s^2+2s \end{bmatrix} = \begin{bmatrix} 1 & 0 \\ 0 & 1 \end{bmatrix} = I_2. \end{aligned}$$

APPENDIX B

Solution of Linear Time-Invariant Differential Equations by Laplace Transforms

B-1 Laplace Transform Definition

The Laplace transform of a time function such as a signal in a system is a function of a complex variable, but it does not depend on the time variable. Suppose $x(t)$ is a signal defined for $t \geqslant 0-$. The Laplace transform of $x(t)$, denoted by $X(s)$, is defined by

$$X(s) \triangleq \int_{0-}^{\infty} x(t)\epsilon^{-st}\,dt \tag{B.1}$$

for those values of s for which the integral exists.[1] The lower limit $0-$ of the integral denotes the limit from below of $-\delta$ for $\delta > 0$ as $\delta \rightarrow 0$. If $x(t)$ has an impulse at $t = 0$, it is included in the integration. We shall denote the operation of taking the Laplace transform of a time function by the symbol $\mathscr{L}$ so that

$$X(s) \triangleq \mathscr{L}[x(t)]. \tag{B.2}$$

The variable t is integrated out in (B.1) so that the Laplace transform $X(s)$ does not depend on t at all but only on a new variable s. For example, if $x(t) = \epsilon^{-3t}$ for $t \geqslant 0$ and zero for $t < 0$, then

$$X(s) = \int_{0-}^{\infty} \epsilon^{-3t}\epsilon^{-st}\,dt = \frac{-\epsilon^{-(s+3)t}}{s+3}\bigg|_{0-}^{\infty} = \frac{1}{s+3} \tag{B.3}$$

[1]For the integral in (B.1) to exist, it is sufficient that

$$\int_{0-}^{\infty} |x(t)|\epsilon^{-\sigma t}\,dt < \infty$$

where $s = \sigma + j\omega$, σ, and ω are real variables and $j = \sqrt{-1}$. The smallest σ (denoted by σ_a) for which the above inequality holds is called the abscissa of absolute convergence.

provided that the real part of s is greater than -3. For $Re[s] \leqslant -3$, the integral does not converge.

It is clear that if the Laplace transform of $x(t)$ is $X(s)$, then the Laplace transform of a sum of signals $x_1(t)+x_2(t)$ is the sum of the Laplace transforms of the individual signals $X_1(s)+X_2(s)$. We need to develop other properties. For example, since we plan to apply the Laplace transform to differential equations, we must investigate the nature of the Laplace transform of the derivative of a signal.

From the definition, the Laplace transform of the derivative of a signal $x(t)$ is (assuming that it is Laplace transformable)

$$\mathscr{L}\left[\frac{dx}{dt}\right] \triangleq \int_{0-}^{\infty} \frac{dx(t)}{dt} \epsilon^{-st}\, dt. \tag{B.4}$$

Integrating by parts, we obtain

$$\int_{0-}^{\infty} \frac{dx}{dt} \epsilon^{-st}\, dt = x(t)\epsilon^{-st}\Big|_{0-}^{\infty} - \int_{0-}^{\infty} x(t)\epsilon^{-st}(-s)\, dt$$

$$= -x(0-) + s\int_{0-}^{\infty} x(t)\epsilon^{-st}\, dt \tag{B.5}$$

so that

$$\mathscr{L}\left[\frac{dx}{dt}\right] = sX(s) - x(0-) \tag{B.6}$$

where $x(0-)$ is the limit of $x(t)$ as t approaches 0 from the left. By a repeated application of the above procedure, we may derive the formula for the Laplace transform of the second derivative of $x(t)$ as

$$\mathscr{L}\left[\frac{d^2x}{dt^2}\right] = s^2X(s) - sx(0-) - x'(0-) \tag{B.7}$$

where $x'(0-)$ is the value of the derivative at $t=0-$. In general, for the nth derivative of $x(t)$, the formula is

$$\mathscr{L}\left[\frac{d^nx(t)}{dt^n}\right] = s^nX(s) - s^{n-1}x(0-) - s^{n-2}x'(0-) \cdots - x^{(n-1)}(0-) \tag{B.8}$$

Example B-1. Suppose a system is modeled by the differential equation

$$\frac{d^2y}{dt^2} + 5\frac{dy}{dt} + 4y = 3\frac{dx}{dt} + 2x \tag{B.9}$$

where $y(t)$ is the output signal and $x(t)$ is the input signal. Taking the transform of both sides of the equation, we have

$$s^2Y(s) - sy(0-) - y'(0-) + 5[sY(s) - y(0-)] + 4Y(s)$$
$$= 3[sX(s) - x(0-) + 2X(s) \qquad \text{(B.10)}$$

or

$$Y(s) = \frac{(3s+2)}{s^2+5s+4}X(s) + \frac{sy(0-) + y'(0-) + 5y(0-) - 3x(0-)}{s^2+5s+4}. \quad \blacktriangle \qquad \text{(B.11)}$$

The output time function $y(t)$ is obtained by taking the inverse transform of $Y(s)$. This may be done in several ways. One way is to expand $Y(s)$ as the sum of simpler expressions each of which is the Laplace transform of a known time function. For such a method, a table of transform pairs is useful. For most applications, this method is the simplest. When the expressions cannot be found in tables, we use the formula

$$y(t) = \frac{1}{2\pi j}\int_{\sigma-j\infty}^{\sigma+j\infty} Y(s)\epsilon^{st}\,ds \qquad \text{(B.12)}$$

where σ is greater than the abscissa of absolute convergence of $Y(s)$. This formula may be established by substituting the Laplace transform integral definition of $Y(s)$ in the integral of (B.12) and interchanging the order of integration [which is allowed if the integral in (B.1) is absolutely convergent]. The integral is a line integral in a complex s-plane.

Instead of taking the Laplace transform of the final differential equation relating an output signal to an input signal, we may take the Laplace transform of the system of integrodifferential equations and then solve for the Laplace transform of the desired output signal. In order to carry this out, we need the formula for the Laplace transform of the integral of $x(t)$ in terms of the Laplace transform of $x(t)$:

$$\mathscr{L}\left[\int_{0-}^{t} x(\tau)\,d\tau\right] = \int_{0-}^{\infty}\int_{0-}^{t} x(\tau)\,d\tau\,\epsilon^{-st}\,dt. \qquad \text{(B.13)}$$

Integrating by parts, we have

$$\mathscr{L}\left[\int_{0-}^{t} x(\tau)\,d\tau\right] = \left[-\frac{1}{s}\epsilon^{-st}\int_{0-}^{t} x(\tau)\,d\tau\right]\Bigg|_{0-}^{\infty} + \frac{1}{s}\int_{0-}^{\infty} x(t)\epsilon^{-st}\,dt. \qquad \text{(B.14)}$$

For values of s for which the left-hand side converges, the first term of the right-hand side vanishes and so

$$\mathscr{L}\left[\int_{0-}^{t} x(\tau)\,d\tau\right] = \frac{1}{s}X(s). \qquad \text{(B.15)}$$

Initial condition terms in the integrodifferential equations describing the behavior of dynamical systems appear as constants. These constants may be multiplied by unit step functions without altering the solutions of the equations. The Laplace transform of $Ku(t)$ is

$$\mathscr{L}[Ku(t)] = \int_{0-}^{\infty} K\epsilon^{-st}\,dt = \left.\frac{K\epsilon^{-st}}{-s}\right|_{0-}^{\infty} = \frac{K}{s} \tag{B.16}$$

Applying the formula for the transform of an integral, since $Kt = \int_0^t K\,d\tau$,

$$\mathscr{L}[Kt] = \frac{1}{s}\left[\frac{K}{s}\right] = \frac{K}{s^2}. \tag{B.17}$$

A short table of transforms for some important time functions is given in Table B.1. Entries 1 and 2 have been derived above. Entry 3 may be derived by induction after observing that the integral of t^{n-1} is $(1/n)t^n$ and applying the formula for the Laplace transform of an integral (B.15). Entry 4 is similar to an example considered previously except that, instead of a general parameter a, 3 was used. Entries 5 and 6 may be derived by expressing the sine and cosine functions as exponential functions with complex components and

Table B.1

	$f(t)$, for $t \geq 0$ $f(t) = 0$ for $t < 0$	$F(s)$
1.	$u(t) = \begin{matrix} 1, & t \geq 0 \\ 0, & t < 0 \end{matrix}$	$\frac{1}{s}$
2.	$tu(t)$	$\frac{1}{s^2}$
3.	$t^n u(t)$, n positive integer	$\frac{n!}{s^{n+1}}$
4.	$\epsilon^{-at}u(t)$	$\frac{1}{s+a}$
5.	$\sin at\, u(t)$	$\frac{a}{s^2+a^2}$
6.	$\cos at\, u(t)$	$\frac{s}{s^2+a^2}$
7.	$\delta(t)$	1

Table B.2

	$f(t)$	$F(s) = \mathscr{L}[f(t)]$
1.	$af(t)u(t)$	$aF(s)$
2.	$[ax_1(t) + bx_2(t)]u(t)$	$aX_1(s) + bX_2(s)$
3.	$\left[\frac{dx}{dt}\right]u(t)$	$sX(s) - x(0-)$
4.	$\left[\int_{0-}^{t} x(\tau)\, d\tau\right]u(t)$	$\frac{1}{s}X(s)$
5.	$\epsilon^{-at}f(t)u(t)$	$F(s+a)$
6.	$f(at)u(t)$	$\frac{1}{a}F\left(\frac{s}{a}\right)$
7.	$\lim_{t\to\infty} f(t)$	$\lim_{s\to 0} sF(s)$, all poles of $sF(s)$ in left-half plane
8.	$f(0+)$	$\lim_{s\to\infty} sF(s)$

then applying the result in entry 4. Entry 7 is immediately obtained by applying the definition in (B.1).

Table B.2 is a similar table of transform pairs, but it applies to simple operations rather than specific simple functions as in Table B.1. Entries 1 to 4 have been discussed above. Entries 5 and 6 are obtained directly from the definition. For entry 5,

$$\mathscr{L}[\epsilon^{-at}f(t)] = \int_{0-}^{\infty} \epsilon^{-at}f(t)\epsilon^{-st}\,dt = \int_{0-}^{\infty} f(t)\epsilon^{-(s+a)t}\,dt \qquad \text{(B.18)}$$

$$= F(s+a)$$

where $F(s)$ is the Laplace transform of $f(t)$.

Example B-2. To determine the Laplace transform of

$$x(t) = \epsilon^{-2t} \sin 5t\, u(t), \qquad \text{(B.19)}$$

we may apply the definition of the Laplace transform to $x(t)$ and carry out the integration. Alternatively, we may use the result of entry 5 in Table B.1 and entry 5 in Table B.2. From Table B.1

$$\mathscr{L}[\sin 5t] = \frac{5}{s^2+25} \tag{B.20}$$

Then applying the result of entry 5 in Table B.2,

$$\mathscr{L}[\epsilon^{-2t}\sin 5t] = \frac{5}{(s+2)^2+25}. \blacktriangle \tag{B.21}$$

Example B-3. Suppose it is desired to obtain the Laplace transform of a signal $x(t)$ which is given by

$$x(t) = \epsilon^{-3t}t^4\,u(t). \tag{B.22}$$

Using entry 3 in Table B.1 and entry 5 in Table B.2, we immediately obtain

$$\mathscr{L}[\epsilon^{-3t}t^4] = \frac{4!}{(s+3)^5} = \frac{24}{(s+3)^5}. \blacktriangle \tag{B.23}$$

Entry 6 in Table B.2 may be derived also by a direct application of the Laplace transform. Thus

$$\begin{aligned}\mathscr{L}[f(at)] &= \int_{0-}^{\infty} f(at)\epsilon^{-st}\,dt \\ &= \int_{0-}^{\infty} f(at)\epsilon^{-(s/a)(at)}\frac{1}{a}d(at) \\ &= \frac{1}{a}\int_{0-}^{\infty} f(\tau)\epsilon^{-(s/a)\tau}\,d\tau = \frac{1}{a}F\left(\frac{s}{a}\right).\end{aligned} \tag{B.24}$$

Example B-4. It is given that

$$\mathscr{L}[\epsilon^{-t}\cos t] = \frac{s+1}{(s+1)^2+1}. \tag{B.25}$$

The problem is to determine the Laplace transform of $\epsilon^{-100t}\cos 100t$. From (B.25) and entry 6 in Table B.2, we have

$$\mathscr{L}[\epsilon^{-100t}\cos 100t] = \frac{1}{100}\frac{\left(\frac{s}{100}+1\right)}{\left(\frac{s}{100}+1\right)^2+1} \tag{B.26}$$

which is obtained by replacing s by $s/100$ in the right-hand side of (B.25) and dividing the whole expression by 100. Entry 6 of Table B.2 is a very important relationship between time scaling and frequency scaling. ▲

Tables B.1 and B.2 are sufficient for determining the inverse transform of a rational function of s. A function $F(s)$ is said to be a rational function of s if it can be expressed as the ratio of two polynomials in s. For example,

$$F(s) = \frac{3s+2}{s^2+5s+4} \tag{B.27}$$

is a rational function of s. The values of s which make the numerator polynomial equal to zero are called *zeros* of $F(s)$. The values of s which make the denominator polynomials of the rational function $F(s)$ equal to zero are called *poles* of $F(s)$. For example, the rational function $F(s)$ in (B.27) has a zero at $s=-\frac{2}{3}$ and poles at $s=-1$ and $s=-4$. In the complex plane, zero locations are denoted by 0 and pole locations are denoted by X.

B-2 Partial Fraction Expansions

Rational functions may be expanded in partial fractions and each term identified with a known inverse transform from a table such as that in Table B.1.

Consider a rational function

$$F(s) = \frac{P(s)}{Q(s)} = \frac{K(s+z_1)(s+z_2)\cdots(s+z_m)}{(s+p_1)(s+p_2)\cdots(s+p_m)} \tag{B.28}$$

First, let us assume that the poles are simple, that is, $-p_1, -p_2, \ldots, -p_n$ are different or distinct. The partial fraction expansion of $F(s)$ in (B.28) when $m < n$ is of the form

$$F(s) = \sum_{k=1}^{n} \frac{K_k}{s+p_k} \tag{B.29}$$

The constant K_k is called the residue of the pole at $-p_k$. The residues may be evaluated as follows. Multiply both sides of (B.29) by $s+p_j$

$$(s+p_j)F(s) = \sum_{k=1}^{n} \frac{K_k(s+p_j)}{s+p_k}. \tag{B.30}$$

If the partial fraction expansion of (B.29) is to be an identity, the equality should hold for all values of s. In particular, (B.30) should hold for $s=-p_j$. However, we note that the terms in the right-hand side of (B.30) are zero except for the jth term which is just K_j. Hence

$$K_j = [(s+p_j)F(s)]_{s=-p_j} \tag{B.31}$$

Equation B.31 may be used for computing K_j for $j = 1, \ldots, n$.

Example B-5. For the $F(s)$ in (B.27), the partial fraction expansion is of the form

$$F(s) = \frac{K_1}{s+1} + \frac{K_2}{s+4}. \tag{B.32}$$

The residues are

$$K_1 = \left. \frac{(3s+2)(s+1)}{(s+1)(s+4)} \right|_{s=-1} = \frac{-3+2}{-1+4} = -\frac{1}{3} \tag{B.33}$$

and

$$K_2 = \left. \frac{(3s+2)(s+4)}{(s+1)(s+4)} \right|_{s=-4} = \frac{-12+2}{-4+1} = \frac{10}{3}. \blacktriangle \tag{B.34}$$

If the rational function has multiple poles, the partial fraction expansion form of (B.29) is not sufficient. We say that a pole has multiplicity k if k poles coincide at the point. For example, if $F(s) = 1/(s+1)^2$, then -1 is double pole. Suppose that $F(s)$ has a pole at $-p_1$ of multiplicity r. Then we write

$$F(s) = \frac{K_{1r}}{(s+p_1)^r} + \frac{K_{1r-1}}{(s+p_1)^{r-1}} + \cdots + \frac{K_{11}}{(s+p_1)} + F_1(s) \tag{B.35}$$

where $F_1(s)$ contains only the partial fraction terms for the other poles. Then

$$(s+p_1)^r F(s) = K_{1r} + K_{1r-1}(s+p_1) + \cdots + K_{11}(s+p_1)^{r-1} + (s+p_1)^r F_1(s). \tag{B.36}$$

Clearly,

$$K_{1r} = [(s+p_1)^r F(s)]_{s=-p_1}. \tag{B.37}$$

Differentiating both sides of (B.36), and setting $s = -p_1$, we obtain

$$K_{1r-1} = \left\{ \frac{d}{ds} [(s+p_1)^r F(s)] \right\}_{s=-p_1}. \tag{B.38}$$

The other coefficients may be obtained by a repeated differentiation of (B.36) and setting $s = -p_1$. In general,

$$K_{1j} = \frac{1}{(r-j)!} \left\{ \frac{d^{(r-j)}}{ds^{(r-j)}} [(s+p_1)^r F(s)] \right\}_{s=-p_1}. \tag{B.39}$$

Similarly, if $-p_2$ is a pole of multiplicity m_2, the pole at $-p_2$ will have m_2 terms associated with it in the partial fraction expansion.

Example B-6. Let us expand

$$F(s) = \frac{s+1}{(s+2)(s+3)^2} \tag{B.40}$$

in partial fractions. Then $F(s)$ is of the form

$$F(s) = \frac{K_1}{s+2} + \frac{K_2}{(s+3)^2} + \frac{K_3}{(s+3)}. \tag{B.41}$$

The coefficients are readily evaluated as

$$K_1 = [(s+2)F(s)]_{s=-2} = \frac{-2+1}{(-2+3)^2} = -1 \tag{B.42}$$

$$K_2 = [(s+3)^2 F(s)]_{s=-3} = \frac{-3+1}{-3+2} = 2 \tag{B.43}$$

and

$$K_3 = \left[\frac{d}{ds}\frac{s+1}{s+2}\right]_{s=-3} = \left[\frac{(s+2)-(s+1)}{(s+2)^2}\right]_{s=-3} = 1 \tag{B.44}$$

so that

$$\frac{s+1}{(s+2)(s+3)^2} = \frac{-1}{s+2} + \frac{2}{(s+3)^2} + \frac{1}{s+3}. \tag{B.45}$$

This is verified to be correct by combining the three terms on the right hand side into one term with a common denominator. Thus

$$\begin{aligned}\frac{-1}{s+2} + \frac{2}{(s+3)^2} + \frac{1}{s+3} &= \frac{-(s+3)^2 + 2(s+2) + (s+2)(s+3)}{(s+2)(s+3)^2} \\ &= \frac{-s^2 - 6s - 9 + 2s + 4 + s^2 + 5s + 6}{(s+2)(s+3)^2} \\ &= \frac{s+1}{(s+2)(s+3)^2}\end{aligned} \tag{B.46}$$

which is correct.

Let the inverse transform operation be denoted by $\mathscr{L}^{-1}$. So the inverse transform of $F(s)$ in (B.45) is

$$\mathscr{L}^{-1}\left[\frac{s+1}{(s+2)(s+3)^2}\right] = \mathscr{L}^{-1}\left[\frac{-1}{s+2}\right] + \mathscr{L}^{-1}\left[\frac{2}{(s+3)^2}\right] + \mathscr{L}^{-1}\left[\frac{1}{s+3}\right]. \tag{B.47}$$

From Table B.1, entry 4, and the linear property of the transform, we recognize that

$$\mathscr{L}^{-1}\left[\frac{-1}{s+2}\right]=-\epsilon^{-2t}, \quad t \geqslant 0 \tag{B.48}$$

and

$$\mathscr{L}^{-1}\left[\frac{1}{s+3}\right]=\epsilon^{-3t}, \quad t \geqslant 0. \tag{B.49}$$

The second term of the right-hand side of (B.47) is not found in the table directly. However, we note from entry 3 that

$$\mathscr{L}^{-1}\left[\frac{1}{s^2}\right]=t, \quad t \geqslant 0 \tag{B.50}$$

and from entry 5, Table B.2, applied to (B.50), we have

$$\mathscr{L}^{-1}\left[\frac{1}{(s+3)^2}\right]=\epsilon^{-3t}t, \quad t \geqslant 0. \tag{B.51}$$

Hence

$$\mathscr{L}^{-1}\left[\frac{s+1}{(s+2)(s+3)^2}\right]=-\epsilon^{-2t}+\epsilon^{-3t}+2t\epsilon^{-3t}. \blacktriangle \tag{B.52}$$

An alternative way for determining the coefficients corresponding to a multiple pole will now be described. Suppose that a rational function $H(s)$ has a pole of multiplicity n at $s=s_0$,

$$H(s)=\frac{N(s)}{D(s)}=\frac{N(s)}{(s-s_0)^n D_1(s)} \tag{B.53}$$

where $N(s)$ and $D(s)$ are polynomials in s, and s_0 is not a root of $D_1(s)=0$. It is assumed that the degree of N is less than the degree of D. If it is not, the polynomial must be divided out before a partial fraction expansion is attempted. Denote $(s-s_0)$ by p so that $s=p+s_0$, and form $(s-s_0)^n H(s)$,

$$(s-s_0)^n H(s)=p^n H(p+s_0)=\frac{N(p+s_0)}{D_1(p+s_0)}. \tag{B.54}$$

Divide the numerator $N(p+s_0)$ by the denominator $D_1(p+p_0)$ with both polynomials written in ascending powers of p. The result is of the form

$$p^n H(p+s_0)=k_0+k_1 p+\cdots+k_{n-1}p^{n-1}+\frac{k_n p^n}{D_1(p+s_0)}. \tag{B.55}$$

Dividing (B.55) by p^n, we obtain

$$H(p+s_0)=\frac{k_0}{p^n}+\frac{k_1}{p^{n-1}}+\cdots+\frac{k_{n-1}}{p}+\frac{k_n}{D_1(p+s_0)} \tag{B.56}$$

so that

$$H(s) = \frac{k_0}{(s-s_0)^n} + \frac{k_1}{(s-s_0)^{n-1}} + \cdots + \frac{k_{n-1}}{(s-s_0)} + \frac{k_n}{D_1(s)} \tag{B.57}$$

If $k_n/D_1(s)$ has other multiple poles, the above process may be applied repeatedly. If $k_n/D_1(s)$ has simple poles only, then the partial fraction expansion is easily obtained as before.

Example B-7. Let us apply the above procedure for obtaining a partial fraction expansion of $F(s)$ in Example B-6. To obtain the coefficients corresponding to the pole at $s=-3$, form

$$(s+3)^2F(s) = \frac{s+1}{s+2}. \tag{B.58}$$

Let $s+3=p$. Then (B.58) becomes

$$p^2F(p-3) = \frac{s+1}{s+2} = \frac{p-2}{p-1}. \tag{B.59}$$

The numerator is divided by the denominator with the polynomials arranged in ascending powers of p:

$$\begin{array}{r|l} & \;\; 2+p \\ -1+p & \overline{-2+p} \\ & \underline{-2+2p} \\ & \quad\; -p \\ & \quad\; \underline{-p+p^2} \\ & \qquad\; -p^2 \end{array}$$

Hence

$$\frac{p-2}{p-1} = 2+p-\frac{p^2}{p-1} \tag{B.60}$$

or

$$F(p-3) = \frac{2}{p^2}+\frac{1}{p}-\frac{1}{p-1} \tag{B.61}$$

and

$$F(s) = \frac{2}{(s+3)^2}+\frac{1}{(s+3)}-\frac{1}{(s+2)} \tag{B.62}$$

which agrees with (B.45). ▲

For the differential equation in (B.9), if the input signal $x(t)$ is a linear combination of exponentials (possibly with complex damping constants) and polynomials of t times exponentials, then the transform $X(s)$ is a rational function so that the entire expression on the right-hand side in (B.11) is a rational function. Hence $Y(s)$ is rational and the partial fraction technique described above may be applied.

As mentioned earlier, we have a choice of transforming a system of integrodifferential equations and solving for the transform of the desired output signal, or we may obtain first the differential equation containing only the desired output signal as unknown without any other unknown system variables appearing and then transform the differential equation. In either case, we have to obtain the inverse transform of the transform of the unknown output by using tables, or by using the integration called for by the inverse transformation formula. Initial conditions must be determined usually from the physical system rather than the system of equations. The procedure of transforming the system of integrodifferential equations before eliminating the other unknown system variables is usually easier than the other method because the initial conditions are most readily available from the system specifications. For example, initial conditions are commonly given by specifying capacitor voltages and inductor currents in electrical networks. These quantities will appear directly in the transformed system of equations. On the other hand, if the system of equations is first reduced to a single differential equation and then transformed, then the initial condition quantities required for solving the problem are the values at $t = 0-$ of the output and input and their derivatives of various orders. These quantities must first be determined from the given initial conditions in terms of capacitor charges and inductor currents if the network is electrical.

Example B-8. Consider the electrical network in Figure B.1.

Suppose that the source $v(t)$ is a constant V and the switch is closed at $t = 0$. Just before the switch is closed, the inductor current is given to be zero and the capacitor voltage v_2 with respect to the ground reference node is $v_2(0-) = V_2$. The manner in which the capacitor attains its charge at $t = 0-$ is not specified. The network is assumed to be linear and time-invariant. First we write a set of differential equations which describe the behavior of the network for $t \geqslant 0$. Applying Kirchhoff's voltage law, we have

$$-v + R_1 i_L + L\frac{di_L}{dt} + v_2 = 0 \tag{B.63}$$

or

$$\frac{di_L}{dt} = -\frac{R_1}{L}i_L - \frac{1}{L}v_2 + \frac{1}{L}v. \tag{B.64}$$

Applying Kirchhoff's current law to the node corresponding to v_2, we have

$$C\frac{dv_2}{dt} + \frac{v_2}{R_2} - i_L = 0 \tag{B.65}$$

or

$$\frac{dv_2}{dt} = \frac{1}{C}i_L - \frac{1}{R_2 C}v_2. \tag{B.66}$$

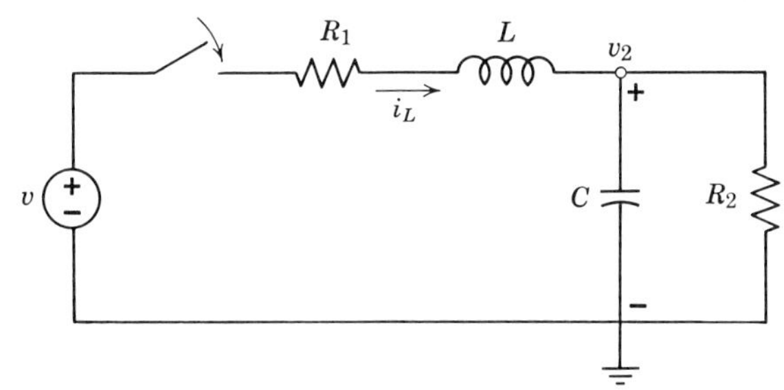

Figure B.1 Linear time-invariant electric network to be analyzed by Laplace transforms.

Equations B.64 and B.66 describe the behavior of the network in terms of $i_L(t)$ and $v_2(t)$. Note that this network may be simulated on an analog computer using two integrators. Since the parameters R_1, L, R_2, and C are assumed to be constant and since v is a constant V for $t \geqslant 0$, the Laplace transforms of (B.64) and (B.66) are

$$sI_L(s) - i_L(0-) = -\frac{R_1}{L}I_L(s) - \frac{1}{L}V_2(s) + \frac{1}{L}\frac{V}{s} \tag{B.67}$$

and

$$sV_2(s) - v_2(0-) = \frac{1}{C}I_L(s) - \frac{1}{R_2C}V_2(s). \tag{B.68}$$

We may use Cramer's rule for solving for $I_L(s)$ and $V_2(s)$. Notice that we need the values for $i_L(0-)$ and $v_2(0-)$.

Solving for the transform of $v_2(t)$, using Cramer's rule, we obtain

$$V_2(s) = \frac{\begin{vmatrix} s + \dfrac{R_1}{L} & i_L(0-) + \dfrac{1}{L}\dfrac{V}{s} \\ -\dfrac{1}{C} & v_2(0-) \end{vmatrix}}{\begin{vmatrix} s + \dfrac{R_1}{L} & \dfrac{1}{L} \\ -\dfrac{1}{C} & s + \dfrac{1}{R_2C} \end{vmatrix}} = \frac{\left(s + \dfrac{R_1}{L}\right)v_2(0-) + \dfrac{1}{C}i_L(0-) + \dfrac{1}{LC}\dfrac{V}{s}}{s^2 + \left(\dfrac{R_1}{L} + \dfrac{1}{R_2C}\right)s + \left(1 + \dfrac{R_1}{R_2}\right)\dfrac{1}{LC}}. \tag{B.69}$$

Suppose $R_1 = R_2 = 100$ ohms, $L = 0.1$ henry, $C = 0.1$ microfarad, $V = 1$ volt, $v_2(0-) = 1$ volt, and $i_L(0-) = 1$ milliamp. Then

$$V_2(s) = \frac{s + 10^4 + 10^4 + (10^8/s)}{s^2 + 2 \times 10^4 s + 2 \times 10^8} = \frac{s(s + 2 \times 10^4) + 10^8}{s(s^2 + 2 \times 10^4 s + 2 \times 10^8)} \tag{B.70}$$

The poles of $V_2(s)$ are given by the roots of

$$s = 0 \tag{B.71}$$

and

$$s^2 + 2 \times 10^4 s + 2 \times 10^8 = 0 \tag{B.72}$$

or

$$s = \frac{-2 \times 10^4 \pm \sqrt{4 \times 10^8 - 8 \times 10^8}}{2}$$

$$= -10^4 \pm j10^4. \tag{B.73}$$

So we may expand $V_2(s)$ in partial fractions as

$$V_2(s) = \frac{A}{s} + \frac{B}{s + 10^4 - j10^4} + \frac{C}{s + 10^4 + j10^4}. \tag{B.74}$$

We now compute the residues

$$A = sV_2(s)\Big|_{s=0} = \frac{10^8}{2 \times 10^8} = \frac{1}{2} \tag{B.75}$$

$$B = (s + 10^4 - j10^4) V_2(s)\big|_{s=-10^4+j10^4} \tag{B.76}$$

$$= \frac{s(s + 2 \times 10^4) + 10^8}{s(s + 10^4 + j10^4)}\bigg|_{s=-10^4+j10^4}$$

$$= \frac{1}{2(1 + j1)}$$

and

$$C = (s + 10^4 + j10^4) V_2(s)\big|_{s=-10^4-j10^4}$$

$$= \frac{s(s + 2 \times 10^4) + 10^4}{s(s + 10^4 - j10^4)}\bigg|_{s=-10^4-j10^4} = \frac{1}{2(1 - j1)} \tag{B.77}$$

From Table B.1, we recognize that the inverse transform of $V_2(s)$ is

$$v_2(s) = \frac{1}{2} + \frac{1}{2(1 + j1)} \epsilon^{(10^4 + j10^4)t} + \frac{1}{2(1 - j1)} \epsilon^{(-10^4 - j10^4)t} \tag{B.78}$$

for $t \geqslant 0$. The above expression for $v_2(t)$ may be simplified into

$$e_2(t) = \tfrac{1}{2}[1 + \epsilon^{-10^4 t} \sqrt{2} \cos (10^4 t - \pi/4)]. \tag{B.79}$$

Instead of transforming (B.64) and (B.66) and then solving for $V_2(s)$, suppose we eliminate $i_L(t)$ first to obtain a single differential equation in $v_2(t)$. From (B.64), we have

$$\frac{di_L}{dt}+\frac{R_1}{L}i_L=-\frac{1}{L}v_2+\frac{1}{L}v\cdot \tag{B.80}$$

But from (B.66),

$$i_L=C\frac{dv_2}{dt}+\frac{1}{R_2}v_2\cdot \tag{B.81}$$

We now form the derivative of both sides of (B.81) and also multiply both sides of (B.81) by R_1/L. Then

$$\frac{di_L}{dt}+\frac{R_1}{L}i_L=\left(C\frac{d^2v_2}{dt^2}+\frac{1}{R_2}\frac{dv_2}{dt}\right)+\frac{R_1}{L}\left[C\frac{dv_2}{dt}+\frac{1}{R_2}v_2\right]\cdot \tag{B.82}$$

Since the left-hand sides of (B.80) and (B.82) are equal, the right-hand sides are equal to each other too, and we have

$$C\frac{d^2v_2}{dt^2}+\frac{1}{R_2}\frac{dv_2}{dt}+\frac{R_1}{L}C\frac{dv_2}{dt}+\frac{R_1}{R_2L}v_2=-\frac{1}{L}v_2+\frac{1}{L}v \tag{B.83}$$

or

$$\frac{d^2v_2}{dt^2}+\left(\frac{R_1}{L}+\frac{1}{R_2C}\right)\frac{dv_2}{dt}+\left(1+\frac{R_1}{R_2}\right)\frac{1}{LC}v_2=\frac{1}{LC}v\cdot \tag{B.84}$$

We may take the Laplace transform of (B.84) and obtain

$$\left[s^2+\left(\frac{R_1}{L}+\frac{1}{R_2C}\right)s+\left(1+\frac{R_1}{R_2}\right)\frac{1}{LC}\right]V_2(s)-sv_2(0-)-v_2'(0-)-\left(\frac{R_1}{L}+\frac{1}{R_2C}\right)v_2(0-)=\frac{1}{LC}\frac{V}{s} \tag{B.85}$$

or

$$V_2(s)=\frac{\dfrac{1}{LC}\dfrac{V}{s}+sv_2(0-)+v_2'(0-)+\left(\dfrac{R_1}{L}+\dfrac{1}{R_2C}\right)v_2(0-)}{s^2+\left(\dfrac{R_1}{L}+\dfrac{1}{R_2C}\right)s+\left(1+\dfrac{R_1}{R_2}\right)\dfrac{1}{LC}}. \tag{B.86}$$

Equation (B.86) differs from (B.69) in that the initial conditions in (B.69) are in terms of $v_2(0-)$ and $i_L(0-)$, whereas in (B.86) we have $v_2(0-)$ and $v_2'(0-)$, the value at $t=0-$ of the derivative of $v_2(t)$. That the two expressions are equal may be verified from the following argument. Equation B.65 is valid for all t and, in particular, it holds for $t = 0-$. Hence

$$\frac{1}{C}i_L(0-) = v_2'(0-) + \frac{1}{R_2C}v_2(0-). \tag{B.87}$$

Replacing $(1/C)i_L(0-)$ in (B.69) by the expression on the right-hand side of (B.87), we obtain (B.86). Hence the two solutions yield exactly the same answer. ▲

Let us now derive entry 7 in Table B.2. First we note that if $F(s)$ is a rational function with poles in the left-half plane only, then the inverse transform of each term in the partial fraction expansion gives rise to a term that is multiplied by an exponentially damped term $\epsilon^{-\alpha_i t}$ where $-\alpha_i$ is the negative real part of the pole $-p_i$. Hence, for $t \to \infty$, the response $f(t) \to 0$. If $F(s)$ has a pole in the right-half plane, then $f(t)$ contains an exponentially growing term and $f(t)$ becomes unbounded for $t \to \infty$. If $F(s)$ has a pole on the $j\omega$ axis other than the origin, then $f(t)$ contains a sinusoidal term and $f(t)$ has no limit for $t \to \infty$. If $F(s)$ has a simple pole at the origin with all the other poles in the left-half plane, then the partial fraction expansion of $F(s)$ contains a term K/s where K is the residue of the pole at the origin. This residue is equal to

$$K = sF(s)|_{s=0} = \lim_{s \to 0} sF(s). \tag{B.88}$$

As $t \to \infty$, $f(t) \to K = \lim_{s \to 0} sF(s)$. For multiple poles at the origin, $f(t)$ will contain terms which grow as $t \to \infty$. Thus we conclude that if $sF(s)$ has only left-half plane poles,

$$\lim_{t \to \infty} f(t) = \lim_{s \to 0} sF(s). \tag{B.89}$$

If $F(s)$ has no pole at the origin, then (B.89) yields $f(t) \to 0$ for $t \to \infty$.

Entry 7 of Table B.2 can be derived also be considering the Laplace transform of the derivative of $f(t)$

$$\int_{0-}^{\infty} \frac{df}{dt}\epsilon^{-st}\,dt = sF(s) - f(0-). \tag{B.90}$$

The transform in (B.90) is defined for all s whose real part is greater than the σ_a of absolute convergence. In particular, suppose that $sF(s)$ has poles in the left-half plane only. Then $\sigma_a < 0$ and (B.90) is defined for $s \to 0$. But, for $s \to 0$, (B.90) yields

$$\int_{0-}^{\infty} \frac{df}{dt}\,dt = \lim_{s \to 0}\,[sF(s)] - f(0-)$$

$$\lim_{t \to \infty} f(t) - f(0-) = \lim_{s \to 0}\,[sF(s)] - f(0-) \tag{B.91}$$

Hence, (B.89) is obtained which is entry 7 in Table B.2. It is known as the final value theorem.

Entry 8 in Table B.2 is known as the initial value theorem. If $f(t)$ has a discontinuity at $t=0$, then df/dt will have an impulse of area $f(0+)-f(0-)$ at $t=0$. The Laplace transform of df/dt may be written as

$$\int_{0-}^{\infty} \frac{df}{dt} \epsilon^{-st} = f(0+) - f(0-) + \int_{0+}^{\infty} \frac{df}{dt} \epsilon^{-st}\, dt. \tag{B.92}$$

But the transform of the derivative df/dt is also given by (B.90). Hence, from (B.90) and (B.92),

$$\int_{0+}^{\infty} \frac{df}{dt} \epsilon^{-st}\, dt = sF(s) - f(0+). \tag{B.93}$$

But if df/dt is Laplace transformable, (B.90) and (B.92) hold for all s with real part greater than σ_a. In particular, (B.93) holds for $s \to \infty$ along the positive real axis. But for $s \to \infty, \epsilon^{-st} \to 0$ for all $t > 0$. Hence

$$\lim_{s \to \infty} \int_{0+}^{\infty} \frac{df}{dt} \epsilon^{-st}\, dt = 0 = \lim_{s \to \infty} sF(s) - f(0+) \tag{B.94}$$

and

$$f(0+) = \lim_{s \to \infty} sF(s), \tag{B.95}$$

provided the limit exists.

B-3 The Second-Order All-Pole Transfer Function

In this section, we discuss the properties of a class of second-order systems. A mechanical system composed of a component that stores potential energy (such as a spring), a component that stores kinetic energy (such as mass), and a component that dissipates mechanical energy (such as a damper) is described by a second-order differential equation. An electrical network with one L, one R, and one C is also described by a second-order differential equation. Even when a system is modeled by a higher-order system, its behavior may be approximately described in terms of quantities that are defined for second-order systems. For example, in investigating the roots of the characteristic equation associated with a differential equation, it sometimes happens that a pair of "dominant" roots has the most significant effect on system behavior, and preliminary analysis or design considerations may be based on a second-order approximation. Even when there is no dominant pair of roots, a partial fraction expansion of a transfer function makes it possible to examine the individual contributions of the terms. Terms corresponding to conjugate poles are grouped together and rewritten as second-order expressions. Thus, properties of second-order transfer functions can be utilized to advantage in dealing with higher-order transfer functions.

Consider the second-order differential equation

$$\frac{d^2y}{dt^2}+2\zeta\omega_n\frac{dy}{dt}+\omega_n^2y=\omega_n^2x \tag{B.96}$$

where $x(t)$ is an input signal, $y(t)$ is an output signal, and ζ and ω_n are constant parameters. If $y(0)=y'(0)=0$, then the Laplace transform of (B.96) yields

$$(s^2+2\zeta\omega_n s+\omega_n^2)Y(s)=\omega_n^2X(s). \tag{B.97}$$

The ratio $Y(s)/X(s)$ is defined as the system transfer function. Denote the transfer function by $H(s)$. From (B.97),

$$H(s)=\frac{Y(s)}{X(s)}=\frac{\omega_n^2}{s^2+2\zeta\omega_n s+\omega_n^2}. \tag{B.98}$$

This transfer function is a rational function, with a zero-degree numerator polynomial and a second-degree denominator polynomial. Hence $H(s)$ has no finite zeros and two finite poles. The poles are given by the roots of

$$s^2+2\zeta\omega_n s+\omega_n^2=0. \tag{B.99}$$

The roots of (B.99) are

$$s=-\zeta\omega_n\pm\sqrt{(\zeta\omega_n)^2-\omega_n^2}. \tag{B.100}$$

If $|\zeta|>1$, the two roots are real and distinct. Furthermore, the roots will be on the negative real axis if $\zeta\omega_n>0$. If $\zeta\omega_n<0$, and $|\zeta|>1$, one of the roots will be on the positive real axis. For any bounded $x(t)$, $y(t)$ will grow without bounds for $t\to\infty$, and the system is said to be unstable. If $|\zeta|=1$, the roots coincide and $H(s)$ has a double pole. Again, for a stable system $\zeta\omega_n$ must be greater than zero. For $|\zeta|<1$, the roots in (B.100) are complex but $\zeta\omega_n$ must be positive in order for the complex roots to be in the left-half plane.

Thus for a stable system, $\zeta\omega_n>0$. There is no loss of generality if ζ and ω_n are taken to be positive individually. Figure B.2 shows the locus of the poles as ζ is varied. For $\zeta<1$, the locus lies on a semicircle of radius ω_n. For $\zeta=1$, the roots coincide at $s=-\omega_n$. For $\zeta>1$, the roots are on the negative real axis.

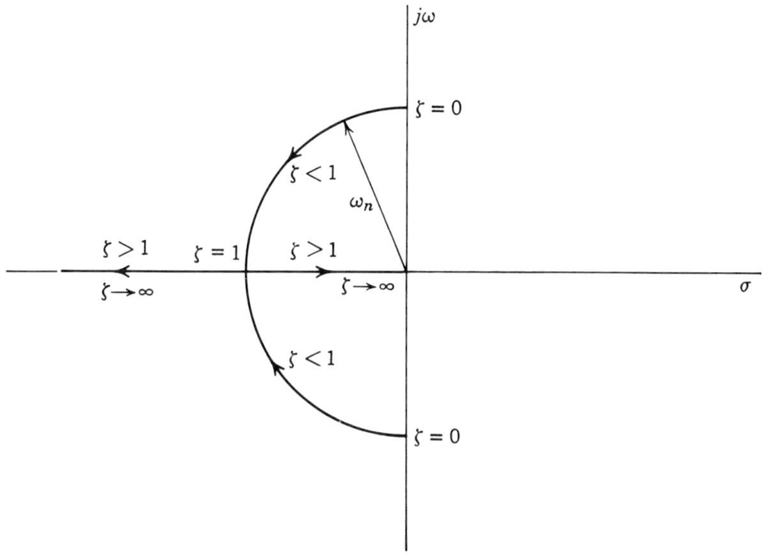

Figure B.2 Locus of roots of (B.99) as ζ varies from 0 to ∞.

Let us investigate the unit step response for the second-order all-pole transfer function of (B.98). The three cases for $\zeta > 1, \zeta = 0$, and $\zeta < 1$ will be treated separately.

Case 1. $\zeta > 1$ (overdamped): Let the two negative real poles be denoted by $-\alpha_1$ and $-\alpha_2$ where $\alpha_2 > \alpha_1$ so that

$$(s+\alpha_1)(s+\alpha_2) = s^2 + 2\zeta\omega_n s + \omega_n^2. \tag{B.101}$$

Hence $\alpha_1\alpha_2 = \omega_n^2$ and $\alpha_1 + \alpha_2 = 2\zeta\omega_n$. For a unit step, $X(s) = 1/s$ and

$$Y(s) = \frac{\alpha_1\alpha_2}{s(s+\alpha_1)(s+\alpha_2)}. \tag{B.102}$$

The partial fraction expansion of (B.102) is

$$Y(s) = \frac{1}{s} + \frac{1}{\alpha_2 - \alpha_1}\left[\frac{\alpha_1}{s+\alpha_2} - \frac{\alpha_2}{s+\alpha_1}\right]. \tag{B.103}$$

Thus

$$y(t) = \left[1 + \frac{1}{\alpha_2 - \alpha_1}(\alpha_1\epsilon^{-\alpha_2 t} - \alpha_2\epsilon^{-\alpha_1 t})\right]u(t) \tag{B.104}$$

as $\zeta \to \infty$, $\alpha_2 \to \infty$ and $\alpha_1 \to 0$. So, for large ζ, α_2 is large and α_1 is close to zero.

Hence, $\alpha_1\epsilon^{-\alpha_2 t}$ is negligible compared to $\alpha_2\epsilon^{-\alpha_1 t}$, and $\alpha_2-\alpha_2 \approx \alpha_2$ so that $y(t)$ is given by

$$y(t) \approx (1-\epsilon^{-\alpha_1 t})u(t) \tag{B.105}$$

which is the response of a first-order system. Figure B.3 shows a typical response for this case. The larger ζ is, the smaller is α_1 and the more sluggish is the response. Also, the slope of $y(t)$ in (B.104) is

$$\frac{dy}{dt} = \frac{\alpha_1\alpha_2}{\alpha_2-\alpha_1}(\epsilon^{-\alpha_1 t}-\epsilon^{-\alpha_2 t})u(t). \tag{B.106}$$

Since $\epsilon^{-\alpha_1 t}-\epsilon^{-\alpha_2 y} > 0$ for all $t > 0$ if $\alpha_2 > \alpha_1$, then dy/dt remains positive for all $t > 0$. At $t=0$, $y(0)=0$. So the response $y(t)$ must be rising for all t, and the maximum value is attained for $t \to \infty$. This maximum is $y_{\text{max}} = 1$.

Case 2. $\zeta = 1$ (critically damped): For this case, the transfer function becomes

$$H(s) = \frac{\omega_n^2}{(s+\omega_n)^2} \tag{B.107}$$

so that the response transfer $Y(s)$ due to a unit step function is

$$Y(s) = \frac{\omega_n^2}{s(s+\omega_n)^2}. \tag{B.108}$$

The partial function expansion of $Y(s)$ in (B.108) is

$$Y(s) = \frac{A}{s}+\frac{B}{s+\omega_n}+\frac{C}{(s+\omega_n)^2}. \tag{B.109}$$

By inspection,

$$A = sY(s)|_{s=0} = 1$$

$$C = (s+\omega_n)^2Y(s)|_{s=-\omega_n} = \frac{\omega_n^2}{-\omega_n} = -\omega_n. \tag{B.111}$$

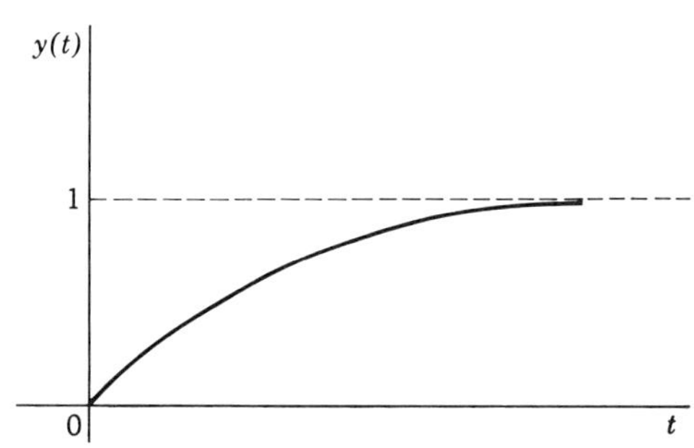

Figure B.3 Step response of overdamped second-order system.

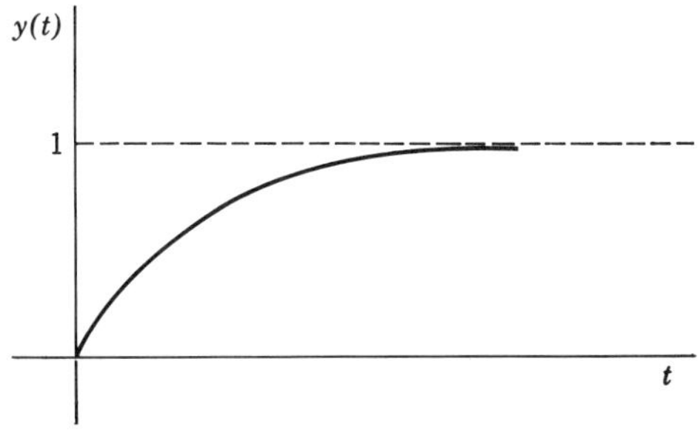

Figure B.4 Step response for a critically damped second-order system.

The coefficient B is obtained from

$$B = \left\{\frac{d}{ds}[(s+\omega_n)^2 Y(s)]\right\}_{s=-\omega_n} = \left\{-\frac{\omega_n^2}{s^2}\right\}_{s=-\omega_n} = -1 \qquad \text{(B.112)}$$

Hence

$$Y(s) = \frac{1}{s} - \frac{1}{s+\omega_n} - \frac{\omega_n}{(s+\omega_n)^2} \qquad \text{(B.113)}$$

and

$$y(t) = [1 - \epsilon^{-\omega_n t} - \omega_n t \epsilon^{-\omega_n t}]u(t) \qquad \text{(B.114)}$$

for $t \geqslant 0$. This response is more sluggish than the corresponding response of a first-order system $y(t) = [1-\epsilon^{-\omega_n t}]u(t)$. The slope of $y(t)$ in (B.114) is

$$\begin{aligned}\frac{dy}{dt} &= [\omega_n \epsilon^{-\omega_n t} + \omega_n^2 t \epsilon^{-\omega_n t} - \omega_n \epsilon^{-\omega_n t}]u(t) \\ &= \omega_n^2 t \epsilon^{-\omega_n t} u(t) > 0. \end{aligned} \qquad \text{(B.115)}$$

Hence, $y(t)$ is always rising, attaining a maximum for $t \to \infty$, which is $y_{\text{max}} = 1$. A typical step response for a critically damped second-order system is shown in Figure B.4. It is faster than an overdamped system.

Case 3. $\zeta < 1$ (underdamped): For the underdamped case, the roots

$$s_1 = -\zeta\omega_n + j\omega_n\sqrt{1-\zeta^2}$$

and

$$s_2 = -\zeta\omega_n - j\omega_n\sqrt{1-\zeta^2}$$

which are complex conjugates of each other. For a unit step input, $Y(s)$ is

$$Y(s) = \frac{\omega_n^2}{s(s+\zeta\omega_n - j\omega_n\sqrt{1-\zeta^2})(s+\zeta\omega_n + j\omega_n\sqrt{1-\zeta^2})}. \qquad \text{(B.116)}$$

Its partial fraction expansion is

$$Y(s) = \frac{1}{s} + \frac{1}{[-2(1-\zeta^2) - j2\zeta\sqrt{1-\zeta^2}](s+\zeta\omega_n - j\omega_n\sqrt{1-\zeta^2})}$$

$$+ \frac{1}{[-2(1-\zeta^2) + j2\zeta\sqrt{1-\zeta^2}](s+\zeta\omega_n + j\omega_n\sqrt{1-\zeta^2})} \tag{B.117}$$

Thus

$$y(t) = \left\{1 - \frac{1}{(2(1-\zeta^2) + j2\zeta\sqrt{1-\zeta^2})} \exp\left[(-\xi\omega_n + j\omega_n\sqrt{1-\zeta^2})t\right]\right.$$

$$\left. - \frac{1}{[2(1-\zeta^2) - j2\zeta\sqrt{1-\zeta^2}]} \exp\left[(-\xi\omega_n - j\omega_n\sqrt{1-\zeta^2})t\right]\right\} u(t). \tag{B.118}$$

Notice that the last two terms in (B.118) are complex conjugates. Let

$$A = \frac{1}{[2(1-\zeta^2) + j2\zeta\sqrt{1-\zeta^2}]}. \tag{B.119}$$

Notice that the sum of conjugates is equal to twice the real part. Hence

$$y(t) = u(t) - 2Re\{\epsilon^{-\zeta\omega_n t} A \epsilon^{j\omega_n\sqrt{1-\zeta^2}t}\} u(t). \tag{B.120}$$

Let us express A in polar form:

$$A = |A|\epsilon^{j\phi}$$

where

$$|A| = \frac{1}{\sqrt{[-2(1-\zeta^2)]^2 + 4\zeta^2(1-\zeta^2)}} = \frac{1}{2\sqrt{1-\zeta^2}} \tag{B.121}$$

and

$$\phi = -\tan^{-1}\frac{2\zeta\sqrt{1-\zeta^2}}{2(1-\zeta^2)} = -\tan^{-1}\frac{\zeta}{\sqrt{1-\zeta^2}} = -\cos^{-1}\sqrt{1-\zeta^2} \tag{B.122}$$

Then

$$y(t) = \left[1 - \frac{\epsilon^{-\zeta\omega_n t}}{\sqrt{1-\zeta^2}} \cos\left(\sqrt{1-\zeta^2}\,\omega_n t + \phi\right)\right] u(t). \tag{B.123}$$

Notice that for $t = 0$, $y(0) = 0$. As t increases from 0, the angle $(\sqrt{1-\zeta^2}\omega_n t + \phi)$ increases from ϕ which is negative until it reaches 90° at which time $y(t) = 1$. For t greater than this, the angle becomes greater than 90° so that the cosine function is negative and the signal $y(t)$ is greater than 1. Thus, for any $0 < \zeta < 1$, there will always be an instant at which $y(t)$ is greater than 1. The final value of $y(t)$ for $t \to \infty$ is 1. The maximum value of y minus the final value is called the overshoot:

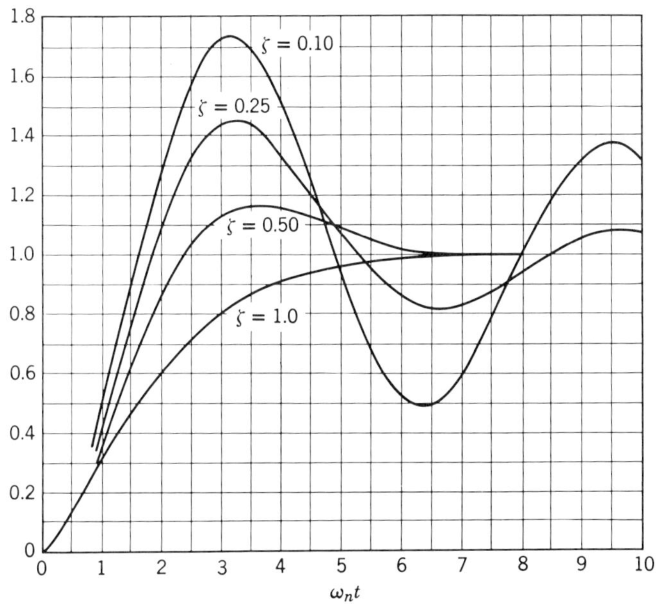

Figure B.5

$$\text{overshoot} = y_{\max} - y_{\text{final}} \tag{B.124}$$

and the percent overshoot is

$$\%\ \text{overshoot} = \frac{\text{overshoot}}{y_{\text{final}}} \times 100. \tag{B.125}$$

Typical response curves for different values of ζ are shown in Figure B.5. The curve crosses the line $y = 1$ every time the angle $(\sqrt{1-\zeta^2}\,\omega_n t + \phi)$ equals $\pi/2$ plus a multiple of π

$$\sqrt{1-\zeta^2}\,\omega_n t + \phi = \frac{\pi}{2} + K\pi. \tag{B.126}$$

Its frequency of oscillation is $\sqrt{1-\zeta^2}\,\omega_n$. The parameter ω_n is called the *undamped natural frequency*. If $\zeta = 0$, the frequency of oscillation is ω_n; otherwise it is $\sqrt{1-\zeta^2}\,\omega_n$. The parameter ζ is called the *damping ratio*.

The percent overshoot is plotted versus ζ in Figure B.6.

In summary of the three cases, for $0 < \zeta < 1$, the step response always has an overshoot, and it is oscillatory. As $\zeta \to 1$ the oscillation frequency decreases, the overshoot decreases, and the response tends to one faster. Then

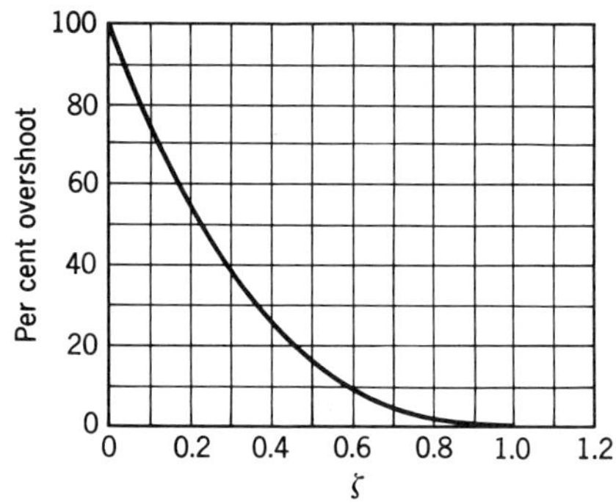

Figure B.6

at $\zeta = 1$, we have critical damping with no oscillation and no overshoot. For $\zeta > 1$, which is the underdamped case, there is no overshoot and no oscillation either, but the response is more sluggish than that for $\zeta = 1$.

Selected References for Collateral Reading

Chapter 1

1. W. E. Wilson, *Concepts of Engineering System Design*, McGraw-Hill Book Company, New York, 1965.
2. H. Chestnut, *Systems Engineering Tools*, John Wiley & Sons, Inc., New York, 1965.
3. R. E. Machol, *Systems Engineering Handbook*, McGraw-Hill Book Company, New York, 1965.
4. W. Gosling, *The Design of Engineering Systems*, John Wiley and Sons, Inc., New York, 1962.
5. D. Meister and G. F. Rabideau, *Human Factors Evaluation in System Development*, John Wiley & Sons, Inc., New York, 1965.
6. J. H. Milsum, "Automatic Control and Medicine," *Proc. Third Congress of IFAC*, London, 1966.
7. V. A. Trapeznikov, "Control, Economy, Technological Progress," *Proc. Third Congress of IFAC*, London, 1966.
8. H. Chestnut, *Systems Engineering Methods*, John Wiley & Sons, Inc., New York, 1967.
9. D. O. Ellis and F. J. Ludwig, *Systems Philosophy*, Prentice-Hall, Inc., Englewood Cliffs, N. J., 1962.
10. J. E. Gibson, *Introduction to Engineering Design*, Holt, Rinehart & Winston, Inc., New York, 1968.
11. *Proceedings of the IEEE*, Special Issue on Transportation, **56**(**4**), April 1968.

Chapters 2 and 3

1. W. A. Lynch and J. G. Truxal, *Signals and Systems in Electrical Engineering*, McGraw-Hill Book Company, New York, 1962.
2. R. H. Cannon, Jr., *Dynamics of Physical Systems*, McGraw-Hill Book Company, New York, 1967.
3. J. L. Shearer, A. T. Murphy, and H. H. Richardson, *Introduction to System Dynamics*, Addison-Wesley Publishing Co., Inc., Reading, Mass., 1967.
4. J. B. Cruz, Jr. and M. E. Van Valkenburg, *Introductory Signals and Circuits*, Blaisdell Publishing Co., Waltham, Mass., 1967.

Chapter 4

1. W. A. Lynch and J. G. Truxal, *Signals and Systems in Electrical Engineering*, McGraw-Hill Book Company, New York, 1962.
2. J. L. Shearer, A. T. Murphy, and H. H. Richardson, *Introduction to System Dynamics*, Addison-Wesley Publishing Co., Inc., Reading, Mass., 1967.
3. G. A. Korn and T. M. Korn, *Electronic Analog and Hybrid Computers*, McGraw-Hill Book Company, New York, 1964.

Chapter 5

1. J. L. Meriam, *Dynamics*, John Wiley & Sons, Inc., New York, 1966.
2. W. T. Thomson, *Introduction to Space Dynamics*, John Wiley & Sons, Inc., New York, 1961.

Chapter 6

1. M. E. Van Valkenburg, *Network Analysis*, Second Edition, Prentice-Hall, Inc., Englewood Cliffs, N.J., 1964.
2. B. C. Kuo, *Automatic Control Systems*, Second Edition, Prentice-Hall, Inc., Englewood Cliffs, N.J., 1967.
3. R. Saucedo and E. E. Schiring, *Introduction to Continuous and Digital Control Systems*, The Macmillan Company, New York, 1968.

Chapter 7

1. P. M. DeRusso, R. J. Roy, and C. M. Close, *State Variables for Engineers*, John Wiley & Sons, Inc., New York, 1965.
2. D. G. Schultz and J. L. Melsa, *State Functions and Linear Control Systems*, McGraw-Hill Book Company, New York, 1967.
3. R. Saucedo and E. E. Shiring, *Introduction to Continuous and Digital Control Systems*, The Macmillan Company, New York, 1968.

Chapter 8

1. B. C. Kuo, *Automatic Control Systems*, Second Edition, Prentice-Hall, Inc., Englewood Cliffs, N.J., 1967.
2. R. Saucedo and E. E. Shiring, *Introduction to Continuous and Digital Control Systems*, The Macmillan Company, New York, 1968.
3. R. J. Schwarz and B. Friedland, *Linear Systems*, McGraw-Hill Book Company, New York, 1965.

Chapter 9

1. R. J. Schwarz and B. Friedland, *Linear Systems*, McGraw-Hill Book Company, New York, 1965.
2. G. J. Thaler and M. P. Pastel, *Analysis and Design of Nonlinear Feedback Control Systems*, McGraw-Hill Book Company, New York, 1962.

3. J. E. Gibson, *Nonlinear Automatic Control*, McGraw-Hill Book Company, New York, 1963.
4. D. McRuer, *Analysis of Nonlinear Control Systems*, John Wiley & Sons, Inc., New York, 1961.

Chapter 10

1. R. Beckett and J. Hurt, *Numerical Calculations and Algorithms*, McGraw-Hill Book Company, New York, 1967.
2. E. I. Jury, *Theory and Application of the z-Transformation*, John Wiley & Sons, Inc., New York, 1964.
3. B. C. Kuo, *Analysis and Synthesis of Sampled-Data Control Systems*, Prentice-Hall, Inc., Englewood Cliffs, N.J., 1963.
4. J. R. Ragazzini and G. F. Franklin, *Sampled-Data Control Systems*, McGraw-Hill Book Company, New York, 1959.
5. P. M. DeRusso, R. J. Roy, and C. M. Close, *State Variables for Engineers*, John Wiley & Sons, Inc., New York, 1965.

Chapter 11

1. K. Ogata, *State Space Analysis of Control Systems*, Prentice-Hall, Inc., Englewood Cliffs, N.J., 1967.
2. R. W. Newcomb, *Concepts of Linear Systems and Control*, Brooks/Cole Publishing Company, Belmont, California, 1968.

Chapter 12

1. D. G. Schultz and J. L. Melsa, *State Functions and Linear Control Systems*, McGraw-Hill Book Company, New York, 1967.
2. G. Newton, L. Gould, and J. Kaiser, *Analytical Design of Linear Feedback Controls*, John Wiley & Sons, Inc., New York, 1957.

Chapter 13

1. D. J. Wilde, *Optimum Seeking Methods*, Prentice-Hall, Inc., Englewood Cliffs, N.J., 1964.
2. D. J. Wilde and C. S. Beightler, *Foundations of Optimization*, Prentice-Hall, Inc., Englewood Cliffs, N.J., 1967.
3. T. L. Saaty and J. Bram, *Nonlinear Mathematics*, McGraw-Hill Book Company, New York, 1964.
4. R. Beckett and J. Hurt, *Numerical Calculations and Algorithms*, McGraw-Hill Book Company, New York, 1967.

THE END

Answers to Selected Problems

Chapter 2

2-6. $t = [J\omega_0/T][1 + \sqrt{1 + (2T\theta_0)/J\omega_0^2}]$

2-8. $3 - \sqrt{5} \leq i \leq 3 + \sqrt{5}$

2-12. $C(v) = \epsilon^{-v}$

2-14. (a) Nonlinear time-invariant.
(b) Linear time-varying.
(c) Linear time-invariant.
(d) Linear time-varying.
(e) Nonlinear time-invariant.
(f) Linear time-varying.
(g) Nonlinear time-varying.
(h) Linear time-varying.

2-19. Area $= 2.2 \times 10^{-3} m^2$; length $= 11.3$ cm.

2-20. $f = -v_2$ Newton.

Chapter 3

3-1. $\dfrac{dv_C}{dt} = \dfrac{1}{C} i_L, \quad \dfrac{di_L}{dt} = -\dfrac{1}{L} v_C - \dfrac{R}{L} i_L + \dfrac{1}{L} v$

3-7. Gears: $\omega_1/\omega_2 = N_1/N_2, \quad \tau_1/\tau_2 = -N_2/N_1$

3-8. $\ddot{v}_1 = \dfrac{1}{M_1}\dot{f} - \dfrac{D}{M_1}\dot{v}_1 - \dfrac{K}{M_1}(v_1 - v_2)$

$$\ddot{v}_2 = \frac{K}{M_2}(v_1 - v_2)$$

3-11. $\dfrac{J_1 J_2}{D}\dddot{\Omega} + (J_1 + J_2)\ddot{\Omega} + \dfrac{KJ_2}{D}\dot{\Omega} + K\Omega = K\omega_{\text{ext}}.$

3-12. $y(t) = K_1\epsilon^{\alpha_1 t} + K_2\epsilon^{\alpha_2 t} + C_1\dfrac{K}{M}$

$$\text{where } \alpha_1 = -\frac{D}{2M} + \sqrt{\left(\frac{D}{2M}\right)^2 - \left(\frac{K}{M}\right)}$$

$$\alpha_2 = -\frac{D}{2M} - \sqrt{\left(\frac{D}{2M}\right)^2 - \left(\frac{K}{M}\right)}$$

$C_1 = \ddot{x}$, $K_1 = -C_1K\alpha_2/M(\alpha_2 - \alpha_1)$

3-15. $f_e = \frac{1}{2}v_1{}^2(dc/dx)$

3-24. Let $\partial y/\partial\phi = \alpha$. Then

$$\ddot{\alpha} + \frac{D}{M}\dot{\alpha} + \frac{K}{M}\alpha = \frac{\partial \ddot{x}}{\partial\phi} - g$$

3-25. For zero initial conditions, $w = \frac{1}{2}Li^2 = i\lambda$.

3-54. For zero initial conditions, $v = \frac{1}{2}Kx^2$.

3-55. For zero initial conditions, $v = \frac{1}{2}K\theta^2$.

3-56. For zero initial conditions, $w = \frac{1}{2}Mv^2$.

3-57. For zero initial conditions, $w = \frac{1}{2}I\omega^2$.

3-58. (a) $y = 1$. (b) $(\dot{\Delta y}) = \Delta x - \Delta y$.

3-59. (a) $y = 1$. (b) $(\dot{\Delta y}) = 10\Delta x - 10\Delta y$.

3-61. $a_{11} = R_1L_2/(M^2 - L_1L_2)$ $\quad a_{12} = (R_2 + R_3)M/(M^2 - L_1L_2)$
$a_{21} = R_1M/(M^2 - L_1L_2)$ $\quad a_{22} = (R_2 + R_3)L_1/(M^2 - L_1L_2)$
$b_1 = L_2/(L_1L_2 - M^2)$ $\quad b_2 = M/(L_1L_2 - M^2)$

Chapter 4

4-1. State equations are

$$\frac{dv_c}{dt} = \frac{1}{c}i$$

$$\frac{di}{dt} = -\frac{1}{L}v_c - \frac{R}{L}i + \frac{1}{L}v$$

4-3. Applying D'Alembert's principle:

$$\ddot{x}_1 = -\frac{K_1 + K_0}{M_1}x_1 - \frac{D_0 + D_1}{M_1}\dot{x}_1 + \frac{K_1 + K_0}{M_1}x_2 + \frac{D_0}{M_1}\dot{x}_2$$

$$\ddot{x}_2 = -\frac{D_2 + D_0}{M_2}\dot{x}_2 + \frac{D_0}{M_2}\dot{x}_1 - \frac{K_0}{M_2}x_2 + \frac{K_0}{M_2}x_1 + \frac{1}{M_2}f(t)$$

Choose $y_1 = x_1$, $y_2 = \dot{x}_1$, $y_3 = x_2$, $y_4 = \dot{x}_2$ as states.
Then draw diagram by inspection of state equations.

4-15. For $z_1 = y$, $z_2 = \dot{y}$, $z_3 = x$, $z_4 = \dot{x}$, equations are

$$\dot{z}_1 = z_2, \quad \dot{z}_2 = -\frac{K}{M}z_1 - \frac{D}{M}z_2 + \frac{K}{M}z_3 + \frac{D}{M}z_4, \quad \dot{z}_3 = z_4, \quad \dot{z}_4 = u = \ddot{x}$$

4-16. Draw diagram from

$$\dot{v}_1 = \frac{1}{R_1C_1}v - \frac{1}{R_1C_1}v_1$$

$$\dot{v}_2 = -\frac{1}{R_2C_2}v_2 + \frac{1}{R_2C_2}v_1$$

$$\dot{v}_3 = \frac{1}{R_3C_3}(v_2 - v_3)$$

4-17. $\dot{\omega}_2 = -\dfrac{D}{J}\omega_2 + \dfrac{n}{J}\tau_1$

Chapter 5

5-1. $\omega_1(t) = \omega_1(0)\cos Kt + [\dot{\omega}_1(0)/K]\sin Kt$
$\omega_2(t) = (T/K^2) + \omega_1(0)\sin Kt - [\dot{\omega}_1(0)/K]\cos Kt$
$\omega_3(t) = \Omega = \text{constant}$
where $K = (J_3 - J)\Omega/J$

5-5. $\mathring{A} = -PA$

5-8. $\alpha(t) = (t/2) - (1/20)(1 - \epsilon^{-10t}),\ t \geq 0$

5-11. $J_v\dot{\omega}_i - \tau'_{di} + \tau_{mi} = 0, \quad i = 1, 2, 3,$
$J_v = J_{vi} + J_w, \quad \tau_{mi} = J_W\dot{\Omega}_i$

Chapter 6

6-1. (a) $\zeta = K/2\sigma\omega_n J = 4J\sigma^2/2\sigma(2\sigma)J = 1$

6-4. Partial answer: $\Delta(s) = 1 + \dfrac{K_c - K_aK_b}{s} - \dfrac{K_aK_d}{s^2}$

6-8. $y(s)/x(s) = \left[\sum_{i=0}^{n} (b_i/s^{n-i})\right] \Big/ \left[1 + \sum_{i=0}^{n-1} a_i/s^{n-i}\right]$

6-10. $1/(RCs + 1)$

6-12. $H(s) = \dfrac{R[(L_2 s/C)+R/C]}{(L_2 s+R)\{L_1 L_2 s^3 + RL_1 s^2 + [(L_1+L_2)/C]s + R/C\}}$

6-16. $y/x = (fc+ac)/(1-ace-ab-dc+abcd)$

6-17. (a) $H(s) = \dfrac{\frac{1}{6}}{s} - \dfrac{\frac{2}{3}}{s+3} + \dfrac{\frac{3}{2}}{s+4}$

(b) $H(s) = 1 - \dfrac{5}{s+2} + \dfrac{5}{s+3}$

(c) $\dfrac{6}{s+2} - \dfrac{3}{(s+2)^2} - \dfrac{6}{s+3}$

6-23. $y_{ss}(t) = 8.28 \sin\left(5t + \dfrac{\pi}{6}\right)$

6-28. $y(t) = (\frac{1}{8}) + (\frac{1}{6})t^2 + (\frac{2}{9})\epsilon^{-3t} - (\frac{3}{8})\epsilon^{-4t},\ t \leqslant 0$
6-33. $H(j\omega) = \sqrt{10}\, j\omega/(1+j\omega/10)^3$
6-36. $H(s) = 18.6(s+2)/s(s+1)$

Chapter 7

7-1. $(R_1/L) + (1/R_2C) > 2/\sqrt{LC}$

7-6. $\mathbf{A} = \begin{bmatrix} 0 & 1 \\ -3 & -4 \end{bmatrix}, \quad \mathbf{b} = \begin{bmatrix} 1 \\ -2 \end{bmatrix}, \quad \mathbf{c} = [1 \quad 0], \quad D = 0$

7-8. $a_{11} = -1, \quad a_{22} = -2, \quad a_{21} = a_{12} = 0$
7-11. For integrator outputs as states, numbering from the output as x_1: $B = 0$,

$$\mathbf{A} = \begin{bmatrix} 0 & 1 & 0 \\ 0 & 0 & 1 \\ -4 & -2 & -3 \end{bmatrix}, \quad \mathbf{C} = [1 \quad 0 \quad 0], \quad D = 0, \quad \phi_{11}(s) = \frac{s^2+3s+2}{s^3+3s^2+2s+4}$$

7-14. $y_1(t) = (\frac{22}{5}) - (\frac{3}{2})\epsilon^{-t} + (\frac{1}{10})\epsilon^{-5t}, \quad t \geqslant 0$
$y_2(t) = 2 - (\frac{1}{2})\epsilon^{-5t} + (\frac{3}{2})\epsilon^{-t}, \quad t \geqslant 0$

7-15. (a) $\phi(t) = \frac{1}{4}\begin{bmatrix} 5\epsilon^{-t} - \epsilon^{-5t} & \epsilon^{-t} - \epsilon^{-5t} \\ -5\epsilon^{-t} + 5\epsilon^{-5t} & -\epsilon^{-t} + 5\epsilon^{-5t} \end{bmatrix}$
7-20. (b) $b_1 = 1, \quad b_2 = -1, \quad b_3 = -3, \quad a_1 = 2, \quad a_2 = 3, \quad a_3 = 4$
7-22. Let $H(s) = y(s)/u(s)$. Then $x_1 = -5x_1 + u, \quad x_2 = -2x_2 + u, \quad y = (\frac{40}{3})x_1 - (\frac{10}{3})x_2$ is one possibility.
7-25. $H(s) = (-\frac{1}{24})/(s+2) - (\frac{8}{15})/(s+5) + (\frac{63}{40})/(s+10)$ (partial answer)
7-30. One possibility is $\dot{x}_1 = -x_1,\ \dot{x}_2 = -3x_2,\ y = 5x_1 + 5x_2,\ x_1(0) = -x_2(0) = 1$

Chapter 8

8-1. (a) $s^2+s+K=0$ (b) $s(2s+K-1)=0$ (c) $s^3+ks^2+1=0$
(d) $s^3+s^2+K=0$ (e) $s^2+(K+2)s+1=0$

8-3. $\det(sI-T^{-1}AT)=\det[T^{-1}(sI-A)T]=\det(sI-A)$

8-5. (a) $K>0$ (e) $K>-2$. Others are not *asymptotically* stable for any values of K.

8-14. $D=1920$.

8-15. $6<K<10$.

8-19. (c) is the only stable system.

8-27. Not AS, but is BIBO stable.

Chapter 9

9-6. Switching lines are the same as in Figure 9.7 except that for the inner two lines, the x_1 intercepts are not at $\pm\delta_1$ but at $\pm(\delta_1+(\tau_c\tau^2/2J)-(K\tau_c\tau/J)$

9-10. Duty cycle $=\frac{1}{2}-\dfrac{1}{2\pi}\left(\sin^{-1}\dfrac{\delta_1}{E}+\sin^{-1}\dfrac{\delta_2}{E}\right)$

9-11. $N(j\omega,E)=\dfrac{2T}{a\pi E}\left[E\sin^{-1}\dfrac{a}{E}+a\left(1-\left(\dfrac{a}{E}\right)^2\right)^{1/2}\right]$

9-12. Limit cycle at $\omega=1, E=1$

9-13. If $a>0$, equilibrium point is stable.

9-15. Equilibrium point is stable if $a_1>0$, $a_3>0$, $a_1a_2-a_3>0$. For $Q=I$, P is a 3×3 symmetric positive definite matrix where

$$P_{11}=\frac{a_3^3+a_1a_3^2+a_3a_1^2+a_1a_2^2-a_2a_3}{2a_3(a_1a_2-a_3)}$$

$$P_{12}=\frac{a_2a_1^2+a_2a_3^2-a_3^2}{2a_3(a_1a_2-a_3)}$$

$$P_{13}=\frac{1}{2a_3}$$

$$P_{22}=\frac{a_1^3+a_1a_3^2+a_1^2a_3+a_1a_3+a_3^2+a_2a_3-a_1a_2+a_3}{2a_3(a_1a_2-a_3)}$$

$$P_{23}=\frac{a_1^2+a_3^2+a_1a_3}{2a_3(a_1a_2-a_3)}$$

$$P_{33}=\frac{a_1+a_3+a_2a_3}{2a_3(a_1a_2-a_3)}$$

Chapter 10

10-1. $x(t_{k+1}) = \dfrac{x(t_k)}{2} + \dfrac{b}{2}\dfrac{1}{x(t_k)}$. The equation is first-order, nonlinear, and time-invariant.

10-2.
```
Read M, T
cM is number of steps
cT is step size
Read X0, (u(I)I = 1, M)
Y1 = 0.0
Y = X0
DO  10  I = 1, M
Y1 = Y + T*FCN(Y, U(I))
cFCN is function subroutine
Y = Y1
Write Y
10 Continue
End
```

10-5. $v_c(t_{k+1}) = \epsilon^{-T/RC} v_c(t_k) + [1-\epsilon^{-T/RC}]u(t_k)$ where $t_k = kT$, $t_{k+1} - t_k = T$

10-8. State equations are

$$c_{k+1} = c_k + v_k$$
$$v_{k+1} = -2v_k + r_k - c_k$$

The difference equation relating r_k to c_k is

$$c_{k+2} + c_{k+1} - c_k = r_k$$

10-9. $c_{k+2} + c_{k+1} + c_k = r_{k+2} + 2r_{k+1} + 2r_k$

10-12. (a) $F(z) = \dfrac{z}{z-1}$

(b) $F(z) = \dfrac{z \sin 1}{z^2 - 2z\cos 1 + 1}$

(c) $F(z) = \dfrac{z(z+1)}{(z-1)^3}$

10-13. (a) $f(n) = 2 - 2(\frac{1}{2})^n$
(b) $f(n) = n(\frac{1}{2})^{n+1}$
(c) $f(n) = 4 + \frac{4}{9}(-1)^n - \frac{31}{9}(\frac{1}{2})^n - \frac{5}{3}(\frac{1}{2})^{n+1}$

10-16. If the input voltage is constant with unit value, the capacitor voltage is

$$v_n = u_n(1 - \epsilon^{-(T/RC)n})$$

10-21. $x_1(n) = 2 - 2(\frac{1}{2})^{n-1}$
$x_2(n) = -1 + (\frac{1}{2})^{n-1}$

10-23. $\dfrac{Y(z)}{U(z)}=\dfrac{z}{z^2-\frac{3}{2}z+\frac{1}{2}}$

The zero is at the origin.
The poles are at $z=1$ and $z=\frac{1}{2}$.

10-24. $\dfrac{Y(z)}{U(z)}=\dfrac{3}{5}\dfrac{z-\frac{2}{3}}{z(z-1)}$

The zero is at $z=\frac{2}{3}$.
The poles are at $z=0$ and $z=1$.

10-25. (a) $\phi_{11}(mT)=2-2^{-m}$
$\phi_{12}(mT)=2-2^{1-m}$
$\phi_{21}(mT)=-1+2^{-m}$
$\phi_{22}(mT)=-1+2^{1-m}$
(b) The answer is the same as in (a).

10-27. (a) $w_{k+1}=TFT^{-1}w_k+TGu_k$
$y_k \quad =CT^{-1}w_k+Du_k$
(b) $H(z)$ remains the same as before.

10-28. $x(t_n)=\phi(t_n,t_0)x_0+\sum_{j=0}^{n-1}\phi(t_n,t_{j+1})G(t_{j+1},t_j)u(t_j)$

where
$$\phi(t_n,t_k)=\prod_{i=k}^{n-1}F(t_{i+1},t_i)$$

10-31. (a) Not stable.
(b) Not stable.

Chapter 11

11-1. (a) $[B \mid AB]$ has rank 2. Therefore the system is completely controllable.
(b) $K_1=\frac{9}{2},\quad K_2=\frac{1}{2}$

11-2. $K_1=\dfrac{-2(\epsilon^{-2}-1)x_1(0)}{\epsilon^{-3}+\epsilon^{-2}-3\epsilon^{-1}+1}+\dfrac{[2(2\epsilon^{-1}-1)(1-\epsilon^{-2})+(\epsilon^{-3}+\epsilon^{-2}-3\epsilon^{-1}+1)]}{(\epsilon^{-1}-1)(\epsilon^{-3}+\epsilon^{-2}-3\epsilon^{-1}+1)}x_2(0)$

$K_2=4\dfrac{(2\epsilon^{-1}-1)x_2(0)+(\epsilon^{-1}-1)x_1(0)}{\epsilon^{-3}+\epsilon^{-2}-3\epsilon^{-1}+1}$

11-4. $K_1=\dfrac{10^5}{1-\epsilon^{.5}}\left[\dfrac{v_2{}^0}{2(1-\epsilon^{-.5})}-\dfrac{v_1{}^0}{(\epsilon-1)}\right]$

$K_2=-\dfrac{10^5 v_1{}^0}{\epsilon^2-\epsilon}-\epsilon^{-1}K_1$

11-5. (a) $x_1 = \theta, \quad x_2 = \dot{\theta}$

$\dot{x}_1 = x_2, \quad \dot{x}_2 = -10x_2 + 100v$

(b) $x_1(k+1) = x_1(k) + 0.063x_2(k) + 0.367v(k)$

$x_2(k+1) = 0.368x_2(k) + 10.0v(k)$

(c) The discrete-time system in (b) is completely controllable.

(d) $K = 5.76, \quad K_2 = -8.19$

(e) Answer is the same as in (d).

11-6. (a) $\dfrac{Y(s)}{R(s)} = \dfrac{b_3 s^2 + b_2 s + b_1}{s^2 + a_2 s + a_1}$

(b) $A = \begin{bmatrix} 0 & 1 \\ -a_1 & -a_2 \end{bmatrix}, \quad B = \begin{bmatrix} 0 \\ 1 \end{bmatrix}, \quad C = [b_1 - a_1 b_3 \quad b_2 - a_2 b_3], \quad D = b_3$

(c) The system is completely state controllable.

(d) The system is output controllable.

(e) The system is completely observable.

11-14. (a) The system is output controllable.

(b) The system is output controllable.

11-16. (a) The system is output controllable.

(b) The system is output controllable.

11-18. (a) The system is completely state controllable.

(b) The feedback system remains completely state controllable for any choice of k_1 and k_2.

(c) The only zero of the transfer function is at $s = -3$ regardless of the values of k_1 and k_2.

11-22. The sensitivity remains less than one provided the plot of $L(j\omega)$ in the complex $L(j\omega)$ plane remains outside a unit circle with center at $-1 + j0$.

11-25. $\dot{x}_1 = -x_1 + u$

$\dot{x}_2 = x_1 - 2x_2$

$\dot{x}_3 = -2x_3 + u$

$\dot{x}_4 = x_3 - x_4$

The system is not completely state controllable.

11-28. $K_1 = -\frac{1}{8}, \quad K_2 = \frac{3}{4}, \quad K_3 = -\frac{3}{2}$

Chapter 12

12-1. $K = 1/\sqrt{\rho}, \quad a = \rho^{-1/4}/2^{1/2}$

12-3. $\rho = \frac{1}{4}, \quad K = \frac{1}{2}, \quad a = 1$

12-5. (a) Use definition of $E(s)$ in terms of $e(t)$, and differentiate with respect to s.
(b) $K=1$
(c) Error is weighted more heavily for large t compared to errors for small t.

Chapter 13

13-1. $x_0 = 10.0$ $\quad x_3 = 5.34$
$x_1 = 7.59$ $\quad x_4 = 5.04$
$x_2 = 6.13$ $\quad x_5 = 5.00$

13-2. $x_0 = -5$ $\quad x_3 = -1.18$
$x_1 = -2.97$ $\quad x_4 = -1.01$
$x_2 = -1.77$ $\quad x_5 = -1.00$

13-12. (a) Choose an arbitrary x_0, and $\gamma > 0$.

(b) Form $2f\left.\frac{\partial f}{\partial x}\right|_{x_i} = 2(x_i^3 - 6x_i^2 - 3x_i + 10)(3x_i^2 - 12x_i + 3)$.

$$\stackrel{\Delta}{=} g_i$$

(c) Form $x_{i+1} = x_i - \gamma g_i$.
(d) If $\gamma|g_i|$ is sufficiently small, stop program. If not, replace x_i by x_{i+1} and go back to (b).

13-13. $I = \dfrac{1}{2K_1K_2} + \dfrac{K_2}{2} + \dfrac{K_1}{2K_2}\rho$

$$\frac{\partial I}{\partial K_1} = \frac{-1}{2K_1^2K_2} + \frac{\rho}{2K_2} \stackrel{\Delta}{=} I_{K_1}$$

$$\frac{\partial I}{\partial K_2} = -\frac{1}{2K_1K_2^2} + \frac{1}{2} - \frac{\rho K_1}{2K_2^2} \stackrel{\Delta}{=} I_K$$

Algorithm:

Guess K_1^0, K_2^0.
(1) Calculate $I_{K_1}{}^i$, $I_{K_2}{}^i$, I^i.
(2) Calculate $K_1^{i+1} = K_1^i - I^i/I_{K_1}{}^i$
$K_2^{i+1} = K_2^i - I^i/I_{K_2}{}^i$.
(3) Check if $\Delta K_1^i = K_1^{i+1} - K_1^i$ and $\Delta K_2^i = K_2^{i+1} - K_2^i$ are sufficiently small.
(4) If yes, terminate. If not, go to (1).

13-14. *Algorithm:*
Guess K_1^0, K_2^0.
(1) Calculate $I_{K_1}{}^0$, $I_{K_2}{}^0$.

(2) Calculate I^i.

(3) Calculate $K_1^{i+1} = K_1^i - I^i/I_{K_1}^0$

$K_2^{i+1} = K_2^i - I^i/I_{K_2}^0$.

(4) Check if $\Delta K_1^i = K_1^{i+1} - K_1^i$

$\Delta K_2^i = K_2^{i+1} - K_2^i$

are sufficiently close to zero.

(5) If yes, terminate. If not, go to (2).

Index

Acceleration, 52
Accelerometer, 52, 151, 235–237
Actuator, 9, 22, 55
  gas jet, 22
  reaction wheel, 55
Additivity, 275, 277
Admittance, 150
Analog, 31, 50, 96
Armature, 64
Attitude control, gas jet, 22, 117, 138–141
  reaction-wheel, 141, 249–252, 269–270, 473–481, 513–514

Bellman, R., 271
Bihovski, M. L., 81, 191
Block diagrams, transfer function, 152–156
Bode plots, 173–185
Body angular velocities, 117–126, 129–130
Body axes, 118
Body rates, *see* Body angular velocities
Bram, J., 491
Brockett, R. W., 422
Butman, S., 406

Canonic phase variables, 423
Capacitance, 28, 36
  incremental, 36
Capacitor, 28, 35
  linear time-invariant, 28
  nonlinear, 35
  time-varying, 35
Center, 226–227
Characteristic equation, 287
Characteristic roots, *see* Eigenvalues
Closed-loop structure, 436
Commutator, 64
Comparison sensitivity, 439
Complete solution, 232–237, 246–252, 274
Compliance, 33
  torsional or rotational, 33
Conductance, 33
Conservation laws, 57–68
Controllability, 398–409
  output, 408
  reactor wheel attitude control system, 413–415
  state, 398
  test, 404
Controller, 398
Convolution, 213–214, 234
  sum, 377
Critically damped response, 546
Cruz, J. B., Jr., 439

D'Alembert's principle, 46
Damper, 17, 20
  rotational or torsional, 20
  translational, 17
Damping constant, 17, 20
  torsional or rotational, 20
  translational, 17
Damping ratio, 549
Dashpot, 17
Decomposition, 275, 276
Delayor, 363
Describing function, 338–341
Design, *see* Systems engineering
Desoer, C. A., 238
Difference equations, 359
Direction cosines, 144–145
Duty cycle, 332

Eigenvalues, 217–218, 222, 245, 289
Eigenvectors, 219, 221
Elements, 11, 13, 27, 34
  electrical, 27
  linear, 13, 27
  mechanical, 13
  nonlinear, 34
  time-invariant, 13, 27
  time-varying, 34

Engineering, *see* Systems engineering
Equality constraint, 472
Equilibrium line, *see* Equilibrium point
Equilibrium point, 74, 214, 225, 342
Equilibrium state, 378
Euclidean norm, 314
Euler angles, 117, 127–131, 133, 140, 143, 145, 210
Euler's dynamical equations, 121, 124, 142–143, 209–210, 269–270
Euler's procedure, 360
Evans, W. R., 299

Feedback, 98, 395–398
Feedback control, 141, 395–398, 421–436, 462–466
Field, 64
Focus, 226–227
Force, 14, 18
Forward path and forward path transfer function, 156–157
Franklin, G. F., 369
Frequency response, 169–188
  Bode plots (straight line approximations or corner plots), 173–185
  from pole-zero diagrams, 185–188
  second order plots, 180, 182

Gas jets, 22, 117, 138–141
Gibson, J. E., 350
Gilbert, E. G., 255, 405, 411, 418
Gimbal-rotor assembly, 133
Gould, L. A., 460
Gradient method, 501–509
Gyro, 117, 131–138, 145
  position, 136, 145
  rate, 136, 145
  rate-integrating, 136, 145
Gyroscope, *see* Gyro

Homogeneity, 275, 276
Hooke's law, 16
Householder, A. S., 467

Impedance, 150
Inductance, 29, 31, 36
  incremental, 36
  mutual, 31
  self, 31
Inductor, 28, 36
  linear time-invariant, 28
  nonlinear, 36
  time-varying, 36
Inertia, 18, 124
  moments of, 18, 124
  products of, 124
  rotational, 18
Inertial axes, 118
Inertial reference, 14
Ingwerson, D. R., 350
Input-output stability, 313
Integrator, 100
Isoclines, 215–216

Johnson, C. D., 428
Jury, E. I., 369, 380

Kaiser, J. F., 460
Kalman, R. E., 399, 409, 428, 466
KCL, *see* Kirchhoff's laws, current law
Kirchhoff's laws, 40, 41, 42
  current law (KCL), 40
  voltage law (KVL), 41
Kokotović, P. V., 191
Kreindler, E., 408
KVL, *see* Kirchhoff's laws, voltage law

Laplace transforms, 148–151, 527–550
Leeds, J. V., Jr., 81
Limit cycle, 329
Linearity, 275–281
Linearization, 73, 135–138, 140, 145, 148
Luenberger, D. G., 467
Lyapunov stability, 341–354

Mason's formula, 156–164, 244, 251
Mass, 14, 52
  seismic, 52
  test, 52
Matrix algebra, 121, 524–526
Mechanical mobility, 150
Meriam, J. L., 118
Minimal polynomial, 289
Minimum-order realization, 255, 265
Modeling, 3, 7, 27, 34, 117
  electrical elements, 27
  linear elements, 27
  mechanical elements, 13
  nonlinear elements, 34
  single-axis gas jet attitude control system, 22
Momentum, 14, 18
  angular, 18
  linear, 14
Morgan, B. S., Jr., 422, 428

Motor, 64, 68
a-c induction, 68
a-c servomotor, 68
d-c, 64

Network, 40, 46, 67
electrical, 40
mechanical, 46
three-port, 67
Newton, G. C., Jr., 460
Newton-Raphson method, 488–501
basic, 488
modified, 492
Newton's law, 118
Node, 40, 225–227
Nontouching loops, 158
Norm, 314, 350
of a matrix, 350
of a vector, 314
Normal coordinates, 405
Nyquist stability criterion, 308–313

Observability, 409–413
of composite systems, 418–421
of modes, 411
of reaction wheel attitude control system, 413–417
Open-loop structure, 436
Overdamped response, 545
Overshoot, 549

Parameter optimization, 454
Parseval's theorem, 438, 459
Partial fraction expansions, 533
Performance index, 455–459
Perkins, W. R., 439
Phase plane, *see* State, plane
Phase portrait, 226
Plant, 398
disturbance input, 398
input, 398
output, 398
Poles and zeros, 185–188, 533
Pontriagin, L. S., 238, 271
Principal axes, 124
Pseudocharacteristic equation, 341

Ragazzini, J. R., 369
Reaction wheel, 117, 141–143
Resistance, 28
Resistor, 27, 35
linear time-invariant, 27
nonlinear time-varying, 35
Resonance, 205
Response, 148, 206
unit impulse, 213, 235, 237, 247
zero input, 213–214, 234, 237, 238, 247–249, 275
zero state, 213–214, 234, 237, 246–249, 275
Riccati equation, 463, 466–471, 484
Rigid body motion, 117
Root-locus, 298–307
Roots of a characteristic equation, *see* Eigenvalues
Rotor, 64
Routh-Hurwitz test, *see* Routh stability test
Routh stability test, 291
Runge-Kutta, 360
Rutman, R. S., 191

Saaty, T. L., 491
Saddle point, 225–227
Sarachik, P. E., 408
Scaling, 109
Schultz, D. G., 350
Sensitivity, 80–82, 188, 191–196, 205
comparison, 439
function, 80, 188, 205
model, 81
network, 82
of nonlinear systems, 354–355
rule for drawing sensitivity block diagram, 191–196
Shilov, G. E., 402
Simulation, 3, 136–137, 143
analog, 96
digital, 362
Sinusoidal steady-state response, 169–173
Sivan, R., 406
Source, 17, 18, 29, 30
controlled or dependent, 30
current, 29
force, 17
torque, 18
velocity, 17, 18
voltage, 29
Spacecraft motion, 117
Spring, 15, 19
linear, 15
torsional, 19
Stability, 287, 291, 308–315, 341–354, 377
asymptotically stable linear time-invariant system, 287
discrete-time, 377
Lyapunov, 341–354

Nyquist criterion, 308–313
relation to poles of transfer function, 315
Routh test for, 291
zero state bounded-input bounded-output (BIBO), 314
State, 15, 75, 100, 206
discrete-time, 375
equations, 209, 374
feedback, 421–436
initial, 207
plane, 214
regulator, 462
transition matrix, 218–219, 227–232, 238–246, 270–274, 375
variables, 15, 75, 100, 206
Stator, 64
Steepest ascent, 505
Steepest descent, 505
Stiffness, 16, 19, 37
coefficient of, 16
incremental, coefficient of, 37
torsional, coefficient of, 19
Superposition, *see* Linearity
Systems, 1–2
analysis, 147
conservative, 59
discrete-time, 359, 482
first order, 212
linear, 209, 268, 275–277
nonlinear, 209–211
piece-wise linear, 323
sampled-data, 359
second order, 217, 227
time-invariant, 209, 275–280
time-varying, 209, 268
zero-input linear time-invariant, 275–278
zero-state linear time-invariant, 275–279
Systems engineering, 1–4

Time-invariance, 275–281
Torques, 18, 139
control, 139
disturbance, 139
Trajectory, 214
Transfer functions, 148–151
block diagrams, 152–155
of discrete-time systems, 369
loop, 157
multiple-input multiple-output, 164
relationship to state, 252
Transition matrix, 218–219, 227–232, 238–246, 270–274, 375

Undamped natural frequency, 549
Underdamped response, 547
Unit impulse response, 213, 235, 237, 247

Variables, 11, 14, 17, 18, 28, 29
across, 11
charge, 28
current, 28
flux linkage, 29
force, 14, 18
momentum, 14
state, 15, 75, 100
through, 11
torque, 18
velocity, 14, 18
voltage, 28
Velocity, 14, 18, 47
angular, 18
relative, 47
translational, 14, 18

Wonham, W. M., 428

Zadeh, L. A., 238
Z-transform, 369–374